UNDERSTANDING AERODYNAMICS

Aerospace Series List

Introduction to UAV Systems, 4th Edition	Fahlstrom and Gleason	August 2012
Theory of Lift: Introductory Computational Aerodynamics with MATLAB and Octave	McBain	August 2012
Sense and Avoid in UAS: Research and Applications	Angelov	April 2012
Morphing Aerospace Vehicles and Structures	Valasek	April 2012
Gas Turbine Propulsion Systems	MacIsaac and Langton	July 2011
Basic Helicopter Aerodynamics, 3rd Edition	Seddon and Newman	July 2011
Advanced Control of Aircraft, Spacecraft and Rockets	Tewari	July 2011
Cooperative Path Planning of Unmanned Aerial Vehicles	Tsourdos et al	November 2010
Principles of Flight for Pilots	Swatton	October 2010
Air Travel and Health: A Systems Perspective	Seabridge et al	September 2010
Design and Analysis of Composite Structures: With applications to aerospace Structures	Kassapoglou	September 2010
Unmanned Aircraft Systems: UAVS Design, Development and Deployment	Austin	April 2010
Introduction to Antenna Placement & Installations	Macnamara	April 2010
Principles of Flight Simulation	Allerton	October 2009
Aircraft Fuel Systems	Langton et al	May 2009
The Global Airline Industry	Belobaba	April 2009
Computational Modelling and Simulation of Aircraft and the Environment: Volume 1 - Platform Kinematics and Synthetic Environment	Diston	April 2009
Handbook of Space Technology	Ley, Wittmann Hallmann	April 2009
Aircraft Performance Theory and Practice for Pilots	Swatton	August 2008
Surrogate Modelling in Engineering Design: A Practical Guide	Forrester, Sobester, Keane	August 2008
Aircraft Systems, 3rd Edition	Moir & Seabridge	March 2008
Introduction to Aircraft Aeroelasticity And Loads	Wright & Cooper	December 2007
Stability and Control of Aircraft Systems	Langton	September 2006
Military Avionics Systems	Moir & Seabridge	February 2006
Design and Development of Aircraft Systems	Moir & Seabridge	June 2004
Aircraft Loading and Structural Layout	Howe	May 2004
Aircraft Display Systems	Jukes	December 2003
Civil Avionics Systems	Moir & Seabridge	December 2002

UNDERSTANDING AERODYNAMICS

ARGUING FROM THE REAL PHYSICS

Doug McLean

Technical Fellow (retired), Boeing Commercial Airplanes, USA

A John Wiley & Sons, Ltd., Publication

Registered office
John Wiley & Sons Ltd, The Atrium, Southern Gate, Chichester, West Sussex, PO19 8SQ, United Kingdom

For details of our global editorial offices, for customer services and for information about how to apply for permission to reuse the copyright material in this book please see our website at www.wiley.com.

Library of Congress Cataloging-in-Publication Data

McLean, Doug (Doug J.)
  Understanding aerodynamics : arguing from the real physics / Doug McLean.
    pages cm
  Includes bibliographical references and index.
  ISBN 978-1-119-96751-4 (hardback)
  1. Aerodynamics. I. Title.
  TL570.M3823 2013
  629.132′3 – dc23

                                    2012032706

A catalogue record for this book is available from the British Library

Print ISBN: 978-1-119-96751-4

Typeset in 10/12pt Times by Laserwords Private Limited, Chennai, India

Contents

Foreword   xi

Series Preface   xiii

Preface   xv

List of Symbols   xix

1   Introduction to the Conceptual Landscape   **1**

2   From Elementary Particles to Aerodynamic Flows   **5**

3   Continuum Fluid Mechanics and the Navier-Stokes Equations   **13**

3.1   The Continuum Formulation and Its Range of Validity   13

3.2   Mathematical Formalism   16

3.3   Kinematics: Streamlines, Streaklines, Timelines, and Vorticity   18

     3.3.1   Streamlines and Streaklines   18

     3.3.2   Streamtubes, Stream Surfaces, and the Stream Function   19

     3.3.3   Timelines   22

     3.3.4   The Divergence of the Velocity and Green's Theorem   23

     3.3.5   Vorticity and Circulation   24

     3.3.6   The Velocity Potential in Irrotational Flow   26

     3.3.7   Concepts that Arise in Describing the Vorticity Field   26

     3.3.8   Velocity Fields Associated with Concentrations of Vorticity   29

     3.3.9   The Biot-Savart Law and the "Induction" Fallacy   31

3.4   The Equations of Motion and their Physical Meaning   33

     3.4.1   Continuity of the Flow and Conservation of Mass   34

     3.4.2   Forces on Fluid Parcels and Conservation of Momentum   35

     3.4.3   Conservation of Energy   36

     3.4.4   Constitutive Relations and Boundary Conditions   37

     3.4.5   Mathematical Nature of the Equations   37

     3.4.6   The Physics as Viewed in the Eulerian Frame   38

     3.4.7   The Pseudo-Lagrangian Viewpoint   40

3.5   Cause and Effect, and the Problem of Prediction   40

3.6	The Effects of Viscosity	43
3.7	Turbulence, Reynolds Averaging, and Turbulence Modeling	48
3.8	Important Dynamical Relationships	55
	3.8.1 Galilean Invariance, or Independence of Reference Frame	55
	3.8.2 Circulation Preservation and the Persistence of Irrotationality	56
	3.8.3 Behavior of Vortex Tubes in Inviscid and Viscous Flows	57
	3.8.4 Bernoulli Equations and Stagnation Conditions	58
	3.8.5 Crocco's Theorem	60
3.9	Dynamic Similarity	60
	3.9.1 Compressibility Effects and the Mach Number	63
	3.9.2 Viscous Effects and the Reynolds Number	63
	3.9.3 Scaling of Pressure Forces: the Dynamic Pressure	64
	3.9.4 Consequences of Failing to Match All of the Requirements for Similarity	65
3.10	"Incompressible" Flow and Potential Flow	66
3.11	Compressible Flow and Shocks	70
	3.11.1 Steady 1D Isentropic Flow Theory	71
	3.11.2 Relations for Normal and Oblique Shock Waves	74
4	**Boundary Layers**	**79**
4.1	Physical Aspects of Boundary-Layer Flows	80
	4.1.1 The Basic Sequence: Attachment, Transition, Separation	80
	4.1.2 General Development of the Boundary-Layer Flowfield	82
	4.1.3 Boundary-Layer Displacement Effect	90
	4.1.4 Separation from a Smooth Wall	93
4.2	Boundary-Layer Theory	99
	4.2.1 The Boundary-Layer Equations	100
	4.2.2 Integrated Momentum Balance in a Boundary Layer	108
	4.2.3 The Displacement Effect and Matching with the Outer Flow	110
	4.2.4 The Vorticity "Budget" in a 2D Incompressible Boundary Layer	113
	4.2.5 Situations That Violate the Assumptions of Boundary-Layer Theory	114
	4.2.6 Summary of Lessons from Boundary-Layer Theory	117
4.3	Flat-Plate Boundary Layers and Other Simplified Cases	117
	4.3.1 Flat-Plate Flow	117
	4.3.2 2D Boundary-Layer Flows with Similarity	121
	4.3.3 Axisymmetric Flow	123
	4.3.4 Plane-of-Symmetry and Attachment-Line Boundary Layers	125
	4.3.5 Simplifying the Effects of Sweep and Taper in 3D	128
4.4	Transition and Turbulence	130
	4.4.1 Boundary-Layer Transition	131
	4.4.2 Turbulent Boundary Layers	138
4.5	Control and Prevention of Flow Separation	150
	4.5.1 Body Shaping and Pressure Distribution	150
	4.5.2 Vortex Generators	150
	4.5.3 Steady Tangential Blowing through a Slot	155

	4.5.4	Active Unsteady Blowing	157
	4.5.5	Suction	157
4.6	Heat Transfer and Compressibility		158
	4.6.1	Heat Transfer, Compressibility, and the Boundary-Layer Temperature Field	158
	4.6.2	The Thermal Energy Equation and the Prandtl Number	159
	4.6.3	The Wall Temperature and Other Relations for an Adiabatic Wall	159
4.7	Effects of Surface Roughness		162

5   **General Features of Flows around Bodies**   **163**

5.1	The Obstacle Effect		164
5.2	Basic Topology of Flow Attachment and Separation		168
	5.2.1	Attachment and Separation in 2D	169
	5.2.2	Attachment and Separation in 3D	171
	5.2.3	Streamline Topology on Surfaces and in Cross Sections	176
5.3	Wakes		186
5.4	Integrated Forces: Lift and Drag		189

6   **Drag and Propulsion**   **191**

6.1	Basic Physics and Flowfield Manifestations of Drag and Thrust		192
	6.1.1	Basic Physical Effects of Viscosity	193
	6.1.2	The Role of Turbulence	193
	6.1.3	Direct and Indirect Contributions to the Drag Force on the Body	194
	6.1.4	Determining Drag from the Flowfield: Application of Conservation Laws	196
	6.1.5	Examples of Flowfield Manifestations of Drag in Simple 2D Flows	204
	6.1.6	Pressure Drag of Streamlined and Bluff Bodies	207
	6.1.7	Questionable Drag Categories: Parasite Drag, Base Drag, and Slot Drag	210
	6.1.8	Effects of Distributed Surface Roughness on Turbulent Skin Friction	212
	6.1.9	Interference Drag	222
	6.1.10	Some Basic Physics of Propulsion	225
6.2	Drag Estimation		241
	6.2.1	Empirical Correlations	242
	6.2.2	Effects of Surface Roughness on Turbulent Skin Friction	243
	6.2.3	CFD Prediction of Drag	250
6.3	Drag Reduction		250
	6.3.1	Reducing Drag by Maintaining a Run of Laminar Flow	251
	6.3.2	Reduction of Turbulent Skin Friction	251

7   **Lift and Airfoils in 2D at Subsonic Speeds**   **259**

| 7.1 | Mathematical Prediction of Lift in 2D | | 260 |

7.2	Lift in Terms of Circulation and Bound Vorticity	265
	7.2.1 *The Classical Argument for the Origin of the Bound Vorticity*	267
7.3	Physical Explanations of Lift in 2D	269
	7.3.1 *Past Explanations and their Strengths and Weaknesses*	269
	7.3.2 *Desired Attributes of a More Satisfactory Explanation*	284
	7.3.3 *A Basic Explanation of Lift on an Airfoil, Accessible to a Nontechnical Audience*	286
	7.3.4 *More Physical Details on Lift in 2D, for the Technically Inclined*	302
7.4	Airfoils	307
	7.4.1 *Pressure Distributions and Integrated Forces at Low Mach Numbers*	307
	7.4.2 *Profile Drag and the Drag Polar*	316
	7.4.3 *Maximum Lift and Boundary-Layer Separation on Single-Element Airfoils*	319
	7.4.4 *Multielement Airfoils and the Slot Effect*	329
	7.4.5 *Cascades*	335
	7.4.6 *Low-Drag Airfoils with Laminar Flow*	338
	7.4.7 *Low-Reynolds-Number Airfoils*	341
	7.4.8 *Airfoils in Transonic Flow*	342
	7.4.9 *Airfoils in Ground Effect*	350
	7.4.10 *Airfoil Design*	352
	7.4.11 *Issues that Arise in Defining Airfoil Shapes*	354
8	**Lift and Wings in 3D at Subsonic Speeds**	**359**
8.1	The Flowfield around a 3D Wing	359
	8.1.1 *General Characteristics of the Velocity Field*	359
	8.1.2 *The Vortex Wake*	362
	8.1.3 *The Pressure Field around a 3D Wing*	371
	8.1.4 *Explanations for the Flowfield*	371
	8.1.5 *Vortex Shedding from Edges Other Than the Trailing Edge*	375
8.2	Distribution of Lift on a 3D Wing	376
	8.2.1 *Basic and Additional Spanloads*	376
	8.2.2 *Linearized Lifting-Surface Theory*	379
	8.2.3 *Lifting-Line Theory*	380
	8.2.4 *3D Lift in Ground Effect*	382
	8.2.5 *Maximum Lift, as Limited by 3D Effects*	384
8.3	Induced Drag	385
	8.3.1 *Basic Scaling of Induced Drag*	385
	8.3.2 *Induced Drag from a Farfield Momentum Balance*	386
	8.3.3 *Induced Drag in Terms of Kinetic Energy and an Idealized Rolled-Up Vortex Wake*	389
	8.3.4 *Induced Drag from the Loading on the Wing Itself: Trefftz-Plane Theory*	391
	8.3.5 *Ideal (Minimum) Induced-Drag Theory*	394
	8.3.6 *Span-Efficiency Factors*	396
	8.3.7 *The Induced-Drag Polar*	397

	8.3.8	The Sin-Series Spanloads	398
	8.3.9	The Reduction of Induced Drag in Ground Effect	401
	8.3.10	The Effect of a Fuselage on Induced Drag	402
	8.3.11	Effects of a Canard or Aft Tail on Induced Drag	404
	8.3.12	Biplane Drag	409
8.4	Wingtip Devices		411
	8.4.1	Myths Regarding the Vortex Wake, and Some Questionable Ideas for Wingtip Devices	411
	8.4.2	The Facts of Life Regarding Induced Drag and Induced-Drag Reduction	414
	8.4.3	Milestones in the Development of Theory and Practice	420
	8.4.4	Wingtip Device Concepts	422
	8.4.5	Effectiveness of Various Device Configurations	423
8.5	Manifestations of Lift in the Atmosphere at Large		427
	8.5.1	The Net Vertical Momentum Imparted to the Atmosphere	427
	8.5.2	The Pressure Far above and below the Airplane	429
	8.5.3	Downwash in the Trefftz Plane and Other Momentum-Conservation Issues	431
	8.5.4	Sears's Incorrect Analysis of the Integrated Pressure Far Downstream	435
	8.5.5	The Real Flowfield Far Downstream of the Airplane	436
8.6	Effects of Wing Sweep		444
	8.6.1	Simple Sweep Theory	444
	8.6.2	Boundary Layers on Swept Wings	449
	8.6.3	Shock/Boundary-Layer Interaction on Swept Wings	464
	8.6.4	Laminar-to-Turbulent Transition on Swept Wings	465
	8.6.5	Relating a Swept, Tapered Wing to a 2D Airfoil	468
	8.6.6	Tailoring of the Inboard Part of a Swept Wing	469
9	**Theoretical Idealizations Revisited**		**471**
9.1	Approximations Grouped According to how the Equations were Modified		471
	9.1.1	Reduced Temporal and/or Spatial Resolution	472
	9.1.2	Simplified Theories Based on Neglecting Something Small	472
	9.1.3	Reductions in Dimensions	472
	9.1.4	Simplified Theories Based on Ad hoc Flow Models	472
	9.1.5	Qualitative Anomalies and Other Consequences of Approximations	481
9.2	Some Tools of MFD (Mental Fluid Dynamics)		482
	9.2.1	Simple Conceptual Models for Thinking about Velocity Fields	482
	9.2.2	Thinking about Viscous and Shock Drag	485
	9.2.3	Thinking about Induced Drag	486
	9.2.4	A Catalog of Fallacies	487
10	**Modeling Aerodynamic Flows in Computational Fluid Dynamics**		**491**
10.1	Basic Definitions		493

10.2   The Major Classes of CFD Codes and Their Applications   493
    10.2.1   Navier-Stokes Methods   493
    10.2.2   Coupled Viscous/Inviscid Methods   497
    10.2.3   Inviscid Methods   498
    10.2.4   Standalone Boundary-Layer Codes   501
10.3   Basic Characteristics of Numerical Solution Schemes   501
    10.3.1   Discretization   501
    10.3.2   Spatial Field Grids   502
    10.3.3   Grid Resolution and Grid Convergence   506
    10.3.4   Solving the Equations, and Iterative Convergence   507
10.4   Physical Modeling in CFD   508
    10.4.1   Compressibility and Shocks   508
    10.4.2   Viscous Effects and Turbulence   510
    10.4.3   Separated Shear Layers and Vortex Wakes   511
    10.4.4   The Farfield   513
    10.4.5   Predicting Drag   514
    10.4.6   Propulsion Effects   515
10.5   CFD Validation?   515
10.6   Integrated Forces and the Components of Drag   516
10.7   Solution Visualization   517
10.8   Things a User Should Know about a CFD Code before Running it   524

References   **527**

Index   **539**

Foreword

The job of the aeronautical engineer has changed dramatically in recent years and will continue to change. Advanced computational tools have revolutionized design processes for all types of flight vehicles and have made it possible to achieve levels of design technology previously unheard of. And as performance targets have become more demanding, the individual engineer's role in the design process has become increasingly specialized.

In this new environment, design work depends heavily on voluminous numerical computations. The computer handles much of the drudgery, but it can't do the thinking. It is now more important than ever for a practicing engineer to bring to the task a strong physical intuition, solidly based in the physics. In this book, Doug McLean provides a valuable supplement to the many existing books on aerodynamic theory, patiently exploring what it all means from a physical point of view. Students and experienced engineers alike will surely profit from following the thought-provoking arguments and discussions presented here.

John J. Tracy
Chief Technology Officer
The Boeing Company
September 2012

Series Preface

The field of aerospace is wide ranging and multi-disciplinary, covering a large variety of products, disciplines and domains, not merely in engineering but in many related supporting activities. These combine to enable the aerospace industry to produce exciting and technologically advanced vehicles. The wealth of knowledge and experience that has been gained by expert practitioners in the various aerospace fields needs to be passed onto others working in the industry, including those just entering from University.

The *Aerospace Series* aims to be a practical and topical series of books aimed at engineering professionals, operators, users and allied professions such as commercial and legal executives in the aerospace industry, and also engineers in academia. The range of topics is intended to be wide ranging, covering design and development, manufacture, operation and support of aircraft as well as topics such as infrastructure operations and developments in research and technology. The intention is to provide a source of relevant information that will be of interest and benefit to all those people working in aerospace.

Aerodynamics is the fundamental enabling science that underpins the world-wide aerospace industry – without the ability to generate lift from airflow passing over wings, helicopter rotors and other lifting surfaces, it would not be possible to fly heavier-than-air vehicles as efficiently as is taken for granted nowadays. Much of the development of today's highly efficient aircraft is due to the ability to accurately model aerodynamic flows using sophisticated computational codes and thus design high-performance wings; however, a thorough understanding and insight of the aerodynamic flows is vital for engineers to comprehend these designs.

This book, *Understanding Aerodynamics*, has the objective of providing a physical understanding of aerodynamics, with an emphasis on how and why particular flow patterns around bodies occur, and what relation these flows have to the underlying physical laws. It is a welcome addition to the Wiley Aerospace Series. Unlike most aerodynamics textbooks, there is a refreshing lack of detailed mathematical analysis, and the reader is encouraged instead to consider the overall picture. As well as consideration of classical topics – continuum fluid mechanics, boundary layers, lift, drag and the flow around wings, etc. – there is also a very useful coverage of modelling aerodynamic flows using Computational Fluid Dynamics (CFD).

Peter Belobaba, Jonathan Cooper, Roy Langton and Allan Seabridge

Preface

This book is intended to help students and practicing engineers to gain a greater physical understanding of aerodynamics. It is not a handbook on how to do aerodynamics, but is motivated instead by the assumption that engineering practice is enhanced in the long run by a robust understanding of the basics.

A real understanding of aerodynamics must go beyond mastering the mathematical formalism of the theories and come to grips with the physical cause-and-effect relationships that the theories represent. In addition to the math, which applies most directly at the local level, intuitive physical interpretations and explanations are required if we are to understand what happens at the flowfield level. Developing this physical side of our understanding is surprisingly difficult, however. It requires navigating a conceptual landscape littered with potential pitfalls, and an acceptable path is to be found only through recognition and rejection of multiple faulty paths. It is really a process of argumentation, thus the "arguing" in the title. This kind of argumentation is underemphasized in other books, in which the path is often made to appear straighter and simpler than it really is. This book explores a broader swath of the conceptual landscape, including some of the false paths that have led to errors in the past, with the hope that it will leave the reader less likely to fall victim to misconceptions.

We'll encounter several instances of serious misinterpretations of mathematical theory that are in wide circulation and of erroneous physical explanations that have found their way into our folklore. In any case where a misconception has been widely enough propagated, the "right" explanation would not be complete without the debunking of the "wrong" one. I have tried to do this kind of debunking wherever it seemed appropriate and have not hesitated to say so when I think something is wrong. This is part of what makes aerodynamics so much fun. It's one of those little perversities of human nature that coming up with a good explanation is much more satisfying when you know there are people out there who have got it wrong. But debunking bad explanations serves a pedagogical purpose as well, because the contrast provided by the wrong explanation can strengthen understanding of the right one.

This effort devoted to basic physical rigor and avoiding errors comes at a cost. We'll spend more time on some topics than some will likely think necessary. I realize some parts of the discussion are long and are not easy, but I hope most readers will find it worth the effort.

We are now well into what I would call the computational era in aerodynamics, made possible by the ever-advancing capabilities of computers. In the 1960s, we began to calculate practical numerical solutions to linear equations for inviscid flows in 3D. In the 1970s, it became economical to compute solutions to nonlinear equations for inviscid transonic flows in 3D and to include viscous effects through boundary-layer theory and viscous/inviscid coupling. By the 1990s, we were routinely calculating solutions to the Reynolds-averaged Navier-Stokes (RANS) equations for full airplane configurations. These computational fluid dynamics (CFD) capabilities have revolutionized aerodynamics analysis and design and have made possible dramatic improvements in design technology. CFD is now such a vital part of our discipline that this book would not be complete if it did not address it in some way. While this is not a book about CFD methods or about how to use CFD, there are conceptual aspects of CFD that are relevant to our focus, and these are considered in chapter 10.

I believe that although we now rely on CFD for much of our quantitative work, it is vitally important for a practicing engineer to have a sound understanding of the underlying physics and to be familiar with the old simplified theories that our predecessors so ingeniously developed. These things not only provide us with valuable ways of thinking about our problems, they also can help us to be more effective users of CFD.

The unusual scope of the book is deliberate. The book is not intended to be a handbook. Nor is it intended as a substitute for the standard textbooks and other sources on aerodynamic theory, as I have omitted the mathematical details whenever the physical understanding I seek to promote can be conveyed without them. This applies especially to the discussion of the basic physics in the early chapters. Those looking for rigorous derivations of the mathematical details will have to look elsewhere. Also, exhaustive scope is not a practical goal. So, for the details on many of the topics treated here, and for any treatment at all of the many topics neglected here, the reader will have to consult other sources. This book is also not intended as an introduction to the subject. Though it would not be impossible for someone with no prior exposure to follow the development given here, some experience with the subject will make it much easier. And while I assume no prior knowledge of the subject, I do assume a higher level of technical sophistication than is often assumed in undergraduate-level texts.

An understanding of the physical basics is more secure if it includes an appreciation of the "big picture," the logical structure of the body of knowledge and the collection of concepts we call aerodynamics. I have tried to at least touch on all of the topics that are so basic that the overall framework could not stand without them. I also devote more attention than most aerodynamics textbooks to the relationships between the parts, to how it all "fits together." Beyond that, several considerations have guided my choice of topics and the kinds of treatment I've given them. One is my own familiarity and experience. Another is my observation of some common knowledge gaps, things that don't seem to be covered well in the usual aero engineering education. But we'll also spend a good part of our time on some of the very familiar things that we tend to take for granted. Our understanding of these things is never so good that it can't benefit from taking a fresh look. We'll put a different spin on some familiar topics, for example, what the Biot-Savart law really means and why it causes so much confusion, what "Reynolds number" and "incompressible flow" really mean, and a real physical explanation for how an airfoil produces lift.

As we'll see in chapter 1, the *subject matter* of aerodynamics consists of physical principles, conceptual models, mathematical theories, and descriptions and physical explanations of flow phenomena. Some of this subject matter has direct practical applications, and some doesn't. We'll spend considerable time on some topics that have no apparent practical import, for example, physical explanations of things for which we have perfectly good quantitative theories and esoterica such as how lift is felt in the atmosphere at large. We'll do these things because they provide general fluid-mechanics insight and because they serve to expand our appreciation of the *cognitive dimension* of the subject, the processes by which we *think about* aerodynamic phenomena and the practical problems that arise from them. They also help us to see how mistaken thinking can arise and how to avoid it. The medical profession in recent years has begun to pay more attention to the cognitive dimension of their discipline, studying how doctors think, in an effort to improve the accuracy of their diagnoses and to avoid mistakes. Doing some of the same would be good for us as well.

Aerodynamics as a subject encompasses a wide variety of flow situations that in turn involve a multitude of detailed flow phenomena. The subject is correspondingly multifaceted, with a rich web of interconnections among the phenomena themselves and the conceptual models that have been developed to represent them. Such a subject has a logical structure of course, but it is not well suited to exposition in a single linear narrative, and there is therefore no ideal solution to the problem of organizing it so that it flows completely naturally as a single string of words. The organization I have chosen is based not on the historical development or on a progression from "easy" concepts to "advanced," but on a general conceptual progression, from the basic physics, to the flow phenomena, and finally to the conceptual models. I have tried to organize the material so that it can be read straight through and understood without the need to skip forward. I have also tried to provide direct references whenever I think referring back to previous chapters would be helpful and to alert the reader when further discussion of a topic is being deferred until later.

The general flow of the book is as follows. First, we take an overview of the conceptual landscape in chapter 1. Then we consider the basic *physics* as embodied in the NS equations in chapters 2 and 3. We turn to the phenomenological aspects of *general flows* in boundary layers and around bodies in chapters 4 and 5. We then enter the more specific realm of *aerodynamic forces* and their manifestations in flowfields to deal with drag in chapter 6 and lift generation, airfoils, and wings in chapters 7 and 8. All of this sets the stage for a bit of a regression into *theory*, with discussions of theoretical approximations and CFD in chapters 9 and 10.

When I started writing I had something less ambitious in mind, something more on the scale of a booklet with a collection of helpful ways of looking at aerodynamic phenomena and a catalog of common misconceptions and how to avoid them. As the project progressed, it became clear that effective explanations required more background than I had anticipated, and the book gradually grew more comprehensive. The first draft in something close to the final form was completed in late 2008 and was reviewed by several Boeing colleagues (acknowledged below). Their feedback was incorporated into a second draft that was used in a 20-week after-hours class for Boeing engineers in 2009. Feedback from class participants and others led to significant revisions for the final draft. As it turned out, the general argumentative approach I've taken to the subject extended to the writing process itself. Many sections saw multiple and substantial rewrites as my thinking evolved.

I gratefully acknowledge the help of many people in getting me through this long process. First, my wife, Theresa, who put up with the many, many weekends that I spent in front of our home computer. Then The Boeing Company, which allowed me to spend considerable company time on the project, Boeing editors, Andrea Jarvela, Lisa Fusch Krause, and Charlene Scammon, who turned my raw Word files and graphics into a presentable draft and helped me take that draft through several revisions, and Boeing graphics artist John Jolley, who redrew nearly half the graphics. Finally, the friends and colleagues without whose help

the book would have been much poorer. Mark Drela (MIT), Lian Ng, Ben Rider, Philippe
Spalart, and Venkat Venkatakrishnan provided very detailed feedback and suggestions for
improvement. Steve Allmaras and Mitch Murray made special CFD calculations just for the
book. My former Boeing colleague Guenter Brune wrote the excellent 1983 Boeing report
on flow topology that introduced me to the topic and served as the basis of much of Section
5.2.3. Another former Boeing colleague, Pete Sullivan, did the CFD calculations plotted in
Section 6.1.5. And many others contributed feedback on various drafts of the manuscript:
Anders Andersson, John D. Anderson (University of Maryland), Byram Bays-Muchmore,
Bob Breidenthal (University of Washington), Julie Brightwell, Tad Calkins, Dave Caughey
(Cornell University), Tony Craig, Jeffrey Crouch, Peg Curtin, Bruce Detert, Scott Eberhardt,
Winfried Feifel, David Fritz, Arvel Gentry, Mark Goldhammer, Elisabeth Gren, Rob Hof-
fenberg, Paul Johnson, Wen-Huei Jou, T. J. Kao, Edward Kim, Alex Krynytsky, Brenda
Kulfan, Louie LeGrand, Adam Malachowski, Adam Malone, Tom Matoi, Mark Maughmer
(Penn State University), the late John McMasters, Kevin Mejia, Robin Melvin, Greg Miller,
Deepak Om, Ben Paul, Tim Purcell, Steve Ray, Matt Smith, John Sullivan (Purdue Univer-
sity), Mary Sutanto, Ed Tinoco, David Van Cleve, Paul Vijgen, Dave Witkowski, Conrad
Youngren (New York Maritime College), and Jong Yu.

Thanks also to the copyright owners who kindly gave permission to use the many graphics
I borrowed from elsewhere. They are acknowledged individually in the figure titles.

Doug McLean,
April 2012.

List of Symbols

Many of the symbols listed below have different meanings in different contexts, as indicated when multiple definitions are given. When an example of usage (a figure or equation) is listed, it is not necessarily the only example.

English Symbols

a	Acceleration
	Speed of sound
A	Streamtube area
	Amplitude of a disturbance in laminar-flow stability theory
A_1, A_2	Coefficients of induced-drag polar (equation 8.3.17)
A_i	Coefficients of sin-series spanloads (equation 8.3.20)
A_θ	Wake momentum area (equation 6.1.6)
AR	Aspect ratio $= b^2/s$
b	Wingspan
b_o	Span between the centers of trailing vortex cores (figure 8.3.3)
B	Constant in law of wall (equation 4.4.10)
c	Airfoil chord
c_{avg}	Wing average chord
$\bar{c}$	Wing average chord (figure 8.3.7)
c_p	Specific heat at constant pressure
c_v	Specific heat at constant volume
C	Cylindrical part of the outer boundary of a control volume (figure 6.1.1)
C_d	Drag coefficient (2D or per unit span)
C_D	Drag coefficient (3D)
C_{Di}	Induced-drag coefficient (3D)
C_{Dimin}	Minimum induced drag coefficient on induced-drag polar (equation 8.3.19)
C_{Dio}	Induced drag coefficient at zero lift (equation 8.3.17)
C_{Do}	Drag coefficient at zero lift (equation 8.3.15)
C_f	Skin-friction coefficient
C_l	Lift coefficient (2D or per unit span)
C_L	Lift coefficient (3D)
C_{Lmax}	Maximum lift coefficient of a 3D wing

C_m	Pitching moment coefficient (2D or per unit span)
C_M	Pitching moment coefficient (3D)
C_n	Sectional normal-force coefficient (figure 8.3.7)
C_N	Total normal-force coefficient (figure 8.3.7)
C_p	Pressure coefficient
$\overline{C_p}$	Smith's canonical pressure coefficient (equation 7.4.1)
C^*	Attachment-line Reynolds number (equation 8.6.15)
$\mathbf{d}$	Viscous drag vector of a propeller blade section (figure 6.1.17)
d_b	Diameter of fuselage (equation 8.3.29)
D	Drag
D_i	Induced drag (equation 8.3.5)
e	Thermodynamic internal energy
	Base of natural logs
	Induced-drag efficiency factor (equation 8.3.14)
e_{NT}	Induced-drag efficiency factor of untwisted version of a wing (equation 8.3.19)
e_o	Oswald efficiency factor (equation 8.3.15)
F	Force
f	Function
g	Genus of a region of a surface (equation 5.2.2)
h	Enthalpy
	Height of a vortex generator
	Height of an excrescence (figure 6.2.1)
	Height above the ground (equation 7.4.2)
h_p	Riblet protrusion height (figure 6.3.6)
H	Total enthalpy
	Boundary-layer shape factor
i	Unit vector in x direction (equation 6.1.1)
	Square root of minus one
I	Index of a region of a surface (equation 5.2.3)
I_{CF}	Cumulative skin-friction integral (equation 6.1.13)
I_{ENS}	Enstrophy integral (equation 6.1.15)
J	Propeller advance ratio $= V/nd_p$
k	Thermal conductivity
	Roughness height
$\mathbf{k}$	Unit curvature vector (figure 4.2.4)
	Unit vector in z direction (equation 5.4.1)
k_s	Equivalent sand-grain height
l	Length
l	Lift per unit span of a wing or propeller blade
L	Length
	Lift in 3D (equation 5.4.1)
L_b	Carry-through lift on fuselage (figure 8.3.15)
L_t	Lift on tail or canard (figure 8.3.17)
L_w	Lift on exposed wing (figure 8.3.15)
M	Mach number
	Propeller shaft torque (equation 6.1.19)

M_{BR}	Wing-root bending moment (equation 8.3.28)
m	Mass
	Exponent in power-law velocity distributions for laminar boundary layers (section 4.3.2)
	Excrescence-drag magnification factor (figure 6.1.13)
	Exponent in Smith's power-law airfoil velocity distributions (figure 7.4.12)
$\dot{m}$	Mass flux of a source or sink (equation 8.3.3)
n	Normal direction
	Propeller revolutions per unit time
n	Unit normal vector (figure 3.3.6)
N	Nodal-point singularity (figure 5.2.6)
p	Pressure
P	Propeller shaft power (equation 6.1.20)
Pr	Prandtl number
q	Dynamic pressure $= 1/2\rho_{ref}\, u_{ref}^2$
R	Ideal gas constant
	Reynolds number (any subscript indicates reference length used)
	Radius from airplane to a point on the ground (figure 8.5.3)
$\overline{R}$	Attachment-line Reynolds number parameter (equation 8.6.16)
R_a	Average of absolute value of roughness height
R_{crit}	Critical Reynolds number, at onset of instability (figure 4.4.3)
r	Radius
	Recovery factor (equation 4.6.4)
r_1	Radius to the maximum-velocity point in a trailing vortex core (figure 8.1.8)
r_2	Radius to the point of effectively zero vorticity in a vortex core (figure 8.1.8)
r_b	Radius of fuselage (figure 8.3.14)
r_c	Radius of vortex core (equation 8.3.6)
s	Arc length
	Riblet spacing
	Wing area
s_p	Propeller disc area
S	Entropy
	Saddle-point singularity (figure 5.2.6)
	Denotes integration over a surface (equation 6.1.1)
T	Temperature
	The Trefftz plane (figure 6.1.1)
	Thrust of a propeller (equation 6.1.18)
t	Time
t	Unit tangent vector
u	Cartesian x-velocity component
	Perturbation velocity in the x direction (figure 7.4.1 (a))
u_τ	Friction velocity $u_\tau = \sqrt{\tau_w/\rho}$
U_∞	Undisturbed freestream velocity in the x direction
v	Cartesian y-velocity component
	Perturbation velocity in the y direction (figure 7.4.1 (a))
V	Velocity vector

V	Velocity magnitude
	Volume
w	Cartesian z-velocity component
	Perturbation velocity in the z direction (equation 8.3.5)
W	Outer function in the law of the wake (equation 4.4.13)
x	Cartesian space coordinate
y	Cartesian space coordinate
z	Cartesian space coordinaten

Greek Symbols

α	Wavenumber in x direction (equation 4.4.1)
	Angle of attack
α_o	Angle of attack at zero total lift (equation 8.3.16)
β	Flow direction angle
	Laminar boundary-layer similarity parameter (figures 4.3.5 and 4.3.6)
	Turbulent boundary-layer similarity parameter (equation 4.3.8)
	Wavenumber in z direction (equation 4.4.1)
δ	Boundary-layer thickness
δ	Boundary-layer thickness
δ^*	Boundary-layer displacement thickness (equation 4.2.13)
δ^*_{loc}	Local δ^* integral (equation 4.2.14)
ϕ	Velocity potential (equation 3.10.1)
γ	Ratio of specific heats, c_p/c_v
Γ	Circulation
Γ_o	Circulation of vortex core (equation 8.3.6)
η	Dimensionless spanwise coordinate on a wing = 2y/b
η	Propeller efficiency = thrust work out/shaft work in (equation 6.1.21)
η_i	Propeller induced efficiency
κ	Streamline curvature (figure 4.2.4)
	Von Karman constant (equation 4.4.10)
Λ	Wing sweep (figure 8.6.1)
λ	Mixing length
μ	Coefficient of shear viscosity
	Propeller torque coefficient (equation 6.1.19)
μ_{eff}	Effective viscosity, sum of viscous and turbulent (equation 6.1.15)
ν	Kinematic viscosity, μ/ρ
π	pi = 3.14159 ...
Π	Constant in the law of the wake (equation 4.4.13)
θ	Boundary-layer momentum thickness (equation 4.2.11)
	Angular coordinate around circular cylinder (figure 5.1.3)
	Flow angles entering and leaving a cascade (figure 7.4.23 (b))
	Dihedral angle (equation 8.3.11)
ρ	Density
σ	Propeller power loading (equation 6.1.20)

τ	Shear stress
	Propeller thrust loading (equation 6.1.18)
ω	Vorticity magnitude
	Frequency (equation 4.4.1)
$\boldsymbol{\omega}$	Vorticity vector
ψ	Transformed spanwise coordinate (equation 8.3.22)

Subscripts

1	In the boundary-layer x direction (equation 4.2.15)
	Denotes conditions upstream of a cascade (figure 7.4.23 (b))
	Denotes conditions upstream of the shock on a transonic airfoil (figure 7.4.31)
2	Denotes conditions downstream of a cascade (figure 7.4.23 (b))
	Denotes conditions downstream of the shock on a transonic airfoil (figure 7.4.31)
3	In the boundary-layer z direction (equation 4.2.15)
b	Of the boundary-layer coordinate system (figure 4.2.2)
c	Pertaining to a vortex core (figure 8.3.3)
	Cross-flow component (figure 4.1.7)
	Airfoil chord
ch	Chordwise (figures 4.3.11 and 4.3.12)
cut	Pertaining to a cut through a wake (figure 6.1.3)
d	Based on diameter
	Drag per unit span
D	Drag
e	At the edge of the boundary layer
f	Friction
i	Incompressible
	Induced, as applied to a propeller
j	Pertaining to a blowing jet (equation 4.5.1)
K	Kinematic (equation 4.6.8)
l	Lift per unit span
L	Based on length L
	Lift
local	Denoting the effective local freestream condition for an excrescence
m	Pitching moment per unit span
n	Connectivity of a 2D domain (equation 5.2.5)
	Direction normal to wake cut (equation 8.3.10)
o	Denotes conditions at the start of an airfoil pressure recovery (equation 7.4.1)
p	At constant pressure
	Propeller
ref	Reference
rms	Root-mean-square
s	Relative to streamwise direction at boundary-layer edge (figure 4.1.8)
sep	Denotes conditions at separation (discussion of figure 7.4.13)
sp	Spanwise (figures 4.3.11 and 4.3.12)

sw	From steamwise at boundary-layer edge to the wall (figure 4.1.8)
t	Turbulent (equation 3.7.5)
	Total (or stagnation) (equations 3.8.3–6)
T	Thermal (equation 4.6.2)
v	At constant volume
w	At the wall
x	Based on x
∞	At infinite distance, far field
	At infinite height from ground (figure 8.3.12)
$\perp$	Perpendicular
perp	Perpendicular to constant-percent-chord lines (figure 8.6.11)
$\parallel$	Parallel

Greek Subscripts

$\delta*$	Based on displacement thickness
μ	Momentum, as in C_μ (equation 4.5.1)
θ	Based on momentum thickness
τ	Friction velocity, in $u_\tau = \sqrt{\tau_w/\rho}$

Superscripts

$\prime$	Fluctuating part
	Independent variable in integration
	Denotes a half singularity at a boundary of a surface (equation 5.2.5)
—	Time average
	Averaged along the length
$\wedge$	Nondimensional
*	Denotes conditions at Mach 1 (equations 3.11.4 and 3.11.5)
	Displacement thickness, when used with δ
+	Turbulent-boundary-layer wall variables (equations 4.4.3 and 4.4.4)
=	Tensor (equation 5.4.1)

Acronyms and Abbreviations

1D	One dimensional
2D	Two dimensional
3D	Three dimensional
BC	Boundary condition
BLC	Boundary-layer control
CFD	Computational fluid dynamics
CF	Cross-flow
CPU	Central processing unit
DES	Detached-eddy simulation
DNS	Direct numerical simulation

ESDU	Engineering Sciences Data Unit
GGNS	General geometry Navier-Stokes
HLFC	Hybrid laminar flow control
LES	Large-eddy simulation
LFC	Laminar flow control
LTA	Lighter than air
NACA	National Advisory Committee for Aeronautics
NASA	National Aeronautics and Space Administration
NLF	Natural laminar flow
NS	Navier-Stokes
ODE	Ordinary differential equation
ONERA	Office National d'Etude et Recherches Aérospatiales
PC	Personal computer
PDE	Partial differential equation
RABL	Reynolds-averaged boundary-layer equations
RANS	Reynolds-averaged Navier-Stokes equations
SA	Spalart-Allmaras
SBVG	Sub-boundary-layer vortex generator
SST	Shear Stress Transport
TKE	Turbulence kinetic energy
TS	Tollmien-Schlichting
URANS	Unsteady Reynolds-averaged Navier-Stokes
VG	Vortex generator
WINGOP	Wing Optimization

1

Introduction to the Conceptual Landscape

The objective of this book is to promote a solid *physical understanding* of aerodynamics. In general, any understanding of physical phenomena requires conceptual models:

> *It seems that the human mind has first to construct forms independently before we can find them in things. Kepler's marvelous achievement is a particularly fine example of the truth that knowledge cannot spring from experience alone but only from the comparison of the inventions of the intellect with observed fact.*
>
> – Albert Einstein on Kepler's discovery that planetary orbits are ellipses

Einstein wasn't an aerodynamicist, but the above quote applies as well to our field as to his. To understand the physical world in the modern scientific sense, or to make the kinds of quantitative calculations needed in engineering practice, requires conceptual models. Even the most comprehensive set of observations or experimental data is largely useless without a conceptual framework to hang it on.

In fluid mechanics and aerodynamics, I see the conceptual framework as consisting of four major components:

1. Basic physical conservation laws expressed as equations and an understanding of the cause-and-effect relationships those laws represent,
2. Phenomenological knowledge of flow patterns that occur in various situations,
3. Theoretical models based on simplifying the basic equations and/or assuming an idealized model for the structure of the flowfield, consistent with the phenomenology of particular flows, and
4. Qualitative physical explanations of flow phenomena that ideally are consistent with the basic physics and make the physical cause-and-effect relationships clear at the flowfield level.

By way of introduction, let's take a brief look at what these components encompass, the kinds of difficulties they entail, and how they relate to each other.

Understanding Aerodynamics: Arguing from the Real Physics, First Edition. Doug McLean.
Images and Text: Copyright © 2013 Boeing. All Rights Reserved. Published 2013 by John Wiley & Sons, Ltd.

The fundamental *physical conservation laws* relevant to aerodynamic flows can be expressed in a variety of ways, but are most often applied in the form of partial-differential equations that must be satisfied everywhere in the flowfield and that represent the local physics very accurately. By solving these basic equations, we can in principle predict any flow of interest, though in practice we must always accept some compromise in the physical accuracy of predictions for reasons we'll come to understand in Chapter 3.

The equations themselves define local physical balances that the flow must obey, but they don't predict what will happen in an overall flowfield unless we solve them, either by brute force numerically or by introducing simplified models. There is a wide gulf in complexity between the relatively simple physical balances that the equations represent and the richness of the phenomena that typically show up in actual flows. The raw physical laws thus provide no direct predictions and little insight into actual flowfields. Solutions to the equations provide predictions, but they are not always easy to obtain, and they are limited in the insight they can provide as well. Even the most accurate solution, while it can tell us *what* happens in a flow, usually provides us with little understanding as to *how* it happens or *why*.

Phenomenological knowledge of what happens in various flow situations is a necessary ingredient if we are to go beyond the limited understanding available from the raw physical laws and from solutions to the equations. Here I am referring not just to descriptions of flowfields, but to the recognition of common flow patterns and the physical processes they represent. The phenomenological component of our conceptual framework provides essential ingredients to our simplified theoretical models (component 3) and our qualitative physical explanations (component 4).

Simplified theoretical models appeared early in the history of our discipline and still play an important role. Until fairly recently, solving the "full" equations for any but the simplest flow situations was simply not feasible. To make any progress at all in understanding and predicting the kinds of flow that are of interest in aerodynamics, the pioneers in our field had to develop an array of different simplified theoretical models applicable to different idealized flow situations, generally based on phenomenological knowledge of the flow structure. Though the levels of physical fidelity of these models varied greatly, even well into the second half of the twentieth century they provided the only practical means for obtaining quantitative predictions. The simplified models not only brought computational tractability and accessible predictions but also provided valuable ways of "thinking about the problem," powerful mental shortcuts that enable us to make mental predictions of what will happen, predictions that are not directly available from the basic physics. They also aid understanding to some extent, but not always in terms of direct physical cause and effect.

So the simplified theoretical models ease computation and provide some degree of insight, but they also have a downside: They involve varying levels of mathematical abstraction. The problem with mathematical abstraction is that, although it can greatly simplify complicated phenomena and make some global relationships clearer, it can also obscure some of the underlying physics. For example, basic physical cause-and-effect relationships are often not clear at all from the abstracted models, and some outright misinterpretations of the mathematics have become widespread, as we'll see. Thus some diligence is required on our part to avoid misinterpretations and to keep the real physics clearly in view, while taking advantage of the insights and shortcuts that the simplified models provide.

We've looked at the roles of formal theories (components 1 and 3) and flow phenomenology (component 2), and it is clear that the combination, so far, falls short of providing us with a completely satisfying physical understanding. Physical cause-and-effect at the local level is clear from the basic physics, but at the flowfield level it is not. Thus to be sure we really understand the physics at all levels, we should also seek *qualitative physical explanations* that make the cause-and-effect relationships clear at the flowfield level. This is component 4 of my proposed framework.

Qualitative physical explanations, however, pose some surprisingly difficult problems of their own. We've already alluded to one of the main reasons such explanations might be difficult, and that is the wide gulf in complexity between the relatively simple physical balances that the raw physical laws enforce at the local level and the richness of possible flow patterns at the global level. Another is that the basic equations define implicit relationships between flow variables, not one-way cause-and-effect relationships. Because of these difficulties, misconceptions have often arisen, and many of the physical explanations that have been put forward over the years have flaws ranging from subtle to fatal. Explanations aimed at the layman are especially prone to this, but professionals in the field have also been responsible for errors. Given this history, we must all learn to be on the lookout for errors in our physical explanations. If this book helps you to become more vigilant, I'll consider it a success.

This completes our brief tour of the conceptual framework, with emphasis on the major difficulties inherent in the subject matter. My intention in this book is to devote more attention to addressing these difficulties than do the usual aerodynamics texts. Let's look briefly at some of the ways I have tried to do this.

The theoretical parts of our framework (components 1 and 3) ultimately rely on mathematical formulations of one sort or another, which leads to something that, in my own experience at least, has been a pedagogical problem. It is common in treatments of aerodynamic theory for much of the attention to be given to mathematical derivations, as was the case in much of the coursework I was exposed to in school. While it is not a bad thing to master the mathematical formulation, there is a tendency for the meaning of things to get lost in the details. To avoid this pitfall, I have tried to encourage the reader to stand back from the mathematical details and understand "what it all means" in relation to the basic physics. As I see it, this starts with paying attention to the following:

1. Where a particular bit of theory fits in the overall body of physical theory, that is, what physical laws and/or ad hoc flow model it depends on; and
2. How it was derived from the physical laws, that is, the simplifying assumptions that were made;
3. The resulting limitations on the range of applicability and the physical fidelity of the results; and
4. The implications of the results, that is, what the results tell us about the behavior of aerodynamic flows in more general terms.

The brief tour of the physical underpinnings of fluid mechanics in Chapters 2 and 3 is an attempt to set the stage for this kind of thinking.

How computational fluid dynamics (CFD) fits into this picture is an interesting issue. CFD merely provides tools for solving the equations of fluid motion; it does not change the conceptual landscape in any fundamental way. Still, it is so powerful that it has become indispensable to the practice of aeronautical engineering. As important and ubiquitous as CFD has become, however, it is not on a par with the older simplified theories in one significant respect: CFD is not really a *conceptual model* at the same level; and a CFD solution is rightly viewed as just a *simulation* of a particular real flow, at some level of fidelity that depends on the equations used and the numerical details. As such, a CFD solution has some of the same limits to its usefulness as does an example of the real flow: In both cases, you can examine the flowfield and see *what* happened, and, of course, a detailed examination of a flowfield is much easier to carry out in CFD than in the real world. But in both CFD and real-world flowfields, it is difficult to tell much about *why* something happened or what there is about it that might be applicable to other situations.

Before we proceed further, a bit of perspective is in order: While correct understanding is vitally important, we mustn't overestimate what we can accomplish by applying it. As we'll see, the physical phenomena we deal with in aerodynamics are surprisingly complicated and difficult to pin down as precisely as we would like, and it is wise to approach our task with some humility. We should expect that we will not be able to predict or even measure many things to a level of accuracy that would give us complete confidence. The best we'll be able to do in most cases is to try to minimize our unease by applying the best understanding and the best methods we can bring to bear on the problem. And we can take some comfort in the fact that the aeronautical community, historically speaking, has been able to design and build some very successful aeronautical machinery in spite of the limitations on our ability to quantify everything to our satisfaction.

2

From Elementary Particles to Aerodynamic Flows

Step back for a moment to consider the really big picture and ponder how aerodynamics fits into the whole body of modern physical theory. The tour I'm about to take you on will be superficial, but I hope it will help to put some of the later discussions in perspective.

First, consider some of the qualitative features of the phenomena we commonly deal with in aerodynamics. Even in flows around the simplest body shapes, there is a richness of possible global flow patterns that can be daunting to anyone trying to understand them, and most flows have local features that are staggeringly complex. There are complicated patterns in how the flow attaches itself to the surface of the body and separates from it (Figure 2.1a, b), and these patterns can be different depending on whether you look at the actual time-dependent flow or the "mean" flow with the time variations averaged out. Even in flows that are otherwise steady, the shear layers that form next to the surface and in the wake are often unsteady (turbulent). This shear-layer turbulence contains flow structures that occur randomly in space and time but also display a surprising degree of organization over a wide range of length and time scales. Examples include vortex streets in wakes and the various instability "waves," "spots," "eddies," "bursts," and "streaks" in boundary layers. Examples are shown in Figure 2.1c–f, and many others can be found in Van Dyke (1982). Such features usually display extreme sensitivity to initial conditions and boundary conditions, so that their apparent randomness is real, for all practical purposes. The "butterfly effect" we've all read about doesn't just apply to the weather; the details of a small eddy in the turbulent boundary layer on the wing of a 747 are just as unpredictable.

How does all this marvelous richness and complexity arise? It is natural to expect that complexity in the flow requires complexity in the basic physics and that complex behavior in the flow must therefore have its origin at a "low level," in the statistical behavior of the molecules that make up the gas or in the behavior of the particles that make up the molecules. But this natural expectation is wrong. Instead, the complexity we see arises from the aggregate behavior of the fluid represented by the continuum equations. In fact, the essential features of everything we observe in ordinary aerodynamic flows could be predicted from the equations for the continuum viscous flow of a perfect gas, that is, the Navier-Stokes (NS) equations, provided we could solve them in sufficient detail.

Understanding Aerodynamics: Arguing from the Real Physics, First Edition. Doug McLean.
Images and Text: Copyright © 2013 Boeing. All Rights Reserved. Published 2013 by John Wiley & Sons, Ltd.

But there are two caveats that must accompany this sweeping claim. The first is that although the NS equations are a high-fidelity representation of the real physics, they are not exact. Imagine comparing a real turbulent flow with the corresponding exact solution to the NS equations, starting at an initial instant in which the theoretical flowfield is exactly the same as the real one in every detail. We would find that the NS solution matches the detailed time history of the real flow only for a short time and then gradually diverges from it. Detailed time histories of flows, however, are rarely of much interest in aerodynamics, where a statistical description of the flow nearly always suffices. In a statistical sense, we expect that a real flow and the corresponding NS solution would be practically indistinguishable.

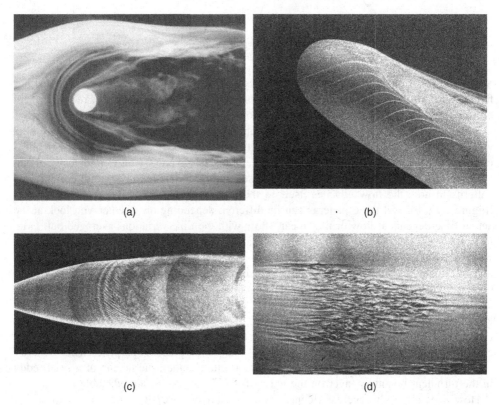

(a)   (b)

(c)   (d)

Figure 2.1   Examples of complexity in fluid flows, from Van Dyke (1982). (a) Horseshoe vortices in a laminar boundary layer ahead of a cylinder. Photo by S. Taneda, © SCIPRESS. Used with permission. (b) Rankine ogive at angle of attack. Photo by Werle (1962), courtesy of ONERA. (c) Tollmien-Schlichting waves and spiral vortices on a spinning axisymmetric body, visualized by smoke. From Mueller, *et al.* (1981). Used with permission. (d) Emmons turbulent spot in a boundary layer transitioning from laminar to turbulent. From Cantwell, *et al* (1978). Used with permission of *Journal of Fluid Mechanics*. (e) Eddies of a turbulent boundary layer, as affected by pressure gradients. Top: Eddies stretched in a favorable pressure gradient. Bottom: Boundary layer thickens and separates in adverse pressure gradient. Photos by R. Falco from Head and Bandyopadhyay (1981). Used with permission of *Journal of Fluid Mechanics*. (f) Streaks in sublayer of a turbulent boundary layer. From Kline, *et al* (1967). Used with permission of *Journal of Fluid Mechanics*

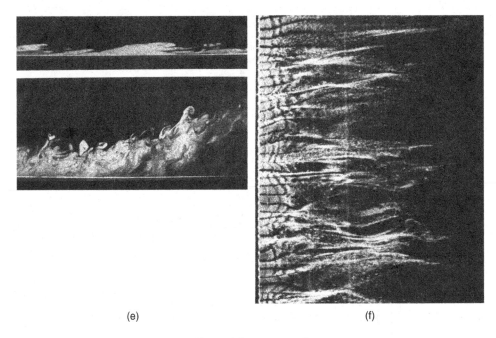

(e)   (f)

Figure 2.1   (*continued*)

The second caveat is that even this less ambitious claim of statistical equivalence is nearly impossible to evaluate quantitatively. For one thing, exact solutions to the NS equations are not practically available for anything but the simplest of flows, and agreement for these simple cases doesn't prove much. For all other flows, especially turbulent flows, we must settle for numerical solutions. Numerical calculations that fully resolve the turbulence down to the last detail have been carried out only for simple flow geometries and relatively low Reynolds numbers. Comparisons of such calculations with the real world, at the level of detailed time histories, would be extremely difficult, if not impossible, and have not yet been attempted. Comparisons of statistical quantities have been encouraging, but for some of the most interesting and revealing quantities, Reynolds stresses, for example, the uncertainties in the experimental measurements are large. Still, in spite of these reservations, I'm confident that the NS equations could in principle predict any phenomenon of interest in practical aerodynamics to an accuracy sufficient for any reasonable engineering purpose. This doesn't mean that they can do so practically, as we'll discuss later in connection with the computational work involved in generating solutions.

The NS equations are, of course, part of a larger system of theory applicable to a hierarchy of physical domains. Figure 2.2 illustrates this by listing the levels of physical phenomena and the corresponding levels of physical theory that deal with them. Here and in the following discussion I use the word "level" both in a conceptual sense, as in levels of physical detail, and in a physical sense, as in levels of physical and temporal scale, from small to large. As already mentioned, the NS equations are an aggregate-level theory applicable to a physical domain separated by several levels from that of the elementary particles that make up the molecules of the fluid. Starting at the lowest level shown and moving upward, each

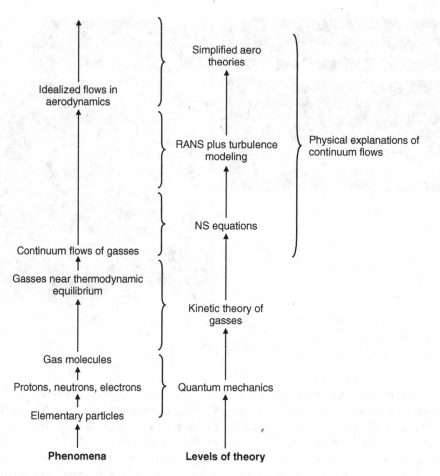

Figure 2.2 Hierarchy of domains of physical theory leading to computational and theoretical aerodynamics

domain or level represents a narrowing of focus, a specialization to a particular situation or class of conditions. We can follow the same logical sequence in deriving the theoretical models, starting with the known properties of elementary particles and eventually reaching the NS equations. The historic development of the theories didn't follow this orderly progression, but the required steps have now been filled in, and the required assumptions and approximations are now understood.

I've already made the claim that the potential for complexity in aerodynamic flows arises at the level of continuum gas flows in Figure 2.2. Is this potential for complexity really independent of what goes on at the lower levels? How can this be? In all of the levels below that of the continuum gas, there is considerable local complexity, and this is reflected in the difficulty of the corresponding mathematical theories. However, adjacent levels are separated by huge gaps in physical scale and very little of the complexity at one level is felt at the next level up. These gaps in scale act a bit like low-pass filters that allow only certain "integrated" effects of the structure at lower levels to be felt at higher levels. For example, the structure of a nitrogen molecule is affected very little by the detailed internal structure of electrons and atomic nuclei. Then, although the electron cloud of a molecule has a complex structure, the details of that internal structure have very little effect on the statistical mechanics of a dilute gas made up of large numbers of molecules. Finally, the continuum properties of gases that really matter to us in aerodynamics are highly insensitive to the details of the molecular motions. For example, the viscous behavior of the fluid under shearing deformation that we assume in the NS equations does not even depend on the fact that the fluid is a gas; it is the same for liquids such as water. As we move up the line toward the NS equations, instead of increasing complexity, we encounter a series of natural simplifications, and surprisingly, these simplifications cost us little loss of physical fidelity for flows that interest us.

So the interesting behavior we see in aerodynamic flows is not inherent in any of the lower levels of the physics. Instead, it emerges in the behavior of the fluid at the continuum level and can be captured in solutions to the NS equations. This is a bit surprising at first, because the physical balances represented by the equations at the local level aren't all that complicated. However, in recent years, studies in the field of complexity science have identified a broad class of seemingly simple systems that can exhibit complex behavior. The NS equations are one example of this kind of system, and solutions (fluid flows) commonly exhibit *emergent behavior* in which great complexity arises from many simple local interactions. The NS equations are, after all, a set of nonlinear partial-differential field equations in space and time, with multiple, interacting dependent variables, and the space of possible solutions is huge. This, combined with the fact that the possible interactions between flow variables at different points in space are sufficiently rich, makes the emergence of complexity in the solutions a natural outcome.

There is also a much more mundane way in which fluid flows must be considered complex, and that is that even some of the simplest flows are not easy to understand or "explain" even qualitatively in a satisfying way without appealing to mathematics. Why is it so difficult? Remember that we are dealing with multiple, interacting flow quantities (dependent variables) and that if we are to understand a flow properly, the behavior of these quantities must be known over some extended spatial domain. Understanding simultaneous behavior over an extended spatial domain is inherently difficult, and the problem of local physics versus global flow behavior contributes to the difficulty. The basic physical laws impose relationships between flow quantities locally, while global behavior is constrained by the requirement that these laws be satisfied everywhere simultaneously. A flowfield is a global phenomenon in which what happens at one point depends to some extent on what happens everywhere else.

Another major difficulty has to do with assigning cause and effect. Here I'm not referring to the general philosophical difficulties of linking causes and their effects. The NS equations are based on Newtonian mechanics and classical thermodynamics, in which clear physical cause-and-effect relationships are assumed. The problem is that when these physical laws are combined in the NS equations, they define relationships between different flow quantities, but they don't define one-way trains of causation. For example, consider Bernoulli's equation, which relates the pressure to the velocity under certain conditions. In trying to explain a flowfield, do we consider the behavior of the velocity to be known and then use Bernoulli's principle to deduce the behavior of the pressure? Or vice versa? In most situations, the right answer is "neither." Cause-and-effect relationships in fluid mechanics tend to be circular, or reciprocal, in the sense that A and B cause each other and are caused by each other at the same time, and often at the same point in space. The upshot of this is that linear explanations assigning one-way cause and effect (A causes B, which in turn causes C) are nearly always wrong. Instead, we must seek explanations of the type that begin with a hypothesis and eventually come back around to consistency with the hypothesis. (If A, then B, then C, which is consistent with A.) We will consider the issue of cause and effect in greater detail with regard to basic fluid mechanics in Section 3.5. Later, we will see examples, such as nonmathematical explanations of the lift on a wing or airfoil in Chapter 7, where trying to force one-way causation has led to errors.

So far, we have considered some of the generic characteristics of continuum flows governed by the NS equations. We've seen how the emergence of complexity in the structure of the flows themselves is a natural outcome of the physics, and we've identified some of the reasons why understanding and explaining flows can be difficult. We've looked at how the NS equations fit into the overall body of physical theory, and we've seen that they represent an aggregate-level theory that involves some simplifying approximations. I've also asserted that within their range of validity the NS equations are a highly accurate representation of reality and that they can in principle predict anything of interest in practical aerodynamics.

Given this essentially complete predictive capability of the NS equations, we could say that they represent all of the "hard science" we should need in aerodynamics. But, of course, the NS equations by themselves are not enough, and there is much more to aerodynamics as a science than what we've seen so far on this brief tour. For the foreseeable future, we won't be calculating an NS solution every time we want to predict or understand what will happen in an aerodynamic flow, and until recently in the history of our discipline, we couldn't have done so anyway. The pioneers in the field had to devise other ways of getting answers: higher level theories that provided both intuitive insight and computational tractability. Such theories are usually derivable at least in part from the NS equations using additional simplifying assumptions, but often also depend on conceptual models of the flows in question. These theories therefore tend to be specialized to particular situations and to have more restricted ranges of applicability than the basic equations. The strategies are several and varied, and we'll consider them in some detail in later chapters. In any case, aerodynamic theory carries a considerable superstructure of higher level theoretical models in addition to the basic NS equations.

Considering this body of aerodynamic theory as a whole, what is its status as science? Like everything else on the conceptual or theoretical side of science, it is "just a theory." This is something the proponents of creationism like to say of Darwinian evolution, implying that it is merely a tentative hypothesis. But of course evolution is much more than a tentative hypothesis, and so is aerodynamic theory. Both have been tested again and again against empirical observations, and have so far always passed the test.

In principle, of course, any theory could be contradicted tomorrow by new evidence and end up needing to be replaced. The NS equations, however, are about as secure as a scientific theory can ever be. In the context of science as a whole, the NS equations purport to be valid only for a narrow range of phenomena, and within this range the NS equations are unlikely ever to be significantly contradicted. The higher level extensions of aerodynamic theory are also not likely ever to be completely overthrown. They, however, typically have even more limited ranges of applicability than the NS equations and are known to be seriously contradicted in common situations that are outside their ranges but still within the realm of aerodynamics. In using the higher level conceptual models of aerodynamics, we must always keep their limitations in mind.

So aerodynamics is on solid ground as a science. But what can we say of its general character? Where does it stand on the deductive/inductive spectrum? In this regard, it has a distinctly split personality. On one hand, we have an all-encompassing theory, the NS equations with a no-slip condition, which is solidly tied to the rest of modern physical theory and from which we can in principle deduce anything of interest in the field. On the other hand, the computational intractability of the equations has greatly limited what we can deduce from first principles in most situations, and much of what we know comes from empirical observations, that is, the inductive approach. This is something we've already discussed in connection with the vital role that phenomenological knowledge plays in our conceptual framework.

The actual physics embodied in our theoretical framework consists of Newtonian mechanics and classical thermodynamics, combined with a mathematical formalism that enables us to bookkeep material properties, forces, and fluxes in a continuous material. Thus when we apply the framework correctly, we are adhering to what I call a Newtonian worldview, in which all effects must have causes that are consistent with Newtonian (or post-Newtonian) scientific principles. As we'll see, some of the errors that can arise in aerodynamic reasoning can be traced to regressions to pre-Newtonian ways of thinking.

3

Continuum Fluid Mechanics and the Navier-Stokes Equations

The Navier-Stokes (NS) equations provide us with a nearly all-encompassing, highly accurate physical theory that can predict practically all phenomena of interest in aerodynamics, including "aerodynamic" flows of liquids such as water. In Section 3.1, we briefly consider the general way in which these equations represent the physics, the assumptions that had to be made to arrive at them, and their range of validity. Then in the sections after that, we delve into the specifics of the equations and what they mean.

3.1 The Continuum Formulation and Its Range of Validity

In the NS formulation, the fluid is treated as a continuous material, or continuum, with local physical properties that can be represented by continuous functions of space and time. These continuum properties, of course, depend on the properties of the molecules that make up the gas or liquid and on the lower level physics of their motions and interactions. However, the continuum properties represent only the integrated effects of the lower level physics, not the details. As I noted in Chapter 2, this provides a representation that is not merely adequate, but highly accurate over a wide range of conditions.

The early historic development of the NS formulation followed an ad hoc approach, assuming continuum behavior a priori and developing a model for the effects of viscosity based on experiments in very simple flows. Much of the hard work involved in this development was devoted to the development of the mathematical formalism that was required to generalize from simple flows to more general ones. We'll touch again on mathematical formalism issues in Section 3.2.

The NS formulation can also be derived formally from the lower level physics, with simplifying assumptions to get rid of the dependence on the details. For gases, the appropriate next lower level to start from is a statistical description of the motion of the molecules and the conservation laws that apply to them, as embodied in the Boltzmann equations. With reference to this kind of derivation, the statement that continuum properties represent only "integrated effects" takes on a literal meaning. We use time-and-space averaging, that is,

Understanding Aerodynamics: Arguing from the Real Physics, First Edition. Doug McLean.

integration, over molecular motions to define the continuum properties of the flow at every point in space and time: the density and temperature of the fluid, and its average velocity. For the definitions of these basic flow quantities, we don't have to make any simplifying assumptions beyond the averaging process itself and that the fluid must be sufficiently "dense" for the averages to "converge." This "convergence" problem is one we'll consider in more detail below.

Although the averaging process gives us rigorous definitions of our basic continuum flow quantities, it doesn't get us all the way to the NS formulation. When we apply the averaging process to the basic conservation laws for mass, momentum, and energy, we get two different types of terms that represent separate sets of phenomena and end up requiring different assumptions:

1. Terms in which only the simple averages defining the continuum density, temperature, and velocity appear explicitly. No further assumptions are needed because these are already the basic variables of the NS formulation. Terms of this type represent the local *time rate of change* of a conserved quantity or the *convection* of a conserved quantity by the local continuum velocity of the flow.
2. Terms that involve averages of products of molecular velocities or products of a velocity component and the kinetic energy. Such terms represent *transport* of a conserved quantity *relative to the local continuum motion* of the flow. The transport of thermal energy is just the heat flux due to molecular conduction. The transport of momentum has the same effect as if a continuum material were under an internal stress and is thus the source of both the local continuum hydrostatic pressure and the additional continuum "stresses" due to viscous effects. The averaging process alone leaves these terms in a form that still depends on statistical details of the molecular motions, and further simplifying assumptions are required to get them into forms that can be expressed as functions of our basic continuum flow variables.

In the NS equations, the terms representing the above transport phenomena have very simple functional dependences on local continuum properties. The hydrostatic pressure is given by an equilibrium thermodynamic relation (an equation of state). The heat flux and the viscous "stresses" are given by gradient-diffusion expressions in which the flux of a conserved quantity is proportional to a gradient of the conserved quantity. Fluids exhibiting the simple behavior of the viscous stresses described in the NS equations are often referred to as *Newtonian*. To get to these simple forms from the general ones that we get from the averaging process requires some simplifying assumptions about the physics. For gases, we must assume the fluid is everywhere locally near thermodynamic equilibrium. This means that the probability distribution functions for molecular velocity that appear in the full transport expressions must be near their equilibrium forms, which in turn requires that significant changes can take place only over length and time scales that are long compared with the mean-free path and time. When these conditions are satisfied, that is, when the local deviations from equilibrium are small, the transport-related terms can be represented very accurately by the simple relationships we use in the NS equations.

The main relationships comprising the NS equations are the basic conservation laws for mass, momentum, and energy. To have a complete equation set we also need an equation of state relating temperature, pressure, and density, and formulas defining the other required gas properties. For aerodynamics applications it is usually a good approximation to assume the ideal gas law, along with a constant ratio of specific heats (γ) and viscosity and thermal conductivity coefficients (μ and k) that depend on temperature only. It seems counterintuitive that the transport coefficients μ and k are well represented as being independent of density at constant temperature, but there is a simple way to understand why this is. As density increases, one might think that the transport coefficients should increase as well because there is more mass per unit volume to transport momentum and thermal energy. However, as density increases, the molecular mean free path decreases, which hinders molecular transport. At the ideal-gas level of approximation, the effects of increasing mass per unit volume and decreasing mean free path exactly offset each other. Thus, practically speaking, the effectiveness of molecular transport depends only on the average speed of the molecules, or the temperature. In some forms of the equations, the local speed of sound ("a") appears, which for an ideal gas also depends only on temperature.

The NS equations, like any field equations, need boundary conditions (BCs). At flow boundaries, where the flow simply enters or leaves the domain, the NS equations themselves determine what combinations of BCs can be imposed and what combinations are required to "determine" the solution in various ways. For boundaries that are interfaces with other materials, for example, gas-solid or gas-liquid interfaces, the NS equations themselves don't fully define the situation, and we need to introduce additional physics. According to theoretical models and experimental evidence, the interaction between most of the liquid and solid surfaces encountered in engineering practice and air at most ordinary conditions is such that the continuum velocity and temperature of the air accommodate almost perfectly to the velocity and temperature of the surface. Thus assuming *no slip* and *no temperature jump* at the "wall," and imposing BCs accordingly, is an extremely good approximation.

A correct physical interpretation of the no-slip BC requires care. In some popular descriptions, the fluid is said to "stick" or "adhere" to the surface. This description is not completely inappropriate, but it is misleading, especially in the case of gasses. Saying that something "adheres" conjures up an image of a bond that can withstand tension as well as shear. Of course a gas cannot be put in tension at all, let alone form a tension-resisting bond with anything else. But the no-slip condition does assume no sliding between the fluid and the solid, so that with regard to shear, the fluid does behave *as if* it were adhering to the surface.

The no-slip condition applies to both liquids and gasses. How it comes about is easier to explain in the case of a gas. While an occasional gas molecule may adhere temporarily to a solid surface (or react chemically with it and remain more permanently), an overwhelming fraction of the molecules that impinge on the surface bounce off. The no-slip condition arises from the nature of these bounces. First, imagine the gas molecules as smooth spheres bouncing specularly off a smooth surface and not losing any tangential momentum in the process. In this case, there would be no shear force exchanged between the surface and the gas, the gas would slip easily along the surface, and there would be no no-slip condition. But on the scale of the molecules, no real surface acts as a smooth surface. All real surfaces consist of atoms similar in size to the gas molecules, and thus even the smoothest is rough on the scale of a gas molecule. And most real surfaces have considerable roughness on larger scales as well. The upshot is that gas molecules impinging on real surfaces bounce off in effectively random directions, which forces the average tangential velocity of molecules near the surface to be very small. Kinetic theory can be used to estimate the effective slip velocity (see White, 1991, Section 1-4.2), showing that in practical situations it is practically zero.

And this must be true even for surfaces that feel slick to the touch, for which our intuition wrongly imagines air being able to slide freely.

Thus our complete physical model consists of the NS equations combined with the no-slip and no-temperature-jump BCs. The range of applicability of this formulation is very broad, and there are only a few applications of practical "aerodynamics" interest where it doesn't apply. Some examples of such exceptions are gas flows at very low densities (for example, very high altitude) and the detailed internal structure of shock waves. Even flows in which ionization, dissociation, or chemical reactions take place are not generally exceptions, because such effects can be incorporated into our continuum formulation by the inclusion of appropriate species-concentration variables, reaction-rate equations, and equations of state. Fortunately for us in aerodynamics, we don't have to deal with the complexities of non-Newtonian liquids, which are important in biological systems and many industrial processes.

So what is it that causes our NS formulation not to apply in the exceptional situations? Is it that very low densities in very high altitude flight, or very small length scales as in the shock-wave problem, cause our averaging process not to converge, a possibility I alluded to earlier? This can happen, but in many cases it is not the cause of "failure." Of course, at a single instant in time, the convergence of a spatial average would require integrating over a large enough volume to include a large number of molecules. Such instantaneous spatial averages might not resolve the internal structure of a shock wave very well, for example, but many flows are close enough to being steady that we can get around this problem and define averages in small spatial volumes by averaging over a sufficiently long period of time. I would think that most of the interesting cases of flight at extreme altitudes or of detailed shock-wave physics can be resolved in this way. In these cases, then, the "failure" of our continuum formulation comes not from the failure of our averaging process to converge, but from the failure of the local-thermodynamic-equilibrium assumption behind our modeling of the "transport" effects when flow gradients become significant on the scale of a mean-free path. Another thing that tends to happen under such conditions is that the errors inherent in the no-slip and no-temperature-jump BC, negligibly small under "ordinary" conditions, become much bigger fractions of the differences in flow quantities in the rest of the field, and these approximations break down as well.

3.2 Mathematical Formalism

Now let's consider some of the issues that arise in casting our formulation of the physics in mathematical terms. Our final formulation will take the form of a set of partial-differential field equations (PDEs), along with some algebraic auxiliary relations. The variables we use, as well as which variables are independent and which are dependent, depend on how we choose to describe the flow. We can choose to describe it in terms of what happens as seen at "fixed" points in space and time, the so-called Eulerian description, or we can choose to define the trajectories of "fixed" parcels of fluid as they evolve in time, the Lagrangian formulation. In the Eulerian description, time and the coordinates in some spatial reference frame, which may or may not be inertial, are the independent variables, and the velocity, pressure, and other state variables of the fluid are dependent. In the Lagrangian description, the independent variables identify the fluid parcels, for example, in terms of their spatial coordinates at an initial instant, and the dependent variables include the spatial coordinates of the parcels at succeeding instants. These two modes of description are in principle equivalent in the sense that they can be used to model exactly the same physics, but they do it in such different ways that they are not practically interchangeable.

For most purposes, the Eulerian framework is more convenient and is therefore the basis for nearly all quantitative work in theoretical aerodynamics and computational fluid dynamics (CFD). A major reason for this is that the Eulerian description provides a much more natural framework for treating steady flows, which are the predominant focus of aerodynamics. All of the higher level conceptual modeling we'll encounter involves the Eulerian description, but we'll still find it helpful to invoke the Lagrangian description in some of our discussion of the basic physical laws.

The time rate of change of any physical quantity (e.g., velocity and temperature) associated with a Lagrangian fluid parcel is called the *Lagrangian derivative* and is usually denoted by the upper case D/Dt. This Lagrangian rate of change is made up of contributions from either or both of two effects, as seen in the Eulerian frame. First, the quantity may be changing with time at the points in space through which the parcel is moving, as reflected in the *unsteady-flow* term $\partial/\partial t$, or the *Eulerian rate of change*. Second, if the parcel is moving with velocity $\mathbf{V}$ through a nonuniform field, it must experience a rate of change $\mathbf{V} \bullet \nabla$ in addition to the unsteady-flow term. In general, then, the Lagrangian derivative is related to derivatives in the Eulerian frame by

$$\frac{D}{Dt} = \frac{\partial}{\partial t} + \mathbf{V} \bullet \nabla. \tag{3.2.1}$$

This transformation has interesting consequences when we apply it to the fluid *velocity*, to determine the *Lagrangian acceleration*. For example, for the simplest case of a 1D steady flow, Equation 3.2.1 applied to the velocity reduces to

$$\frac{Du}{Dt} = u\frac{\partial u}{\partial x}. \tag{3.2.2}$$

From this we see that a given material acceleration Du/Dt requires a large spatial gradient $\partial u/\partial x$ when u is small, but only a small $\partial u/\partial x$ when u is large. This is a consequence of a Lagrangian fluid parcel's motion through the velocity field. In Section 3.4.6, we'll look at how the convective acceleration appears in the convective terms in the momentum equation, and in Sections 4.1.2 and 4.2.1 we'll consider it again in the context of boundary-layer flows.

A factor that complicates the mathematics is that some of the quantities we must deal with are vectors and tensors. The velocity is a vector, and the equation for conservation of momentum is a vector equation. In 3D, this results in three variables and three of our equations, which is a pretty straightforward thing to grasp intuitively.

What is less intuitive is the problem of representing the transmission of forces by "contact" between adjacent fluid parcels. Physically speaking, of course, these forces are a result of momentum transfer by molecular motions, but in the continuum formulation the integrated effects of the motions of many molecules are represented as apparent internal stresses in the fluid, or forces per unit area of the boundary of a parcel. We'll encounter this idea again in our discussion of the momentum equation in Section 3.4.2. The mathematical problem we face is the general problem of representing the state of stress in a continuous material. First, we have to get used to the idea of imaginary boundaries separating adjacent parcels of material. Then we must visualize how any two adjacent parcels exert equal-and-opposite stresses on each other across their common bounding surface. Our description must be able to define the state of stress at any point in the fluid in such a way that it gets the right value for the opposing forces for any orientation of the imaginary boundary. The stress is a force per unit area (a vector) that depends on the orientation of an imaginary dividing surface, which can be defined by the direction of the normal to the surface (another vector).

The stress is thus a tensor. An entire field of mathematics, tensor analysis, was developed just to provide rigorous means for mathematical manipulation of such quantities, not just in continuum mechanics, but in other branches of physics as well, and along with it came powerful shorthand notations for expressing the manipulations. Tensor notation provides the least error-prone way to deal with the stress terms and convection terms in our equations, especially when it comes to deriving the many terms that arise when the equations are transformed to different coordinate systems. Such manipulations can be done without tensor notation, but avoiding errors becomes much more difficult. With or without tensor notation, however, such manipulations quickly become exercises in manipulating symbols, and it can be difficult to maintain any grasp of the physical meaning. We'll try to reach a physical understanding of the most important aspects of the viscous stresses by using very simple flow situations as examples in Section 3.6.

So far, we have talked about the NS equations only in their local or differential form, which is the form that will relate most directly to most of our succeeding discussions. However, in some applications, a more global view of the flow suffices and can be easier to deal with. For these situations, we have the control-volume form of the equations, in which the equations have been integrated over a volume and the surfaces bounding the volume. The control-volume equations are "exact" in the sense that there is no loss of accuracy relative to the differential equations, but they are "simplified" in the sense that they can tell us only what happens to integrated quantities and nothing about how the local quantities are distributed over the volume and bounding surfaces. We'll use the control-volume approach to calculate viscous drag in Section 6.1.4 and lift-induced drag in Section 8.3.2, and we'll consider the general approach further in Chapter 9.

In ordinary solutions to the NS equations, all flow quantities are continuous and differentiable, even through shocks (see more about shocks in Section 3.11.2). This brings some powerful mathematical facts to our aid, without our having to introduce any "physics" at all, which brings us to the topic of the next section.

3.3   Kinematics: Streamlines, Streaklines, Timelines, and Vorticity

Kinematic descriptions are basic to all our attempts to understand flowfields. Obviously, we must understand the kinematic structure of a flow before we can understand the underlying dynamics. The kinematic structure of a flowfield is constrained to have certain characteristics because the velocity field is a continuous vector field.

3.3.1   Streamlines and Streaklines

Two kinematic concepts we often appeal to are *streamlines* and *streaklines*. Streamlines are simply 3D space curves, defined as being everywhere parallel to the velocity vector. A streakline is also a 3D space curve, but is defined by the locations of a string of Lagrangian fluid parcels that all passed through a given "point of origin" somewhere upstream in the flowfield. (We introduced the Lagrangian description of the flowfield in Section 3.2, and we'll define what a Lagrangian fluid parcel means more precisely in Section 3.4.) The point of origin for a streakline is usually taken to be a fixed point in space, but it needn't be; it can be allowed to move with time. A streamline is obviously a mathematical construct that can be defined only by "solving" a mathematical problem, that is, "construct a curve everywhere parallel to a given vector field." A streakline, on the other hand, can be "realized," at least approximately, in real flows that are marked by a passive contaminant such as dye in liquids or smoke in air.

If the flow is steady, streamlines and streaklines from fixed points of origin will coincide and will be the same as individual *particle paths,* that is, the paths of individual Lagrangian parcels. Even in steady flows, however, interesting issues can arise in the interpretation of flow patterns. Figure 3.3.1, for example, compares a streamline pattern constructed from a steady-flow CFD solution with the corresponding nominally steady streakline pattern marked by dye in a water tunnel. The two patterns should of course be the same, and on close inspection they agree about as well as a CFD calculation and an experiment can be expected to. But at a glance, the two images give very different impressions. In the CFD solution, the separation at about 60% chord and the formation of a closed separation bubble stand out clearly, while in the water-tunnel photo the separation is evident only if you look very carefully. Part of the problem is that the field of view of the photo doesn't show the whole length of the separation bubble. I must admit that I didn't realize that this flow separates ahead of the trailing edge until I saw the CFD streamlines. And I'm in good company: Van Dyke (1982) published this same photo with the comment that the flow "appears to be unseparated."

If the flow is unsteady, the situation is much more complicated, and streamlines, streaklines, and particle paths will generally all be different. Looking at the pattern formed by any one of them gives an incomplete and usually misleading picture of the flow. Figure 3.3.2 shows how different the unsteady flow in the wake of a circular cylinder looks in terms of streaklines (a) and streamlines (b). Timelines (c), which we'll define in Section 3.3.3, also present a very different picture.

3.3.2 *Streamtubes, Stream Surfaces, and the Stream Function*

The concept of a *streamtube* is one that is usually applied only to steady flows. The definition of a streamtube starts with a closed curve in the flowfield, as illustrated in Figure 3.3.3a. Steady streamlines or streaklines passing through all points on the curve define a surface that forms the boundary of a curvilinear tube. Because the bounding surface is parallel to the velocity vector, no continuum fluid parcel passes through it. In a steady flow, according to the principle of continuity, which we'll discuss in the next section, the mass flux in a streamtube is the same at any cross section along its length. In a 2D flowfield, we could still define a streamtube the same way we did in 3D, using a closed curve to define the boundary, but a more useful definition is to allow the closed curve defining the streamtube to degenerate into two points, so that the streamtube becomes a 2D layer of flow defined by one streamline through each point, as illustrated in Figure 3.3.3b.

The bounding surface of a streamtube is a special case of the more general concept of a *stream surface,* which also is usually applied only to steady flows. The space curve from which a stream surface originates needn't be a closed curve, and a stream surface needn't

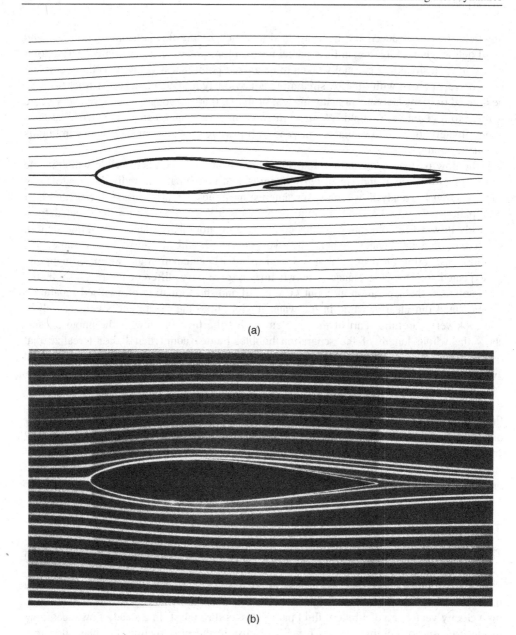

(a)

(b)

Figure 3.3.1 Streamlines and streaklines in the entirely laminar steady flow around a NACA 64A015 airfoil at zero incidence, R = 7,000. (a) Streamlines in a steady-flow CFD solution showing a separation bubble starting at about 60% chord. Laminar RANS calculation by Steven R. Allmaras. (b) Streaklines marked by dye released from upstream in a water tunnel. The streaklines closest to the trailing edge apparently consist of dye streaming forward from the closure region of the separation bubble, beyond the right edge of the photo. Aft of mid-chord there are variations in streakline spacing that are not present in the CFD solution. Photo by Werle, 1974, courtesy of ONERA

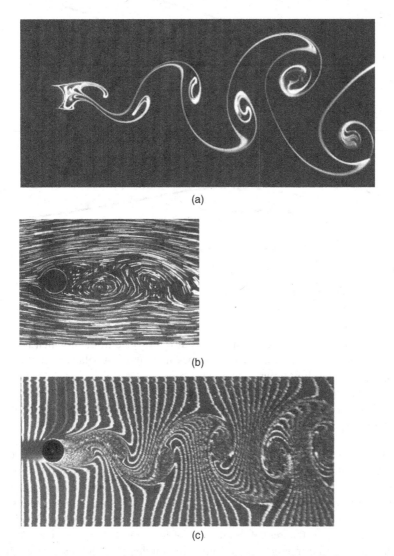

(a)

(b)

(c)

Figure 3.3.2 Unsteady wake (von Karman vortex street) of a circular cylinder at Reynolds numbers in the range 136–140. Photos by S. Taneda, © SCIPRESS. Used with permission. (a) Streaklines marked by dye introduced at the cylinder surface. (b) Streamlines approximated by short time exposure of suspended particles. Used with permission. Photos by S. Taneda, © SCIPRESS. Used with permission. (c) Timelines marked by hydrogen bubbles from a pulsed wire upstream. Photos by S. Taneda, © SCIPRESS. Used with permission

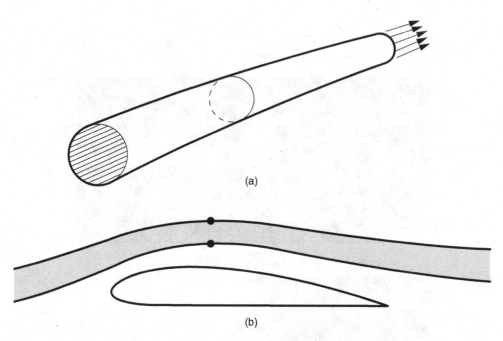

Figure 3.3.3 Illustrations of streamtubes. (a) As a compact streamtube defined by a closed contour in a 3D flow. (b) As a sheet of flow defined by two points in a 2D flow

form a closed tube. A general stream surface is also a surface through which no continuum fluid parcel passes. In 3D flows, stream surfaces that start out relatively flat can become highly contorted as the flow progresses downstream.

The *stream function* is a concept that applies only to 2D flows. Considering any two points A and B in a 2D flow, the mass flux across any curve joining the two points depends only on the locations of the points and on time, provided the flow is either incompressible or steady. (For example, for the two points in Figure 3.3.3b, the mass flux across any contour joining the points is the mass flux in the shaded streamtube.) Thus if we fix point A, the mass flux defined in this way for all other points B defines a single-valued function we call a stream function. It follows then that the stream function is constant along streamlines and that the difference in its value between two streamlines is the mass flux in the streamtube bounded by them. The stream function was used more frequently in the past than it is now. It was often used in earlier theoretical discussions of incompressible flows (see Section 3.10) and was sometimes used in numerical methods for solving the NS equations in 2D.

3.3.3 Timelines

Another useful kinematic concept is that of timelines, which are usually considered most useful in 2D flows, though they can be defined in any flow, steady, or unsteady. The definition starts with the marking of a string of Lagrangian fluid parcels arrayed across the flow at some initial instant. A timeline is then the space curve defined by that same string of parcels at some future instant. Timelines are most useful when defined in sets of multiple lines whose initial instants are separated by equal time intervals. In real flows, timelines can be realized approximately by passive-contaminant markers, usually emanating from a fine wire stretched across the flow. In air, the wire is coated with oil, and a pulsed electric current in the wire produces brief puffs of smoke, marking cross-stream lines that convect downstream. In water, electric pulses can produce lines of tiny hydrogen or oxygen bubbles that mark the flow. In Figure 3.3.2c, we saw timelines in the flow past a circular cylinder.

Figure 3.3.4 shows an example of timelines in a turbulent boundary layer, illustrating a key aspect of timelines in turbulent flows. In a fully turbulent boundary layer, the turbulent velocity fluctuations are not large fractions of the mean velocity, and as a result, the younger timelines near the left edge of the photo remain ordered and build up distortions slowly, looking as if they were in a smoother flow than that in the rest of the photo. As the flow progresses from left to right, the distortions accumulate until, in the right half of the photo, the timelines that are entirely inside the boundary layer appear as a chaotic jumble. In this fully turbulent flow, the timeline picture gives the misleading impression that the intensity of the turbulent motions is increasing from left to right. In steady flows, timeline patterns tend to be simpler and less prone to misinterpretation, as we'll see later in the case of a circular cylinder in ideal potential flow in Figure 5.1.3c.

3.3.4 The Divergence of the Velocity and Green's Theorem

The fact that the velocity is a continuous and differentiable vector field also means that the usual theorems of vector analysis apply. Some constrain the physics in ways that can simplify our task considerably.

Green's theorem relates the divergence of the velocity to a surface integral:

$$\iiint \nabla \bullet \mathbf{V} \mathrm{d}v = \iint \mathbf{V} \bullet \boldsymbol{n} \mathrm{d}s, \tag{3.3.1}$$

where the triple integral is over any volume occupied by the fluid, and the double integral is over the surface that encloses the volume. This is a key ingredient in derivations of the control-volume forms of the equations.

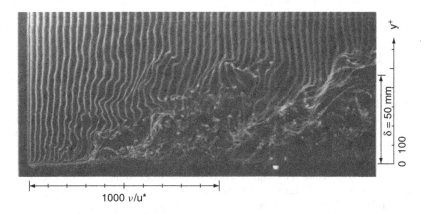

y^+

$\delta = 50\ \mathrm{mm}$

0 100

1000 v/u^*

Figure 3.3.4 Timelines in a turbulent boundary layer in water, marked by hydrogen bubbles from a pulsed wire at the left edge of the photo. From Y. Iritani, N. Kasagi and M. Hirata © (1980). Used with permission

3.3.5   Vorticity and Circulation

Relationships involving the *vorticity* are extremely useful for purposes both of conceptualizing and doing quantitative calculations. In this section, we'll concentrate on the kinematics of vorticity and its relationship to the velocity field. We'll encounter the *Biot-Savart* law, which leads naturally to the idea of vortex *induction,* and we'll take pains to come to a correct understanding that induction is not a dynamic phenomenon, as implied by the way people often talk about it, but is instead a strictly kinematic concept. The real dynamical aspects of vorticity are yet to come, in Sections 3.6 and 3.8 and in the discussion of lift in Chapters 7 and 8.

The vorticity is just the curl of the velocity:

$$\omega \equiv \nabla \times \mathbf{V}, \qquad\qquad (3.3.2)$$

from which it follows by a basic vector-calculus identity that the vorticity is divergence free:

$$\nabla \bullet \omega = 0. \qquad\qquad (3.3.3)$$

If we know the vorticity at a point in a flowfield, we know something about how the velocity varies in the neighborhood of that point, but we don't know everything. In a constant-density flow, the deviations in velocity in the neighborhood of a point can be expressed as the sum of two parts: a deformation velocity field and a solid-body rotation with angular velocity $\omega/2$, a result known as *Helmholtz's first theorem* (see Milne-Thomson, 1966, Section 3.22). Examples of how these two components of motion look in isolation and in combination in 2D flow are illustrated in Figure 3.3.5. In each sketch, a square fluid

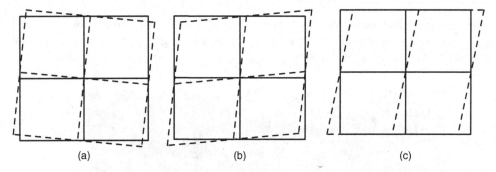

     (a)   (b)   (c)

Figure 3.3.5 Illustrations of the effects of solid-body rotation and deformation on initially square fluid parcels in 2D flow. The square and perpendicular lines bisecting it at an initial instant are shown as solid lines, and the same lines anchored in the fluid are indicated at a later instant by dashed lines. (a) A pure solid-body rotation in which the two bisecting lines have rotated in the same direction by the same amount. (b) A pure irrotational deformation in which the vorticity, and therefore the average angular velocity of the fluid, are zero. This is reflected in the fact that the bisecting lines have rotated by equal-and-opposite amounts. (c) Components (a) and (b) added together. The result is a simple shearing motion in which the horizontal bisecting line hasn't rotated at all, and the vertical bisecting line has rotated twice as far as in either (a) or (b). The average angular velocity, and thus the vorticity, are the same as in (a)

parcel and perpendicular lines bisecting it at an initial instant are shown as solid lines, and the same lines anchored in the fluid are indicated at a later instant by dashed lines. Because we're interested in velocity deviations, we anchor our reference frame to the fluid at the center of the square and show the center as not moving. In Figure 3.3.5a we have a pure solid-body rotation in which the two bisecting lines have rotated in the same direction by the same amount. In Figure 3.3.5b we have a pure deformation in which the vorticity, and therefore the average angular velocity of the fluid, are zero. This is reflected in the fact that the bisecting lines have rotated by equal-and-opposite amounts. In Figure 3.3.5c we have added components Figure 3.3.5a,b together. The result is a simple shearing motion in which the horizontal bisecting line hasn't rotated at all, and the vertical bisecting line has rotated twice as far as in either Figure 3.3.5a or b. The average angular velocity, and thus the vorticity, are the same as in Figure 3.3.5a. These examples are 2D, but they are indicative of what these effects would look like in 3D in planes perpendicular to the vorticity vector.

The deviation velocity components comprising the solid-body rotation part of the motion are in a plane perpendicular to the vorticity vector, but there is no such constraint on the deformation part. Note that the vorticity does not determine the deformation part of the field, but that the vorticity and the solid-body-rotation part of the field are proportionally related. Thus at any point where the local velocity field has a solid-body-rotation component to it, the vorticity must be nonzero. Likewise, if the vorticity is zero, there is no solid-body rotation component, and because of this, flows with zero vorticity are often called *irrotational*.

Much of our theorizing in subsequent sections will make use of the interplay between vorticity and the *circulation,* which is defined as the line integral of the velocity around a closed contour. The crucial relationship is given by Stokes's theorem:

$$\Gamma = \oint \mathbf{V} \bullet \mathbf{t}\, dl = \iint_s \boldsymbol{\omega} \bullet \mathbf{n}\, ds, \qquad (3.3.4)$$

where the double integral is over any continuous surface that is bounded by the contour and is piecewise smooth over its entire area, as shown in Figure 3.3.6. For Equation 3.3.4 to hold, the integrands need merely to be integrable; they needn't be continuous. Thus creases in the surface are allowed, because they occupy zero area, or a subset of zero measure, as the mathematicians would say. Likewise, kinks in the closed contour are allowed.

What Stokes's theorem says, in words, is that the circulation around a closed contour is equal to the *flux* of vorticity through the contour. When we discuss dynamics in Section 3.8, we'll find this relation useful for drawing conclusions about the persistence of the circulation, or lack of circulation, and often we'll be able to do so without any actual calculation.

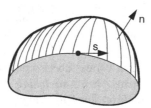

Figure 3.3.6   Illustration of a closed contour and bounded surface to which Stokes's theorem applies

3.3.6 The Velocity Potential in Irrotational Flow

If the flow is irrotational (zero vorticity everywhere), we can derive a further useful result from Equation 3.3.4, which now states that the circulation around any closed contour must be zero. We start by noting that for any two points A and B in the field, we can define many different closed paths from A to B and back to A. Because the line integral of the velocity around any of these paths must be zero, the line integral along one path segment from A to B must be the negative of the line integral along the other path segment from B to A, and because this must be true for any choice of the two path segments, both line integrals must be independent of the path taken. Now if we fix point A, the line integral from A to any other point B is a single-valued scalar function of the location of point B, and it is easy to show that the velocity vector must be equal to the gradient of that function. Thus whenever the velocity field is *irrotational,* it can be expressed as the gradient of a scalar function we call a *velocity potential* ϕ:

$$\mathbf{V} = \nabla\phi. \tag{3.3.5}$$

The existence of a velocity potential can greatly simplify the analysis of inviscid flows by way of *potential-flow theory,* which we'll discuss further in Section 3.10. In the above discussion, we assumed the simplest situation, in which the vorticity is zero everywhere. However, many of the situations in which potential-flow theory is applied are not so simple. Many practical flows are effectively irrotational everywhere except for isolated concentrations of vorticity. Potential-flow theory can still be applied in the irrotational parts of these flows, but special treatments are required to account for the presence of the isolated vorticity. Examples include the jumps in velocity potential that must be allowed across vortex sheets, as discussed in Section 3.3.7, and the special treatment required in the potential-flow theory for 2D airfoils, for which the region of irrotational flow is not simply connected, as we'll discuss further in Section 7.1.

3.3.7 Concepts that Arise in Describing the Vorticity Field

We have a variety of concepts that are useful for thinking about how vorticity is distributed in the flowfield. The first ones we'll consider are applicable to the usual realistic situation in which vorticity is continuously distributed.

Anywhere that the vorticity is nonzero, we can define a *vortex line* as a curve in space that is parallel to the vorticity vector, just as a streamline is parallel to the velocity vector. So a vortex line in the vorticity field is analogous to a streamline in the velocity field, and just as we extended the concept of a streamline to define a streamtube, we can extend the concept of a vortex line to define a *vortex tube.* By definition, the flux of vorticity across the bounding surface of a vortex tube is zero. This, combined with Equation 3.3.3, means that the flux across any cross-section of the tube, anywhere along its length, is the same.

The fact that the vorticity flux in a vortex tube is constant dictates the changes in vorticity magnitude that must accompany *vortex stretching.* If the cross-sectional area of a vortex tube decreases, either in time or along the length of the tube, the strength of the vorticity (the magnitude of the vorticity vector) must increase. For a section of vortex tube containing a given amount of fluid, a reduction in cross-sectional area usually requires an increase in length, or a stretching. (It definitely requires it if the fluid density is constant, for reasons we'll discuss in Section 3.4.1 in connection with the conservation of mass.) Thus the stretching of a vortex tube usually increases the local vorticity magnitude.

A *vortex filament* is a vortex tube whose cross section has a maximum dimension that is infinitesimally small. The cross-sectional area of a vortex filament is thus also infinitesimally small, but it is still assumed to vary along the length of the filament, so that the filament can still satisfy the definition of a vortex tube. For a vortex filament, the flux of vorticity across a cross-section reduces to the product of the vorticity magnitude and the cross-sectional area, which is called the *intensity* of the filament. Note that this definition of the intensity as the flux of vorticity through an infinitesimal area is different from other concepts of intensity you may be familiar with, for example, the intensity of a light beam, which is defined as the energy flux per unit area. The result that the intensity of a vortex filament is constant along its length is *Helmholtz's second theorem* (see Milne-Thomson, 1966, Section 9.31). This conservation of intensity means that a vortex filament cannot end anywhere inside the fluid domain and must either form a closed loop (vortex loop) or end on the boundary of the domain.

Depending on the nature of the boundary, there will be constraints on how vortex filaments or vortex lines can end there. First, consider the special case of an isolated vortex filament that is surrounded by irrotational flow. If the flow is steady, and the boundary is an interface across which the fluid cannot flow, such a vortex filament can intersect the boundary only in the normal direction. This is so because there must be an essentially circular flow pattern in the neighborhood of the filament, in planes perpendicular to the filament, as we'll see in Section 3.3.8. This would violate the no-through-flow condition at the boundary if the filament were not normal to the boundary. Further, if the boundary is a stationary solid surface at which the no-slip condition applies, the velocity components in planes perpendicular to the filament must vanish at the wall, and the vorticity magnitude must go to zero. Thus an isolated vortex filament cannot end at all at a solid surface with a no-slip condition.

In the more general case of distributed vorticity, vortex lines may intersect a no-through-flow boundary at which there is slip, and the intersection need not be in the normal direction. On the usual kind of stationary surface with no slip, the situation is much more constrained. Because the tangential velocity is zero on the surface, the component of vorticity normal to the surface must be zero everywhere on the surface. Then if the magnitude of the vorticity is nonzero, the vortex lines must be tangent to the surface. In the viscous flow about a stationary body, this applies practically everywhere on the surface. The only exceptions are isolated singular points of separation or attachment, which we'll discuss further in Section 5.2.2 and which are places where the magnitude of the vorticity on the surface is zero. At such a point, a vortex line may intersect the surface in the normal direction, but it does so with a "whimper," because the normal component of the vorticity must still go to zero where the line intersects. Thus vortex lines can intersect a no-slip surface only at isolated singular points. It is sometimes erroneously stated that vortex lines cannot intersect a no-slip surface at all, without acknowledgment of the above exception (as pointed out by Saffman, 1992; Section 1.4).

So we see that when vortices approach a solid no-slip surface anywhere other than at an isolated singular point, the vortex lines must turn to avoid intersecting the surface, and in doing so they often become part of the vorticity in a viscous boundary layer on the surface. Figure 3.3.7 illustrates one such situation, the "inlet vortex" that often forms when an engine

Figure 3.3.7   An inlet vortex marked by water droplets. Because all but one of the vortex lines cannot end on a solid surface with a no-slip condition, the vortex lines must spread out in the boundary layer on the ground as illustrated. The possible spiral component to the vortex lines' alignment is omitted here for clarity. Modified from original photo by Alastair Bor. Used with permission

inlet is near the ground, in this case made visible by water droplets. Lines were added to the photo to illustrate how the vortex lines must spread out in all directions along the ground in this situation. (There is often a spiral component to the vortex lines' alignment, which is omitted here for clarity.) Though this vortex is not a single thin filament, it is fairly concentrated and is surrounded by mostly irrotational flow. Just outside the boundary layer on the ground, the vortex is close to perpendicular to the ground, roughly in keeping with our conclusion above that an isolated vortex filament must approach a no-throughflow boundary in the normal direction.

Now let's consider concepts that were developed for idealized models of flows with highly concentrated vorticity. Concentrations of vorticity in limited regions are important features in some flows we'll study later. For example, as we'll see in Chapter 8, the vorticity in the wake behind a lifting wing starts out concentrated in a relatively thin *shear layer* and ends up concentrated in two isolated, more or less axisymmetric *vortices,* all surrounded by practically irrotational flow. In conceptual models of such flows, these vortical structures are often idealized as mathematically thin concentrations, with the shear layers idealized as *vortex sheets,* and the vortices as *line vortices.* These idealized entities carry vorticity fluxes that are finite even though the vorticity is concentrated in a region with zero cross-sectional area. The vorticity distribution must therefore be singular, or infinite, at the location of the sheet or line. For a vortex sheet, we must generally integrate over a cut through a finite width of sheet to find a finite flux of vorticity, though the area we have integrated over is still zero, because the sheet is infinitely thin. For a line vortex, we need only integrate over a single cut through the line (a point) to find a finite flux. There is a formal mathematical theory that makes all of this rigorous, but we won't go into it here because the concepts can be understood well enough without it.

A line vortex looks superficially similar to the vortex filament we defined earlier, but there are crucial differences. The cross-sectional area of a line vortex is zero, while that of a filament is infinitesimal, and the vorticity flux of a line vortex is finite, while that of a filament is infinitesimal. We must also take care not to confuse a line vortex, which is a singular distribution of vorticity, with a vortex line, which is simply parallel with the vorticity vector, usually in fields in which vorticity is continuously distributed.

A line vortex in a 2D planar flow is often called a *point vortex*, because it must be a straight line that extends to infinity in both directions perpendicular to the 2D plane and therefore appears as a single point in the 2D plane. The line vortex is one of the elementary singularities that can be used as a building block to construct solutions in potential-flow theory, as we'll discuss further in Section 3.10. In more general flows, a line vortex would usually be curved, a situation that raises a special problem. At any point on a line vortex where the curvature of the vortex is nonzero, the fluid velocity at that point on the vortex, in a direction perpendicular to the vortex, must be infinite. This makes it impossible to determine a realistic velocity at which such a vortex line will be convected by the flow. (Convection of vorticity is discussed in Sections 3.6 and 3.8.) In real flows, the vorticity is spread out continuously and has finite magnitude, and such infinite velocities do not occur.

3.3.8   Velocity Fields Associated with Concentrations of Vorticity

We've just seen that highly concentrated vorticity is often idealized as a vortex sheet or a line vortex. With Stokes's theorem in hand, we are in a position to determine the nearfield velocity distributions that must accompany these idealized distributions of vorticity, as illustrated in Figure 3.3.8.

A vortex sheet in 2D flow is shown in Figure 3.3.8a. By applying Stokes's theorem to a small closed contour inclosing a short section of the sheet, we see that there must be a jump in velocity magnitude across the sheet equal to the local vorticity strength, or vorticity per unit distance along the sheet in the direction perpendicular to the vorticity vector. In this 2D case, the vorticity vector is perpendicular to the plane of the paper, and the distance along the sheet is measured in the flow direction. The physical flow corresponding to this idealized vortex sheet is a shear layer with the velocity jump spread across a finite thickness, as shown in Figure 3.3.8b.

In a 3D flow, the velocity jump across a vortex sheet, in a vector sense, must still be perpendicular to the vorticity vector. A common situation in aerodynamics, as we'll see in the modeling of wing flows in 3D in Chapter 8, is to have a sheet with no jump in velocity magnitude, only in direction. In this case, the jump in the velocity vector is perpendicular to the vorticity vector, which is parallel to the direction of the mean of the velocity vectors on the two sides of the sheet, as sketched in Figure 3.3.8c. It is easy to show that if the vorticity vector were not parallel to the mean of the two velocity vectors, there would have to be a jump in velocity magnitude.

Vortex sheets of the kind sketched in Figure 3.3.8c are often modeled in 3D potential-flow theory. It is clear from the definition of the velocity potential in Section 3.3.6 that the jump in the velocity vector requires a jump in the velocity potential as well.

If a physical shear layer is effectively thin, that is, if the flow changes across the layer are much faster than changes in other directions, the velocity jump will be approximately equal in magnitude and perpendicular to the integral of the vorticity across the layer. This is

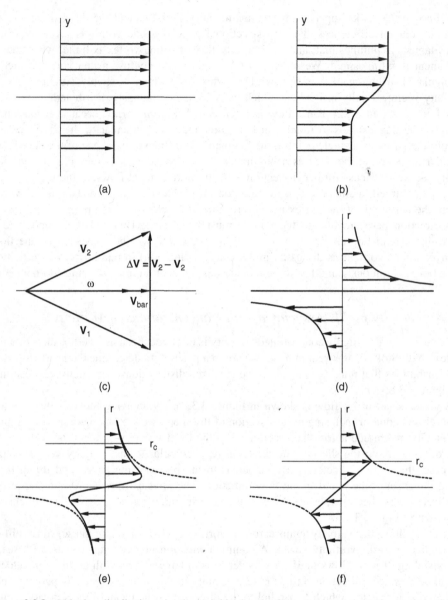

Figure 3.3.8 Velocity distributions in the neighborhood of common concentrations of vorticity. (a) An idealized vortex sheet in 2D. (b) A physical shear layer in 2D with a finite thickness. (c) Plan view of a vortex sheet in 3D in the common situation of no jump in velocity magnitude, only direction. The velocity jump ΔV is perpendicular to the vorticity ω, which is parallel to the mean velocity V_{bar} (d) An idealized line vortex. (e) A physical vortex with a core of finite radius. (f) The Rankine vortex: an idealization of a physical vortex in which the vorticity is constant in a circular core and zero outside the core

an observation we'll find helpful in some of our thinking about boundary layers and wakes in later chapters.

Our next example is the idealized line vortex shown in Figure 3.3.8d. For purposes of this discussion, we'll assume the line is locally straight, so that we don't have to deal with the problem of the infinite perpendicular velocity of a curved line vortex that we mentioned in Section 3.3.7. The circulation on any circular contour of radius r centered on the vortex line must be the same, provided the contour encloses no other vorticity. We conclude that the circumferential velocity must go as 1/r as shown and that the circumferential velocity is singular on the vortex line itself. In the corresponding physical vortex shown in Figure 3.3.8e, the vorticity is spread over a *vortex core* with a radius r_c and a circumferential velocity distribution that depend on the flow process that produced the vortex. Outside of the core, the vorticity is zero, the circulation is constant, and the velocity goes as 1/r, just as it did for the ideal line vortex. An idealization that lies between the line vortex of Figure 3.3.8d and the physical vortex of Figure 3.3.8e is the Rankine vortex, shown in Figure 3.3.8f. In the Rankine vortex, the vorticity is constant throughout a circular core of radius r_c and zero outside the core. The motion within the core is thus a solid-body rotation with velocity proportional to r, as shown. The Rankine vortex is an ingredient in one of the theories of induced drag we'll consider in Section 8.3.

The point vortex and Rankine vortex are idealized flow structures that can be sustained only in inviscid flow. In the real world, viscosity would diffuse them, forming a physical vortex core as shown in Figure 3.3.8e.

3.3.9   The Biot-Savart Law and the "Induction" Fallacy

Now we come to the more general question of what we can say globally about the velocity when the distribution of vorticity is known. Let's start with the general problem of determining velocity from the vorticity outright, in a mathematical sense. It turns out that inverting the definition of vorticity (Equation 3.3.2) does not entirely determine the velocity field, but determines it only to within an unknown additive part that must be irrotational. The Biot-Savart law expresses the solution for the part of the velocity field that is determined. It can be expressed in three forms: applicable to vorticity distributed continuously through a volume, vorticity concentrated in thin sheets, or vorticity concentrated in line vortices.

For the Biot-Savart law to hold, the following assumptions must be met (see Milne-Thomson, 1966):

1. The fluid fills all of space.
2. The fluid is at rest at infinity, with the velocity magnitude at large distances dying off at least as $1/r^2$.

For the ways we typically want to use the Biot-Savart law, these assumptions are not restrictive. If there is a solid body in the flowfield, we can assume that it is filled with fluid at rest that is separated from the external field by a vortex sheet that represents the surface of the body and provides the velocity jump from the interior to the exterior. A constant farfield velocity, as we often have in steady-flow aerodynamics, is not a problem because the constant velocity can be removed via a Galilean transformation.

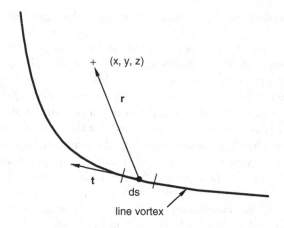

Figure 3.3.9   Definitions of geometric quantities appearing in the Biot-Savart law

The simplified theoretical models for flows around airfoils and wings that we'll discuss in Chapters 7 and 8 are generally idealized inviscid models in which the vorticity is assumed to be concentrated in thin sheets, and we often discretize the sheets into line vortices. Thus the form of Biot-Savart that we use most often is the form for vorticity distributed as a line vortex:

$$\mathbf{V}(x, y, z) = \frac{\Gamma}{4\pi} \int \frac{\mathbf{t} \times \mathbf{r}}{r^3} \, dl, \qquad (3.3.6)$$

where Γ and $\mathbf{t}$ are the strength and the tangent vector of the line vortex, the radius vector $\mathbf{r}$ is defined as illustrated in Figure 3.3.9, and the integration is over all line vortices in the field.

The Biot-Savart law is of course useful for quantitative calculations, but just the qualitative idea that knowing the vorticity at one point allows us to infer something about the velocity at another point is valuable in itself. It is in fact one of our most powerful conceptual tools for reasoning about flowfields. However, as powerful as this idea is, it can be a mixed blessing because it frequently leads to confusion regarding cause and effect.

The problem arises because the vorticity is the "input" and the velocity is the "output" in Equation 3.3.6, and it is common practice to refer to the velocity inferred from the vorticity as the *induced velocity*. Because of this, it's just too easy to think of the vorticity as somehow "causing" the part of the velocity that it "determines." But, of course, this kind of thinking is wrong. In the absence of significant gravitational or electromagnetic body forces, there is no action at a distance in ordinary fluid flows. Significant forces are transmitted only by direct contact between adjacent fluid parcels. So there is no way a vortex at point A can directly "cause" a velocity at some remote point B, and terms such as "caused by" and "induced" and even "due to" misrepresent the physics. We must be careful to remember that Biot-Savart is just a calculus relation between a vector field and its curl, and that in fluid mechanics it doesn't reflect a direct physical cause-and-effect relationship.

This is a crucial point that cannot be overemphasized, and yet it has received surprisingly little emphasis in the literature. It is interesting to look at what other authors have had to say about it. Sears (1960) on p. 1.12 has a short paragraph that mentions the lack of a "mechanism" by which a vortex "induces velocities" remotely, but he doesn't provide further discussion other than to point out that it's just kinematics resulting from the assumption of irrotational flow. Batchelor (1967) on p. 87 states that the vorticity can be said to "produce" or "induce" the "velocity distribution in the surrounding fluid," but then goes on to say that this does not imply a "mechanical cause and effect," but only that the velocity distribution is the "solenoidal velocity whose curl has the specified value everywhere and which is therefore associated with the given distribution of vorticity." Milne-Thomson (1966) on p. 167 defines the "induced velocity" as "the velocity field that coexists with a given distribution of vorticity and vanishes with it." But he also makes the unequivocal statement that the vorticity and the induced velocity "*occur together* but neither can properly be said to *cause* the other" (emphasis his). Durand (1967a) introduces the term "induced velocity" on p. 135 and warns the reader that he will continue to use it even though the implied causation isn't real, saying "it would presumably be more correct to say ... that the field velocity is consistent with, or correlative to, the existence of the vortex."

Just as Equation 3.3.6 seems to imply that the vorticity "causes" the velocity, it also gives the misleading impression that individual elements of a vortex filament somehow make their own separate "contributions" to the velocity. Milne-Thomson (1966) on p. 171 is unequivocal on this point as well, saying that "This impression must be guarded against; otherwise an improper physical picture may be imagined" and pointing out that only the velocity given by the complete result of the integration can be "asserted to have physical reality." This is correct, of course. Still, in our discussion of the flow around a 3D wing in Chapter 8, we'll find it instructive to think about separate "contributions" to the velocity "associated with" partial portions of the vorticity configuration. In such discussions, we must always be on our guard against attributing any direct physical causation to these associations.

We aerodynamicists have contributed to our own confusion by using the terms "induced velocity" and "induction" much too freely. This terminology comes from another field where the Biot-Savart law applies, that is, classical electromagnetics, in which it is said that the magnetic field is "induced" by the electric current. In electromagnetics, the terminology is appropriate because there is supposed to be real action at a distance taking place, for which the term "induction" is physically appropriate. In fluid mechanics, however, there is no direct causal link. We know that vorticity is produced, convected, and diffused in ways we'll discuss in Section 3.6, so we know why the vorticity in our flowfields has to be there: It is there more as a manifestation of the overall flow pattern than as a cause of it.

So, when we want to infer something about a flow pattern without solving the equations of motion, and we already know the general arrangement of the associated vortices, it can be very helpful to appeal to the vorticity and Biot-Savart. The simplified wing and induced-drag theories we'll discuss in Sections 8.2 and 8.3 make use of this approach for quantitative calculations as well. We should make full use of the insights and computational shortcuts that Biot-Savart provides, but we should avoid terms such as "caused," "induced," "induction," and "due to." Then when we want to explain *why* a flow pattern exists, we must appeal to the real physics, that is, to the local force balance among fluid parcels.

This completes our discussion of the aspects of vorticity that can be deduced just from kinematic considerations. In Section 3.8, we'll consider vorticity further, looking at aspects that arise from dynamics.

3.4 The Equations of Motion and their Physical Meaning

My approach in this section is not to derive the equations but to try to give brief, intuitive explanations of what the various terms in the equations mean and to look at some of the general things we can infer from the equations themselves about the behavior of flows.

Our basic equations are expressions of conservation laws for mass, momentum, and energy. These laws can be described most directly and understood most easily in the Lagrangian reference frame, in which we describe the flow in terms of the trajectories of "fixed" parcels of fluid as they evolve in time. However, as I argued in Section 3.2, the Eulerian reference frame, in which we describe the flow as it streams past points in a spatial reference frame not tied to the fluid, is ultimately the preferred choice for conceptual and quantitative purposes. The approach I'll follow here is to briefly discuss what the conservation laws mean in the Lagrangian frame and then proceed to a discussion of how they are expressed in the Eulerian frame.

In both the Lagrangian and Eulerian frames, we will be considering what happens to elemental volumes of fluid, just differently defined in the two cases. Deriving our conservation laws in the form of PDEs involves a formal procedure of taking the limit as the dimensions of our fluid parcels go to zero. We won't go through the details of that procedure in this discussion, but the reader should keep in mind that fluid parcels in either reference frame should be thought of as arbitrarily small.

Lamb (1932) defines a fixed Lagrangian parcel of fluid as containing the same fluid particles, and only the same fluid particles, for all time. The bounding surface of our parcel must therefore move with the fluid in such a way that no fluid particles pass through it. This is, of course, an idealization that makes sense only in our imaginary continuum world. In the real world, molecules will always be diffusing across such a boundary in both directions, and the best we can do is to make the boundary follow the average motion of the fluid in such a way that it has no net flux of material across it. In either way of looking at it, our parcel will always have the same amount of material in it and have no net flux of material across its bounding surface. Sweeping mass diffusion under the rug in this way works fine for single-species fluids or multispecies fluids in situations where relative species concentrations remain constant. If relative species concentrations vary significantly, defining a Lagrangian fluid parcel becomes problematic. For now, we'll ignore this minor limitation on the Lagrangian description and continue our discussion.

As we noted earlier, we have conservation laws for mass, momentum, and energy. Why do we have them for these quantities and not for others, such as the pressure or the viscous stresses? It is because mass, momentum, and energy are quantities whose conservation is required by elementary physics and thermodynamics, and the other quantities are not. Our conserved quantities are also physically tied to the fluid material in such a way that they are carried along, or *convected*, with it. Thus by definition these convected quantities are carried along with our Lagrangian fluid parcels. The amount of such a quantity in a parcel can change only if some physical process acting inside the parcel or at its boundaries accounts for the change. Our conservation laws simply quantify this accounting. Here is a brief description of what they mean within the Lagrangian framework.

3.4.1 Continuity of the Flow and Conservation of Mass

By our very definition of a fluid parcel in the Lagrangian description, we have implicitly enforced conservation of mass within a parcel. However, the equation that explicitly enforces conservation of mass must do more than that. The continuity equation relates the fluid density at all points to the volume occupied, so as to satisfy two requirements:

1. Mass is conserved within each Lagrangian parcel, as required by the definition of the parcel and
2. There are no voids between Lagrangian parcels, nor do adjacent parcels overlap. The entire volume occupied by fluid must be considered to be filled with Lagrangian parcels that conserve mass.

The physical interpretation of the continuity equation in the Lagrangian description is very simple: As the volume of a parcel of fluid changes, the fluid density must change so as to keep the mass of the parcel constant.

Although the basis for the continuity equation is physical (requirements 1 and 2 above), the requirements it imposes on the flow are not as direct in a cause-and-effect sense as those imposed by other equations. For example, in the conservation of momentum (Section 3.4.2), forces directly cause accelerations; and in the conservation of energy (Section 3.4.3), work done in compressing the fluid directly causes a rise in temperature. The continuity equation is different in this regard. It can be tempting to think that a change in streamtube area "causes" a change in velocity, as a result of continuity. But of course the more direct cause of a change in velocity is unbalanced forces applied to fluid parcels. So compared with the momentum and energy equations, the continuity equation is less directly tied to the dynamics and is more like a kinematic constraint imposed on the flowfield.

3.4.2   Forces on Fluid Parcels and Conservation of Momentum

In the Lagrangian reference frame, conservation of momentum is imposed explicitly in the form of Newton's second law, F = ma. Our Lagrangian fluid parcel has fixed mass, and its *acceleration* is the result of the sum of the forces acting on it. As we saw in Section 3.2, this is a vector relation that in the general case requires a vector equation, or equivalently, three scalar equations, for its expression.

There can be external body forces (gravitational and electromagnetic) acting on the parcel, but in aerodynamics these are usually negligible, and we are concerned only with the forces exerted on the surface of the parcel by adjacent parcels. These surface forces that adjacent parcels exert on each other must be equal-and-opposite across the shared boundary, according to Newton's third law. They are the same apparent internal fluid "stresses" we discussed in Section 3.1 in our consideration of the issues of modeling the flow as a continuum. There we saw that it is valid to view them as distributed stresses only in the idealized continuum world, and that in the real world they are just apparent stresses that are the result of momentum transferred relative to the average flow by molecular motion. In any case, from now on we'll think of them as if they were actual stresses.

In Section 3.2, we discussed representing these stresses as a tensor. This is convenient for mathematical manipulation, but for purposes of physical understanding, thinking in terms of force vectors is more intuitive. If we contract the stress tensor with the unit vector normal to the imaginary boundary between parcels, we get a vector representing the force per unit area acting across the boundary. Further, we can resolve this vector into a component perpendicular to the boundary and a component parallel. In the NS equations, the perpendicular component is assumed to be the local hydrostatic pressure (*static pressure*, for short). The parallel component is called the shear stress and is due entirely to the effects of viscosity.

The pressure is one of the most fundamental quantities in continuum fluid mechanics, but understanding it intuitively isn't trivial. At a single point in space, the normal stress on any imaginary boundary containing the point is the same regardless of the orientation of the boundary. Thus we have the idea that the pressure at a point, a scalar quantity, "acts equally in all directions," not an easy concept to grasp. The difficulty of expressing the concept in easily understood terms has led some commentators to errors, such as Anderson and Eberhardt's (2001) description of the static pressure as "the pressure measured parallel to the flow." This contradicts two aspects of pressure as correctly understood: The *definition* of the pressure is independent of the flow, and pressure acts equally in all directions. It is a bit easier to grasp pressure intuitively in terms of its effect on a small but finite fluid parcel, which in a field of constant pressure is pushed inwardly by the surrounding fluid equally in all directions. Understanding the shear stress intuitively entails similar difficulties, which we'll defer to Section 3.6, where we'll discuss in some detail how the shear stress arises and now it is represented in the NS equations.

For the surface stresses to contribute any acceleration to the parcel, the *vector sum* of the stresses acting on all the parcel's faces must be nonzero, that is, there must be an *unbalanced* force left over. Stresses on opposite sides of a parcel by definition act in opposite directions, and if their magnitudes are the same, they cancel. The normal stresses in a field of constant pressure, for example, would cancel each other out, and there would be no unbalanced force. For there to be an unbalanced force, the magnitudes of the stresses on opposite sides of a parcel must be different, and for this to happen the pressure or the viscous stress must be nonuniform. Thus the unbalanced force depends not on the stress itself but on a gradient of the stress, which in the case of the pressure is simply ∇p. This generally requires nonuniform motion of the fluid. We'll look at some examples of how this works in the case of viscous stresses in Section 3.6. In any case, because the forces depend on the motion of the parcel and the motions of the parcel's neighbors, the cause-and-effect relationship between the stresses and the velocities is circular, which complicates our task and is a topic we'll consider in more detail in Section 3.5.

Because the momentum equation governs a parcel's acceleration, determining the parcel's velocity requires integration of the equation. We'll see in Section 3.8.4 how an integration of the momentum equation for the steady flow of an inviscid fluid leads to Bernoulli's equation, one of our most useful special-purpose flow relations.

3.4.3 Conservation of Energy

The principle of conservation of energy is just the first law of thermodynamics, which states that the rate of change of the energy stored by the material in our Lagrangian fluid parcel is equal to the rate at which energy is added to it from outside, in the form of heat added and/or mechanical work done. Only two parts of this would be new to a student of elementary thermodynamics. One is that the motion of the parcel is an important part of the picture, and thus the bulk kinetic energy of the parcel must be included as one of the forms of stored energy to be accounted for. The other is that viscous forces, not just the pressure, provide an avenue by which mechanical work can add to the energy.

Heat can be added to or subtracted from the parcel both by electromagnetic radiation absorbed or emitted within the parcel or by molecular conduction across the parcel's boundary. Note that radiation to or from the interior of a parcel is a volume-proportional or "body" effect, while conduction across the boundary of a parcel is a "surface" effect, and that in aerodynamics it happens that only the surface effect is usually significant. The mechanical work done on the parcel is done by the same forces we considered in momentum conservation. Again, in aerodynamics the external body forces are usually negligible, and we are concerned only with forces exerted on the parcel by adjacent parcels. But the effects of these internal fluid stresses on energy conservation are more complicated than their effects on momentum. In momentum conservation, we had to consider only the net force on the parcel. In energy conservation, the net force is important as well: Acting over the distance moved by the center of mass of the parcel, the net force contributes to changes in the bulk kinetic energy of the parcel. But there is more. If the parcel deforms, either volumetrically or in shear, parts of the parcel's boundary move relative to the parcel's center of mass, and significant work can be done on the parcel that way as well. The pressure, acting through compression or expansion, heats, or cools the parcel, and the viscous stresses heat the parcel, a process called viscous dissipation, which we'll consider further in Sections 3.6 and 6.1.1.

Turbulence raises interesting issues with regard to conservation of energy. We often think of turbulent flows and model them theoretically in terms of time averages, in which the unsteady turbulence motions have been averaged out, an approach we'll discuss in detail in Section 3.7. In the time-averaged flowfield, the kinetic energy of the turbulence is a form of energy that must, in principle, be accounted for. However, in many flow situations the production and dissipation of turbulence kinetic energy (TKE) are roughly in local equilibrium, and TKE can be neglected. We'll discuss this further in Section 3.7.

3.4.4 Constitutive Relations and Boundary Conditions

We've just taken a brief look at what the three basic conservation laws mean in the Lagrangian reference frame. Whether we implement these laws in the Lagrangian frame or the Eulerian, they provided us with five equations, and we have eight unknowns. Our unknowns are three space coordinates (Lagrangian) or velocity components (Eulerian) and five local material and thermodynamic properties: pressure, density, temperature, and the coefficients of molecular viscosity and thermal conductivity. We therefore need three additional constitutive relations to complete the system. For aerodynamics applications, these relations are usually taken to be the ideal-gas equation of state relating the pressure, density, and temperature; the Sutherland law defining the viscosity as a function of temperature only; and Prandtl's relation for the thermal conductivity.

The complete NS system provides all the internal-to-the-fluid physics we need. At the boundaries of our flow domain, the BCs we need to apply depend on the type of boundary. At flow boundaries, we need to invoke no additional physics, and the NS equations themselves determine what BCs are permissible or required, depending on the flow situation. At any boundary that is an interface with another material (often referred to as a "wall"), additional physical considerations are needed to define the BC. We saw in Section 3.1 that under most conditions where the continuum equations apply, the no-slip and no-temperature-jump BCs are appropriate.

3.4.5 Mathematical Nature of the Equations

As we've just seen, our system of equations consists of five field PDEs and three algebraic constitutive relations, with eight unknowns in all. The equations are of mixed hyperbolic/

elliptic type in space, so that the solution depends on conditions on the entire boundary of the domain. Numerical solutions can be "marched" forward in time, but not in space. The equations are nonlinear, so that solutions cannot generally be obtained by superposition of other solutions. Even a steady-flow solution cannot be obtained by a single matrix-inversion operation, but must be approached by time-marching or some process of iteration. These are issues we'll discuss in greater detail in connection with CFD methods in Chapter 10.

Solutions to the NS equations are sometimes nonunique, for example, when more than one steady-flow solution exists for the same body geometry, as we'll see in the case of some airfoils at high angles of attack in Section 7.4.3. Generally, solutions without turbulence exist mathematically, but in most situations at high Reynolds numbers they are dynamically unstable and are not to be found in nature. The instabilities and other mechanisms that can lead to the appearance of turbulence in solutions to the equations are discussed in Section 4.4.

Because of the above general difficulties, analytic solutions to the NS equations are known for only a few simple cases with reduced dimensions and constant fluid properties, and even then only in limiting situations in which the inertia terms can be neglected. For example, there are effectively 1D solutions for steady, fully developed flow in planar 2D or circular-cross-section ducts or pipes and 2D solutions for flow around a circular cylinder or sphere in the limit of low Reynolds number. For some idealized situations at high Reynolds numbers, boundary-layer theory provides approximate solutions to the 2D NS equations that require only the solution of an ordinary differential equation (ODE) in 1D, as we'll see in Section 4.1. For more general flows, numerical solutions are our only option, unless we can make simplifying assumptions.

3.4.6   The Physics as Viewed in the Eulerian Frame

In the Eulerian description, we track what happens as fluid flows past points in a given spatial reference frame. So now, instead of tracking what happens to fixed parcels of fluid, as we did in the Lagrangian description, we track what happens in infinitesimal elements of volume imbedded in our spatial coordinate system. These Eulerian volume elements have fluid continuously streaming through them and across their bounding surfaces. This is, of course, the same streaming motion that was part of the flow when we described it in the Lagrangian frame. We are just seeing it now in a different reference frame, and the difference in vantage point requires us to treat the convection process differently when we implement our conservation laws. In the Lagrangian formulation, convection is accounted for implicitly by our definition of a fixed fluid parcel, and our conservation equations have no terms representing convection across the boundaries of a fluid parcel, because there is none by definition. In the Eulerian formulation, where there is generally a flux of fluid across the boundaries of our volume elements, the convection process must appear explicitly in the form of additional terms in the equations.

Mathematically, the additional terms arise when we replace the time derivatives in the Lagrangian equations with their Eulerian equivalents, using Equation 3.2.1. In the Eulerian equations that result, convection effects are represented by terms that arise from the $\mathbf{V} \cdot \nabla$ term on the right-hand side. To see how this works, consider, for example, the x component of the momentum of a Lagrangian parcel of volume dV, which is given by $\rho\, u\, dV$.

Applying Equation 3.2.1 to this quantity gives

$$\frac{D}{Dt}(\rho u d\mathrm{V}) = \frac{\partial}{\partial t}(\rho u d\mathrm{V}) + \mathbf{V} \bullet \mathbf{\nabla}(\rho u d\mathrm{V}). \tag{3.4.1}$$

The second term on the right-hand side represents the convection of momentum in the Eulerian x-momentum equation in its rawest form. There is another form often seen in the literature, in which the density is taken outside the derivative, and the relationship to the Lagrangian acceleration Du/Dt is clearer. To derive this other form, we must invoke conservation of mass, which in its Lagrangian form simply states that the mass of a Lagrangian parcel doesn't change with time. Applying Equation 3.2.1 to that gives

$$\frac{D}{Dt}(\rho d\mathrm{V}) = \frac{D\rho}{Dt}(d\mathrm{V}) + \rho \frac{D}{Dt}(d\mathrm{V}). \tag{3.4.2}$$

Using the product rule to expand the derivatives on the right-hand side of Equation 3.4.1 and invoking Equation 3.4.2, we get the Lagrangian rate of change of x momentum, per unit volume:

$$\frac{1}{d\mathrm{V}}\frac{D}{Dt}(\rho u d\mathrm{V}) = \rho \frac{\partial u}{\partial t} + \rho \mathbf{V} \bullet \mathbf{\nabla} u. \tag{3.4.3}$$

The last term on the right-hand side of Equation 3.4.3 is just the product of density and the convective acceleration that we introduced in Section 3.2 and that we looked at for the special case of 1D flow in Equation 3.2.2.

Now let's look at the convection process in more detail. One thing that almost goes without saying is that across a shared boundary between two parcels, convection is reciprocal. It is a kind of analog to Newton's third law in mechanics, which states that the mutual forces exerted by two bodies in contact with each other must be equal and opposite, because there is nothing at the interface that can support an unbalanced force. At a shared boundary between two fluid parcels, there is nothing that can add to or subtract from the flux of a conserved quantity, so the flux leaving one parcel must be equal to the flux entering the other. In our general NS formulation, we end up not having to enforce this reciprocity explicitly, because it is guaranteed by the continuity of all of our flow variables. There are specialized theories in which we allow surfaces of discontinuity, such as when we model shocks in solutions to the inviscid versions of the equations. In these cases, we must introduce additional equations to explicitly enforce the applicable conservation relationships across the discontinuity.

The physical interpretation of the convection terms in our conservation equations is straightforward. When the rate of convection into a volume element is not balanced by the convection out, convection becomes a source of the conserved quantity and must be taken into account in the conservation law. The convection terms thus represent the *net* rate at which a conserved quantity is being convected into or out of a volume element. In the conservation of mass, this net convection is the only contribution to the time rate of change of the total mass inside the element. When the flow is steady, this means that the flux of mass into the element must equal the flux out. The same thing applies to larger volumes than just a local parcel. For a steady-flow streamtube, as we defined it in the previous section, this means that the mass flux through any surface that cuts across the tube must be the same. Net convection is an important part of the balance in the conservation of momentum and energy too, but we must also account for the contributions from forces applied to the fluid (momentum and energy) and from heat conduction (energy only).

Contributions to the momentum and energy balances from external sources such as gravitational or electromagnetic forces exerted on the fluid, or heat transmitted to and from the fluid by absorption and emission of radiation, can be accounted for in a straightforward manner in our formulation. Likewise for exchanges of force or energy between parcels within the fluid that are not in direct contact with each other, which would be internal to the flow but nonlocal effects. In aerodynamics, however, such external effects and internal nonlocal effects are usually negligible, so that the only effects that remain to be represented in our equations are those that are transmitted by direct parcel-to-parcel contact. This leaves us with just the interparcel forces represented by the apparent internal stresses, and the heat fluxes due to conduction, that are exchanged between adjacent parcels of fluid. These quantities, as we have already seen, are not physically tied to the fluid material and are not convected with it. They are not affected by changes in the velocity of our reference frame, and they look the same in the Eulerian frame as they did in the Lagrangian.

So in the usual situation that prevails in aerodynamics, the only significant transmission of forces within the fluid is between adjacent parcels of fluid. Convection effects in an Eulerian frame are similar in the sense that they too act only between adjacent Eulerian parcels. Thus there is no mechanism in our usual aerodynamic flows for any kind of exchange of "force at a distance," and therefore no remote "induction" or other such effects. Although the Biot-Savart law seems to imply a kind of remote induction effect, as we saw in Section 3.3.9, it is a fallacy to think of the velocity at one point as being "induced" or "caused" by the vorticity at another point. This is just one example of the difficulties associated with assigning cause and effect in fluid mechanics. In spite of the difficulties, we will try to make some sense of cause and effect in Section 3.5.

3.4.7 The Pseudo-Lagrangian Viewpoint

So far in this section, we've discussed the equations of motion and the physical conservation relationships they represent, as viewed in both the Lagrangian and Eulerian reference frames. The two reference frames are generally kept distinct, and theoretical models or quantitative calculations generally use either one frame or the other, usually the Eulerian. However, in qualitative discussions the distinction is not always so clear cut. This is especially so in discussions of steady flows, for which time should enter into the description of the flow only in the Lagrangian frame, not the Eulerian. In spite of this, in many discussions in the literature and some in this book, you will notice that time-like terminology is used, even though the reference frame used is not explicitly Lagrangian in the sense of following individual Lagrangian fluid parcels. For example, a statement such as "As the flow approaches the leading edge of the airfoil, it is deflected upward" is distinctly time-like in the image it evokes, but it is not strictly Lagrangian, because it refers to the progress of some imprecisely defined macroscopic body of fluid instead of a small fluid parcel. In a sense, this is sloppy terminology, but the meaning is usually clear enough that no harm is done, and more rigorous alternatives are often awkward.

3.5 Cause and Effect, and the Problem of Prediction

A basic characteristic of mathematical theories, both in fluid mechanics and other branches of physics, is that governing equations by themselves are not predictive. Only solutions to the equations can make predictions with any "reach" in space and time. Just knowing this, however, doesn't tell us how difficult it will be to predict things in fluid mechanics, either computationally or mentally, or how difficult it will be to understand the flow phenomena that can be predicted. To try to gain insight into these complicated issues, I'll start by considering the nature of cause and effect, which will bring us naturally back to the problem of prediction.

To understand a fluid flow, or to explain it, we would like to establish cause-and-effect relationships between what happens at different points in space, between the different flow quantities (for example, velocity and pressure), and between the flowfield and the forces and heat fluxes that it exchanges with its environment. One of the main reasons fluid mechanics is such a difficult discipline is that these cause-and-effect relationships are complicated, which can make it difficult to explain things in a satisfying way, as we'll see below. And in Section 3.3.9, we saw an example of another kind of difficulty, where, in connection with the Biot-Savart law and the idea of induced velocity, it was easy to assign cause and effect incorrectly.

Let's start our discussion of this issue by looking at what determines the motion of an individual parcel of fluid. According to Newton's second law, the acceleration that a Lagrangian fluid parcel experiences as it moves through the field is proportional to the net force exerted on the parcel. As we noted in Section 3.4.7, in aerodynamic flows there are usually no significant "body forces" acting over long distances, and the only forces we need to consider are those exerted by parcel-to-parcel contact, that is, by the pressure and the viscous stresses. So the motion of our Lagrangian parcel is influenced directly only by the resultant of the forces exerted by all the adjacent parcels. And the motions of the adjacent parcels are influenced in turn by the forces exerted by the parcels adjacent to them, and so on throughout flowfield. If all of these forces were known a priori, predicting all of the motions would be trivial, but of course we can't know the interparcel forces a priori because they depend on the motions of the parcels. And the motions depend on the forces, and around and around it goes. We're dealing with the motions of many parcels, all interacting with their neighbors, and as a result, we effectively have circular cause-and-effect between the forces and the motions. Regarding Newton's second law, we tend to think of the force as the "input" and the motion as the "output," but that way of thinking is not consistent with the situation we face in continuum fluid mechanics, where circular cause and effect is a fact of life.

Because the cause-and-effect relationship between interparcel forces and parcel motions is circular, the relationship between the global flowfield and the integrated forces it exchanges with its environment is circular as well. We might be tempted to ask whether the lift on a wing is there as a result of the flowfield, or vice versa, that the flowfield is there as a result of the force, but the answer is "both." The interaction between the integrated force and the flowfield is mutual, and the cause-and-effect relation is necessarily circular.

All of this circularity of cause and effect is reflected in the governing equations, which define implicit relationships between the independent variables, not explicit one-way causation relationships. Tracking even these implicit relationships intuitively is made more difficult by the fact that they are mediated simultaneously by more than one physical principle, which is reflected in the fact that there is more than one equation to deal with. The momentum equations relate the velocity components to the interparcel forces, but are not sufficient by themselves to determine either of them. To have a fully determined system, we must also include the energy and continuity equations and the auxiliary relations. To predict what happens in a flowfield, we must solve the whole system of equations.

We've just seen that to predict the interactions between the interparcel forces and parcel motions, we can't just look at the equations themselves, we must solve the equations. The same thing applies when we consider the interactions between what happens at one point in space and what happens at other points throughout the flowfield. Parcel motions are the direct result of interparcel forces, but they are at the same time constrained by the principle of continuity, that is, the requirement that Lagrangian fluid parcels do not overlap each other or open gaps between them. Lagrangian parcels must therefore move in concert, not with their positions relative to each other rigidly fixed, but varying in a coordinated way as the fluid deforms. Thus we can't determine the individual motions separately, but must determine them collectively by solving the equations.

Now we've looked at two major aspects of cause and effect: relationships between forces and motions, and relationships between what happens at different points in space. Both have reinforced what we already knew, that is, that the governing equations by themselves don't predict what will happen in a flowfield and that prediction requires solving the equations. What has also become clear is that there is a wide gap between our basic physical laws as embodied in the equations of motion, and the flow phenomena that the laws govern. The physical balances represented by the equations are relatively simple to understand, and the direct interactions represented are only local. A flowfield, on the other hand, is a global, collective phenomenon that can be exceedingly complex. The problem we face is that of bridging two disparate regimes: equations governing local interactions, on the one hand, and flowfields embodying complicated collective behavior on the other, with a huge gap in potential complexity in between.

This gulf between local physics and global behavior is at the root of a fundamental difficulty of prediction in fluid mechanics. To predict from first principles what will happen in a flowfield, we must determine a flow pattern that satisfies the equations everywhere simultaneously, that is, we must solve the equations. The wide gulf we must bridge to do this is reflected in the difficulty of the equations we must solve: a set of PDEs in multiple spatial dimensions and time, with multiple dependent variables (flow quantities) to be determined. Even in situations where a qualitative description of the flow would suffice, we humans are not well equipped mentally to do the required "solving" in our heads, and "mental" predictions based solely on the basic physics are usually not reliable. In all but the simplest flow situations, there are too many possible flow patterns, and determining which one will prevail requires a degree of quantitative precision not attainable by mental means. In most situations that are complex enough to be of practical interest, a "first principles" prediction requires a detailed numerical solution to the equations.

In principle, as I argued in Section 3.1, a numerical solution of the full NS equations should be able to predict any flow of interest in aerodynamics. Strictly speaking, this sweeping claim can be true only if we assume our idealized solution process can find multiple solutions in cases where they exist, as, for example, in the lift-curve hysteresis of some airfoils and wings, which we'll discuss in Section 7.4.3. In practice, the possible existence of multiple solutions is only one of our worries, however. In practical calculations, we cannot resolve the effects of turbulence directly and must model them, and our inability to model them accurately compromises the accuracy of our predictions to varying degrees depending on the situation, an issue we'll consider further in Section 3.7 and encounter several more times in Chapters 7–10.

Even with the shortcut of turbulence modeling, solutions to the NS equations are not always readily done, and even when a solution is available, the information it provides has limits. A solution can tell us *what,* for example, by determining a flow-separation pattern, or *how much,* as in determining lift or drag, but it provides very little information as to *why.* Thus we will still often seek qualitative predictions and explanations of flow phenomena. For these purposes, we must generally rely to some extent on experience, that is, on our knowledge of the *phenomenology* of general flow patterns in a variety of situations. Where starting from scratch with nothing but first principles to rely on usually gets us nowhere unless we resort to computation, simply predicting the likely flow pattern based on phenomenological knowledge of flows in similar situations at least gives us a starting point from which to build a more detailed explanation. This is a mental approach that comes so naturally that we tend to take it for granted and to forget that this is what we are actually doing. Of course, any explanation we construct from there must be true to the physics. It is especially important not to force one-way causation where it doesn't fit. Explanations should correctly acknowledge circular cause-and-effect relationships and avoid the temptation to oversimplify.

A qualitative "analysis" constructed along these lines would start with a conjecture as to the basic flow pattern, drawn from our phenomenological experience base, followed by a sketch of the velocity field (qualitative velocity magnitudes and directions). If large regions of the flow are expected satisfy the conditions for a steady Bernoulli equation to hold (see Section 3.8.4), it should then be possible to sketch out a qualitative pressure field based on the velocity magnitudes. Then the qualitative flow curvatures can be assessed for consistency with the pressure gradients. Some mental "iteration" may be required to find velocity and pressure patterns that are consistent with the physics. Of course, the quantitative resolution of this kind of "analysis" is very low and not generally sufficient to distinguish whether the conjectured flow pattern was correct or not. The approach is really suited only to generating explanations of things we already know actually happen. We'll explore some arguments of this type in Chapters 7 and 8, as applied to airfoils, wings, and the generation of lift.

In this section, we've identified two major characteristics of fluid mechanics that make prediction difficult: the circular nature of cause and effect and the great gulf between local physical laws and global flow behavior. Our task in much of what follows will be to make some headway in the face of the daunting challenge that these difficulties pose.

3.6 The Effects of Viscosity

Air and water are "simple fluids" in the sense that they cannot resist deformation in the same way a solid can. When a solid body is subjected to a steady force that tries to deform it in a way that does not change the volume, the solid can resist by assuming a steady deformation, usually small. A simple fluid cannot put up such steady-state resistance. Another way of saying this is that the resistance of a simple fluid to such deformation vanishes as the *rate* of deformation vanishes. The upshot is that when a fluid is at rest, all of the off-diagonal terms in the apparent-stress tensor, that is the tangential or shear stresses, must be zero. At rest, only the normal stresses represented by the terms on the diagonal can be nonzero, and the three of them must be equal.

The viscous behavior of a fluid is defined by how the apparent internal stresses respond to a deformation that changes with time. In general, a nonzero rate of deformation will result in nonzero shear stresses, and the normal stresses will no longer be equal. For gases, the description of this response that we use in the NS equations can be arrived at through statistical analysis of the motion of the molecules (kinetic theory of gases) and an assumption of small deviations from thermodynamic equilibrium. The NS formulation can also be arrived at through a strictly continuum approach. We'll not go into the mathematical details of this continuum derivation here, but the assumptions and their consequences are worth enumerating. The assumptions are:

1. The deviatoric stress tensor (the stress tensor with the pressure subtracted out) is a linear function of the rate-of-deformation tensor. At this level of generality, the factor of proportionality (the "viscosity") is a fourth-order tensor. The direct response of the pressure to the rate of deformation (the "bulk viscosity" effect) is left as a separate issue and is usually assumed negligible.
2. The fluid behaves isotropically, meaning that the response of the stress in a parcel to a given deformation field is independent of the orientation of the parcel.

The consequences are:

1. Because of the exclusion of the pressure from consideration and the assumption of isotropic behavior, the viscosity of the fluid can be represented by a single material property μ.
2. A simple shearing deformation $\partial u/\partial y$ results in a resisting shear stress $\mu \partial u/\partial y$. We'll look in greater detail at what this means a little later.
3. In the special case of a uniform fluid (constant density and viscosity), the net viscous force on a fluid parcel due to the shear stresses on all of its faces is proportional to the curl of the vorticity. Thus the viscous shear stresses can affect the motion only if the vorticity is nonuniform (and therefore not everywhere zero). Even in more general flows in which density and viscosity vary, there must usually be significant vorticity for viscosity to exert a net force on a fluid parcel.
4. A corollary of item 3, again for uniform fluids, is that irrotational flows (flows with zero vorticity) that satisfy the inviscid equations also satisfy the NS equations. This is not because the viscous stresses are zero, but because the stresses are distributed so as not to exert unbalanced forces on fluid parcels.

The derivation of these consequences is a nontrivial exercise in tensor analysis, of which a clear account is given in Chapters 1 and 3 of Batchelor (1967).

The range of applicability of the resulting NS formulation is wider than we might expect. The linear relationship between the stresses and the rate of deformation, which we would expect to be a good approximation for small rates, seems to hold quite accurately for rates as large as we commonly encounter in applications. The dependence of the stresses on the instantaneous gradient of the velocity requires slow change on the scale of the mean-free time, but this restriction is practically never an issue because the mean free time is extremely short. And with the exception of high-frequency acoustic waves and the internal structure of shock waves, the bulk-viscosity effects are negligible, and only the shear stresses are significant. We can usually ignore any deviation of the pressure from what is given by the equilibrium equation of state.

Recall from Section 3.1 that the continuum viscous stresses are really just apparent stresses representing transport of molecular momentum by random molecular motion relative the mean flow. In the simple linear relationship described in item 2 above, this transport of molecular momentum is represented as diffusion that is proportional to a gradient. Likewise, the transport of heat by conduction is represented by a simple linear diffusion relationship. Both of these relationships work in the direction such that they try to reduce the nonuniformity that drives the transport, which is the direction we would expect based on the second law of thermodynamics.

First, consider how this works for the viscous stresses. The viscous shear stresses act like ordinary friction in that the work done against them is dissipated irreversibly into heat. We can see this by considering a small parcel of fluid undergoing a shear deformation.

In a reference frame moving with the center of mass of the parcel, we would see a local distribution of relative velocity as sketched in Figure 3.6.1, which would result in the stresses shown, as imposed on the parcel by its surroundings. The forces on the parcel faces are in the same direction as the relative velocities, so that the work done on the parcel by the surroundings is positive. Because the parcel has no average translational kinetic energy in this frame, and rotational kinetic energy must be negligible if the parcel is small enough, the work done can add only to the internal energy of the parcel. In other words, the work is dissipated irreversibly into heat.

The linear relationship for the diffusion of heat by conduction guarantees that heat is always conducted in the direction from higher temperature to lower temperature. Thus our linear relationships for viscous shear stress and thermal conduction are both consistent with the second law of thermodynamics in that they never contribute to decreases in the total entropy of the system.

Now let's take a very elementary look at how the shear stresses typically affect fluid motion. In the simple steady-flow situations illustrated in Figure 3.6.2, the effects of viscosity are sufficiently isolated from other complicating effects that they can be easily

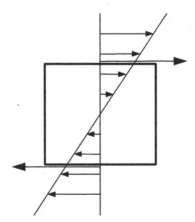

Figure 3.6.1 Typical velocities relative to the center of mass of a parcel (short arrows), and resulting viscous stresses exerted by the surrounding fluid on the top and bottom of the parcel (long arrows). The stresses and the relative velocities are in the same direction, so that the work done on the parcel by its surroundings is positive and is dissipated into heat

understood. For simplicity, we'll assume the viscosity μ is constant. Figure 3.6.2a,b show 2D flows confined between parallel walls with velocity distributions as shown, assumed to be unchanging in the x direction, so-called *Couette flows*. Figure 3.6.2c shows an "external" flow, bounded on only one side by a wall. In all cases, a no-slip condition is enforced at walls. The small square in each diagram represents a typical fluid parcel, and the arrows illustrate the balance of horizontal forces: pressure forces on vertical faces, and shear forces on horizontal faces. Note that the forces on opposing faces are in opposite directions and mostly offset each other. A net force on the parcel requires a change in the magnitude of the force between one opposing face and the other. For a sufficiently small parcel, net forces are proportional to gradients of the stresses, that is, the pressure gradient $\partial p/\partial x$ and the shear-stress gradient $\partial \tau/\partial y$, where $\tau = \mu \partial u/\partial y$ is the shear stress.

In Figure 3.6.2a, the lower wall is stationary, and the upper wall moves to the right at a constant velocity. The motion is a simple shear with a constant velocity gradient $\partial u/\partial y$ from bottom to top. The shear stress $\tau = \mu \partial u/\partial y$ is also constant, so that the typical parcel shown has equal and opposite stresses on its top and bottom faces that contribute no net force in the x direction. If the pressure is uniform in the x direction, it contributes no net force either, and the parcel therefore has no acceleration, which is consistent with our assumption that velocity is constant in x.

In Figure 3.6.2b, both walls are stationary, but we assume there is flow from left to right. The simplest velocity profile we can assume that satisfies the no-slip condition is a parabola with the maximum at midchannel. Now the stresses imposed on the top and bottom faces of our typical parcel are different in magnitude, and the parcel experiences an unbalanced force proportional to the difference, which is in turn proportional to the second derivative of the velocity. For the parabolic profile we've assumed, $\partial \tau/\partial y$ is thus constant and negative across the channel. A constant negative pressure gradient $\partial p/\partial x$ in the x direction can balance these unbalanced shear forces and allow an unaccelerated flow consistent with our assumption that velocity is constant in x. This flow requires pressure decreasing in the x direction to sustain it and is thus an idealized prototype for flows driven by pressure drops in long ducts or pipes.

Figure 3.6.2c is an idealization of external flow past the surface of a solid body. Here we have only one wall with flow streaming past it, and the flow is assumed to have uniform

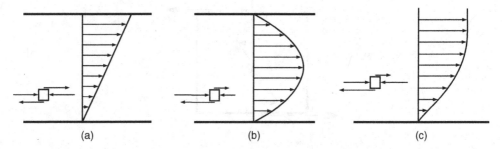

(a)   (b)   (c)

Figure 3.6.2 Idealized flow situations illustrating how viscous stresses affect fluid motion. (a) Couette flow with one moving wall and no pressure gradient. The shear stresses balance and the pressures balance. (b) Couette flow with stationary walls and a pressure gradient. The difference in shear stresses is balanced by the difference in pressures. (c) External flow along one wall (boundary-layer flow). The unbalanced shear stresses result in slowing of the fluid parcel and thickening of the boundary layer

velocity except close to the wall. We assume the pressure is constant in x, but that the velocity close to the wall can vary. Qualitatively the velocity profile will be as shown, typical of flow in "wall boundary layers," which are quite important in aerodynamics and which we'll take up in much greater detail in Chapters 4 and 5. $\partial \tau / \partial y$ is negative, as it was in Figure 3.6.2c, though it is not constant in this case because the velocity profile is not a simple parabola. As in Figure 3.6.2b, the unequal shear stresses on the top and bottom of a typical parcel exert an unbalanced force on the parcel, but in this case there is no pressure gradient to balance it, and the parcel slows down as it moves from left to right. As this happens to all parcels in the layer as they move downstream, the velocity profile is "stretched" away from the wall, and the boundary layer becomes thicker.

The view we have just taken of these simple viscous-flow examples is based on conservation of momentum and on thinking of the viscous stresses as real continuum stresses. It is important to remember, however, that the continuum stresses are only apparent and that they really reflect transport of momentum by the motion of molecules relative to the average (continuum) motion of the fluid. So the view we have just taken is really the *momentum transport* view.

It is also instructive to view viscous effects in terms of the *transport of vorticity*. If we take the curl of the momentum equation, we get the *vorticity equation* governing the creation, convection, stretching, and diffusion of vorticity. The vorticity equation for constant-density flow can be found in many books (Batchelor, 1967; White, 1991, for example). Other specialized forms are discussed in the meteorology literature. The general form is harder to find, but at the time of this writing, Wikipedia had a good description under "Vorticity equation."

An examination of the vorticity equation for flows with variable properties shows that there are only two ways that vorticity can be created or destroyed in the interior of a flow: by the pressure gradient acting in the presence of nonuniform density or by an external body force that is nonconservative. As a result, under most conditions of importance in aerodynamics, the only creation of vorticity that can take place in the interior of a flow is through shocks of nonuniform strength. However, unless a shock is quite strong, the vorticity it generates is very weak. Thus for purposes of this discussion, we can assume that only boundaries can serve as sources or sinks of vorticity and that transport in the interior involves only convection, stretching, and diffusion. In the special case of zero viscosity, there is no diffusion, only convection and stretching, and if there is any vorticity, it remains attached to the same fluid parcels and is carried with them as they move through the field. In Section 3.8, we'll discuss two of Helmholtz's theorems that deal with this situation.

A simple boundary-layer flow like that in Figure 3.6.2c raises interesting vorticity-transport issues. In 2D flow, the vorticity cannot be changed through vortex stretching, and in constant-property flow, there can be no production of vorticity within the volume, only convection and diffusion. We'll look at the resulting vorticity "budget" of a 2D constant-property boundary layer in some detail in Section 4.2.4. There we'll see that the vorticity gradient can diffuse vorticity either away from the wall or toward the wall, and that the wall can thus act as either a source or sink for vorticity. The vorticity that is convected along within the boundary layer can be thought of as having been "created" at the wall upstream, through the action of the no-slip condition and the shear stress at the wall, at locations where the wall was acting as a source. Farther from the wall, the vorticity gradient weakens, the rate of diffusion decreases, and eventually the flow becomes effectively irrotational far from the wall.

This is a general pattern that characterizes flows past bodies when the viscosity is small. Near the solid surface, there is formed a viscous, vortical boundary layer outside of which the flow remains effectively irrotational, as if it were inviscid. Note that no matter how small the viscosity is, a viscous boundary layer will form along the surface as long as the no-slip condition prevails, and there will always be at least this thin layer where the effects of viscosity are important. We'll consider this general flow structure further, including more details of the behavior of boundary layers, in Chapters 4 and 5. Also note that "small viscosity" is usually referred to as "high Reynolds number," where the inverse of the Reynolds number (1/R) is a dimensionless parameter that multiplies the viscous terms in the nondimensional form of the momentum equations. We discuss the Reynolds number and what it means in some detail in Section 3.9.2.

3.7 Turbulence, Reynolds Averaging, and Turbulence Modeling

At the high Reynolds numbers typical of engineering applications, boundary layers usually transition from laminar to turbulent before reaching the back of the body, and the viscous wakes behind bodies are always turbulent. Turbulence is a complicated beast that is highly random but at the same time displays a surprising degree of organized structure over ranges of length and time scales that depend on the situation. In Chapters 4 and 5, we'll look at some of the details of turbulent structures and their consequences in boundary layers and wakes. In this section, our focus will be on the effects that turbulence has on the flow that are important in applications and how these effects can be practically handled in the framework of the NS equations. For computational purposes, our main concern will be to avoid having to solve for the details of the turbulent motions, which is generally impractical. This will lead us to the problem of *turbulence modeling.* For the kinds of effects we generally wish to model, we'll find that we're more interested in the local statistical properties of the turbulence than in the organized structure, and in fact, that our increasing knowledge of organized structures has not contributed much to our quantitative prediction capabilities.

Turbulence in boundary layers and wakes is of course unsteady, but in most applications the unsteadiness is not of direct engineering interest because the length scales associated with the turbulent motions are typically small compared with the dimensions of the body. As a result, the fluctuations in the integrated forces on the body are also very small, and in most cases we will be interested only in the time-averaged properties of the turbulent flowfield. (Buffeting due to large-scale unsteadiness, usually associated with flow separation, airplane cabin noise coming from the surface-pressure fluctuations of the external boundary layer, and "airframe noise," mostly from deployed landing gear and flaps, are some exceptions.)

Defining the time-averaged properties of a flow by averaging out the turbulent motions is very similar to defining continuum properties by averaging out molecular motions. In both types of averaging, we can still treat globally unsteady flows, provided there is sufficient separation between the global time scales we wish to resolve and the random-motion time scales we wish to average out. Because turbulent time scales are typically many orders of magnitude longer than molecular-collision time scales, the requirement to keep the time scales separate is potentially more limiting for turbulence averaging than for molecular. In most applications, however, we'll be looking at time-averaged flows that are nominally steady, so that this is not a problem. In any case, the result of the averaging process is an imaginary continuum flow, or *mean flow,* in which we can identify the same kinematic features as in a real flow, such as streamlines, streamtubes, and timelines, and all of the same kinematic rules apply. And Lagrangian and Eulerian fluid parcels have the same meaning in the mean flow as they would in any other continuum flow.

The analogy between averaging turbulent motions and averaging molecular motions is a close one, not just in terms of the mathematics, but also in terms of the effects of the physical processes that are averaged out. In Section 3.1, we saw how molecular motions relative to the average (continuum) motion of the fluid transport momentum and heat and that averaging over these motions defines the static pressure as well as the apparent continuum stresses due to viscosity and the heat flux due to conduction. Turbulent motions relative to the mean flow also transport momentum and heat and can do so even more effectively than molecular transport does. Time averaging the turbulent motions gives rise to apparent steady stresses that we call *Reynolds stresses* and to an average heat flux due to turbulent transport. These apparent stresses and heat fluxes are directly analogous to the viscous stresses and molecular heat conduction that we studied in Section 3.1. It is important to remember that both the turbulent and molecular versions of these stresses and fluxes are *apparent* and that in both cases they are actually due to transport by random unsteady processes that have been averaged out.

Reynolds stresses affect the motion in the same ways as the apparent viscous stresses do, that is, in the ways we discussed in Section 3.6:

1. Gradients in the stresses produce unbalanced forces on mean-flow fluid parcels, contributing to the parcels' accelerations.
2. The unbalanced forces on fluid parcels result in the diffusion of mean-flow vorticity, usually in the direction away from the body surface into the field.
3. The apparent turbulent stresses result in dissipation into heat just as molecular viscous stresses do. This dissipation is not as direct as that due to viscous stresses, but the end result is much the same.

It is helpful to think of the turbulent dissipation process as consisting of two separate processes taking place simultaneously: production and dissipation of TKE. First, the work done by the mean deformation against the turbulent stresses feeds directly into the kinetic energy of the turbulent motions. This turbulence production is nearly always positive, like molecular viscous dissipation. Second, the turbulent motions themselves contain local, unsteady velocity gradients that produce molecular viscous stresses that dissipate the TKE directly into heat (turbulence dissipation). It happens that practically everywhere in most flows, production and dissipation are roughly in local equilibrium, and the end result is as if the work done against the turbulent stresses were dissipated directly into heat. The fact that the energy dissipated by turbulent stresses goes through the intermediate stage of the kinetic energy of turbulence is therefore irrelevant for most practical purposes. For example, for purposes of computing the flowfield, it has been shown that TKE does not contribute significantly to the energy balance that determines the temperature and density distributions and so can be ignored in CFD calculations (see Cebeci and Bradshaw, 1984).

Given that the basic physical effects of turbulence and molecular viscosity are so similar, in all of the preceding discussion of viscous effects, and in the discussion in the remainder of the book, terms such as "shear stress" and "dissipation" can be taken to refer to either the molecular or turbulent variety.

Of course, there are also important differences between turbulent and laminar flows. Outside the sublayer in a turbulent boundary layer, or in a wake, the apparent turbulent stresses are much larger than the viscous stresses, and the direct contribution of turbulence to viscous drag is therefore typically large. Thus keeping the flow laminar over at least part of the body surface can be a powerful means of drag reduction (discussed further in Section 6.3.1). In most practical applications, however, turbulent flow over at least part of the surface is actually beneficial because of its greater resistance to flow separation, as we'll see in Section 4.1.4.

These are the qualitative physical effects of turbulence on the mean flowfield. How are these effects represented in the time-averaged equations? Think back for a moment to how this worked in the case of molecular transport. In Section 3.6, we saw that the continuum viscous stresses in the NS equations are represented in terms of simple velocity-derivative expressions such as $\mu \partial u/\partial y$. This simple linear dependence of the stress on an instantaneous velocity derivative very accurately represents the effects of the molecular-motion physics that was "thrown out" in the averaging process that leads to the continuum equations. And not only is the NS model for momentum transport very simple, it introduces only one new fluid property, the viscosity coefficient μ, and it can usually be determined with sufficient accuracy as a function of the local static temperature alone; likewise for molecular heat conduction. Unfortunately, the physics thrown out in averaging over the turbulent motions cannot be so simply or accurately modeled. However, the practical importance of the effects of turbulence demands that we make the attempt, which leads us to the difficult problem of *turbulence modeling*.

It is worth taking a brief look at how the turbulence-modeling problem arises formally. First, we'll look at how the raw turbulent stresses arise and are expressed in the equations, and then we'll consider some of the ways they are modeled. We start with the full, unsteady NS equations and apply the process of *Reynolds averaging*. Each dependent variable is decomposed into a time-averaged part and a left-over fluctuating part that contains all of the turbulent time dependence. For the velocity component u, for example, we have

$$u = \overline{u} + u'. \tag{3.7.1}$$

We substitute expressions like this for all the dependent variables in the equations, and then time-average the equations. To simplify the result, we make use of identities that apply to the averaging integration, such as

$$\overline{\overline{u}} = \overline{u}$$
$$\overline{u'} = 0, \tag{3.7.2}$$

and appeal to the continuity equation, which applies separately to the mean and fluctuating parts of the flowfield. For constant-property flow, the process is relatively simple, and the details can be found in Schlichting (1979) or White (1991). For compressible flow, things are more complicated, and further assumptions have to be made, the details of which are given in Cebeci and Bradshaw (1984). For purposes of this discussion, we'll look at the forms that arise in incompressible flow. In the momentum equation, for example, the convection terms involve products of velocity components, which, when decomposed as in Equation 3.7.1 and simplified yield components of the *Reynolds shear stress* that look like

$$\tau_{xy} = -\rho \overline{u'v'}, \tag{3.7.3}$$

and components of the *Reynolds normal stress* that look like

$$\tau_{xx} = -\rho\overline{u'u'}.\tag{3.7.4}$$

The resulting time-averaged equations with these terms included are called the *Reynolds-averaged Navier-Stokes* (RANS) equations, and the time-averaged flow is what we have been calling the *mean flow*. The apparent turbulent stress terms like those given above are unknowns in these equations, and they represent the turbulent-motion physics that was thrown out in the averaging process. The addition of the unknown Reynolds stresses to our problem requires us to include additional equations embodying what we call a *turbulence model*. The Reynolds stresses depend on what's happening in the flowfield in complicated ways, and as we observed earlier, they cannot be "modeled" nearly as simply or as accurately as the viscous-stress terms are "modeled" in the NS equations.

Remember that our objective is to avoid having to solve the full, unsteady NS equations for the details of the turbulent motions in what is called direct numerical simulation (DNS), which generally requires too much computational work to be practical. By introducing the RANS equations plus a turbulence model, we accomplish this objective, but not by "simplifying" the equations. As we'll see below, the addition of a turbulence model actually makes the equation set more complicated than the unsteady NS equations. Instead, the way the turbulence-modeling approach saves computation is by enabling a lower resolution numerical solution to suffice. Resolving the turbulent motions in a DNS solution would require very fine spatial grids in all directions in the turbulent regions and very small time steps. And a 3D spatial grid would be required even for a nominally 2D flow, because turbulent motions always require 3D resolution. Spalart (2000) has estimated that a DNS calculation for a full large-airplane configuration at flight Reynolds number will require 10^{16} grid points and $10^{7.7}$ time steps to reach a statistically steady solution. The computing power that would make such a calculation practical is not likely to be available until around 2080. On the other hand, for the resolution of just the mean flow using the RANS equations with a turbulence model, a much coarser grid suffices, especially in directions parallel to the turbulent shear layers, and a full-airplane simulation requires only about 6×10^7 grid points and 10^3 iterations or pseudo-time steps to converge to a steady solution, which became practical around 1990. As of 2008, such a 3D solution could be computed in 40 000 seconds of clock time on a PC cluster using 64 CPUs. A RANS solution for a nominally 2D flow requires only a 2D grid, and as of 2008 a solution for the flow around an airfoil on a grid with 30 000 points could be obtained in 1400 seconds on a PC using a single CPU.

Before we consider the various strategies for modeling the Reynolds stresses, let's look a little further at the nature of the physical problem we're dealing with. We've already pointed out the general similarities between molecular transport and turbulent transport of momentum. At a more detailed level, however, the two processes are very different, and it is in the differences that we will find the reasons why turbulence modeling is such a difficult problem.

To explore the differences, let's start by reviewing the nature of molecular momentum transport. Consider one of the horizontal faces of a tiny Eulerian fluid parcel immersed in the fluid at a location in the flow where the mean velocity is horizontal with a value of $\overline{u}$, as illustrated in Figure 3.7.1a. There is no net flux of mass across the face, because the face is parallel to the continuum flow velocity, but there is random motion of molecules up and down relative to the mean flow, and therefore through the face from both above and below.

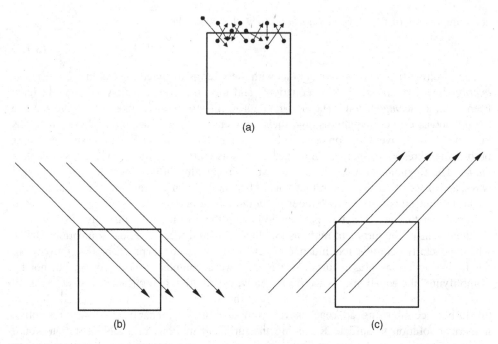

Figure 3.7.1 Transport of momentum across the top face of a small fluid parcel. (a) By molecular motions. (b) and (c) By turbulent motions, seen at different instants in time

These molecules carry with them, on average, horizontal velocities slightly different from $\bar{u}$, reflecting, on average, the conditions where they experienced their last collisions, typically very small distances (on the order of the mean-free-path) above and below the face. The molecular shear stress τ_{xy} in response to a positive velocity gradient $\partial u/\partial y$ is a result of the transport, on average, of a slight excess of momentum from above the face and a slight deficit from below the face. The key feature of molecular momentum transport is that it is a result of molecular interactions over very short ranges and can therefore be represented by a simple, local model, as in the NS equations.

Now consider the same mean-flow situation as we just did for the molecular shear stress, but assume that there is now also a turbulent shear stress τ_{xy} given by $\rho\overline{u'v'}$ Again there is no mean velocity perpendicular to the face of our fluid parcel, but at any instant the perpendicular continuum velocity component v' will usually be nonzero, as illustrated in Figure 3.7.1b. In these illustrations, v' is shown as uniform over the parcel face because the parcel is assumed small compared to distances over which v' varies. In the case of molecular transport, we had molecules crossing the face in both directions at once, arriving from very short distances since their last collisions. In the case of turbulent transport, we have macroscopic masses of fluid sloshing across the parcel face from one direction and then the other, carrying with them some "memory" of conditions where they came from, conditions that depend on the complex turbulent structures (e.g., eddies, vortices, bursts, and sweeps) of which the sloshing motion is a part. The nature of the memory that the sloshing fluid masses carry is complicated because each part of the mass has its motion modified continuously by interaction with the surrounding fluid up until the instant that that part crosses the parcel face. The distances over which the sloshing carries some influence can be as large as the thickness of the turbulent shear layer or as small as the dimension of the smallest turbulent structure, but in any case many orders of magnitude longer than a molecular mean-free-path. Thus although the Reynolds stress $\overline{\rho u'v'}$ is a local statistical property of the turbulent motion, it depends directly on flow structure that is not only complex, but is much less local than the simple velocity gradient $\partial u/\partial y$ that determines the viscous stress.

So in the Reynolds-averaging process, we've thrown out some very nonlocal turbulence effects, and our task in turbulence modeling is to represent these effects in the context of the RANS equations, a set of local PDEs. The additional equations that are introduced in implementing various turbulence models tend to be mostly local algebraic or differential equations. Some models also include explicit nonlocal information, but generally only of a very low-level variety, such as the distance to the nearest solid surface, or some measure of the thickness of the turbulent shear layer as reflected in the mean-velocity profile. Otherwise, nonlocal effects are felt only through the behavior of turbulence-modeling variables such as "TKE," "eddy viscosity," "dissipation rate," and so on, that depend on the flow solution. The local PDEs that govern the nonlocal behavior of such variables can represent only a limited repertoire of effects such as "generation," "convection," "diffusion," "dissipation," and "history" that at best are pale reflections of the complicated physics that was thrown out. This might seem like a recipe for failure, but so far, for obtaining practical predictions of turbulent flows, there seems to be no realistic alternative to this general approach to turbulence modeling. We just have to be realistic in our expectations and understand that our predictions will often be far from perfect.

Ideally, we would like a turbulence model to possess the following:

1. Universality, or the ability to reasonably represent any turbulent flow without requiring different empirical content for each category of flow;
2. Invariance, by which the model gives equivalent results regardless of the reference frame in which a problem is posed; and
3. Accuracy, sufficient for all reasonable engineering purposes.

With regard to universality, currently available models tend to do well only on relatively narrow classes of flows. Models developed to do well on boundary layers and wakes do less well on highly 3D or highly curved flows, and so forth. Some models, but not all, meet the invariance requirement. With regard to accuracy, many models do very well on simple flows such as free shear layers and flat-plate boundary layers, for which much of the models' empirical content was derived, but do much less well on flows with strong pressure gradients or flow curvature. The effects of flow separation and of three-dimensionality in boundary layers are particularly difficult to model accurately, as we'll see in Sections 4.4.2 and 8.6.2. In RANS solutions for practical flow situations, the turbulence modeling nearly always constitutes the most serious limitation on the physical fidelity of the simulation.

There are two main distinctions to be made between the various modeling strategies:

1. Whether the Reynolds stress is defined explicitly by the model ("Reynolds-stress models"), or indirectly in the form of an eddy-viscosity or mixing-length relationship ("eddy-viscosity models" and "mixing-length models") and

2. The number and type of additional equations that constitute the model, over and above the mean-flow equations. In "algebraic" models, the additional equations are not differential (ODE or PDE). "One-equation" or "n-equation" generally refers to the number of additional PDEs in a model.

Eddy-viscosity modeling is based on an assumption proposed by Boussinesq in 1877 (see White, 1991), that the turbulent shear stress can be expressed in a manner analogous to the molecular shear stress in the NS equations, that is,

$$\tau_t = -\rho \overline{u'v'} = \mu_t \frac{\partial \overline{u}}{\partial y}, \qquad (3.7.5)$$

where μ_t is the "turbulent" or "eddy" viscosity. When applied to an actual flow, Equation 3.7.5 simply defines μ_t as a function of the local statistics of the turbulent motions, that is, the $\overline{u'v'}$ correlation. When used for making flow predictions, as part of a turbulence model, Equation 3.7.5 shifts the problem from that of modeling τ_t to that of modeling μ_t. The advantage of this is that μ_t can be simpler to model than τ_t, at least in the simple flows on which many models are based. A disadvantage is that there are limited regions in some flows in which the real μ_t is not well-behaved, and eddy-viscosity models will miss some of the details of such flows. Practical eddy-viscosity models are usually kept simple. For example, in 3D boundary-layer flows, where shear stresses in two directions must be calculated, the eddy viscosity is usually assumed to be isotropic, that is, the same in both directions, an assumption that is seriously violated in many flows, as we'll discuss further in Section 4.4.2.

The mixing-length assumption proposed by Prandtl (1925) is closely related and is usually expressed in terms of the eddy viscosity:

$$\mu_t = \rho \ell^2 \left| \frac{\partial \overline{u}}{\partial y} \right|, \qquad (3.7.6)$$

where ℓ is the mixing length. When applied to a real flow, this equation is simply a definition of the mixing length ℓ, just as Equation 3.7.5 was just a definition of μ_t. As part of a turbulence model, Equation 3.7.6 shifts the modeling problem again, from that of modeling μ_t to that of modeling ℓ. Some of the earliest modeling ideas for turbulent boundary layers were more naturally expressed in terms of ℓ than μ_t, an issue discussed in White (1991, Section 6.7).

Early mixing-length and eddy-viscosity models were cast in terms of algebraic equations and made use of rudimentary nonlocal information such as the distance to the wall and a measure of the thickness of the shear layer. In general CFD grids, such nonlocal information is not always easy to define unambiguously and can be expensive to compute. Reducing or eliminating the need for nonlocal information in more recent models has generally dictated moving from algebraic equations to PDEs.

A common strategy for generating additional equations for PDE-based models is to take "moments" of the momentum equation, that is, to take one or more of the components of the full unsteady momentum equation and to multiply each component of the equation by one of the velocity components, and to Reynolds-average the result. Depending on what combination of equations and velocity components is used, this can yield a conservation equation for TKE or a transport equation for Reynolds shear stress. The new equation is effectively an equation *for* the unknown second-order turbulence correlation, that is, the TKE or the Reynolds stress, but it contains new unknowns in the form of third-order correlations such as pressure-velocity correlations and triple-velocity correlations. Such new equations shift the modeling problem from that of modeling the Reynolds stress to that of modeling the third-order correlations. The process can be repeated to shift the modeling to higher orders, but there seems to be little to gain. The appearance of adding more "physics" to a model simply by adding more equations using this type of operation is an illusion. There is only one physical (vector) equation involved, and generating higher moments of it does not add physics; it is just algebraic manipulation. The only way this approach could lead to models with higher accuracy would be if the behavior of the higher-order correlations turned out to be less dependent on the particular flow situation, thus giving the model greater "universality," but there is little evidence so far that this is the case.

Another approach that has been surprisingly successful is to invent a totally ad hoc "transport" equation for a quantity such as eddy viscosity or mixing length that is not actually governed by a physical conservation law. The model by Spalart and Allmaras (1994, the "SA model") that is used in many RANS codes is an example of such a model, and for many types of flow, its performance is comparable to that of many more complicated models.

Viscous flows in most practical applications involve turbulence, and quantitative prediction thus requires turbulence modeling. Turbulence models and flow predictions using turbulence models will be discussed several more times in the course of the book. In Section 6.2.2, we'll look at incorporating surface-roughness effects into turbulence models; in Section 7.4, we'll use calculations made using turbulence models to illuminate several aspects of airfoil flows; and in Section 8.6.2, we'll discuss the turbulence-modeling issues related to swept wings.

3.8 Important Dynamical Relationships

In Section 3.3, we discussed the many things that can be deduced about the nature of flowfields based just on kinematics and on the fact that the velocity is a continuous vector field with only isolated discontinuities and singularities allowed in idealized models of the flow. In this section, we delve into the general relationships that can be deduced based on dynamics, with and without viscosity.

3.8.1 Galilean Invariance, or Independence of Reference Frame

Something we all take for granted as part of our Newtonian worldview is that the phenomena of fluid mechanics are physically the same, regardless of what reference frame we choose to view them in. Any fluid flow may be described with equal validity whether the reference frame is inertial (not accelerating or rotating) or is accelerating and/or rotating. Of course, this idea applies to classical mechanics in general, and a formal justification of it requires knowledge of Newton's laws. But there was some intuitive realization of it before Newton, as for example in the writings of Galileo and Leonardo da Vinci (see Anderson, 1997, p. 24).

For most work in fluid mechanics we use an inertial reference frame. How the idea of independence of reference frame is reflected in the mathematics in this case is worth at least a cursory look. First, the NS equations and the general statements of the BCs used with them are *invariant* under Galilean transformation, which means that the forms of the equations and BCs are *unchanged* by a transformation from one inertial frame to another. However, the phenomena the equations represent are not invariant in the same sense. They don't *look* the same in different reference frames; they are just *physically equivalent* when the effects of the different viewpoints have been accounted for. What we mean by "accounted for" is that things have been appropriately transformed in going from one frame to another. The descriptions of the same flow in two different reference frames are transformed versions of each other. So while the equations are *invariant,* the flowfields they govern, which are solutions to the equations, are merely *transformable.*

In some applications such as propellers or other turbomachines, the flow may be steady and simpler to describe in the reference frame of the rotating object. Then the effects of the rotating reference frame must be accounted for in the equations by inclusion of appropriate rotation terms. The equations are not invariant with respect to the rotational motion of the reference frame, but they can be *transformed* to account to the rotational motion.

The fact that we can look at aerodynamic phenomena in different reference frames and know that they are effectively the same comes in handy in many ways. For example, it is basic to the similarity between wind-tunnel testing and flight. We'll look at other aspects of dynamic similarity in Section 3.9. And in developing physical understanding, it is often helpful to look at different aspects of a phenomenon in different reference frames.

3.8.2   *Circulation Preservation and the Persistence of Irrotationality*

In Section 3.3.5, we defined the circulation as the line integral of the velocity around a closed contour and saw that it is equal to the total vorticity piercing any capping surface bounded by the contour, by virtue of Stokes's theorem. The circulation can reflect the dynamics in ways that make it a useful quantity in numerous situations; for example, as we'll see in Section 7.2, it is related to lift through the Kutta-Joukowski theorem.

It is often useful to be able to make general statements about whether circulation is preserved or not. Considering the circulation around a closed material contour that moves with the flow, under what conditions can we expect that it is preserved, that is, that it remains constant in time? In other words, under what conditions is the Lagrangian rate of change of the circulation equal to zero? This can be written as

$$\frac{D\Gamma}{Dt} = \frac{D}{Dt} \oint \mathbf{V} \bullet \mathbf{t}\, dl = 0. \tag{3.8.1}$$

Note that we can reverse the order of differentiation and integration in Equation 3.8.1, so that it becomes the statement that the line integral of DV/Dt around a closed contour is zero, which is true only if DV/Dt is the gradient of a scalar function, called an *acceleration potential.* (See the analogous argument in Section 3.3.6 regarding the velocity potential in irrotational flow.) By invoking the momentum equation, we can show that this requires:

1. No net viscous force on any fluid parcel on the contour. This requires either that the viscosity is zero, or that the fluid is uniform and the flow is irrotational. But the second of these options is of little interest, because requiring irrotationality means that Equation 3.8.1 is satisfied trivially.
2. Any external body force is the gradient of a potential, such as gravity.

3. The fluid is barotropic, that is, the density is a function of pressure only. This is commonly satisfied only if the fluid is homocompositional and entropy is constant in the flowfield, conditions that in aerodynamics often apply to inviscid flow in the absence of shocks. But zero viscosity is not strictly required for this condition to be met. A uniform fluid meets this requirement by definition, even if it is viscous.

This result is sometimes referred to as *Kelvin's circulation theorem.* In some references (for example, Milne-Thomson, 1966), condition 3 is not mentioned explicitly, but it is still implicitly assumed.

Circulation preservation of this kind supports the persistence of irrotationality in many situations. This is easy to see if we think of a flow that starts out irrotational, say upstream of its approach to a body or other source of disturbance. In this initially irrotational state, the circulation must be zero around any reducible closed contour. Then if the flow is circulation preserving, any circulation must remain zero as the flow progresses. The vorticity must also remain zero, since if any nonzero vorticity appeared, there would be nonzero circulation around some closed contour, which would violate our assumption of circulation preservation. So circulation preservation means that a flow that starts out irrotational will remain irrotational downstream.

Preservation of irrotationality is especially simple in the special case of a uniform fluid (constant density and viscosity). A uniform fluid satisfies condition 3 above by definition, and an irrotational flow of a uniform fluid produces no net forces on fluid parcels even if the viscosity is nonzero (see Section 3.6), so that it also satisfies condition 1. For the kinds of compressible flows we are likely to encounter in practice, variations in fluid properties tend to be slow enough that condition 1 is effectively, though not exactly, satisfied in regions of irrotational flow, and condition 3 is satisfied in regions where the flow is isentropic.

In Section 3.6, we discussed the general flow pattern that characterizes external flow past bodies at high Reynolds numbers (small viscosity), in which a vortical boundary layer develops close to the body surface, surrounded by essentially irrotational flow that behaves as if it were inviscid. What we have seen here regarding the persistence of irrotationality is consistent with that picture. Ordinarily, you should expect that when the flow approaching a body from upstream is irrotational, significant vorticity will appear in the flowfield only where it has diffused and convected from the body surface or where it has been produced by a strong, curved shock (see Crocco's theorem, Section 3.8.5 and the discussion of shocks in Section 3.11).

3.8.3 *Behavior of Vortex Tubes in Inviscid and Viscous Flows*

In Section 3.3.7, we introduced the concepts of vortex tubes and vortex filaments, and we saw that the intensity of a vortex filament is constant along its length (Helmholtz's second theorem). By noting that we can consider a vortex tube as a bundle of vortex filaments, we can define the intensity of a vortex tube as the sum of the intensities of the filaments within the tube. (For a vortex tube of finite cross section, this "sum" is actually an integral, because vortex filaments are of infinitesimal size.) Now we assume that the flow is circulation preserving in the sense defined in Section 3.8.2, that is, that the conditions for Kelvin's circulation theorem are met. In this instance, we are by definition concerned with flows with nonzero vorticity, so the only option that satisfies condition 1 for circulation preservation is for the fluid to be *inviscid*. Then two dynamics-based theorems apply:

Helmholtz's third theorem: As a vortex tube evolves in time, fluid parcels cannot cross the bounding surfaces of the tube, and as a result the tube is always made up of the same fluid parcels.

Helmholtz's fourth theorem: The intensity of a vortex tube is constant regardless of how the tube moves around.

A proof that makes use of Stokes's theorem is given by Milne-Thomson (1966). In a rigorous sense, it might seem that these two theorems are not very informative or useful, because they deal with vorticity in inviscid flows, while in most cases of interest in aerodynamics the only way for vorticity to be there is through the effects of viscosity. But if we view the inviscid case to which they apply as being the limiting case for flows with small viscosity, we can get some useful insight from them. Basically, what Helmholtz's third theorem tells us is that for small viscosity, the natural tendency of a vortex tube is to remain anchored to the same material tube, and that vorticity migrates into or out of that material tube only through viscous diffusion. In Section 3.6, we discussed the vorticity equation and the transport of vorticity by convection and diffusion. Helmholtz's third theorem provides further insight into what it means for vorticity to be transported mainly by convection. Helmholtz's fourth theorem tells us something that's a bit harder to visualize: that vorticity migrates into or out of a vortex tube only by viscous diffusion.

The assumption of small viscosity is not always as limiting as it might seem. Nearly all of the vorticity in aerodynamic flows is generated at solid surfaces, and in the thin viscous or turbulent boundary layers on those surfaces, the effects of viscosity are not small. So in these parts of the flow, the insights provided by these theorems are not very applicable. But when the vorticity is convected into the wake away from the body, the viscous and turbulent stresses are rapidly reduced, and the above insights become more useful, as, for example, in Section 8.1, when we consider the vortex wake behind a 3D wing.

3.8.4 *Bernoulli Equations and Stagnation Conditions*

We've seen that effectively frictionless flow conditions are frequently encountered in practice, particularly outside the viscous boundary layer that develops near the surface in flows around bodies. Under these conditions, with a couple of additional restrictions, Bernoulli's principle defines a simple relationship between the pressure and the velocity magnitude. This is analogous to the many situations in classical particle mechanics in which the total force exerted on a particle can be expressed as the gradient of a potential that is a function of position only, a condition that often applies in the absence of friction effects. In these situations, the sum of the "potential energy" and kinetic energy is constant, which leads to a simple relation between the potential function value and the particle's velocity magnitude.

To determine the requirements for the analogous situation to exist with regard to the net force on a fluid parcel, we can either look directly at the energy equation or take a line integral of the momentum equation. Either way, we find that if:

1. There is no net viscous force on fluid parcels;
2. external body forces are negligible;

3. there is no heat conduction; and
4. the flow is steady,

then the total enthalpy

$$H = \frac{V^2}{2} + e + \frac{p}{\rho} \tag{3.8.2}$$

is constant along streamlines, where e is the thermodynamic internal energy per unit mass. Note that under these conditions p/ρ acts like a form of potential energy. Note also that conditions (1) and (3) amount to specifying the absence of entropy-generating processes (assuming bulk-viscosity effects are also negligible), so that entropy must also be constant along streamlines.

Now we introduce the concept of stagnation conditions, which can be defined for general flows that don't necessarily satisfy the assumptions for Equation 3.8.2. From any point in a general flow, we can imagine bringing the fluid parcel passing that point to rest (stagnation, or zero velocity) by an imaginary steady-flow process in which the viscous forces and heat conduction that may be active in the general flow are "turned off" and for which Equation 3.8.2 therefore applies. Because the imaginary stagnation process is isentropic, the stagnation values of T, p, and ρ are uniquely defined, and we'll designate them by the subscript t for "total." In general flows, these stagnation conditions need not be constant, either along or across streamlines. It is important to keep in mind that the stagnation temperature, pressure, or density is not generally a real temperature, pressure, or density, but is only an imaginary construct defined by an imaginary process. The exception is a real stagnation point in a steady flow, where the stagnation condition can actually be realized.

Returning our attention to flows that satisfy Equation 3.8.2 and introducing stagnation quantities, Equation 3.8.2 becomes

$$H = \frac{V^2}{2} + e + \frac{p}{\rho} = e_t + \frac{p_t}{\rho_t} = \text{constant} \tag{3.8.3}$$

along streamlines, where all of the subscript t quantities are constant along streamlines as well. Now if we assume a perfect gas and impose the constant-entropy condition, we have

$$e + \frac{p}{\rho} = c_p T = \frac{\gamma}{\gamma - 1} RT = \frac{\gamma}{\gamma - 1} \frac{p}{\rho}, \tag{3.8.4}$$

and

$$\frac{p}{p_t} = \left(\frac{\rho}{\rho_t}\right)^\gamma, \tag{3.8.5}$$

so that Equation 3.8.3 becomes (see National Advisory Committee for Aeronautics (NACA) Report 1135, Ames Research Staff, 1953)

$$\frac{\gamma}{\gamma - 1} \frac{p_t}{\rho_t} \left(\frac{p}{p_t}\right)^{\frac{\gamma-1}{\gamma}} + \frac{V^2}{2} = \frac{\gamma}{\gamma - 1} \frac{p_t}{\rho_t} \tag{3.8.6}$$

along a streamline, where the subscript t quantities are constants for that streamline. We'll call this the *steady, compressible Bernoulli equation* for frictionless, nonconducting flow of a perfect gas. If the density is constant, that is if the flow can be assumed effectively *incompressible,* in that the density change in response to any pressure change in the flowfield is a sufficiently small percentage (see Section 3.10), this reduces to

$$\frac{\rho V^2}{2} + p = p_t \qquad (3.8.7)$$

along a streamline.

To recap, the Bernoulli principle can be simply stated: For steady, frictionless flow with no body forces and no heat conduction, the total enthalpy is constant along streamlines. Because such flows are also generally isentropic along streamlines, the total pressure and temperature are also constant along streamlines. How the "Bernoulli constant" and the stagnation conditions vary from streamline to streamline, however, depends on how the flow was established, which brings us to our next topic.

3.8.5 Crocco's Theorem

Assume that a flow satisfies the conditions given above for the Bernoulli principle to apply, so that Equation 3.8.2 holds along streamlines. If we assume further that the fluid is homo-compositional, that is, that it consists everywhere of the "same stuff," as would be the case if the relative species concentrations in a mixture of gasses were constant, then a standard thermodynamic relationship between T, S, e, p, and ρ (one of the familiar "TdS" or "tedious" equations, see Liepmann and Roshko, 1957) applies throughout the flowfield. Combining this with the momentum equation (remember that the flow is assumed steady and frictionless), we get

$$\nabla H = T \nabla S + \mathbf{V} \times \boldsymbol{\omega}, \qquad (3.8.8)$$

which is known as *Crocco's theorem.*

We saw earlier that it is common in external flows around bodies for the flow to be "established," that is, to come from upstream, in an irrotational state. If it continues to satisfy the requirements for Crocco's theorem to apply in the part of the flowfield outside the viscous boundary layers near the body surface, the flow will remain irrotational, in which case we can see that constant H implies constant S and vice versa. Thus in the effectively inviscid parts of many external flows, the total enthalpy, our "Bernoulli constant," will usually be constant from streamline to streamline, as well as along streamlines. If such a flow passes through a curved shock (see Section 3.11.2), however, the situation downstream of the shock will be more complicated. The portions of the shock through which different streamlines pass will have different obliqueness, and S will therefore vary from streamline to streamline, though H will still be constant. In Equation 3.8.8, the RHS will be zero, and the varying S will require nonzero vorticity. Thus a curved shock inevitably produces some vorticity downstream, though it is typically very weak unless the shock is very strong.

For a more detailed discussion of circulation preservation, Bernoulli's principle, and Crocco's theorem, readers would do well to consult Chapter 3 of Batchelor (1967). I hope those who do will accuse me at worst of oversimplifying and not of plagiarizing.

3.9 Dynamic Similarity

Using flows around subscale models to simulate what would happen at full scale has been a major tool in aerodynamic development since the Wright brothers built their first wind tunnel. Subscale model testing has made it practical to explore many more design options and a wider range of flight conditions than could be explored by flight testing alone, and the savings in time, money, and test pilots' lives have been incalculable. Advances in CFD in recent years have made it possible for CFD to replace some subscale model simulation, but it will be a long time before computation replaces all of it.

In practice, the similarity between a subscale flow simulation and the real thing is rarely even close to perfect. However, defining the theoretical requirements for a perfect simulation will tell us what factors are important and provide at least some basis for anticipating how accurate our actual simulations are likely to be.

A perfect simulation requires that the model geometry and the model flow be geometrically scaled versions of the real things, with all of the detailed forces and accelerations and other internal processes reproduced in a properly scaled fashion. Presumably if we knew everything we needed to know to be able to set up such a simulation correctly, we would also know the scaling factors for converting quantities such as pressures and integrated forces from model scale to full scale. The early pioneers were able to deduce some of the important scaling factors from a combination of intuition and experience, such as the scaling of forces with surface or cross-sectional area times the square of the velocity, but they were not always aware of the more subtle requirements for similarity. With our advanced understanding of the governing equations, we are in a position to derive these requirements with some rigor.

The basic idea is that if we "nondimensionalize" the governing NS equations and BCs appropriately for both situations (model and full scale), the two flows will be similar if they represent the same solution to the dimensionless equations. Of course, this can be expected to happen only if the dimensionless equations themselves, and the dimensionless BCs, are the same for both situations. This sounds simple enough, but there are subtleties involved, as we'll see below.

To derive the dimensionless equations, we start by replacing each dimensional quantity in the equations with a product of a dimensional reference constant and a dimensionless version of the original variable. The reference constants are simply quantities appropriate to the flow situation, such as freestream flow properties (e.g., velocity and density) and a reference length that characterizes the body, and we naturally have to use equivalent definitions for these constants for both flow situations. We make these substitutions likewise for all of the independent spatial coordinates and all of the dependent flow variables. Then for each equation, we choose a group of the reference constants having the right combined dimensions so that when we divide the entire equation by that group, the equation becomes dimensionless. There is no unique right choice of the group of constants to use for this nondimensionalizing step, but the resulting equations are of course equivalent, no matter what group is chosen.

The conventional choice is based on the observation that inviscid flows have no inherent scale dependence, and that it is therefore convenient to choose the nondimensionalizing group so that the nonviscous terms in the dimensionless equations end up looking just as they do in the original equations, with no scale-dependent coefficients multiplying them. For the momentum equations, this means dividing by $\rho_{ref}\, u_{ref}^2/L_{ref}$, and for the energy equation in total-enthalpy form by $u_{ref}\, p_{ref}/L_{ref}$. In our final set of dimensionless equations, the nonviscous convective terms look the same as they did in the dimensional equations, while the pressure gradient, viscous-transport, and heat-conduction terms have dimensionless "scaling" or "similarity" parameters multiplying them.

Now we see that for the dimensionless equations to be the same for both the simulation and for full scale, the similarity parameters multiplying the pressure and transport terms must have the same values in both cases.

Now let's look specifically at the dimensionless parameters that arise. We form dimensionless versions of all the basic variables in the equations, designating them by putting "hats" over them. We nondimensionalize the basic variables as shown in Table 3.9.1.

We invert these definitions (e.g., $x = \hat{x}L$) and substitute them into the form of the equations valid for a perfect gas with constant specific heats. We also make use of the specific heat relationship $c_p = \gamma R/(\gamma - 1)$, where γ is the ratio of specific heats and R is the gas constant, and of the fact that the speed of sound is given by $a = (\gamma p/\rho)^{1/2}$. In the resulting dimensionless equations, the continuity equation and the convective terms in the momentum equation and in the energy equation in total-enthalpy form look exactly as they did in the original dimensional equations. Terms in which dimensionless similarity parameters appear are as shown in Table 3.9.2, where $M_\infty = u_\infty/a_\infty$ is the freestream Mach number, $R_L = \rho_\infty u_\infty L/\mu_\infty$ is the Reynolds number, and $Pr = \mu c_p/k$ is the Prandtl number.

Based on this, we might expect perfect similarity if we match γM_∞^2, R_L, and Pr. But things aren't quite so simple with regard to γ and M_∞. One of our primary variables in the energy equation is $\hat{H}$, which is defined in terms of other dimensionless variables as

$$\hat{H} = \frac{\gamma}{\gamma - 1}\hat{T} + \frac{\gamma M_\infty^2}{2}\hat{u}^2. \qquad (3.9.1)$$

Table 3.9.1 Basic variables

Type of quantity	Example
Spatial coordinates	$\hat{x} = x/L$
Velocities	$\hat{u} = u/u_\infty$
Local fluid state variables p, ρ, T, μ	$\hat{p} = p/p_\infty$
Total enthalpy	$\hat{H} = H\rho_\infty/p_\infty$

Table 3.9.2 Terms appearing in dimensionless equations

Equation	Type of term	Example
Momentum	Pressure gradient	$\dfrac{1}{\gamma M_\infty^2}\dfrac{\partial \hat{p}}{\partial \hat{x}}$
Momentum	Viscous force	$\dfrac{1}{R_L}\dfrac{\partial}{\partial \hat{y}}\left(\hat{\mu}\dfrac{\partial \hat{u}}{\partial \hat{y}}\right)$
Energy	Heat conduction	$\dfrac{1}{R_L}\dfrac{\partial}{\partial \hat{y}}\left(\dfrac{\hat{\mu}}{Pr}\dfrac{\partial \hat{H}}{\partial \hat{y}}\right)$
Energy	Viscous dissipation and thermal conduction	$\dfrac{\gamma M_\infty^2}{R_L}\left(1 - \dfrac{1}{Pr}\right)\dfrac{\partial}{\partial \hat{y}}\left(\hat{\mu}\hat{u}\dfrac{\partial \hat{u}}{\partial \hat{y}}\right)$

So the internal energy (first term) and kinetic energy (second term) in the total enthalpy depend on γ in different ways, and true similarity requires that we match γ and M_∞ separately.

To summarize, similarity parameters that must be matched for perfect similarity are:

Ratio of specific heats	$\gamma = c_p/c_v$
Mach number	$M_\infty = u_\infty/a_\infty$
Reynolds number	$R_L = \rho_\infty u_\infty L/\mu_\infty$
Prandtl number	$Pr = \mu c_p/k$

Note that some of these are fluid properties (γ and Pr) and some (M_∞ and R_L) depend in addition on the reference flow conditions, in this case the far field. Now let's consider the major scaling and similarity issues in detail: compressibility effects, the effects of viscosity, scaling of pressure forces, and the consequences of failing to match the similarity requirements.

3.9.1 Compressibility Effects and the Mach Number

Throughout the flowfield the interaction between the pressure and density for compressible flows is governed by implicit relationships built into multiple equations, and γ and M_∞ correspondingly appear in several places. We saw above that similarity requires us to match γ and M_∞ separately.

Of the flow-dependent parameters, Mach number has the most straightforward physical interpretation. It determines how large the pressure changes in the flowfield will be, relative to the absolute pressure, and thus how important pressure-induced density changes will be, a topic we'll consider further in Sections 3.10 and 3.11. The Prandtl number is important with regard to thermal effects in boundary layers, and we'll consider it further in Section 4.6.2.

3.9.2 Viscous Effects and the Reynolds Number

The Reynolds number is also widely familiar, but its proper interpretation is more subtle. Reynolds number is sometimes described as measuring the "ratio of inertial forces to viscous forces," which is incorrect, or the "relative importance of inertial forces and viscous forces" in the flow, which is not as bad, but not quite right either. Actually, the term $1/R$ multiplies the viscous terms in the momentum and energy equations, but it does not directly reflect the relative magnitudes of the forces locally. In any external flow with a Reynolds number above about 10^2, there will be part of the flow outside the boundary layer where the ratio of inertial forces to viscous forces will be practically infinite and part of the flow deep in the boundary layer where it will be practically zero. Thus the ratio of local forces is obviously not what the Reynolds number directly determines.

What the Reynolds number does determine is how "fast," relative to the flow velocity, momentum will be diffused in the cross-stream direction by viscosity or turbulence and thus how thick the boundary layer will grow relative to the dimensions of the body. As Reynolds number increases, the diffusion of momentum becomes relatively slower, and the boundary layer will be thinner. This increases the velocity gradient inside the boundary layer, which offsets the reduction in the $1/R$ multiplier in front of the viscous terms.

The result is that the variation across the boundary layer in the relative importance of viscous forces stays essentially the same, regardless of the Reynolds number: At the "bottom" of the boundary layer, viscous and pressure forces dominate, while at the "edge" of the boundary layer viscous forces are small, and inertial and pressure forces dominate. Thus the idea that the Reynolds number measures "relative importance" makes sense only if we apply it to global viscous "effects" as opposed to local viscous "forces." As the Reynolds number increases, the boundary layer becomes thinner and has less effect on the rest of the flow. It is only in this indirect sense that the Reynolds number measures the relative importance of viscous effects.

We've just used a simple physical argument to explain why the effects measured by the Reynolds number are subtle. The corresponding mathematical argument is less intuitive and starts with the observation that the viscous terms involve the highest order derivatives of the velocity and therefore play a unique role in the equations. If we set 1/R to zero, the viscous terms vanish, and we have the inviscid equations. We might expect, then, that for small values of 1/R, a viscous solution could be obtained as a small perturbation to a corresponding inviscid solution. This expectation is wrong, however, because when the 1/R terms are dropped to obtain the inviscid equations, the order of the equations is reduced, and with it the number of allowable BCs. The BC we must give up to allow nontrivial inviscid solutions is the no-slip condition. To obtain a viscous solution from the corresponding inviscid solution, the no-slip condition must be reinstated, and there will always be a region close to the wall (the boundary layer) where the perturbation imposed on the inviscid solution by viscosity is not small. In mathematical terms, this is an example of a *singular perturbation*, and the kind of behavior we've been discussing is characteristic of singular-perturbation problems. In Section 4.2.1, we'll discuss how the implementation of the singular-perturbation idea leads to a rigorous derivation of boundary-layer theory.

So far, we've seen that the way the characteristics of a flow depend on the Reynolds number is complicated by the fact that a change in the 1/R multiplier in front of the viscous terms tends to be offset by a change in boundary-layer thickness. This complication is compounded by additional subtleties. The location on the body where the boundary layer transitions from laminar to turbulent usually depends on Reynolds number, and this can strongly affect the development of the boundary layer and thus the rest of the flow. Though the variation of boundary-layer thickness with Reynolds number in laminar and turbulent boundary layers is similar qualitatively, it is significantly different quantitatively. The location of laminar-to-turbulent transition strongly affects both the profile drag in the attached-flow regime and the resistance of the boundary layer to separation downstream. Such effects can be very important in airfoil flows, as we'll see in Section 7.4, and in bluff-body drag, which we'll discuss in Section 6.1.6.

3.9.3 Scaling of Pressure Forces: the Dynamic Pressure

When we divide the momentum equations by $\rho_{ref} u_{ref}^2 / L_{ref}$, the resulting dimensionless convection terms look just as they did in their original dimensional form. For example, the original dimensional term

$$\rho u \frac{\partial u}{\partial x} \tag{3.9.2}$$

becomes

$$\hat{\rho}\hat{u}\frac{\partial \hat{u}}{\partial \hat{x}},$$ (3.9.3)

where $\hat{\rho} = \rho/\rho_{ref}, \hat{u} = u/u_{ref}, \hat{x} = x/L_{ref}$

When we nondimensionalized the equations to derive the similarity parameters, we nondimensionalized p by p_∞. But this nondimensionalization of the pressure isn't very convenient for dealing with low-speed flows. If, instead, we take our reference pressure equal to what is called the *dynamic pressure* $q = 1/2\rho_{ref} u_{ref}^2$, the original dimensional pressure-gradient term:

$$\frac{\partial p}{\partial x}$$ (3.9.4)

becomes

$$\frac{1}{2}\frac{\partial \hat{p}}{\partial \hat{x}},$$ (3.9.5)

where $\hat{p} = p/q$. We introduced the factor of 1/2 in the definition of our reference pressure because it arises when we integrate the momentum equation to obtain Bernoulli's equation, as we saw in Section 3.8.4, and q defined in this way is equal to the difference between local static pressure and total pressure in the incompressible, or low-Mach-number case, as in Equation 3.8.7. (A discussion of what "incompressible" means is coming up in Section 3.10.)

Normalizing pressure differences by q is very convenient because it removes the dependence on density and velocity in incompressible flows, and it has therefore become second nature to aerodynamicists. And q is used as a reference for compressible flows as well, though outside of the low-Mach regime, it no longer removes all of the dependence on velocity, or Mach number.

3.9.4 Consequences of Failing to Match All of the Requirements for Similarity

In most subscale flow simulations we must tolerate a substantial mismatch in one or more of the similarity parameters. How does the accuracy of a simulation suffer as a result? In this regard, the first thing to note is that the parameters are not all of equal importance and that their relative importance depends on the situation.

In flows with sufficiently low Mach number throughout the field, matching Mach number is not important, while in flows where local Mach numbers approach or exceed one, matching Mach number is crucial. Fortunately, Mach number is formally scale independent and therefore relatively easy to match regardless of scale. In cases where Mach number is important, we can usually avoid having to tolerate a serious mismatch.

Unfortunately, we can't say the same of the Reynolds number, which varies directly with scale. For a given flow condition, a reduction in scale (our main reason for doing subscale simulation) entails a reduction in Reynolds number. Testing at elevated pressures can alleviate the problem but can't generally eliminate it. At high subsonic speeds, testing at total pressures above about 5 atm is impractical because air loads on a typical airplane model become too high, even for a model constructed of solid steel. The only way the full-scale Reynolds number of a large airplane can be matched on a small-scale model in a wind tunnel is through a combination of elevated pressure (up to 5 atm) and very low temperature (to $-250\,^\circ\text{C}$), called *cryogenic pressure testing*.

Such testing is very expensive and accounts for only a miniscule fraction of all wind-tunnel flow simulation. Most other subscale testing is done at Reynolds numbers roughly an order of magnitude below full scale. As we saw previously, the dependence of a flow on the Reynolds number is complicated, with contributions from a number of physical factors. The sacrifice in simulation accuracy due to a mismatch in the Reynolds number varies greatly depending on the situation, but it is usually serious. A great deal of work has been devoted to the problem of understanding, minimizing, and accounting or correcting for the "Reynolds-number effects" associated with testing at sub-full-scale Reynolds numbers, with mixed success (see Bushnell, 2006, for a review).

In our derivation of the terms in which similarity parameters appear (Table 3.9.2), we assumed a perfect gas. The idea of a perfect dynamic similarity is problematic if we consider the more complicated behavior of real gases over wide ranges of conditions. In the general case, the pressure and temperature, as well as the composition of the gas, would have to be matched, in which case there could be no difference in density, and no difference in scale would be allowed. Fortunately, air under conditions of interest to us in aerodynamics behaves very much like a perfect gas, with practically constant γ and Prandtl number, so that mismatches in parameters other than the Mach and Reynolds numbers tend to have very minor effects (see White, 1991). For example, because of the nonlinear form of the Sutherland law for the dependence of viscosity on temperature, perfect similarity would require matching temperature. However, because the temperature variation within a given flow is usually not large on an absolute scale, the effect of even a large mismatch in temperature at a given Reynolds number is very small.

3.10   "Incompressible" Flow and Potential Flow

In early theoretical work in fluid mechanics, the fluid was generally assumed to have constant density regardless of the pressure, and thus to be "incompressible," an idealization that greatly simplifies the mathematics and is actually justified at sufficiently low Mach numbers. The term "incompressible," however, is a potential source of confusion. Literally speaking, saying that something is "incompressible" means that it cannot be compressed. In this sense, the term is only approximately applicable to liquids and not at all applicable to gases. Yet even with reference to gases, we often talk loosely about "incompressible flow." This is sloppy terminology. We don't mean what the term literally says (i.e., that the flow cannot be compressed). What we really mean is that the flow happens not to be getting compressed significantly in this particular situation and that it therefore behaves *as if* it were incompressible.

In gases at low Mach numbers, a flow can act as if it were incompressible, in that we can make very accurate predictions using equations in which we have assumed that the density is constant. But how can constant density be a reasonable assumption for gases, which are by definition highly compressible? Even in flows at low Mach numbers, where velocity differences, pressure differences, and density differences are all small, the density differences are of the same order as the pressure differences. How can the pressure changes be important and the density changes not be? The answer lies in the different roles that pressure and density play in affecting the motion. In the application of Newton's second law to a fluid parcel, only a pressure *difference* can apply an unbalanced force, while the density itself provides the resisting inertia. In the momentum equation, this is reflected in the fact that the pressure appears inside a derivative (gradient operator), and the density does not. While the small velocity and pressure differences are crucial players in the momentum balance, the small density differences have a comparatively much smaller effect on the flow.

Note that to justify treating the flow as incompressible we do not have to assume small disturbances in the sense of small velocity changes relative to the freestream velocity. The changes in velocity can be large relative to freestream, as long as the velocity itself remains small relative to the speed of sound.

So the equations of motion for an "incompressible" fluid are valid for flows of highly compressible gases, in the limit of low Mach number. If we also assume that the flow is steady and inviscid, and that the onset flow is irrotational, we have met the conditions for classical *potential flow* or *ideal flow* theory. As we saw in Section 3.8.2, a constant-density inviscid flow is circulation preserving, so that irrotational onset flow means the flow will be irrotational everywhere. Then, as we saw in Section 3.3.6, the velocity field can be represented as the gradient of a scalar potential function:

$$\mathbf{V} = \nabla\phi, \tag{3.10.1}$$

and the continuity equation is Laplace's equation for ϕ:

$$\nabla^2\phi = 0. \tag{3.10.2}$$

Assuming the pressure is related to the velocity by the "incompressible" Bernoulli equation (Equation 3.8.7) with a constant total pressure, any flow that satisfies Equation 3.10.2 satisfies both the continuity and momentum equations. Under these conditions, the energy equation is not needed, and Equation 3.10.2 with appropriate BCs represents the complete system describing the inviscid, irrotational flow of a constant-density fluid. The problem has been reduced to a single linear equation with a single scalar as the dependent variable.

Because the incompressible potential-flow equation is linear, solutions can be constructed by superposition of more than one known solution. The velocity potentials for elementary building-block flows around simple singularities are often used for this purpose. Some examples are illustrated in Figure 3.10.1. A *vortex* Figure 3.10.1a can be used to produce circulation, and *sources* or *sinks* Figure 3.10.1b can be used to simulate the thickness effect of a body. The combination of a source and sink of equal strength results in flow pattern Figure 3.10.1c. In the limit as the separation distance between them goes to zero, keeping the product of the source/sink strength and the separation distance constant, we have a *doublet* Figure 3.10.1d. Combining a doublet with a uniform flow produces a pattern Figure 3.10.1e in which the outer part is the potential flow around a circular cylinder. And many other combinations are possible.

For 2D flows, conformal mapping provides solutions for flows over convex and concave corners. Knowing the character of these solutions is useful just for thinking about the local effects of corners in general, and in Section 4.3.2 we'll see that they provide the outer-flow BCs for a family of special solutions to the boundary-layer equations. The basic idea is that a simple analytic mapping can transform the x axis into a line that is bent at the origin through

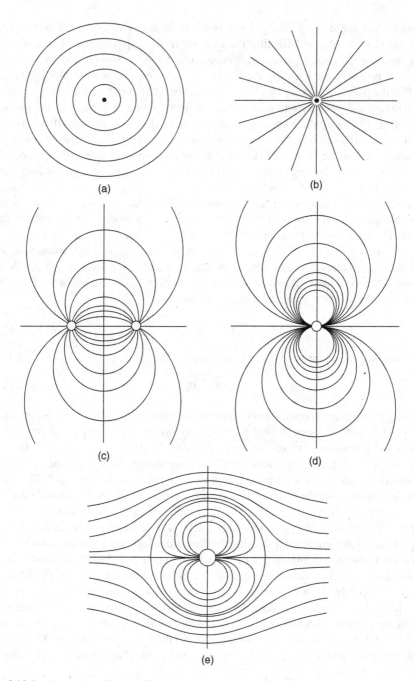

Figure 3.10.1 Examples of streamline patterns associated with simple potential-flow singularities that can be used in constructing other potential-flow solutions by superposition. (a) Vortex. (b) Source or sink. (c) Combination of source and sink of equal strength. From Durand, (1932). Used with permission of Dover Publications, Inc. (d) Doublet. From Durand, (1932). Used with permission of Dover Publications, Inc. (e) Combination of a doublet and a uniform flow yielding a circular cylinder. From Durand, (1932). Used with permission of Dover Publications, Inc.

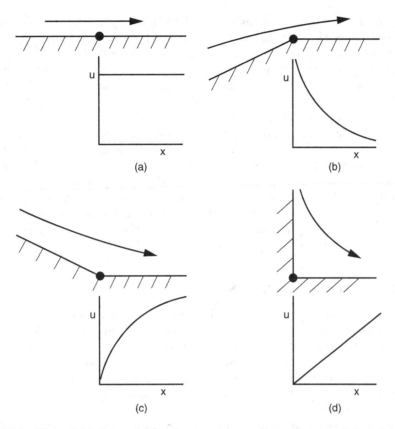

Figure 3.10.2 Illustrations of potential flows over concave and convex corners that can be produced by conformal mapping of the uniform flow in the upper half plane. (a) Uniform flow. (b) A convex corner: $-1 < \beta < 0$. (c) A concave corner: $0 < \beta < 1$. (d) A 90-degree concave corner: $\beta = 1$

an angle $\pi\beta/2$. The uniform flow in the upper half plane, illustrated in Figure 3.10.2a, is then transformed into corner flows like those illustrated in Figure 3.10.2b–d. The velocity distribution along the surface is given by $u \sim x^m$, where $m = -\beta/(\beta - 2)$.

Velocity distributions that go with the different types of corners are illustrated in Figure 3.10.2. In Figure 3.10.2b, we see that a convex corner produces infinite velocity at the corner and decelerating flow away from it. In Figure 3.10.2c, a concave corner produces zero velocity (a stagnation point) at the corner. Combine this flow with its reflection about the x axis, and we have the flow off of a wedge-shaped trailing edge. This demonstrates that in potential flow, an airfoil with a nonzero trailing edge angle must have a trailing-edge stagnation point. In Figure 3.10.2d, we have the special case of $\beta = 1$, a 90° corner with a stagnation point and linear acceleration of the flow away from the corner. Combine this flow with its reflection about the y axis, and we have the prototype for flow attaching itself to a smooth wall at a stagnation point. Conformal mapping can also be used to transform solutions for very simple shapes such as circles into solutions for more-complicated shapes. With the conformal-mapping approach, analytic solutions are possible for a wide variety of shapes.

We can obtain numerical solutions for 2D flows either by discretizing the conformal-mapping problem or by the "panel method," that is solving for a distribution of unknown singularity strength along a set of discrete panels representing the body surface. When the panel method is extended to 3D flows, not only the surface of the body must be "paneled," but also any vortex wakes that exist in the field, which, as we saw in Section 3.3.8, can be idealized as surfaces of discontinuity in the velocity field.

As we saw in Sections 3.6 and 3.8.2, external flows around bodies at high Reynolds numbers remain effectively irrotational outside of a thin viscous boundary layer that develops near the surface. Potential flow is therefore not a bad approximation for many of these flows and can often provide very reasonable predictions of pressure distributions and integrated forces at low Mach numbers. The pioneers of theoretical aerodynamics took full advantage of this, and before large-scale computation of nonlinear flow problems became practical, much of our quantitative prediction capability depended on incompressible potential-flow theory, most often implemented in the form of panel methods. It remains a powerful tool both for conceptualizing and for quantitative calculations.

3.11   Compressible Flow and Shocks

"Compressibility" deals with changes in fluid density in response to changes in pressure. In the previous section, we discussed how these changes are insignificant at low Mach numbers. As Mach number increases, however, we can no longer ignore the density changes, and when the local Mach number exceeds one, their effects on the flowfield become so strong that the character of the flow changes drastically, as reflected for example in the response of the flow to changes in cross-sectional area, which is essentially reversed between the subsonic and supersonic regimes.

There are small-disturbance flow regimes, both subsonic and supersonic, in which the effects of compressibility are treated approximately, in the context of linear equations. These theories have such narrow ranges of applicability, however, that they are usually not very useful, and they have been marginalized in recent years by the availability of higher-fidelity computational methods. We'll touch on the linear theories again in Chapter 9 when we survey broad classes of theoretical approximations. In this section, we explore what happens when we attack the problem of compressibility head on, without assuming that disturbances are small.

In compressible flows, the changes in density bring with them changes in local temperature and in all of the fluid properties that are affected by temperature. These changes have significant local effects on the flow, and accurate flow prediction requires solving for the altered local fluid properties in detail. As we discussed in Section 3.1, the viscosity, thermal conductivity, and local speed of sound are usually assumed to depend on temperature only.

A general feature of most transonic and supersonic flows is the appearance of compression shocks, regions in which significant amounts of deceleration and compression (pressure increase) take place over very short distances, on the order of the mean free path of the gas molecules. The extremely large gradients inside a shock are beyond the strict range of validity of the NS equations. As we've already noted in Section 3.1, the NS equations don't predict the internal structure of a shock accurately, but because the equations also enforce the integrated conservation laws, they do capture the "jump" conditions across shocks correctly. Shocks tend to form in most transonic and supersonic flows because sustained deceleration and compression in supersonic regions is usually not possible, especially when the deceleration goes all the way from supersonic to subsonic. There are exceptional situations in which sustained gradual deceleration can occur, but they are not common in practical situations. In most ordinary situations, infinitesimal compression waves in a supersonic region tend to pile up on each other and to coalesce into a shock. Details on this process can be found in Shapiro (1953) or Liepmann and Roshko (1957).

Shocks in steady flows can be locally perpendicular (normal) or oblique to the flow. Figure 3.11.1 illustrates several examples of steady flowfields with shocks. In Figure 3.11.1a we have a transonic airfoil flow, with a "bubble" of supersonic flow over the upper surface, terminated by a weak shock that is generally curved and slightly oblique. This global flow pattern is typical of jet airliner wings in high-speed subsonic cruise (The simplified sketch omits the details of the interaction between the shock and the boundary layer, which we'll discuss in Section 7.4.8). In Figure 3.11.1b we have flow over a sharp wedge or cone at a sufficiently high supersonic freestream Mach number to produce an *attached oblique shock* propagating from the nose. In this case, the flow downstream of the shock is still supersonic. In Figure 3.11.1c we have supersonic freestream flow past a blunt-nosed body with a detached, curved *bow shock* standing off the nose. In this case, there is a "bubble" of subsonic flow downstream of the part of the shock that is close to normal, and supersonic flow downstream of the more oblique parts of the shock. Of course, more complicated patterns are possible, especially in 3D, with shocks from different parts of a body intersecting, or with shocks being affected by expansion waves from other parts of the body. In Section 3.11.2, we'll discuss the local relationships the flow through a shock must obey.

With or without shocks, compressible flow is inherently more complicated than incompressible flow. The full system of equations for compressible flow, even if assumed inviscid, is nonlinear and entails all the difficulties we discussed in Section 3.5. Except for a very few simple cases, analytic solutions for flows in more than one dimension are not possible, and the prediction of a multidimensional flowfield generally requires large-scale computation.

But all is not lost. There is still a great deal that can be learned about steady multidimensional compressible flows without resorting to computation, based on the combination of continuous 1D flow theory and the relations governing normal and oblique shock waves. Most real flows are not close to being 1D, and the 1D flow theory/shock relations are therefore not globally predictive, but they are still powerful tools because they provide simple ways to calculate useful relationships between the various flow quantities, applicable along streamlines of steady multidimensional flows. Their general usefulness is thus analogous to that of Bernoulli's equation for incompressible flow.

3.11.1 Steady 1D Isentropic Flow Theory

If the flow is steady, the flow in any given slender streamtube in a flowfield is effectively 1D. If we assume that the flow is inviscid and isentropic, as it usually would be outside the boundary layer near solid surfaces, and that the fluid is a perfect gas, then the 1D equations of motion can be integrated analytically. Note that these assumptions satisfy the requirements for the compressible Bernoulli equation (Equation 3.8.6) to hold, so that stagnation conditions are constant along our streamtube. Remember also that if we further assume that the fluid is homocompositional, such that Crocco's theorem (Equation 3.8.8)

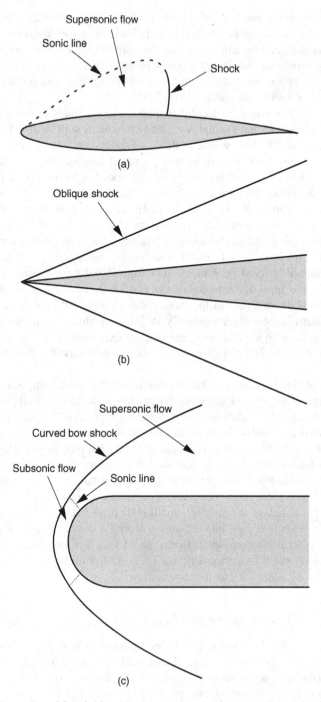

Figure 3.11.1 Examples of flowfields with compression shocks. (a) A transonic airfoil flow with a bubble of supersonic flow over the upper surface, terminated by a shock. (b) Supersonic flow over a sharp wedge or cone, with an attached oblique shock. (c) Supersonic flow over a blunt-nosed body with a detached bow shock

holds, and that the onset flow is irrotational, then the stagnation conditions will be the same in all of the streamtubes in the field, at least upstream of all shocks. Downstream of any shock, the stagnation pressure is reduced, as we'll see next.

The result of the steady 1D flow theory for a perfect gas is a set of relationships defining dimensionless ratios of flow variables. This set of relationships can be expressed in a variety of ways. An exhaustive compilation is given in NACA Report 1135 (Ames Research Staff, 1953). Derivations can be found in Shapiro (1953) and Liepmann and Roshko (1957).

In a commonly used form of these relations, the local temperature, pressure, and density are related to their stagnation values, the streamtube area and flow velocity are related to conditions that would apply at Mach one, and everything is expressed in terms of local Mach number:

$$\frac{T}{T_t} = \left(1 + \frac{\gamma - 1}{2}M^2\right)^{-1} \tag{3.11.1}$$

$$\frac{p}{p_t} = \left(1 + \frac{\gamma - 1}{2}M^2\right)^{\frac{-\gamma}{\gamma - 1}} \tag{3.11.2}$$

$$\frac{\rho}{\rho_t} = \left(1 + \frac{\gamma - 1}{2}M^2\right)^{\frac{-1}{\gamma - 1}} \tag{3.11.3}$$

$$\frac{A^*}{A} = \left(\frac{\gamma + 1}{2}\right)^{\frac{\gamma + 1}{2(\gamma - 1)}} M \left(1 + \frac{\gamma - 1}{2}M^2\right)^{-\frac{\gamma + 1}{2(\gamma - 1)}} \tag{3.11.4}$$

$$\left(\frac{V}{a^*}\right)^2 = \frac{\gamma + 1}{2}M^2 \left(1 + \frac{\gamma - 1}{2}M^2\right)^{-1}, \tag{3.11.5}$$

where A is the streamtube area, a is the speed of sound, and the star refers to conditions that would apply at Mach 1. These relations are plotted in Figure 3.11.2, and they highlight several very useful facts regarding steady inviscid flow of a perfect gas:

1. Streamtube area is at a minimum, and therefore mass flux per unit area is maximum, at a local Mach number of one. Thus for a flow to pass from a subsonic condition to supersonic, it must pass through Mach one at a local minimum in streamtube area. If geometric constraints force the flow area to go through a minimum at some location, as is the case in flows in confined passages or "ducts," a maximum-mass-flux flow condition at Mach one at that location is often referred to as "choked." To produce a steady supersonic flow in a duct, as in a rocket exhaust or a supersonic wind tunnel, a choked nozzle somewhere upstream is generally required.
2. A change in streamtube area has opposite effects in subsonic and supersonic flows. An area increase slows a subsonic flow down and increases the pressure and temperature but does the opposite to a supersonic flow.
3. Dimensionless *ratios* of flow quantities depend only on Mach number and γ. Changing the *level* of pressure or temperature has no effect on dimensionless ratios if Mach number is kept the same.

Steady, 1D flow analysis can be extended to include effects of bulk-averaged heating or cooling, or viscous friction. Treating such effects in a 1D, bulk-averaged sense can

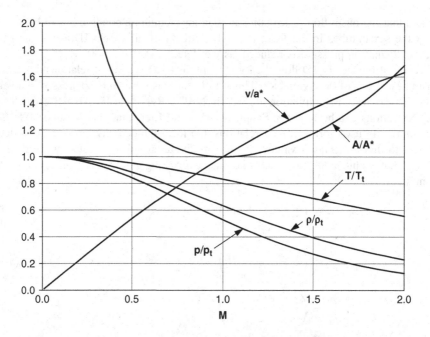

Figure 3.11.2 1 D isentropic-flow relations

be a reasonable approximation for flows in ducts that are sufficiently slender relative to their length. The inclusion of these effects means that entropy and total pressure will no longer be constant. A flow with simple viscous friction and no heat transfer will still have constant total temperature, but a flow with heat transfer will not. Analytic solutions such as those leading to Equations 3.11.2–3.11.5 are possible for constant streamtube area and simple friction or simple heat transfer. Numerical solution is generally required in cases with variable streamtube area and/or combined effects. Equations applicable to these situations are derived in Shapiro (1953, Chapters 6–8)

3.11.2 Relations for Normal and Oblique Shock Waves

In solutions to the NS equations, compression shocks appear as continuous compressions in which all flow quantities vary smoothly. The predicted shock thickness of a bit over one mean free path is roughly correct, but the detailed distributions of flow quantities are not, because the NS equations don't model the molecular transport terms accurately when changes take place over such short distances. The "jumps" in flow quantities across a shock are accurately predicted because they don't depend on the internal structure of the shock, and the NS equations enforce all the relevant physical conservation laws that must be obeyed in the total jump across the shock.

Note that most practical numerical solutions to the NS equations are done on grids that are far too coarse to resolve the correct thickness of a shock. Numerical dissipation smears shocks over a distance of at least several grid intervals, which is typically several orders of magnitude larger than the actual thickness of a shock, and the degree of smearing increases when the shock front is oblique to the grid. Still, except to the extent that the smearing encroaches on flow gradients upstream and downstream, jumps across the shock can be predicted accurately in spite of the smearing.

The "shock relations" are a set of analytic formulas describing the effects of shocks in terms of the jumps in flow quantities from upstream to downstream. In this discussion, we will consider only steady flows, but the results of the steady theory can in principle be transformed to apply to unsteady flow because the flow in the neighborhood of a shock can be rendered locally steady by a Galilean transformation that cancels the motion of the shock normal to itself. The shock relations are "1D" in that they describe what happens only to an individual streamtube as it goes through the shock, and in this sense they are similar to the 1D isentropic flow relations we described in Section 3.11.1. The only multidimensional consideration that comes into play is turning of the streamtube through the shock, which must be accounted for when the shock is oblique to the local flow direction. Because jump conditions through a shock do not depend on the detailed internal structure of the shock, the shock relations can be derived in an essentially inviscid framework.

To derive the shock relations, we enforce conservation of mass, momentum, and energy across the shock discontinuity. Conservation of the component of momentum tangential to the shock requires that the jump in tangential velocity is zero. Because the tangential velocity is the same upstream and downstream, it can be removed by a Galilean transformation, and any shock can be viewed as a normal shock. The derivation is therefore carried out in two stages. First we derive the jump conditions for a normal shock, and then we derive the jump conditions for an oblique shock by transforming the normal-shock relations (i.e., by reversing the Galilean transformation that removed the tangential velocity).

For a normal shock, once we enforce conservation of mass, momentum, and energy; assume a perfect gas; and stipulate that entropy cannot decrease, the relationships between downstream and upstream flow quantities are uniquely defined. The resulting relations are plotted in the form of dimensionless ratios in Figure 3.11.3. A shock obeying these relations is often referred to a *Rankine-Hugoniot* shock (see Liepmann and Roshko, 1957). The following general characteristics of normal shocks are noteworthy:

1. The Mach number jump is always from supersonic to subsonic, with accompanying increases in pressure, temperature, and density, and a decrease in velocity.
2. Total temperature and total enthalpy are unchanged, but because passage through a shock is a dissipative process, there is a loss in total pressure and an increase in entropy. A jump from subsonic to supersonic (an "expansion shock") would require a decrease in entropy and thus violate the second law of thermodynamics.
3. The "strength" of a shock, as indicated by the loss in total pressure, increases rapidly with upstream Mach number. In applications where low drag is important, really strong shocks are to be avoided by design, and the strength of relatively weak shocks becomes an important issue. An important fact to remember in that regard is that for upstream Mach numbers near 1.0 (the weak shock limit), the total-pressure loss goes as $(M - 1)^3$.
4. Because of the loss of total pressure, the minimum area, A^*, through which the flow can be forced, is increased by passage through the shock.
5. Dimensionless ratios between downstream and upstream are unique functions of either upstream or downstream Mach number and γ.

Of course, the normal-shock relations apply to oblique shocks as well, provided we resolve the velocity and Mach number in the direction normal to the shock. So the basic physics

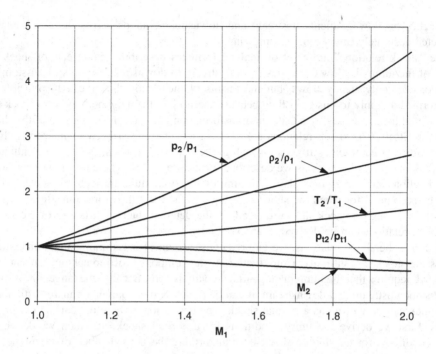

Figure 3.11.3   Normal-shock relations

of an oblique shock is no more complicated than that of a normal shock. With oblique shocks, however, there is a surprising degree of apparent additional complexity, introduced by what amounts to only one additional free parameter in the problem (the tangential velocity). Across an oblique shock, the normal velocity is reduced, and the tangential velocity is unchanged, so that the flow is turned, or deflected, through some angle as it passes through the shock. For a given upstream Mach number, which must be supersonic, there is a maximum deflection angle. For any deflection angle less than the maximum, there are two possible shock inclination angles and two different possible downstream flow conditions. For deflection angles close to the maximum, both of the possible downstream conditions are subsonic, as would be the case with a normal shock, but for deflection angles less than the maximum by more than a fraction of a degree, one of the possible downstream conditions is supersonic. These complexities are best understood through study of the various oblique-shock charts in Shapiro (1953) or NACA Report 1135 (Ames Research Staff, 1953).

The normal and oblique shock relations we've just discussed are essentially 1D, but they apply in multidimensional flows. When supersonic flow is present in steady-flow solutions of the multidimensional inviscid equations (the Euler equations), shocks are usually also present and appear as surfaces imbedded in the field, across which flow quantities are discontinuous. Flow quantities can also be discontinuous across imbedded vortex sheets, or singularities can exist at discontinuities in the slope or curvature of bounding surfaces, but everywhere else in the field flow quantities are continuous and differentiable, and the inviscid equations of motion are satisfied. The locations and shapes of imbedded shock surfaces are not known or specified a priori; all that can be specified is that if a shock arises as part of the solution, the jump across it must satisfy the shock relations. In mathematical jargon, solutions with such imbedded surfaces of discontinuity are called "weak" solutions. Note that the Euler equations, which are satisfied in the regions of continuous flow, are not sufficient to determine the solution if any shock discontinuity is present and that the shock relations must be imposed explicitly across the discontinuity. In numerical solutions, shock discontinuities can either be represented explicitly (*shock fitting*) or approximated automatically in a somewhat smeared-out way in the numerical grid (*shock capturing*). We'll discuss these options further in Section 10.4.1.

4

Boundary Layers

In Section 3.6, we saw that the basic effect of viscosity is to exert shear stress, and that when that is combined with a no-slip condition at a solid surface, it leads to vorticity that originates at the surface, is convected downstream, and diffuses outward. (In Section 4.2.4, we'll look at this in detail and see that some parts of a surface act as sources of vorticity, and others act as sinks, but that the net result is that there is always some vorticity convected along next to a solid wall.) In Section 3.8.2, we saw that irrotational flow tends to remain irrotational until viscosity has had a chance to work on it. The upshot is that flows around bodies at high Reynolds numbers acquire a natural global structure: an inner vortical region where viscosity is important, consisting of the *boundary layer* next to the surface and the *wake* downstream, surrounded by an *outer flow* that is effectively irrotational and that behaves as if it were inviscid. This basic pattern was not generally understood until Prandtl (1904) explained it and proposed his approximate theory for the flow in the boundary layer.

In this chapter, we'll take a detailed look at the physics of the flow in the boundary layer, in preparation for a more global discussion of the whole flowfield in Chapter 5. This ordering of the discussion is convenient because it turns out that the boundary layer and the outer flow interact only through a relatively simple set of boundary conditions at their interface, and everything of interest in the physics of the boundary layer can be discussed in general terms without our having to know much about the outer flow. On the other hand, our later discussion of the global flow will refer often to what goes on in the boundary layer. Boundary-layer specialists like to joke (only half joke, actually) that the boundary layer is the most important part of the flow because it's the only part that touches the body.

When the Reynolds number is sufficiently high, the boundary layer remains relatively thin, unless it separates from the body "prematurely," that is, ahead of the tail of a body or the trailing edge of a wing. Within the thin attached boundary layer, Prandtl's simplified versions of the Navier-Stokes (NS) equations often apply, along with special boundary conditions through which the boundary layer and the outer inviscid flow interact. These equations and the many methods for solving them constitute the subdiscipline of fluid mechanics called *boundary-layer theory*. The terminology can be confusing, for example, when it is sometimes implied that the viscous flow near the surface ceases to be a "boundary-layer flow" when it fails to satisfy the assumptions of boundary-layer theory. We won't take that

Understanding Aerodynamics: Arguing from the Real Physics, First Edition. Doug McLean.
Images and Text: Copyright © 2013 Boeing. All Rights Reserved. Published 2013 by John Wiley & Sons, Ltd.

narrow view here. Most of the physical boundary-layer phenomena we'll discuss in this chapter don't depend on the validity of boundary-layer theory. However, the theory does provide helpful insights into why boundary layers behave as they do, and it is sufficiently accurate, enough of the time, that it is useful for quantitative analysis as well. And, of course, the theory played a huge role in the history of our understanding of viscous effects at high Reynolds numbers.

In this chapter, we'll limit our attention to steady boundary-layer flows. We'll get to the specifics of the theory in Section 4.2, after discussing the general physical aspects of 2D and 3D boundary-layer flows in Section 4.1. In much of the literature, 2D boundary layers and 3D boundary layers are treated as separate universes. To emphasize the common ground these two universes share, we'll take a different tack here, and in the discussion in Sections 4.1 and 4.2, we'll keep 2D and 3D integrated. Topic by topic, we'll introduce ideas in 2D and then discuss what has to be added to go to 3D. Then, in the remaining sections, we'll deal with transition and turbulence, control of flow separation, heat transfer, compressibility, and surface roughness.

4.1   Physical Aspects of Boundary-Layer Flows

The boundary layer is a thin sheet of flow close to the surface, so it's natural to imagine following its progress along the surface and using time-like terminology to describe its development even when the flow is actually steady. And we'll use this kind of time-like terminology to refer to the boundary layer as a whole, even though fluid parcels flowing along at different distances from the surface move at different speeds and thus experience different transit times. This is an example of the "pseudo-Lagrangian viewpoint" we discussed in Section 3.4.7. Imagining a steady flow in this way is especially appropriate in the case of a boundary layer, because, as we'll see in Section 4.2, a boundary layer flow has the character of an initial-value problem. You'll note that many discussions of boundary-layer flows, including those in this chapter, use time-like terminology freely.

4.1.1   The Basic Sequence: Attachment, Transition, Separation

Let's look first at the major milestones that mark the development of a 2D boundary layer as it flows along the surface of a body. The development starts where the flow first attaches at or near the front of the body and ends where the boundary layer finally leaves the body and becomes part of the viscous wake. In between, the boundary layer will usually undergo transition from laminar to turbulent, except in special cases at low Reynolds number, and it may also undergo intermediate separation and reattachment.

The least complicated type of 2D attachment occurs at a stagnation point, at or near a blunt leading edge, as shown in Figure 4.1.1a. The velocity u_e just outside the boundary layer initially increases linearly in both directions from such a stagnation point, and the boundary layer in the neighborhood is always laminar. In the region of linear acceleration, the boundary-layer development is described by one of the special similarity solutions to the equations, as described in Section 4.3.2. The thickness of the boundary layer remains constant until u_e deviates from its initial linear distribution.

If the leading edge is sharp, the situation is not always simple. For supersonic flow or for one particular angle of attack in subsonic flow, the flow can attach directly to the

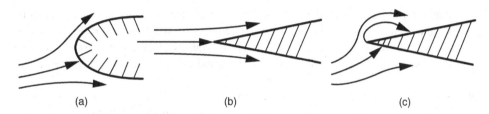

Figure 4.1.1 Types of initial boundary-layer attachment in 2D. (a) At or near a blunt LE (b) Directly on a sharp LE. (c) Near a sharp LE

sharp leading edge, as shown in Figure 4.1.1b. At other angles of attack in subsonic flow, the attachment will be either above or below the leading edge, as in Figure 4.1.1c, with local behavior around the attachment point like that of Figure 4.1.1a, and there will be a separation at the leading edge with at least a short bubble of recirculating flow downstream.

The final separation of the boundary layer from a 2D body can take place either from a sharp trailing edge, as in Figure 4.1.2a, or from a smooth part of the surface, as in Figure 4.1.2b. Separation from a smooth surface raises interesting physical issues that we'll discuss further in Section 4.1.4. Between the initial attachment and the final separation, a boundary layer usually undergoes transition from laminar to turbulent, a process we'll discuss in some detail in Section 4.4.1.

Separation with subsequent reattachment can take place from a sharp leading edge, as in Figure 4.1.1c; from a smooth surface, as in Figure 4.1.3a,b; or from a sharp corner at the edge of a "cove," as in Figure 4.1.3c. In a smooth-wall separation bubble like that in Figure 4.1.3a, the reattachment is usually precipitated by transition to turbulent flow, with

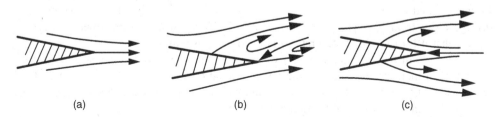

Figure 4.1.2 Types of final 2D boundary-layer separation in 2D. (a) Separation from sharp TE. (b) Single separation from smooth surface and into the wake. (c) Double separation from smooth surface and into the wake

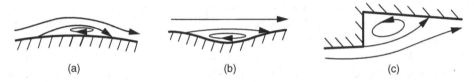

Figure 4.1.3 Types of boundary-layer separation with subsequent reattachment in 2D. (a) Laminar separation from a smooth surface with reattachment triggered by transition to turbulent flow. (b) Laminar separation and laminar reattachment in a dip in the surface. (c) Separation at the sharp edge of a "cove"

its associated rapid increase in shear stress. For reattachment to happen without transition, the separation bubble would generally need to occupy a dip in surface, as in Figure 4.1.3b, because reattachment of a laminar boundary layer requires a favorable pressure gradient, as would occur as the flow exits the dip. Pressure gradients also play an important role in smooth-wall separation, as we'll see below.

Note that the word *separation* connotes two different aspects of a flow's topological structure. First, separation in general involves some flow leaving the surface and forming a shear layer that is at least somewhat "separated" from the surface. Second, a line of separation on the surface divides the surface into "separate" regions, from which the flow along the surface converges from different directions (By "flow" we are referring either to flow a short distance off the surface or, loosely, to the surface shear stress). This is an idea we'll return to in our discussion of 3D separation in Section 4.1.4.

3D boundary-layer flows run the gamut from nearly 2D at one extreme to very different from 2D at the other. In some situations, we can view a 3D flow in cross sections and see the same patterns of attachment and separation that we saw in 2D flows in Figures 4.1.1–4.1.3. But the additional dimension in 3D makes many other patterns of attachment and separation possible as well.

Note that in 2D flows the attachment and separation points are actually lines that extend to plus and minus infinity in the "spanwise" direction. In 3D flows, we can have singular points of attachment and separation, and these are often accompanied by what we commonly call attachment lines and separation lines that can form complicated patterns on the surface of the body. These terms are very useful for discussions of flow patterns, but we should note that they aren't mathematically precise. In real flows around finite bodies, the flow structures that we commonly call attachment lines and separation lines are not uniquely defined lines or curves in the mathematical sense. They are actually bands of finite width, loosely defined by either strong flow convergence toward the body (attachment) or strong diverge away from it (separation), but for practical purposes, they are often sufficiently narrow that it isn't grossly inaccurate to refer to them as "lines."

This issue will become clearer when we discuss what the boundary-layer velocity field looks like in more detail locally in the neighborhood of separation in Section 4.1.4. Then we'll discuss the global topology of points and "lines" of attachment and separation on the body surface in some detail in Section 5.2.3.

4.1.2   General Development of the Boundary-Layer Flowfield

Now let's look at the general features of the boundary-layer velocity field, time-averaged if the flow is turbulent. The velocity in the boundary layer is nearly parallel to the surface. It varies relatively slowly along the surface, but much more rapidly in the direction normal to the surface, in a distribution called the *velocity profile*.

Typical velocity profiles for 2D laminar and turbulent flow are shown in Figure 4.1.4. In both cases, the velocity starts at zero at the surface, in keeping with the no-slip condition, and gradually approaches a distribution consistent with an inviscid outer flow. In the first-order theory that we'll discuss in Section 4.2.1, and therefore in this discussion, we'll ignore the slope of the inviscid velocity distribution in the outer flow and assume that the slope of the boundary-layer profile goes to zero. The boundary between the

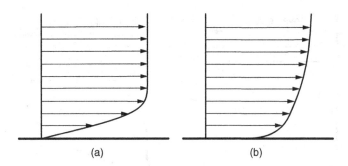

Figure 4.1.4 Typical 2D boundary-layer velocity profiles. (a) Laminar. (b) Turbulent

viscous boundary layer and the effectively inviscid outer flow is often referred to as the boundary-layer "edge," though in fact it is indistinct. Any definition of the boundary-layer edge is thus arbitrary. The usual choice is the point at which the velocity reaches 99% of its outer-flow value, and the term boundary-layer thickness usually refers to the distance from the surface to this 99% point. Because the slope of the velocity profile is small there, the determination of boundary-layer thickness in this way is sensitive to small errors in the determination of the velocity profile, especially in the turbulent case, and the difficulty is compounded if the distribution of velocity in the outer flow has significant slope.

In a laminar constant-property boundary layer the viscosity μ is constant, so that the shear stress in 2D is just proportional to $\partial u/\partial y$. Given the shape of a laminar velocity profile in Figure 4.1.4a, we can see that the shear stress starts at zero at the outer edge of the boundary layer and increases as the wall is approached. The value of $\mu \partial u/\partial y$ at the wall itself is the shear stress τ_w transmitted from the flow to the surface and is referred to as the *skin friction*. This terminology is a bit misleading, because it evokes an image of relative motion between the fluid and the surface, as in mechanical friction. But remember that with the no-slip condition at the surface, there is no relative motion, and that this is "friction" only in the sense that a shear force is exerted on the surface.

In a compressible boundary layer or a turbulent boundary layer the relationship between τ and $\partial u/\partial y$ is more complicated, but τ_w is still proportional to $\partial u/\partial y$ at the wall. Though a turbulent boundary layer is typically much thicker than a laminar one, the turbulent layer has a much larger gradient $\partial u/\partial y$ at the wall and thus much higher skin friction. We'll consider turbulent boundary-layer flow in detail in Section 4.4.2 and compressible flow in Section 4.6.

A streamwise pressure gradient $\partial p/\partial x$ causes additional acceleration that affects the shape of the velocity profile. Because $\partial p/\partial x$ tends to be nearly the same across the thickness of a boundary layer, it contributes nearly the same incremental material (Lagrangian) acceleration to fluid parcels regardless of distance from the wall. In Equation 3.2.2, we saw that a given Lagrangian acceleration requires larger $\partial u/\partial x$ when u is small than when u is large. Thus the bottom of the boundary layer, where u is small, responds to $\partial p/\partial x$ with a more rapid spatial rate of change of u (larger $\partial u/\partial x$) than does the outer part of the boundary layer. As a result, a positive pressure gradient reduces the velocity at the bottom of the profile more than at the top, as shown in Figure 4.1.5a, and pushes the flow closer to separation, which we'll discuss in Section 4.1.4. Because of this association with the approach to separation,

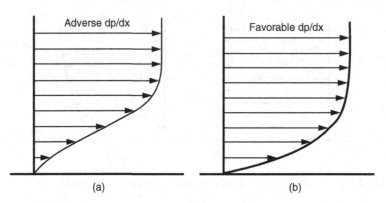

Figure 4.1.5   Effects of pressure gradient on 2D laminar-boundary-layer velocity profiles. (a) Adverse (positive) pressure gradient. (b) Favorable (negative) pressure gradient

a positive pressure gradient is often called an *adverse pressure gradient*. A negative, or *favorable pressure gradient* has the opposite effect, as illustrated in Figure 4.1.5b. With these changes in profile shape come an increase in skin friction in a favorable gradient and a decrease in an adverse gradient.

In Section 3.6, we discussed the general effects of viscosity and used a notional boundary-layer velocity profile as an example in Figure 3.6.2c. There we noted that the negative second derivative $\partial^2 u/\partial y^2$ corresponds to a negative shear-stress gradient $\partial\tau/\partial y$, which constitutes a net viscous force on fluid parcels tending to slow the parcels down. The boundary conditions on $\partial u/\partial y$ require that every boundary-layer velocity profile have at least one region of negative $\partial^2 u/\partial y^2$. And in fact, in the usual situation in most boundary-layer flows, negative $\partial^2 u/\partial y^2$ and negative $\partial\tau/\partial y$ dominate, with the result that fluid parcels in the boundary layer slow down, and the boundary layer grows thicker as it flows along. Of course net viscous forces are not the only thing affecting this *boundary-layer growth;* the pressure gradient also plays a role. An adverse pressure gradient tends to slow fluid parcels down and hasten the growth the boundary layer, while a favorable pressure gradient tends to slow the growth down and can even reverse it. In Section 4.2.2, we'll see how these effects are quantified in terms of the integrated momentum balance.

The tendency toward positive boundary-layer growth is usually quite pronounced. A boundary layer starting at a stagnation point as in Figure 4.1.1a starts with nonzero thickness, and it is not uncommon for the thickness to grow by a couple orders of magnitude over the length of a body. Streamlined bodies usually have regions of adverse pressure gradient over their aft portions that contribute strongly to the overall boundary-layer growth.

The general tendency of viscous forces to slow fluid parcels down in the boundary layer has an important exception, and that is at the bottom of a boundary layer in an adverse pressure gradient. Note that the velocity profile in the adverse pressure gradient in Figure 4.1.5a has a positive second derivative $\partial^2 u/\partial y^2$ close to the wall and a negative second derivative farther out, with an inflection point in between. When we get to the quantitative theory, our discussion in connection with Equation 4.2.7 will explain why this is, but for now the important point is that positive $\partial^2 u/\partial y^2$ at the bottom of the boundary layer corresponds to a positive shear-stress gradient $\partial\tau/\partial y$, which constitutes a net viscous force on fluid parcels pushing them along in the flow direction rather than impeding them.

This "favorable" viscous force is the main mechanism by which the flow at the bottom of the boundary layer resists being slowed in an adverse pressure gradient and thus resists separation, which we'll discuss further in Section 4.1.4.

A boundary layer flow must obey conservation of momentum, both locally and in an integrated sense, something we'll discuss in detail in Section 4.2.2. One result of this is that a boundary layer maintains a "memory" of what it was subjected to upstream, and this memory typically persists over some distance downstream. As an example, consider two turbulent boundary-layer flows, A and B, that are subjected to the same outer flow and differ only in that flow B is subjected to a short patch of surface roughness near the upstream end, that is not present in flow A. For reasons we'll discuss in Section 6.1.8, the roughness in flow B will increase the skin friction locally and thicken the boundary layer, relative to flow A. Conservation of momentum requires that the additional boundary-layer thickness in flow B persists downstream for some distance, but not forever. Downstream of the roughness patch, the skin friction in flow B will be lower than that in flow A because of the increased boundary-layer thickness. The boundary layer thickness in flow B will therefore grow more slowly and asymptotically settle back toward the thickness in flow A. Thus when a boundary-layer flow is perturbed in some way, it "remembers" the perturbation and then gradually "forgets." We'll see a computational example that illustrates this effect in Figure 6.2.4 in connection with a more detailed discussion of surface roughness.

Note that although the idea of "memory" is applicable in boundary-layer flow, "premonition" is not. In attached boundary-layer flow, "influence" is heavily skewed in the downstream direction, that is, flow conditions at one location along the surface strongly influence what happens downstream but have only very weak influence upstream. In Sections 4.2.1 and 4.2.2, we'll see that in the idealized theory for both 2D and 3D flows the direct upstream influence is predicted to be zero.

In 3D boundary layers, the velocity profile takes on an additional dimension, becoming a two-component vector function. An instructive way of visualizing a 3D velocity profile is to resolve it into components parallel and perpendicular to the outer flow, as shown in Figure 4.1.6. The component in the direction of the outer flow at the local boundary-layer edge is called the *streamwise profile*, and it looks qualitatively like the velocity profile in a 2D boundary layer, as in Figures 4.1.4 and 4.1.5. The component in the direction perpendicular to the local outer flow is called the *cross-flow profile* and is different from the streamwise profile in several ways. First, the cross-flow velocity goes to zero by definition at the boundary-layer edge. Second, cross-flow profiles are of different types depending on what "drives" them. If the wall is in motion so as to provide a shearing action in the cross-flow direction, as for example, on a propeller spinner, the cross-flow profile appears as in Figure 4.1.7a and is said to be *shear-driven*. If the wall is stationary, the cross-flow velocity must go to zero there, and the cross-flow profiles appear as in Figure 4.1.7b,c. Cross-flow of this type is said to be *pressure driven*, because a pressure gradient in the cross-flow direction is required to set it in motion. And, of course, it is possible for cross-flow to be shear driven and pressure driven simultaneously.

Given the essentially inviscid momentum balance that pertains in the outer flow, a pressure gradient in the cross-flow direction requires outer-flow streamline curvature in the cross-flow direction. So pressure-driven cross flow is always associated with situations in which there is outer-flow streamline curvature in the cross-flow direction. Because the cross-flow direction is parallel to the local body surface, "curvature in the cross-flow direction" refers

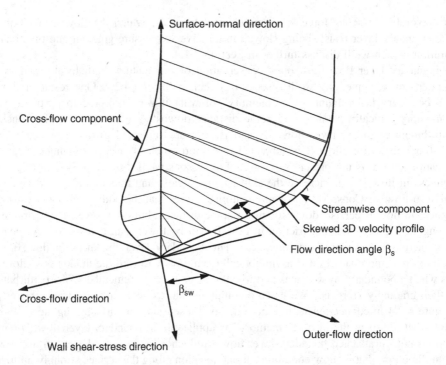

Figure 4.1.6   Isometric view of a 3D velocity profile and its resolution into streamwise and cross-flow components

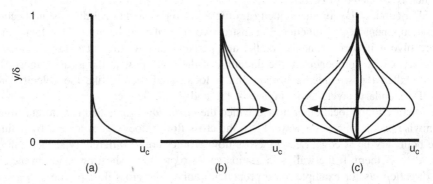

Figure 4.1.7   Profiles of cross-flow velocity u_c in a 3D boundary layer. (a) Shear-driven cross flow produced by motion of the wall. (b) Pressure-driven cross flow increasing. (c) Pressure-driven cross-flow profiles reversing, including one profile of the cross-over type

to the curvature of the streamline as viewed in the local tangent plane of the body surface, as distinct from the part of the curvature that is due to the curvature of the surface itself. We'll explore what this entails in greater detail when we address the theory in Section 4.2.1. The simple way to think of it is that it is the "lateral" component of the outer-flow streamline curvature as seen in a local "plan view" that is associated with pressure-driven cross flow in a 3D boundary layer.

The effect of the cross-flow pressure gradient on the flow is similar in some ways to the effect of a streamwise pressure gradient that we discussed above in connection with 2D flow in Figure 4.1.5. Like a streamwise pressure gradient, the cross-flow pressure gradient tends to have nearly the same strength regardless of depth in the boundary layer, and it has stronger effects on velocity gradients in the low-velocity fluid deep in the boundary layer than it does in the outer flow. These effects can take the form of rapid changes in flow direction, as, for example, when a boundary layer with little cross-flow flows into a region with a strong cross-flow pressure gradient. In this situation the cross-flow rapidly increases, as in Figure 4.1.7b, and the streamline curvature deep in the boundary layer is much greater than in the outer flow.

Just as we saw in 2D, these effects of pressure gradient in 3D are resisted by viscosity. As we saw for a 2D boundary layer, the tendency of an adverse pressure gradient to slow the flow is resisted by viscous forces produced by the positive second derivative of the velocity profile close to the wall (see Figure 4.1.5a). In a 3D boundary layer with a cross-flow pressure gradient, the tendency of the cross-flow to increase as in Figure 4.1.7b is resisted by viscous forces produced by the negative second derivative of the cross-flow profile typically spanning a region starting at the wall and including the peak of the cross-flow profile. The growth of the cross-flow profile often stops when these two tendencies come into equilibrium.

Of course, inertia also plays a role in the evolution of the cross-flow profile. If the cross-flow pressure gradient disappears, the cross-flow profile lags behind and persists for some distance downstream. If the cross-flow pressure gradient reverses sign, the cross-flow profile reverses first at the bottom of the boundary layer and goes through an intermediate stage with a profile of the cross-over type illustrated in Figure 4.1.7c. Because it is a transient state accompanying a reversal in sign, the cross-flow velocity magnitudes associated with a cross-over profile tend to be small.

The distribution of flow direction in the boundary layer can be expressed in terms of a *direction profile,* measured by the flow angle β_s relative to the outer-flow direction. The direction profile consistent with a 3D velocity profile like that of Figure 4.1.6 is shown in Figure 4.1.8. In the limit as y approaches zero, this flow direction is the same as the direction of the shear stress at the wall, β_{sw}, as indicated in Figure 4.1.6.

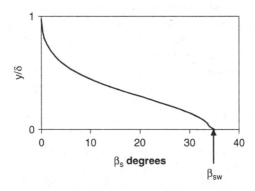

Figure 4.1.8 The *direction profile* in a 3D boundary layer, in terms of βs, the flow angle relative to that at the edge

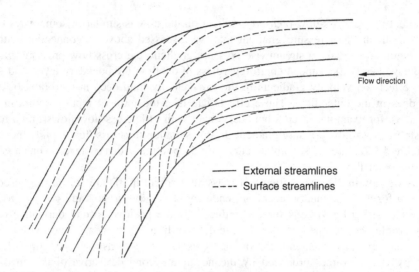

Figure 4.1.9 Skin-friction lines and outer-flow streamlines in a 3D turbulent boundary layer on a flat wall of a curved duct of rectangular cross-section, plotted from measurements by Vermeulen (1971)

Curves constructed parallel to the wall-shear-stress direction are called *skin-friction lines*, or *limiting streamlines*, or even *wall streamlines*, though this last term is misleading, because the velocity at a stationary wall is zero, and no real streamline can be defined. Figure 4.1.9 shows skin-friction lines and outer-flow streamlines in a 3D turbulent boundary layer on a flat wall of a curved duct of rectangular cross-section. The turning of the outer flow is accompanied by a radial pressure gradient that forces the flow deep in the boundary layer to turn inward much more strongly than the outer flow does, as we would expect based on Figure 4.1.6.

The presence of cross flow in a 3D boundary layer often significantly affects the momentum transport and thus the development of the flow compared with what it would be in a 2D boundary layer subjected to the same streamwise pressure distribution. In a 2D boundary layer, the streamwise momentum deficit is convected in only one direction: It comes from upstream and is carried downstream. In a 3D boundary layer, convection is in the direction of the local flow, which varies through the boundary layer, and cross-flow thus plays a direct role in the development of the flow, by transporting momentum "laterally." Convergence or divergence of the flow in the boundary layer also plays an important role. Figure 4.1.10 illustrates what convergence and divergence look like in cross-flow profiles at locations that are a short distance apart in the cross-flow direction. Note that convergence and divergence don't require a change in sign of the cross-flow velocity, just an increase or decrease. The cross-flow gradient associated with convergence or divergence affects the velocity component normal to the wall through continuity, and thus affects momentum transport indirectly.

The cross-flow gradient also transports mass, which can alter the displacement effect of the boundary layer, as we'll see in Section 4.1.3. And the additional degree of freedom in 3D boundary-layer flowfields opens up the possibilities regarding how the flow can separate from the surface, as we'll see in Section 4.1.4.

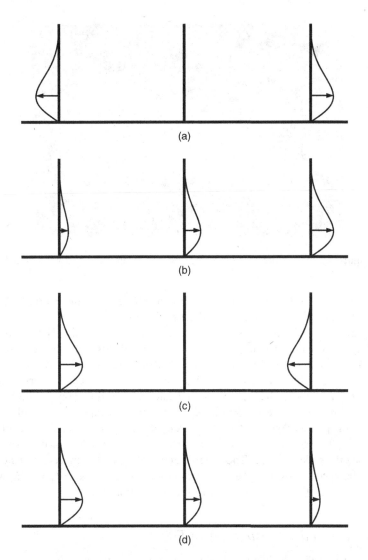

(a)

(b)

(c)

(d)

Figure 4.1.10 Examples of cross-flow convergence and divergence. (a) Divergence to either side of a location with zero cross flow. (b) Divergence in cross-flow all of one sign. (c) Convergence to either side of a location with zero cross-flow. (d) Convergence in cross-flow all of one sign

The oil-flow-visualization technique is often used in wind-tunnel testing to visualize the skin-friction lines. Oil is applied to the model surface, and the shear stress drags it along, forming streaks in the direction of the shear-stress lines. The streaky pattern can be made visible by pigment in the oil or by fluorescent dye illuminated by ultraviolet light. Figure 4.1.11 is an example of a fluorescent-oil-flow photo of a swept wing in a wind tunnel. (We'll discuss the specifics of swept-wing boundary layers in Section 8.6.2.) In general, such photos are useful for diagnosing 3D boundary-layer separation patterns, which we'll discuss

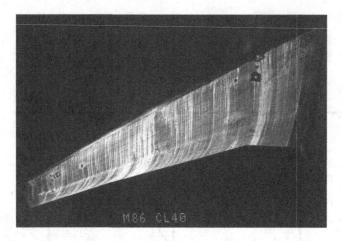

Figure 4.1.11   Fluorescent oil-flow photo of a swept wing in a wind tunnel

in detail in Sections 4.1.4, 5.2.2, and 5.2.3. It can also give an indication of the kind of cross-flow convergence or divergence we discussed above.

As useful as it is, however, oil-flow visualization has a downside: The streaks can give the misleading impression that they represent the general flow direction over the surface. We should always keep in mind when looking at oil-flow pictures that the streaks represent only the surface-shear-stress direction and that the flow direction can be very different only a very short distance above the surface, as indicated by the direction profile Figure 4.1.8. The change in direction above the surface can be especially rapid in a turbulent boundary layer in a strong pressure gradient, as, for example, near the trailing edge of the wing in Figure 4.1.11. The general flow over the surface there doesn't turn outboard nearly as strongly as the oil streaks indicate. The exaggerated turning of the streaks in oil-flow pictures often leads observers to overestimate the importance of "spanwise flow" in swept-wing boundary layers, something we'll discuss further in Section 8.6.2.

4.1.3   Boundary-Layer Displacement Effect

Flow separation like that depicted in Figure 4.1.2b obviously alters the effective shape of the body as seen by the outer flow. What is less obvious is that even an attached boundary layer has a displacement effect, though it tends to be much more subtle than the effect of separation.

We've seen how the no-slip condition and viscous diffusion act in combination to slow the flow in the boundary layer, compared with a corresponding inviscid flow. It follows from the reduced velocity that streamtubes within the boundary layer are thicker than they would have been in the inviscid flow. As a result, the flow outside the boundary layer is displaced outward, away from the body, relative to what would happen in the inviscid case. The effect of this attached-flow displacement effect on the outer flow varies widely depending on how sensitive the outer flow is to small changes in the effective shape of the body. For example, boundary-layer displacement can have dramatic effects on airfoil pressure distributions in transonic flow, as we'll see in Section 7.4.8. But even when the effect on the surface pressures is subtle, it can make a significant contribution to the viscous drag of a body. We'll look at this contribution to pressure drag further in general in Sections 6.1.3 and 6.1.5, and specifically in the case of airfoils in Section 7.4.2.

Note that we just described the displacement effect as the displacement of the outer flow relative to the ideal inviscid flow that would adhere to the contour of the body. In the viscous case, the displaced outer flow acts like a fictitious inviscid flow around a body whose contour has been displaced outward by some amount. In the theory, this fictitious inviscid flow is called the equivalent inviscid flow. Outside the boundary layer and viscous wake, the equivalent inviscid flow is the same as the actual outer flow, while in the region occupied by the boundary layer in the real flow, it is an inviscid extrapolation of the outer flow.

The most widely familiar measure of the displacement effect is the displacement thickness, δ^*, which varies along the surface, mostly increasing in the flow direction, and reflects how far outward the boundary layer has displaced the effective surface as seen by the equivalent inviscid flow. We'll look at using δ^* as way to quantify the effect in Section 4.2.3. In 2D flow, the slowing of the flow in the boundary layer generally results in positive δ^*, except in cases with strong cooling at the surface, which can produce a sink effect that overrides the slowing effect.

So, if we shift the body contour outward by the distance δ^*, we define a new body, the equivalent body seen by the equivalent inviscid flow. But what determines how large δ^* must be to have the required effect? First, the equivalent body surface must meet the flow-tangency requirement, that is, it must be a stream surface of the equivalent inviscid flow. But this requirement by itself is not sufficient to determine a unique value of δ^* because the equivalent inviscid flow has an infinity of stream surfaces where a solid surface could be placed, all presumably having the same effect on the remainder of the outer flow. How do we choose which of these possible stream surfaces is the "right" one for our purposes? To illustrate the problem, Figure 4.1.12 shows schematically the streamlines of the equivalent inviscid flow around a 2D airfoil-like body. It is clear that there is only one streamline that splits at a closed "leading edge" and becomes two streamlines that pass the body on opposite sides (shown as heavier lines than the others). If we choose streamlines farther from the body than these, the equivalent body must extend upstream to infinity, and if we choose streamlines closer than these, the equivalent body must spew mass out of part of its leading edge. These other options would work, in principle, but they'd be inconvenient, to say the least.

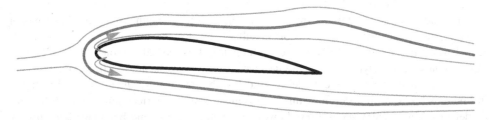

Figure 4.1.12 Schematic of streamlines of the *equivalent inviscid flow* around a 2D airfoil-like body. Only the streamline indicated by the heavier curve forms an equivalent body that is closed at the front and is therefore the preferred streamline to define the *displacement thickness* of the boundary layer of the actual viscous flow. Distances from the body are exaggerated for clarity

The preferred equivalent-outer-flow streamline forms a blunt nose that mimics the shape of the body. The displacement thickness is therefore defined and nonzero at the stagnation point. We'll look at how this works in more detail in Section 4.2.3.

Thus the preferred choice of equivalent-outer-flow streamline to define δ^* is the one for which the equivalent inviscid flow has no mass-flow missing (because the equivalent body extends upstream to infinity) and has no extra mass-flow spewing from its leading edge. With this definition of δ^*, the equivalent inviscid flow between the δ^* surface and any point outside the boundary layer has the same mass flux as the actual viscous flow does between the wall and the same point. In Section 4.2.3, this will be our basis for defining δ^* in 2D as a function of the local velocity profile.

Now note that the equivalent-inviscid-flow streamlines defining δ^* behind the body "neck down" somewhat, but never close off. This reflects the fact that the viscous wake downstream of the body always carries some velocity deficit and therefore retains some displacement effect.

The simple mass-flux argument we made using Figure 4.1.12 to define δ^* in 2D flow doesn't apply in general in 3D. In 3D flows, cross flow can have a major influence on the displacement effect and can decouple it from the local streamwise velocity profile, so that while 2D δ^* can be inferred from the local velocity profile, 3D δ^* cannot. Cross-flow convergence can "pile up" fluid in the boundary layer and increase the displacement thickness. Likewise, cross-flow divergence carries fluid away laterally and decreases the displacement thickness. The general slowing of the flow in the boundary layer still tends to produce mostly positive δ^*, in the absence of strong cooling. In limited regions of strong cross-flow divergence, however, δ^* can be negative. In this situation, the boundary-layer fluid that has been carried away laterally by the divergence must be replaced by the outer flow, and it then appears to the outer flow as if the effective body contour has been locally carved away rather than thickened. Of course, the fluid that is carried away from a region of divergence has to go somewhere, and as a result, regions of negative δ^* are always flanked by areas of unusually large positive δ^*.

The general interpretation of the δ^* surface as an effective solid-wall boundary for an equivalent inviscid flow is a useful mental model, but it is sometimes leads to misunderstanding. The δ^* surface represents an effective solid wall only for the flow situation that produced the boundary-layer flow with that particular distribution of δ^*. In a different flow situation, say because some part of the body geometry elsewhere changes, the entire flow will change, including the δ^* surface, and the flow will not respond as if the original δ^* surface were a solid surface.

A misunderstanding in this regard has arisen in connection with wind-tunnel half-model testing. This is the testing of a model of half of an airplane mounted on a solid wind-tunnel wall that is supposed to enforce a symmetry-plane boundary condition, so that the flow is equivalent to symmetrical flow around a full model. In the idealized situation, the nominal symmetry plane of the model coincides with a flat wall of the tunnel, as illustrated in Figure 4.1.13a. In inviscid flow, this arrangement provides perfect simulation of the full-model flow. In a real viscous flow, however, the symmetry-plane boundary condition is rendered imperfect by the boundary layer on the tunnel wall. An idea that is intuitively appealing, and that has frequently been put into practice, is that using a "standoff" spacer to move the model off the wall by a distance equal to δ^* of the empty-tunnel boundary layer, as illustrated in Figure 4.1.13b, is the right thing to do on physical grounds, since the δ^* surface

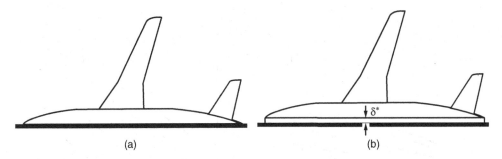

Figure 4.1.13   Schematic illustrations of a half model mounted on the wall of a wind tunnel. (a) Nominal symmetry plane of the model coincides with a flat, solid wall of the tunnel. (b) A spacer is used such that the nominal symmetry plane of the model is spaced away from the wall at the empty-tunnel δ^* surface, which is often erroneously thought to provide a better reflection plane

should be the effective location of the solid wall. But this is erroneous physical reasoning. Although the empty-tunnel δ^* represents an effective solid-wall boundary condition for the empty tunnel, it doesn't do so in the presence of flow changes introduced by the model. A computational fluid dynamics (CFD) study (Milholen, Chokani, and McGhee, 1996) looked at a range of standoff heights and several strategies for controlling the tunnel-wall boundary layer. The results indicated that the best simulation of full-model conditions should be achieved when no model offset is used, and the wall boundary layer is thinned by tangential jet blowing just upstream of the model. (The authors did not calculate the case of zero height, but extrapolation of their results indicates that it would be best.)

4.1.4   Separation from a Smooth Wall

So far, we've seen what 2D separation from a smooth wall looks like topologically in Figure 4.1.2b. Now we'll look at the physics of separation. First, what determines whether a boundary layer separates, and if so, where it separates?

Whether the boundary layer separates or stays attached is determined by what happens to the low-velocity fluid at the bottom of the boundary layer. In an attached-flow region, all of the fluid is moving in the general direction of the outer flow, which, if we take the outer-flow direction to be positive, means that the velocity in the entire boundary layer is positive, except at the wall itself, where it must go to zero. This, of course, requires the slope of the velocity profile to be positive at the wall. Just downstream of separation, there must be a region of reverse flow next to the wall, which requires a negative slope of the velocity profile at the wall. So going from attached flow to separated flow requires a decrease in the slope of the velocity profile at the wall from positive to negative, and the slope must go through zero at the separation point. This sequence is easiest to see and understand in the case of laminar flow, as sketched in Figure 4.1.14a.

As we saw in Section 4.1.2, an attached boundary layer usually thickens as it flows downstream, as viscosity and the no-slip condition act in concert to decelerate the flow. The slope of the velocity profile at the wall thus naturally tends to decrease gradually. But viscosity and the no-slip condition are not sufficient to make the slope go through zero and become negative as in Figure 4.1.14a. For that, an adverse pressure gradient (rising

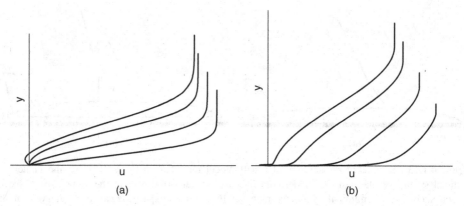

Figure 4.1.14 Progression of velocity profiles in a 2D boundary layer going through separation. (a) Laminar flow. (b) Turbulent flow

pressure) is also required. How do we know that an adverse pressure gradient is needed? Note that for the negative velocity slope at the wall to appear, there must be a region in which the velocity profile is concave "forward" (positive $\partial^2 u/\partial y^2$) somewhere within the boundary layer. When such a region appears, it generally appears first adjacent to the wall, as it does in the sequence sketched in Figure 4.1.14a. In a laminar boundary layer, or at the bottom of a turbulent boundary layer, positive $\partial^2 u/\partial y^2$ means that the shear stress gradient $\partial\tau/\partial y$ is also positive. In our discussion of the theory in connection with Equation 4.2.7, we'll establish that positive $\partial\tau/\partial y$ at the wall requires an adverse pressure gradient.

So it takes an adverse pressure gradient to cause separation of the boundary layer. But an adverse pressure gradient also activates a mechanism by which the boundary layer tends to resist separation, enabling it to remain attached at least for some distance into the region of adverse pressure gradient. As we saw above, an adverse pressure gradient results in a positive $\partial\tau/\partial y$ close to the wall. Of course, $\partial\tau/\partial y$ is the net viscous force on a fluid parcel (see the discussion in Section 3.6 and the flow examples in Figure 3.6.2), and positive $\partial\tau/\partial y$ thus constitutes a net viscous force opposing the pressure gradient. In effect, the fluid at the bottom of the boundary layer experiences a "favorable" viscous force. One way to think of it is to imagine the higher-velocity fluid farther from the surface acting through the viscous stress to drag the lower-velocity fluid along, fighting against the pressure gradient that is trying to slow the fluid down.

Boundary-layer separation thus involves a tug-of-war between the adverse pressure gradient and an opposing viscous force. At any given station along a surface subjected to an adverse pressure gradient, the favorable viscous force will generally be overmatched, and you'll see the pressure gradient winning the tug-of war, slowing the fluid near the wall, and reducing the velocity slope at the wall. How far the boundary layer will be able to persevere into the adverse pressure gradient before it separates depends on the *rate* at which the pressure gradient wins and the velocity slope at the wall decreases. Although the favorable viscous force is generally overmatched locally, its presence is vital. As we'll see below, if it weren't for the favorable viscous force, a boundary layer starting into an adverse pressure gradient would separate immediately. With the favorable viscous force, the rate of approach to separation is finite, and the distance from the onset of the adverse gradient to separation is nonzero. Until the slope of the velocity profile at the wall is brought to zero, the boundary layer remains attached, just as the corresponding inviscid flow would under the same conditions. Separation occurs only when the adverse pressure gradient has acted over a long enough distance to produce reversal of the velocity profile. How long that distance is depends on a number of factors that we'll discuss in Section 7.4.3 in connection with the maximum lift of airfoils.

Thus we've established viscosity as a source of resistance to separation, which seems contradictory because we also tend to think of viscosity as one of the main causes of separation. Without viscosity and the no-slip condition, a flow can remain attached over the entire length of a body, surviving the adverse pressure gradient all the way to an aft stagnation point without separating. But with viscosity and the no-slip condition, there must be a boundary layer with zero velocity at the surface, which introduces the possibility of separation when the flow encounters an adverse pressure gradient. So separation is a possibility only because viscosity and the no-slip condition have introduced a viscous velocity profile. On the other hand, once an adverse pressure gradient sets in, viscosity is the only source of resistance to separation. If you turned off the viscous stresses at the start of the adverse pressure gradient (i.e., switched to Euler equations with the incoming boundary-layer velocity profile as the upstream boundary condition), the flow would separate immediately. The flow near the wall, with near-zero velocity, has near-zero capacity to proceed into a pressure rise without the favorable viscous effect that we discussed above. So viscosity is both an enabler of separation and a source of resistance to separation. The key to this seeming contradiction is that the net viscous force is just $\partial \tau / \partial y$, which is negative in most of the boundary-layer flowfield, slowing the fluid parcels down. Then at the start of an adverse pressure gradient, $\partial \tau / \partial y$ switches to positive at the bottom of the boundary layer and helps that part of the flow overcome the pressure gradient, at least for a while. This favorable viscous effect acts only in the bottom part of the boundary layer in an adverse pressure gradient, while viscosity everywhere else has an adverse effect.

A laminar boundary layer cannot withstand much of a pressure rise without separating (see White, 1991, section 4-2), because the favorable viscous force that resists separation comes only from molecular shear stress, which tends to be small. The amount of pressure rise that can be withstood by a turbulent boundary layer, on the other hand, is much greater, because the favorable net viscous force is much stronger. At first glance it's tempting to think that this is simply because the eddy viscosity and the turbulent shear stress are so much larger than their molecular counterparts (we discussed turbulent shear stress and the eddy viscosity in Section 3.7), but the correct explanation is more complicated. If the eddy viscosity were simply larger, but uniform throughout the boundary layer, we would have the equivalent of a laminar boundary layer at a lower Reynolds number, and separation resistance would be no greater. The key is that a turbulent boundary layer has a thin *sublayer* next to the wall, in which the eddy viscosity is effectively zero, and that the eddy viscosity increases rapidly with distance from the wall outside this sublayer. We'll look in some detail at the physics of the sublayer and the role it plays in the greater separation resistance of turbulent boundary layers in Section 4.4.2.

In the regions of pressure rise that frequently occur in practical flows around bodies, a turbulent boundary layer is usually required if separation is to be avoided. Streamlined bodies, which we'll discuss in greater detail in Sections 5.2 and 6.1.6, must generally have a region of pressure rise at the rear, often referred to as a pressure recovery, and turbulent flow is generally required to prevent premature separation there. Even so-called laminar-flow airfoils (Section 7.4.6) are generally designed to have laminar flow over only part of the airfoil chord, with the boundary layer transitioning to turbulent before it tries to proceed too far into the region of the pressure recovery.

Avoiding premature separation is important in many applications. In external flows, there are the pressure recoveries on airfoils and other streamlined bodies. In internal flows, ducts that serve to provide pressure recovery are often called diffusers, and they are important in propulsion inlets, wind tunnels, and many other flow systems. In such applications, the designer's objective is often to maximize the recovery that can be achieved in a given length or to minimize the length for a given recovery. In this regard, the performance of a flow device with a pressure recovery is strongly dependent on the details of the pressure distribution in the recovery region. We'll consider this issue in some detail in connection with the maximum lift of airfoils in Sections 7.4.3 and 7.4.4, and in Section 4.5 we'll look at general strategies for delaying or preventing separation.

Now let's look further into what happens as a 2D flow approaches separation. Velocity profiles of flows going through separation are illustrated in Figure 4.1.14, for laminar flow and turbulent flow. We've already noted that the basic flowfield topology requires that downstream of separation there be reverse flow close to the wall, which requires negative $\partial u/\partial y$ at the wall, and that the boundary between attached flow and separated flow is where $\partial u/\partial y$ at the wall goes through zero. The surface shear stress and the skin-friction coefficient are therefore zero at the separation point, but we must remember that this applies only in 2D flow.

Where $\partial u/\partial y$ goes through zero at the separation point is easy to see in plots of laminar velocity profiles, as in Figure 4.1.14a. In plots of mean (time-averaged) velocity profiles in turbulent flow, separation doesn't stand out so clearly. A turbulent velocity profile at separation can give the appearance of still having a large positive $\partial u/\partial y$ at the wall, because $\partial u/\partial y$ can drop to zero over a very short distance from the wall, not visible on the scale of a plot like Figure 4.1.14b. This is related to the existence of the thin sublayer we mentioned above, and which we'll discuss in detail in Section 4.4.2. And there are other ways in which the turbulent case is complicated. Turbulent separation is of course unsteady on length and time scales related to the boundary-layer turbulence, and often on longer time scales as well. This means that turbulent separation is marked by two thresholds: the first appearance of intermittent reverse flow, followed downstream by reversal of the mean flow. The special complexities associated with separation in turbulent flow are discussed in detail by Simpson (1989).

Separation in 3D flow is also driven by the pressure gradient, but not just by its streamwise component. We can still think of separation as involving the reversal of one component of the velocity close to the wall, but it needn't be the outer-flow-streamwise component. And a line along which separation takes place, which we'll call a separation line, needn't be perpendicular to the local outer flow, as it would be in 2D.

In some situations, in fact, a separation line can be closer to parallel to the outer flow than to perpendicular. In such cases the separation is sometimes called a "cross-flow separation" (see Hirsch and Cebeci, 1977, for example), because it can be accompanied by reversal of the cross-flow profile. This is unfortunate terminology because it implies that cross-flow reversal defines the separation. Actually, cross-flow reversal occurs in many situations not even remotely associated with separation, and even in cases of so-called cross-flow separation, it isn't the cross-flow reversal that defines the separation location. Cross-flow separations are just situations in which the cross-flow reversal happens to occur very close to the actual separation location.

So separation in 3D is not generally defined by a reversal of either the streamwise or the cross-flow velocity profile, or by zero C_f in either the streamwise or the cross-flow direction. But then what is it defined by? Separation in general involves some flow leaving the surface and forming a shear layer that is at least somewhat "separated" from the surface, and we'll look at some of the separated-flow structures that arise in 3D in Section 5.2.2. But separation should also have a telltale signature on the surface. In 2D, that signature is zero C_f. What is the corresponding signature in 3D?

In Section 4.1.1, we noted that a line of separation on the surface divides the surface into regions from which the flow just off the surface is converging from different directions. In the limit as we approach the wall, the direction of the flow just off the surface defines the direction of the surface shear stress, or the direction of the skin friction lines, which we defined in Section 4.1.2. A 3D separation line is thus a skin-friction line flanked by other skin-friction lines converging toward it from different directions, as in Figure 4.1.15. Although the magnitude of C_f is not zero, the component of C_f perpendicular to the separation line is zero, just as is was in 2D. But zero perpendicular C_f isn't sufficient as a definition of the separation line because it is satisfied on every other skin-friction line as well. And the fact that other skin-friction lines converge toward it doesn't suffice either. So we must look at other aspects of the direction field on the surface to see what it is that makes the separation line different.

Looking at the global pattern on the surface, we see that what distinguishes a 3D separation line is the longer term "history" of the skin-friction lines converging toward it: The skin-friction lines converging toward the separation line from opposite sides "arrive" from locations on the surface that are far apart. Thus I propose as a working definition of a separation line that skin-friction lines converging toward it from opposite sides have *different regions of origin.*

This isn't a mathematically rigorous definition, but we can use a math-like argument to elaborate on what it means. Consider two points close together on the surface, both of them on the same side of a separation line, and consider the skin-friction lines that "arrive" at these points from "upstream," as illustrated in Figure 4.1.16a. The path followed by each of these skin-friction lines can be thought of as being a kind of mathematical function in which the points on the surface map into the surface streamlines arriving at the points.

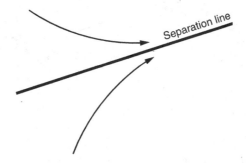

Figure 4.1.15 Local view of a 3D separation line with other skin-friction lines converging toward it

This function can be said to be continuous in the sense that skin-friction lines 1 and 2 can be made arbitrarily close together anywhere along their length if we make points 1 and 2 sufficiently close together. In a strict mathematical sense, this kind of continuity also applies to two points on opposite sides of a separation line, as sketched in Figure 4.1.16b, but there is a practical difference. Crossing the separation line, the rate of change of the "function" is much larger than it is elsewhere, as you can see qualitatively by comparing parts (a) and (b) of Figure 4.1.16. Practically speaking, the separation line is a *near-discontinuity in the region of origin of the skin-friction lines,* which is the most precise way of stating my proposed working definition of a separation line in 3D.

A swept wing is one major application for which it is important to remember that separation in 3D is not generally characterized by zero C_f. We'll look at swept-wing separation patterns and at how the region-of-origin definition of separation applies to them in Sections 5.2.3.2 and 8.6.2.

The region-of-origin way of defining 3D separation leads naturally into a discussion of the distinction between the two major types of separation lines illustrated in Figure 4.1.17. The separation line in Figure 4.1.17a divides a region where the boundary layer is fed by "clean" outer flow originating upstream of the body, from a region within a "closed" separation bubble. This is often referred to as a *closed separation*. If the bubble's footprint on the surface ends at a sharp trailing edge, as on a wing, the separation line can be an actual discontinuity in the region of origin of the skin-friction lines. The other major type is the separation line in Figure 4.1.17b, which is flanked on both sides by boundary layers fed by "clean" outer flow. This type is often called an *open separation*. Note that although the open separation doesn't divide regions of the surface fed by different kinds of flow, it is still a near-discontinuity in the region of origin of the skin-friction lines.

Although Figure 4.1.17 is a good representation of the general distinction between open and closed separation, there is one detail that is not quite right. It shows adjacent skin-friction lines joining the separation line tangentially but firmly, where skin-friction lines in actual flows do so only asymptotically. This is an issue we'll discuss further in Section 4.2.5.

The idea of pressure recovery that we discussed in connection with 2D separation is also relevant to 3D separation, but in 3D it is not just the streamwise component of the pressure gradient that is important. 3D effects can either increase or reduce the amount

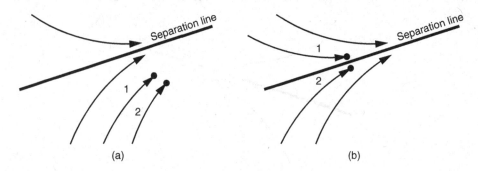

(a)   (b)

Figure 4.1.16 Illustrations of the concept of the *region of origin* of skin-friction lines, as applied to two typical points on the surface. (a) Two points to one side of a separation line. (b) Two points flanking a separation line

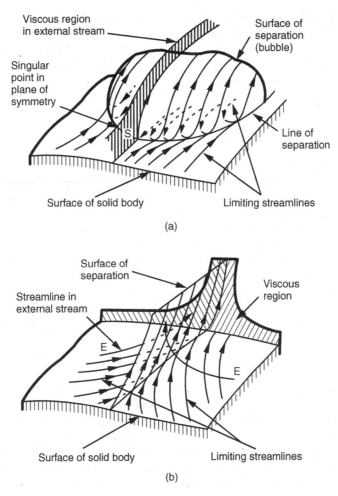

Figure 4.1.17 Illustrations of the two major types of 3D separation lines. (a) Closed type ("bubble"). After Maskell, (1955). From Peake and Tobak, (1980). Published by NASA. (b) Open type ("free shear layer"). After Maskell (1955). From Peake and Tobak (1980), published by NASA

of recovery that can be withstood without premature separation. To take one important example, wing sweep generally reduces the amount of pressure recovery that the boundary layer can withstand, in absolute terms. We'll look at CFD calculations illustrating this effect in Section 8.6.2.

4.2 Boundary-Layer Theory

The idea of dividing the flow around a body, for theoretical purposes, into an outer inviscid flow and an inner viscous flow, both governed by simplified equations, was introduced by Prandtl (1904) and has been extensively developed in the years since. For decades, this approach provided the only means for making quantitative predictions of viscous flows at high Reynolds numbers. Even in the 1970s and 1980s, when CFD predictions of transonic inviscid flows became practical, calculations based on boundary-layer theory provided the only economical means for accounting for viscous effects. In Chapter 10, we'll discuss some of the methods by which coupled CFD solutions are obtained for inner and outer flow regions. In this section, we'll concentrate on the theory and what we can learn from it about the flow in the boundary layer itself.

4.2.1   The Boundary-Layer Equations

Prandtl's, 1904 derivation of the equations of motion for the flow in the boundary layer was based on physical reasoning and an order-of-magnitude analysis applied to the 2D NS equations. The boundary-layer equations can also be derived by a more formal procedure based on the method of matched asymptotic expansions (Van Dyke, 1964). In this procedure, the flow is divided into overlapping inner and outer regions, and the NS equations are expanded in terms of a small parameter ($R^{-1/2}$ for laminar flow). The same boundary-layer equations arrived at by Prandtl arise as the equations governing the first-order solution for the inner-region flow. The equations that arise for the outer-region flow are inviscid, as was also proposed by Prandtl. This is an interesting example of sound physical intuition being reinforced much later by rigorous mathematical analysis. The more rigorous later theory has provided a basis for going beyond the original theory, as, for example, in higher order elaborations of boundary-layer theory, and asymptotic analyses of flows in regions where the first-order boundary-layer approximations break down, for example, at trailing edges of plates and airfoils.

To see how Prandtl's original 2D boundary-layer equations arise and how they relate to the corresponding NS equations, consider the boundary-layer flow developing along the surface of a body as shown in Figure 4.2.1. The flow is described in a curvilinear coordinate system with x along the body surface and y perpendicular to it, but because of the thinness of the boundary layer, we ignore the curvature of the surface and of the coordinate system. The boundary layer (the region in which the effect of viscosity is significant) is assumed to have a small thickness $\delta(x)$, and x derivatives of flow quantities are assumed to be much smaller than y derivatives. The 2D NS equations for constant-property, steady, laminar flow are given below. The strikethroughs indicate the terms that were eliminated by Prandtl's order-of-magnitude analysis to yield the boundary-layer equations:

x momentum:

$$u\frac{\partial u}{\partial x} + v\frac{\partial u}{\partial y} = -\frac{1}{\rho}\frac{\partial p}{\partial x} + \frac{\mu}{\rho}\left(\cancel{\frac{\partial^2 u}{\partial x^2}} + \frac{\partial^2 u}{\partial y^2}\right) \tag{4.2.1}$$

y momentum:

$$u\cancel{\frac{\partial v}{\partial x}} + v\cancel{\frac{\partial v}{\partial y}} = -\frac{1}{\rho}\frac{\partial p}{\partial y} + \frac{\mu}{\rho}\left(\cancel{\frac{\partial^2 v}{\partial x^2}} + \cancel{\frac{\partial^2 v}{\partial y^2}}\right) \tag{4.2.2}$$

which reduces to

$$\frac{\partial p}{\partial y} = 0 \tag{4.2.3}$$

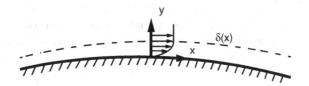

Figure 4.2.1   2D boundary layer developing on the surface of a body, and the coordinate system used in the 2D boundary-layer equations

Continuity:

$$\frac{\partial u}{\partial x} + \frac{\partial v}{\partial y} = 0. \tag{4.2.4}$$

For Reynolds-averaged turbulent flow, we would have to add a Reynolds-stress term to represent turbulent transport in the momentum equation, and then to have a complete set of equations, we would have to incorporate a turbulence model, as discussed in Section 3.7. For compressible flow (variable properties), we would also need to include the thermal energy equation, an equation of state, and an equation defining μ as a function of other fluid properties (usually temperature only). The momentum and continuity equations for incompressible flow will suffice for purposes of this discussion. Note that Equation 4.2.3 stipulates that, according to this set of first-order boundary-layer assumptions, the pressure p is constant in the y direction and is thus effectively a function of x only. The pressure in the x-momentum equation is thus imposed unchanged across the entire thickness of the boundary layer and is, in effect, an environmental condition imposed on the boundary layer from the outside.

So, what does the above simplification accomplish? Remember from Section 3.2 that the NS equations are *hyperbolic-elliptic* in space, which means that the solution at any one point depends on the solution everywhere else. This, combined with the nonlinearity of the equations, effectively precluded the economical computation of numerical solutions for most purposes, even in 2D, until the 1980s. The boundary-layer equations, on the other hand, are *parabolic* because all second derivatives with respect to x have been eliminated, which means that they in effect represent an initial-value problem in x. Starting with an initial condition (initial velocity profile) at an initial x station, we can determine the solution in a one-pass marching sequence from upstream to downstream, and the required computational effort is much less than that required for a corresponding NS solution. Economical numerical solutions to the full boundary-layer equations were obtained in the 1960s.

And further simplifications apply in special cases. There are special situations in incompressible laminar flow, which we'll consider in Section 4.3.2, for which the pioneers of boundary-layer theory were able to find *similarity transformations* that reduce the boundary-layer equations from partial-differential field equations (PDEs) to an ordinary differential equation (ODE), greatly reducing the effort required to generate numerical solutions.

Even in more general situations in which the similarity transformations don't apply, incompressible laminar flows can often be simplified in another way. Equations 4.2.1–4.2.4 can be transformed into a dimensionless form in which the Reynolds number appears only in the definition of the transformed vertical velocity v and not explicitly in the equations themselves (White, 1991, Section 4.2). Thus the dimensionless development of an incompressible laminar boundary layer can be independent of Reynolds number, provided it starts with an initial condition that is consistent with the Reynolds number. A common situation that meets this requirement is when the boundary layer starts at a stagnation point of the 2D outer flow. The 2D stagnation-point boundary layer is one of the similarity situations we alluded to above, and it scales in the right way with Reynolds number so that the development of the rest of the flow downstream will be independent of Reynolds number even if it is nonsimilar. One consequence of this is that incompressible laminar separation depends only on the pressure distribution, not on the Reynolds number, provided the flow starts at a stagnation point.

In Section 4.1.4, we discussed how separation is determined by a tug-of-war between the viscous shear-stress gradient $\partial\tau/\partial y$ and the pressure gradient dp/dx. So how can separation in laminar flow be independent of Reynolds number? Doesn't a change in Reynolds number change the viscous stress? Well, yes, but the change in boundary-layer thickness compensates for it so that the balance between the shear-stress gradient and the pressure gradient remains the same. To see how this works, consider two flows in which the body shape and the flow quantities ρ and u_∞ are the same, but the viscosity μ is different. At comparable locations in the boundary layers in the two flows (same station on the body, at the half-way point in the boundary-layer thickness, for example), we would have $\tau \sim \mu$, if $\partial u/\partial y$ were the same. But $\partial u/\partial y$ is not the same. In a laminar boundary layer that displays the kind of global Reynolds-number independence we're talking about, the boundary-layer thickness $\delta \sim \mu^{0.5}$, so that $\partial u/\partial y \sim \mu^{-0.5}$, and $\tau = \mu\partial u/\partial y \sim \mu^{0.5}$. Then taking $\partial\tau/\partial y$ introduces another factor of $\mu^{-0.5}$, so that $\partial\tau/\partial y$ is independent of μ.

The flow around a circular cylinder is a classic example of this kind of behavior in which laminar separation takes place at a fixed location in the pressure distribution, independent of Reynolds number. Assuming ideal potential flow as the outer-flow input (the effect of the separated wake is not accounted for), an early series solution predicted separation at the 108.8° location on the cylinder (see Schlichting, 1979). It has since been realized that the series solution converges poorly near separation, and more recent numerical solutions give separation at 104.5° (see White, 1991).

In the world governed by the parabolic boundary-layer equations, "information" is directly transmitted in the downstream direction but not the upstream direction. If the pressure distribution $p(x)$ imposed on the boundary layer is held fixed, the flow at any one point within the boundary layer depends on, but has no direct influence on, the flow at points upstream, while the flow at one point influences the flow everywhere downstream. Upstream influence can happen only indirectly through interaction with the outer inviscid flow, which would be felt through changes in $p(x)$. There are two important points to note about this. First, the complete asymmetry between the directions (influence travels downstream but not upstream) reflects the fact that the boundary layer is largely a dissipative viscous flow and is therefore irreversible. Second, the absence of direct upstream influence is an idealization resulting from the theoretical assumptions: the neglect of the streamwise viscous diffusion term in the x-momentum Equation 4.2.1 and the assumption that pressure is an imposed environmental condition that doesn't vary with y. These assumptions don't hold exactly in any real flow. But they are good approximations in many situations, and even in many situations in which they are significantly violated locally, the effects of the violation tend to be localized. For example, if a disturbance in the form of a small bump on the surface is introduced into a real boundary-layer flow, it will, of course, have some direct influence on the flow upstream, primarily through its disturbance pressure field, which will vary in y. But its significant upstream influence will be confined to within a few bump lengths or heights of the bump itself, and if the bump is small, it's direct upstream influence will be limited to a short distance. Thus the idea that direct upstream influence in boundary-layer flows is limited, but that downstream influence can be far reaching, is a useful insight. Of course, this applies only to boundary layers that remain thin and attached to the surface. If the boundary layer separates from the surface, its upstream influence through the outer flow becomes much stronger.

While the boundary-layer equations represent an initial-value problem in x, they also require boundary conditions, some of which can vary with x. At the inner boundary, the usual solid-wall no-slip boundary condition is $u = v = 0$ at $y = 0$, just as it was for the NS equations. For the normal velocity v, this inner condition is all that we can impose, because the highest order y derivative is $\partial v/\partial y$. The outer boundary conditions on the rest of the solution are a little more complicated. We've already observed that the pressure is effectively one boundary condition imposed on the boundary layer from the outside, but we still must apply a condition to the tangential velocity u. Note that $\partial u/\partial y$ should tend toward zero for large y, so that the x-momentum Equation 4.2.1 reduces to the 1D inviscid momentum (Euler) equation:

$$\rho u \frac{\partial u}{\partial x} = -\frac{dp}{dx},$$
(4.2.5)

with the pressure independent of y as stipulated in Equation 4.2.3. Thus for large y, u should tend toward a value u_e (x) that is independent of y and consistent with Equation 4.2.5. We can either impose p(x) as the "outer" boundary condition and let u_e (x) "fall out" as an implicit result of Equation 4.2.5, or we can impose u_e (x) as the boundary condition and use Equation 4.2.5 explicitly to determine a consistent p(x) to impose in Equation 4.2.1. In any case, we will usually use either p(x) or u_e (x) as an outer-flow matching condition, that is, we will require it to match the distribution for some outer inviscid flow. Solving the boundary-layer equations with either p(x) or u_e (x) as an explicit boundary condition is referred to as the *direct mode*. Solving the equations with a boundary-layer flow variable such as the displacement thickness (which we'll define in Section 4.2.3) or the skin-friction coefficient C_f imposed and allowing p(x) and u_e (x) to "fall out" is referred to as an *inverse mode*.

The terms in the streamwise momentum Equation 4.2.1 have straightforward physical interpretations easily related to our physical discussion of Section 4.1.2. The two convective-acceleration terms represent the steady-flow part of the Lagrangian acceleration of fluid parcels as they pass through the flowfield, in the manner we discussed in Sections 3.2 and 3.4.6, and the pressure gradient and shear-stress terms represent the internal fluid-stress gradients that provide the acceleration.

Some further discussion of the order-of-magnitude argument is called for here. At first, it might seem surprising that neither of the convective acceleration terms was eliminated in going from the NS equations to the boundary-layer equations. After all, the boundary layer is a thin region, and because v in the boundary-layer coordinate system must be zero at the wall, v will be small everywhere in the boundary layer. So why can't we neglect $v\partial u/\partial y$? It turns out that although v is small, $v\partial u/\partial y$ and $u\partial u/\partial x$ are of the same order, as Prandtl's order-of-magnitude analysis showed.

To see how this comes about, consider what determines the magnitude of v in a boundary layer. Although v must of course obey a vertical momentum balance, the momentum balance isn't actually operative in determining v, because all of the terms in the y-momentum Equation 4.2.2 are negligible in the boundary-layer version, Equation 4.2.3. Instead, to determine v we must resort to the continuity equation. Integrating the continuity Equation 4.2.4 in the y direction, noting that $v = 0$ at $y = 0$, we get

$$v = -\int_0^y \frac{\partial u}{\partial x} \, dy'. \qquad (4.2.6)$$

So in boundary-layer theory, the vertical velocity v is of the same order as $y\partial u/\partial x$. Furthermore, it is proper to think of v as being primarily a result of the combination of continuity and $\partial u/\partial x$, and to think of the small $\partial p/\partial y$ in the vertical momentum balance as adjusting to accommodate the v distribution that continuity imposes.

Now because v is of the same order as $y\partial u/\partial x$, we must keep both convective terms in the x-momentum equation. And so even in a boundary-layer flow, $u\partial u/\partial x$ isn't the only significant contributor to the Lagrangian acceleration. Because of the $v\partial u/\partial y$ term, the actual Lagrangian acceleration of fluid parcels in a boundary layer is typically considerably smaller, in an absolute-value sense, than $u\partial u/\partial x$. Even though v is small, it makes an important contribution to the boundary-layer's streamwise momentum balance.

At the bottom of the boundary layer, the relative importance of the terms in the momentum balance is different from what it is at the edge. We saw above that according to Equation 4.2.5, the momentum balance at the edge of the boundary layer involves only the pressure gradient and the longitudinal acceleration. At the bottom of the boundary layer, the situation is very different, and the gradients of the shear stress and the pressure dominate. In the limit as the wall is approached, the convective terms in the $x-$momentum Equation 4.2.1 vanish, and the equation reduces to

$$\frac{dp}{dx} = \mu\frac{\partial^2 u}{\partial y^2} = \frac{\partial \tau}{\partial y}. \qquad (4.2.7)$$

This equation applies only to a very limited region at the bottom of the boundary layer, but it still provides interesting insights. It requires, for example, that in an adverse (positive) pressure gradient $\partial^2 u/\partial y^2$ and $\partial \tau/\partial y$ must be positive at the wall. We looked at the implications of this in the physics of flow separation in Section 4.1.4. Because the second derivative is always negative in the outer part of the layer, this means that the velocity profile of a laminar boundary layer in an adverse pressure gradient must always have an inflection point. We'll see in Section 4.4 that this has implications for the stability of the laminar boundary layer.

The 3D boundary-layer equations are analogous to the 2D Equations 4.2.1–4.2.4, with two velocity components, u and w, parallel to the surface, in place of just u, and an additional component of momentum to be accounted for. The momentum Equation 4.2.1 thus becomes a two-component vector equation, or two equations. And in place of just one coordinate, x, along the surface, we must now have two, x and z. If the body surface has compound curvature, the x-z coordinate system that is laid out in the surface must be curvilinear, and coordinate metrics and curvature terms must be introduced. And it is often convenient to make the coordinate system in the surface nonorthogonal. In this respect, the 3D boundary-layer equations are no different from most implementations of the 3D

NS equations, and tensor notation provides the least error prone way to derive equations in curvilinear, nonorthogonal coordinates. For the boundary-layer equations, details can be found in the book by Nash and Patel (1972).

The x-z coordinate system for the 3D boundary-layer equations can be laid out arbitrarily on the body surface. The velocity profiles described in an arbitrary coordinate system can look quite different from the streamwise and cross-flow profiles that we considered earlier. Figure 4.2.2 shows what the velocity profiles of Figure 4.1.6 would look like in an arbitrary boundary-layer coordinate system x_b, z_b.

In some situations, particular choices of coordinate alignment can make life much easier. For example, if a numerical solution is to be generated in a marching sequence, it may be necessary to align one coordinate or the other roughly in the dominant flow direction. Sometimes it may be advantageous to align one of the coordinate families with the streamlines of the outer flow, a choice referred to as *streamline coordinates*, though this is not often done in practice. Sometimes one coordinate line is aligned with a line along which the initial conditions for the boundary-layer flow are known or can be easily generated as solutions to the plane-of-symmetry boundary-layer equations, simplified versions of the 3D equations that we'll discuss in Section 4.3.4. Examples of this would be to align one coordinate line with a plane of symmetry of a body or the leading-edge attachment line on a swept wing, as shown in Figure 4.2.3.

In Section 4.1.2, we discussed in physical terms how the presence of cross flow affects the momentum balance in a 3D boundary layer, and how cross flow is related to the pressure gradient in the cross-flow direction. We also saw how a pressure gradient in the cross-flow

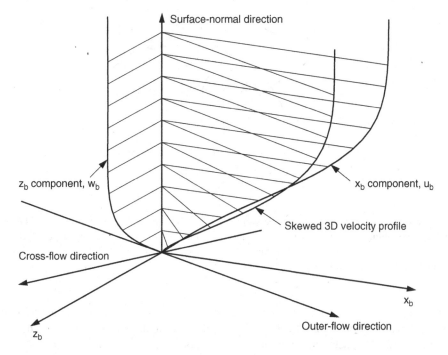

Figure 4.2.2 Isometric view of the 3D boundary-layer velocity profiles of Figure 4.1.6, as resolved in an arbitrary 3D boundary-layer coordinate system x_b, z_b

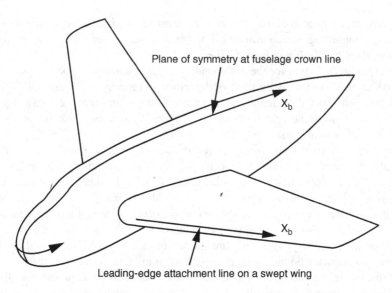

Figure 4.2.3 Examples of aligning one coordinate of a 3D coordinate system with a line along which plane-of-symmetry equations apply

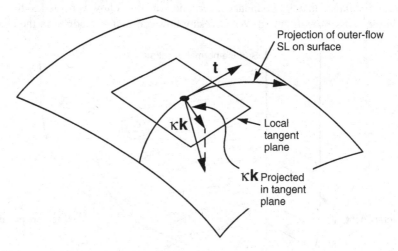

Figure 4.2.4 Illustration of outer-flow streamline curvature. The unit tangent vector to the local outer-flow streamline projection is **t,** and the *unit curvature vector* **k** is given by $(1/\kappa)$ dt/ds. The component of the *curvature vector* $\kappa\mathbf{k}$ projected in the local tangent plane is the component that is active in producing cross flow in a 3D boundary layer

direction requires outer-flow streamline curvature in the cross-flow direction. What this means is illustrated in Figure 4.2.4. The outer-flow streamline projected onto the body surface forms a 3D space curve that at any given point has a unit tangent vector **t**. When the *curvature* dt/ds (the of change of **t** along the curve) is nonzero, we can define a *unit curvature vector* $\mathbf{k} = (1/\kappa)$dt/ds. Then the *curvature vector* $\kappa\mathbf{k}$ can be decomposed into components perpendicular and parallel to the local tangent plane of the body surface. The component perpendicular to the local tangent plane is due to the curvature of the surface and is not active in cross-flow production. The cross-flow pressure gradient that is active in producing cross flow is proportional to the component of $\kappa\mathbf{k}$ parallel to the local tangent plane of the body surface.

Another way to look at the active component of the curvature is to project the outer streamline into the tangent plane instead of into the surface itself. The perpendicular component of $\kappa\mathbf{k}$, which is the one due to surface curvature, is then lost (zero), but the parallel component is the same as before. So the curvature we're interested in is the curvature of the outer streamline as viewed in the local tangent plane. For a curve that lies in a curved surface, the mathematical term for this part of the curvature is *intrinsic curvature*. So pressure-driven cross flow is always associated with situations in which the intrinsic curvature of the outer-flow streamlines is nonzero.

When a curve lying in a surface has zero intrinsic curvature it is called a *geodesic* of the surface. Longitude lines and great-circle routes on a spherical globe are examples of geodesics of a spherical surface. Sedney (1957), using the 3D boundary-layer equations in streamline coordinates, showed that if the outer-flow streamlines are geodesics of the body surface, the 3D boundary layer will have no tendency to generate cross flow. While this sounds like a very general result, the only common examples of it are axisymmetric and 2D planar flows.

Earlier in this section, we discussed how a solution to the 2D boundary-layer equations at one station along the surface depends only on what happens upstream and influences only what happens downstream, when the imposed pressure p(x) is held fixed. The same principle applies in solutions to the 3D boundary-layer equations, but we must generalize the meaning of "upstream" and "downstream" in a particular way. For the 3D boundary-layer solution along a particular column normal to the surface at a point P, when the imposed pressure $p(x_b, z_b)$ is held fixed, the upstream zone of dependence and the downstream zone of influence are curvilinear wedge-shaped regions defined by the widest range of streamline directions passing through the column, as shown in Figure 4.2.5 (see Wang, 1971). The flow at every point on the column at P depends on everything inside the upstream zone of dependence and influences everything in the downstream zone of influence. In the lateral regions outside of both of these zones, the flow neither directly affects, nor is directly affected by, the flow at points along the column at P. As was the case in 2D, this all-or-nothing dependence/influence situation is an idealization resulting from the theoretical assumptions. The earlier comments regarding indirect influence through interaction with

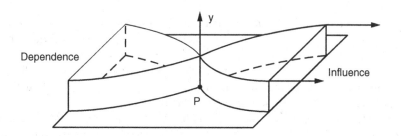

Figure 4.2.5 Zones of dependence and influence in the 3D boundary-layer equations

the outer flow and the effects of local disturbances like small bumps on the surface apply in 3D as they did in 2D. In 3D we have the additional effect of viscous diffusion in the lateral cross-stream direction, which is ignored in the idealized theory but which can have direct effects in the real world that are significant over short distances, on the order of the boundary-layer thickness. Thus it would be reasonable to think of direct dependence and influence in the real world as a kind of blurred version of the picture in Figure 4.2.5.

4.2.2 Integrated Momentum Balance in a Boundary Layer

Looking at the streamwise momentum balance in the boundary layer in an integrated sense is informative for purposes of physical understanding and is the basis for a whole class of boundary-layer prediction methods, the *integral methods*. If we take the 2D momentum Equation 4.2.1, substitute Equation 4.2.5 for the dp/dx term, and integrate in the y direction from $y = 0$ to any height h that is everywhere outside the boundary layer, we obtain

$$\int_{y=0}^{h} \left(u \frac{\partial u}{\partial x} + v \frac{\partial u}{\partial y} - u_e \frac{du_e}{dx} \right) dy = -\frac{\tau_w}{\rho}. \tag{4.2.8}$$

Then if we integrate the continuity Equation 4.2.4 from $y' = 0$ to $y' = y$ and substitute for v in Equation 4.2.8, recognizing that

$$\tau_w = \mu \frac{\partial u}{\partial y} \bigg|_{y=0} = \frac{C_f \rho u_e^2}{2}, \tag{4.2.9}$$

we obtain the von Karman momentum integral equation:

$$\frac{d\theta}{dx} = \frac{C_f}{2} - (H + 2) \frac{\theta}{u_e} \frac{du_e}{dx} = \frac{C_f}{2} + (H + 2) \frac{\theta}{\rho_e u_e^2} \frac{dp}{dx}, \tag{4.2.10}$$

where

$$\theta = \int_0^{\infty} \frac{\rho u}{\rho_e u_e} \left(1 - \frac{u}{u_e} \right) dy \tag{4.2.11}$$

is the boundary-layer *momentum thickness,* and

$$H = \frac{\delta^*}{\theta} \tag{4.2.12}$$

is called the *boundary-layer shape factor*, and δ^* is the displacement thickness, a measure of the displacement effect of the boundary layer, which we'll discuss below (Equation 4.2.13). The second term on the right-hand side defines the effect of the pressure gradient and is shown in two alternate forms, one using du_e/dx and the other using dp/dx directly. The two forms are related by the Euler momentum Equation 4.2.5. A version of Equation 4.2.10 valid for compressible flow can be found in Schlichting (1979).

The momentum thickness is a measure of the flux of momentum deficit and can be related to drag in ways we'll discuss in Section 6.1.4. The flux of momentum deficit isn't the easiest of concepts to appreciate intuitively, but it leads to the most direct way to express the integrated momentum balance, as in Equation 4.2.10.

As we'll see in Section 4.4, Equation 4.2.9 holds for turbulent flow as well as laminar, so that the momentum integral Equation 4.2.10 is valid for turbulent flow, even though we derived it from the field equations for laminar flow.

The momentum integral equation quantifies and illuminates several important relationships in boundary-layer flows. It clearly shows how the growth of the boundary layer, represented by $d\theta/dx$ in Equation 4.2.10, depends on the skin friction and the pressure gradient, relationships we discussed in physical terms in Section 4.1.2. It dictates that the growth rate $d\theta/dx$ will be positive unless the pressure-gradient term is sufficiently negative (i.e., there is a sufficiently strong favorable pressure gradient) to offset the skin-friction contribution. It also shows how the effect of the pressure gradient depends on the shape of the velocity profile, through the shape factor, H. Referring to the laminar velocity profiles with low and high H sketched in Figure 4.2.6, we can see that the higher H is, the higher will be the proportion of the flux of momentum deficit that is accounted for by low-velocity fluid. As we argued in Section 4.1.2, low-velocity fluid responds to the pressure gradient with a larger local velocity gradient $\partial u/\partial x$. A high-H boundary layer is thus more susceptible to the effect of a pressure gradient, an effect that Equation 4.2.10 quantifies in terms of $d\theta/dx$.

The momentum integral equation has four independent variables: u_e, θ, H, and C_f. Generally at least one of these must be imposed as a "boundary condition." Solving the equation with u_e imposed is referred to as the *direct mode*, just as it was in solving the field Equations 4.2.1–4.2.4. And just as before, solving with any of the other three imposed instead is referred to as an *inverse mode*. Whatever mode the equation is solved in, it has three unknowns, and two additional *closure relationships* are therefore required in order to have a complete system. One relation is almost always a *skin-friction law* that defines C_f as a function of other boundary-layer quantities. In many methods, the other closure relation requires making an assumption about the velocity profiles, usually in the form of a family of profile shapes with a range of shape factor H. The skin-friction law and the other closure relation will of course depend on whether the flow is laminar or turbulent. A wide variety of integral boundary-layer methods have been developed for both laminar and turbulent flow. Two of the best-known examples are the lag-entrainment method of Green, Weeks, and Brooman (1977) for turbulent flow, and the laminar and turbulent methods developed by Drela and Giles (1986) for airfoil calculations. White (1991) discusses some of the earlier methods.

Three-dimensional integral boundary-layer equations can be derived in the same way, starting with the 3D boundary-layer momentum equations in general curvilinear,

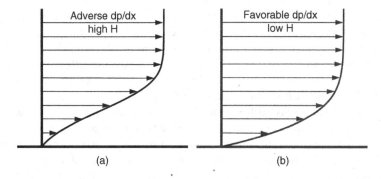

Figure 4.2.6 2D laminar velocity profiles illustrating low and high shape factor H

nonorthogonal coordinates. There are then two equations corresponding to Equation 4.2.10, one for each in-surface coordinate direction. Because there are two surface-parallel velocity components carrying momentum, the equations contain two "displacement" thicknesses and four momentum thicknesses. There are also two skin-friction coefficients to solve for. Compared with 2D, we thus have five more basic unknowns and only one more basic equation, so more closure relations are required. Tying the additional integral thicknesses together generally requires making empirical assumptions about the velocity profiles. The usual approach is to develop relations based on empirical models for the streamwise and cross-flow velocity profiles and then to transform the relations into the general coordinate system. The resulting equations are very complicated. The 3D integral method by P.D. Smith (1972) is one of the better-known examples.

4.2.3 The Displacement Effect and Matching with the Outer Flow

In early applications of boundary-layer theory, p(x) or u_e (x) was usually assumed to be known a priori from a potential-flow solution for inviscid flow around the body, the boundary-layer equations were solved in the direct mode, and any influence of the boundary layer on the outer flow was ignored. Later development of the rigorous theory showed that this is all that is required for the formal first-order theory. But in many practical applications, the effect of the boundary layer on the pressure distribution is significant, and the first-order theory is not enough.

In higher fidelity viscous-inviscid matching calculations, the objective is to impose the boundary layer's displacement effect on the outer flow and thus to have the outer-flow calculation provide a prediction of the boundary-layer's effect on the surface pressure. Note that the correction for displacement is formally a second-order effect (see White, 1991, Section 4.11), but that when we introduce it we still usually ignore all other second-order effects. Interactive schemes that impose the displacement effect always require some iteration, with both the boundary-layer and the outer-flow solutions repeated several times.

The most obvious matching strategy is simple repeated cycling: Calculate the inviscid flow, then input the resulting u_e (x) to a boundary-layer calculation, and then add the resulting δ^* to the body shape, and repeat to the inviscid calculation. This scheme works some of the time, but it doesn't always converge, especially if the boundary-layer calculation gets close to separation. Many other schemes have been developed in attempts to achieve more robust convergence, to extend matching calculations into regions of separation, and to reduce computing time. In these schemes, the inviscid flow and the boundary layer are calculated either in series or in parallel with each other, or simultaneously in systems where the boundary-layer and outer-flow equations have been combined. In various nonsimultaneous schemes, the boundary-layer equations may be solved in the direct mode, an inverse mode, or a combination. In all cases, the matching of p(x) or u_e(x) is either imposed explicitly as a condition on each boundary-layer solution or is sought as an outcome of the iterative coupling. We'll discuss such schemes further in Chapter 10.

In Section 4.1.3, we discussed the displacement effect and the idea of a displacement thickness in qualitative terms. We also discussed the concept of an *equivalent inviscid flow,* which has the same velocity field and pressure distribution outside the boundary layer as the actual flow, but remains inviscid and irrotational across all or part of the near-wall region occupied by the boundary layer in the actual flow. Implementing this concept

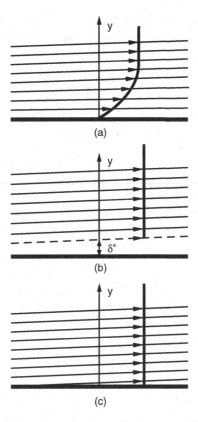

Figure 4.2.7 Illustration of the 2D boundary-layer displacement effect. (a) Actual boundary-layer flow. (b) Equivalent inviscid flow with displacement thickness. (c) Equivalent inviscid flow with surface transpiration

requires altering the boundary conditions on the outer inviscid flow solution so that it matches the desired equivalent inviscid flow. There are two ways to do this: (i) to apply the usual no-through-flow boundary condition on an "effective" body contour that has been moved outward by the displacement thickness or (ii) to apply a nonzero normal velocity, or *transpiration,* on the actual body contour. These ideas are illustrated in Figure 4.2.7.

In Section 4.1.3, we saw that in 2D flows the most convenient value for the displacement thickness δ^* can be defined by requiring that the equivalent inviscid flow between $y = \delta^*$ and any value of y outside the boundary layer, as shown in Figure 4.2.7b, have the same mass flow as the actual flow, as shown in Figure 4.2.7a. The result is that δ^* is given by an integral of the local velocity profile:

$$\delta^* = \int_0^\infty \left(1 - \frac{\rho u}{\rho_e u_e}\right) dy. \tag{4.2.13}$$

Given this definition, the imaginary *displacement surface* at $y = \delta^*$ is a streamline of the equivalent inviscid flow and is therefore a surface along which the equivalent inviscid flow would satisfy a solid-wall boundary condition.

But this is not our only choice for a definition of δ^*. As we noted in Section 4.1.3, we can apply a solid-wall boundary condition at any nearby streamline of the equivalent inviscid flow, and it will have the same effect on the rest of the flow. Thus the value of δ^* defined by Equation 4.2.13 isn't the only choice that would have the effect we seek. However, as we saw in Figure 4.3.12, the drawback to the other choices for δ^* is that the displacement surface won't have a simple, closed shape at the nose of the body. As a practical matter, this is a major drawback, and alternative definitions of 2D δ^* are never used in practice. But the alternative derivation is useful because it shows the way toward a definition of δ^* that works in 3D.

The alternative derivation of 2D δ^* that captures the other possible choices is based on flow tangency and the continuity equation. Recall from Section 4.1.3 that the purpose of defining δ^* is to produce the right flow inclination v/u_e in the equivalent inviscid flow where it contacts the displacement surface. We can calculate $v(y)$ in both the real flow and the equivalent inviscid flow by integration of the continuity equation in the y direction. For the real flow, we start with $v = 0$ at $y = 0$ and integrate outward to a point outside the boundary layer. For the equivalent inviscid flow, we start with v matching the real flow outside the boundary layer and integrate inward to any y inside the boundary layer, thus defining $v(y)$ for the equivalent inviscid flow. The δ^* surface we seek to define must be tangent to the streamlines of the equivalent inviscid flow and thus must have a slope $d\delta^*/dx$ equal to $v(y)/u_e$. The result is an equation for δ^*:

$$\frac{d}{dx}(\rho_e u_e \delta^* - \rho_e u_e \delta^*_{loc}) = 0, \tag{4.2.14}$$

where δ^*_{loc} is defined by the local integral given by Equation 4.2.13. Equation 4.2.14 is singular at a stagnation point, but useable elsewhere. When Equation 4.2.14 is integrated in x, it defines δ^* only to within a constant of integration. In effect, the value of δ^* at any initial x station other than the stagnation point is free to be chosen arbitrarily.

At the stagnation point of attachment, we seem to have a problem: Equation 4.2.14 is singular there, and because $u(y) = u_e = 0$, Equation 4.2.13 doesn't seem to define anything either. But the 2D stagnation-point boundary layer is one of the laminar similarity solutions we'll discuss in Section 4.3.2, and in the neighborhood of the stagnation point δ^*_{loc} is constant, including in the limit as the origin is approached. So $\delta^* = \delta^*_{loc}$ is the definition of δ^* that works at the stagnation point and is consistent with Equation 4.2.14 elsewhere.

In 3D flow, the mass-flux argument leading to Equation 4.2.13 doesn't generally apply. Instead, we must use the same matching and tangency analysis that led to Equation 4.2.14, extended to 3D flows (see Nash and Patel, 1972). The result is a PDE in two dimensions on the body surface, involving two local integrals of the velocity profiles, which in Cartesian coordinates is given by

$$\frac{\partial}{\partial x}(\rho_e u_e \delta^* - \rho_e q_e \delta^*_1) + \frac{\partial}{\partial z}(\rho_e w_e \delta^* - \rho_e q_e \delta^*_3) = 0 \tag{4.2.15}$$

where q_e is the edge-velocity magnitude ($q_e^2 = u_e^2 + w_e^2$), and

$$\delta^*_1 = \frac{1}{\rho_e q_e} \int_0^\infty (\rho_e u_e - \rho u) \, dy \tag{4.2.16}$$

$$\delta^*_3 = \frac{1}{\rho_e q_e} \int_0^\infty (\rho_e w_e - \rho w) \, dy. \tag{4.2.17}$$

The 3D δ^* equation is hyperbolic, and solving it numerically must follow the usual rules for hyperbolic equations. Initial conditions (initial values of δ^*) must be defined at upstream boundaries, in a manner analogous to defining a starting value of δ^* in 2D in Equation 4.2.14. Three-dimensional analogs to starting at a 2D stagnation point are starting at a singular point of attachment, as at the blunt nose of a fuselage, or at an attachment line, as at the leading edge of a swept wing. A similarity solution applies at a singular point of attachment, and at a leading edge attachment line, we can usually use the plane-of-symmetry boundary layer equations discussed in Section 4.3.4. Both of these situations define starting values of δ^* analogous to starting with $\delta^* = \delta^*_{\text{loc}}$ in 2D.

We mentioned earlier that an alternative to using δ^* to produce the displacement effect is to use an equivalent *transpiration boundary condition* applied at the actual body surface. We derive this condition by applying the same process of integrating the continuity equation that led to Equations 4.2.14 and 4.2.15. The 2D result is

$$\rho_w v_w = \frac{d}{dx}(\rho_e u_e \delta^*_{loc}), \tag{4.2.18}$$

and the 3D result for Cartesian coordinates is

$$\rho_w v_w = \frac{\partial}{\partial x}(\rho_e u_e \delta^*) + \frac{\partial}{\partial z}(\rho_e w_e \delta^*), \tag{4.2.19}$$

or, given Equation 4.2.15,

$$\rho_w v_w = \frac{\partial}{\partial x}(\rho_e q_e \delta^*_1) + \frac{\partial}{\partial z}(\rho_e q_e \delta^*_3). \tag{4.2.20}$$

Imposing this as a boundary condition on the outer flow simulates the boundary-layer displacement effect by enforcing the flow situation illustrated in Figure 4.2.7c. Note that because this is a boundary condition on the outer inviscid flow, ρ_w is the density of the inviscid flow at the wall, not the viscous boundary-layer flow.

4.2.4 The Vorticity "Budget" in a 2D Incompressible Boundary Layer

In Section 3.6, we discussed the *vorticity equation* governing the creation or destruction, convection, stretching, and diffusion of vorticity. In a 2D planar flow, only the transverse component of the vorticity can be nonzero, and there is no vortex stretching. Furthermore, if the density is constant, there can be no creation or destruction of vorticity in the interior of the field. With these restrictions, there is only convection and diffusion, and the vorticity equation looks like a conservation equation for a passive scalar (the transverse component of the vorticity). We should keep in mind, however, that it isn't an independent physical equation in its own right, but really just a sort of rearrangement of the momentum equation.

In a 2D, incompressible, steady flow, the conservation of vorticity in control-volume form is quite simple. It reduces to a balance between two fluxes integrated over the boundaries of the control volume: the convective flux $\omega \mathbf{V} \bullet \mathbf{n}$ and the diffusive flux $\nu \partial \omega / \partial n$, where ν is the kinematic viscosity, and ω is the vorticity component in the transverse direction, the only component that can be nonzero.

In Figure 4.2.8, we set up a control volume for a boundary-layer flow in the x direction and indicate the relevant diffusive flux to or from the wall and the convective fluxes through the

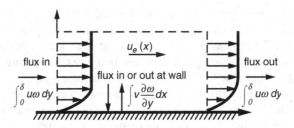

Figure 4.2.8 Control volume for tracking the vorticity "budget" of a 2D incompressible boundary layer

upstream and downstream boundaries. The diffusive fluxes at the upstream and downstream boundaries are negligible. Even if flow passes through the outer boundary, there is no diffusive or convective vorticity flux there because the vorticity is effectively zero outside the boundary layer. So the diffusive flux in or out at the wall must balance the difference between the convective flux in (upstream) and out (downstream).

The diffusive flux of vorticity to the wall is given by

$$v\frac{\partial \omega}{\partial y} = v\frac{\partial^2 u}{\partial y^2}, \qquad (4.2.21)$$

and for a thin boundary-layer flow, the convective flux reduces to

$$\int_0^\delta u\omega dy \approx -\int_0^\delta u\frac{\partial u}{\partial y}dy = -\int_0^{u_e} udu = \frac{u_e^2}{2} \qquad (4.2.22)$$

So we see that the net vorticity flux in a thin boundary layer is given approximately by half the square of the edge velocity. This is analogous to the result that the integrated vorticity in a thin shear layer is approximately equal to the velocity jump across the shear layer, which we discussed in Section 3.3.8.

Now we see that the vorticity balance requires flux to or from the wall only when u_e is changing with x. The wall acts as a source or sink for vorticity depending on the rate of change of u_e. Increasing u_e requires vorticity flux from the wall, or negative $\partial^2 u/\partial y^2$, while decreasing u_e requires vorticity flux to the wall, or positive $\partial^2 u/\partial y^2$. This is of course consistent with our finding in Equation 4.2.7, based on the momentum equation, that $\partial^2 u/\partial y^2$ at the wall is proportional to dp/dx.

A flat-plate boundary layer, for which u_e is constant in x, is an interesting special case. When u_e is constant, there is no diffusive vorticity flux to or from the wall, only a constant integrated convective flux along the flow direction. Of course, the vorticity in the field becomes more diffuse as the boundary layer thickens.

4.2.5 Situations That Violate the Assumptions of Boundary-Layer Theory

Remember that in deriving Prandtl's boundary-layer equations, it was assumed that the boundary layer is thin and that x derivatives are much smaller than y derivatives. Any boundary-layer situation in which the x derivatives become comparable to the y derivatives obviously violates these assumptions. Figure 4.2.9 illustrates a few common examples of

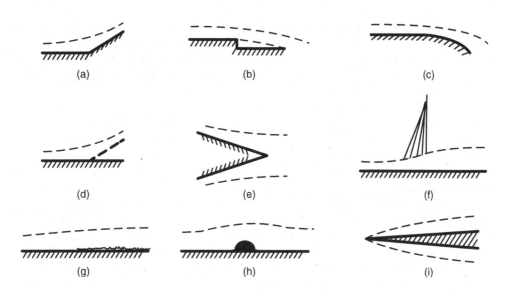

Figure 4.2.9 Illustrations of situations that violate the assumptions of first-order boundary-layer theory. (a) Corners. (b) Steps. (c) Large longitudinal curvature. (d) Separation from a smooth wall. (e) Separation from trailing edge. (f) Interaction with shock. (g) Sudden change in surface roughness. (h) Protuberances. (i) Sharp leading edge

flows that constitute boundary layers in the physical sense but violate the assumptions of first-order boundary-layer theory.

Situations such as these can raise interesting issues for boundary-layer theory. For example, separation from a smooth wall (Figure 4.2.9d) is something we discussed from a physical point of view in Section 4.1.4, where we saw that the skin-friction coefficient C_f goes through zero at a separation point in 2D flow. Because the negative values reached by C_f downstream of separation are generally quite small in magnitude, in the real world governed by the full NS equations C_f usually goes through zero with a small slope (Figure 4.2.10a). Direct-mode solutions to the first-order boundary-layer equations don't predict this behavior correctly at all. The C_f calculated in the direct-mode goes through zero at separation with infinite slope (Figure 4.2.10b), a square-root singularity that in the context of boundary-layer

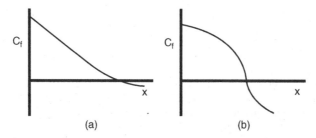

Figure 4.2.10 Illustrations of how C_f goes through zero at separation in 2D. (a) Small slope according to NS equations. (b) Infinite slope (Goldstein singularity) according to direct-mode BL equations

theory is called the Goldstein singularity. Numerical methods for solving the equations in the direct mode generally fail before reaching the singularity. Solving the equations in any of the inverse modes removes the singularity but does not alleviate the problem that x derivatives in the neighborhood of separation are usually still large enough to violate the boundary-layer-theory assumptions.

Continuing the solution downstream of the separation point raises another set of difficulties having to do with dependence and influence. Upstream of separation, a one-direction marching sequence can be used in the numerical scheme, because the solution at any point depends only on the solution upstream. Downstream of separation, the reversed flow at the bottom of the boundary layer convects information from the general downstream direction, and for the numerical scheme to deal with this properly, something more complicated than one-direction marching is required.

Separation from a sharp trailing edge (Figure 4.2.9e) raises a different set of issues. For this situation, an approximate correction to boundary-layer theory has been developed, called triple-deck theory (see White, 1991, for a summary). Of course, the full NS equations handle all of these situations "exactly," and without any of the previous theoretical difficulties, just a large increase in required computational effort.

Separation from a smooth wall in 3D raises the same issues for direct-mode solutions to the boundary-layer equations as it does in 2D. The component of C_f perpendicular to the separation line goes through zero with a square-root (Goldstein) singularity just as 2D C_f does. This means that adjacent skin-friction lines converge *to a distinct separation line* parabolically, rather than just asymptotically converging *toward each other,* as they do in solutions to the NS equations. This distinction is illustrated in Figure 4.2.11. Some illustrations of the topology of separation, such as the one in Figure 4.1.17, show the separation line in the form of a distinct, singular line, as in Figure 4.2.11b, even though separation in real flows takes the form in Figure 4.2.11a.

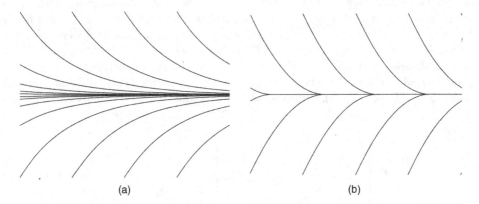

(a)   (b)

Figure 4.2.11   Skin-friction lines in neighborhood of a 3D separation line. These are the 3D analogs of the 2D C_f distributions of Figure 4.2.10. In 3D, the component of C_f perpendicular to the separation line follows qualitatively the same pattern as in 2D. (a) Continuous variation according to NS equations (asymptotic convergence). (b) Singular behavior according to direct mode BL equations (parabolic convergence)

4.2.6   Summary of Lessons from Boundary-Layer Theory

To summarize the conclusions we've gleaned from boundary-layer theory:

1. Because a boundary layer tends to remain thin, the pressure variation across it tends to be small and is ignored in the first-order theory. The pressure plays the role of an environmental condition imposed on the boundary layer from outside, that is, by the outer inviscid flow.
2. The local flow in the boundary layer depends on everything that happens upstream and influences everything that happens downstream.
3. The local boundary layer's influence on what happens upstream is much less direct. If the boundary layer is attached, it has little direct influence upstream (none in the first-order theory), and only a weak indirect influence through the outer flow. If the boundary layer separates, its upstream influence through the outer flow becomes much stronger.
4. The development of a 2D laminar, incompressible boundary layer, expressed in dimensionless terms, is independent of Reynolds number, provided the boundary layer starts at a stagnation point.

Except for point (4), the above conclusions apply to both 2D and 3D boundary layers.

4.3   Flat-Plate Boundary Layers and Other Simplified Cases

In this section, we look at flat-plate flows and a number of other situations in which boundary-layer behavior is simplified in one way or another. These idealized cases provide valuable physical insights that are not as easily seen in more general situations. And in many situations, they are close enough to reality to provided useful quantitative information.

4.3.1   Flat-Plate Flow

When the outer-flow pressure and velocity are constant in 2D flow, the flow development predicted by boundary-layer theory is particularly simple. This idealized situation is referred to as flat-plate flow because a thin flat plate aligned with the flow actually produces something very close to it, except in the neighborhood of the leading edge.

Two-dimensional incompressible laminar flow at constant pressure is one of the similarity situations we mentioned in Section 4.2.1, and the similarity solution is credited to Blasius (see Schlichting, 1979). Treating the corresponding incompressible turbulent flow requires empiricism to account for the effects of turbulence, which we'll consider in more detail in Section 4.4.2.

Expressions for the average skin-friction coefficient $\overline{C_f}$ over the length of a flat plate are widely used in preliminary drag estimation, as we'll see in Section 6.2.1. A question mark that hangs over any such formula has to do with the leading-edge region and what details had to be glossed over there. If we start the boundary layer at zero thickness and ignore the singularity of the boundary-layer equations in that situation, not to mention the fact that the assumptions behind the equations aren't valid there, the result from the Blasius solution for incompressible laminar flow is

$$\overline{C_{fi}} = 1.328 R_L^{-\frac{1}{2}}.$$ (4.3.1)

The most widely used corresponding formula for incompressible turbulent flow is the Prandtl-Schlichting relation (see Schlichting, 1979):

$$\overline{C_{fi}} = 0.455(\log R_L)^{-2.58}.$$ (4.3.2)

These laminar and turbulent $\overline{C_{fi}}$ relations are plotted for a wide range of length Reynolds number in Figure 4.3.1. First, note that in both laminar and turbulent flow $\overline{C_{fi}}$ decreases with increasing Reynolds number, in spite of the fact that the boundary-layer thickness, as a fraction of body length decreases and $\partial u/\partial y$ at the wall (in dimensionless terms) increases. The rate of increase of dimensionless $\partial u/\partial y$ at the wall is not enough to cancel the direct effect of the Reynolds-number increase. Also note that laminar $\overline{C_{fi}}$ decreases faster than turbulent $\overline{C_{fi}}$, becoming a smaller and smaller fraction of turbulent $\overline{C_{fi}}$.

The curves for cases with transition in Figure 4.3.1 were calculated for idealized "instantaneous" transition, in which the boundary layer is assumed to become turbulent immediately at the assumed transition location, and the turbulent boundary layer is assumed to start with the same momentum thickness reached by the laminar boundary layer at the transition point. (We'll see immediately below how θ and $\overline{C_f}$ are related, in Equation 4.3.3.) Note that when the length Reynolds number exceeds the assumed transition Reynolds number, and the boundary layer becomes turbulent on the downstream part of the plate, the average C_f increases rapidly at first and then asymptotically approaches the all-turbulent curve from below.

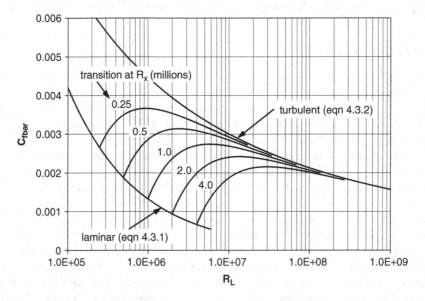

Figure 4.3.1 Average skin-friction coefficients for incompressible laminar, transitional, and turbulent flow on a flat plate

Because the pressure gradient is zero in flat-plate flow, only the skin friction contributes to the growth in momentum thickness along the plate. θ is then just proportional to the cumulative average $\overline{C_f}$, as can be seen in the momentum-integral Equation 4.2.10, which reduces to

$$\overline{C_f}(x) = \frac{2\theta}{x} = \frac{2R_\theta}{R_x}. \tag{4.3.3}$$

This result is valid for either laminar or turbulent flow and is not limited to incompressible flow.

The local C_f is often also of interest, and we can obtain expressions for it by differentiating $x\overline{C_{fi}}$, where $\overline{C_{fi}}$ comes from Equations 4.3.1 and 4.3.2. The result for incompressible laminar flow is

$$C_{fi} = 0.664 R_L^{-\frac{1}{2}} = \frac{1}{2}\overline{C_{fi}}, \tag{4.3.4}$$

and for incompressible turbulent flow:

$$C_{fi} = \overline{C_f}(L = x) - 0.51(\log R_x)^{-3.58}. \tag{4.3.5}$$

These relations are plotted in Figure 4.3.2. The first thing to note about local C_{fi} is that it decreases with increasing R and is therefore less than $\overline{C_{fi}}$ (for flat-plate flow; in more general situations, this is not always the case). The cases with transition are the same ones for which $\overline{C_{fi}}$ was plotted in Figure 4.3.1, but when viewed in terms of C_{fi} in Figure 4.3.2, they look very different. After transition, local turbulent C_{fi} jumps well above the all-turbulent curve.

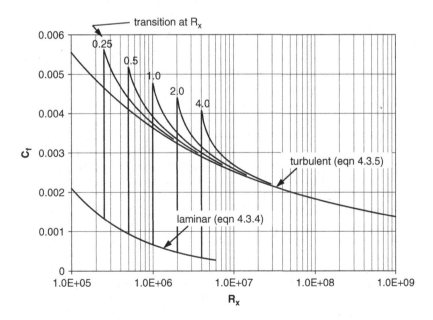

Figure 4.3.2 Local skin-friction coefficients for incompressible laminar, transitional, and turbulent flow on a flat plate

This is because the laminar run results in a thinner boundary layer (smaller θ) at the start of the turbulent boundary layer than would be there in the all-turbulent case. Thus the local-C_{fi} curves in Figure 4.3.2 approach the all-turbulent curve from above, while the $\overline{C_{fi}}$ curves of Figure 4.3.1 approach it from below because the average includes the low-C_f laminar run.

The above formulas for $\overline{C_{fi}}$ and C_{fi} as functions of x are not always accurate for real-world flat-plate flows because the development of a real flow near the leading edge doesn't generally follow the idealized flat-plate assumptions. Away from the leading edge, real-world flows with constant pressure should settle into agreement with the local-C_f formula, but with a shift in the apparent origin in x. In the "fully developed" flat-plate boundary layer the local relationship between C_{fi} and R_θ is independent of what happened near the leading edge. For laminar flow, we can derive this formula by inverting Equation 4.3.4 (expressing x as a function of C_{fi}) and substituting it into Equation 4.3.3, with the result:

$$C_{fi} = 0.664^2 R_\theta^{-1}. \qquad (4.3.6)$$

For turbulent flow, Equation 4.3.5 cannot be inverted in closed form, and the turbulent counterpart to Equation 4.3.6 must be determined by numerical interpolation of tables of $C_{fi}(x)$ from Equation 4.3.5 and R_θ (x) from Equations 4.3.2 and 4.3.3. The resulting $C_{fi}(R_\theta)$ relations are plotted in Figure 4.3.3.

Compressibility and/or heat transfer can alter flat-plate skin friction significantly. For cases with heat transfer, numerical calculations are generally required. For the case of an adiabatic wall, the situation is simpler, and many approximate theories have been developed, with predictions that vary widely (see White, 1991). Experimental data from numerous sources also show considerable scatter. White states that the curve labeled "Frankl-Voishel" in his Figure 7.22 fits "typical data" fairly well. A simple formula that approximately fits this

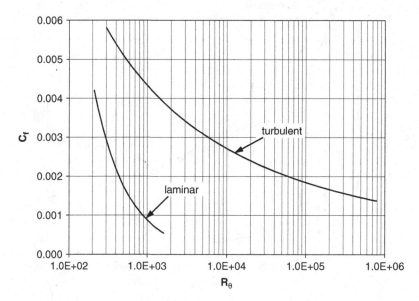

Figure 4.3.3 Local skin-friction coefficients for incompressible laminar and turbulent flow on a flat plate, as functions of R_θ

curve, over the range of Mach numbers below 2.0, is given by

$$\frac{\bar{C}_f}{\bar{C}_{fi}} = \frac{1}{1 + 0.13M_e^2},\qquad (4.3.7)$$

which is plotted in Figure 4.3.4. Note that at Mach 1.0, compressibility reduces $\overline{C}_f$ by only about 10%.

4.3.2 2D Boundary-Layer Flows with Similarity

Similarity in a boundary-layer flow refers to a situation in which the velocity profile in nondimensional form, with an appropriately scaled vertical coordinate, is the same for all x. Strictly speaking, similarity exists only in laminar flow, but there are examples of near-similarity in turbulent flow.

For incompressible laminar flow, it can be shown that similar solutions exist when $u_e \sim x^m$. In such cases, a similarity transformation can be defined that reduces the boundary-layer Equations 4.2.1 and 4.2.4 to a single ODE. A derivation is given in Schlichting (1979). The ODE has solutions for a range of values of m, which are usually referred to in terms of the value of the similarity parameter $\beta = 2m/(m+1)$. Negative values of β correspond to adverse pressure gradients, with a boundary layer in a constant state of incipient separation (zero C_f for all x) corresponding to $\beta = -0.19884$. Zero β corresponds to a flat-plate flow (constant u_e, the Blasius solution), and positive β corresponds to favorable pressure gradients. Velocity profiles for three representative values of β are plotted in Figure 4.3.5.

The laminar similarity solutions are often referred to as wedge flows because the outer-flow velocity distribution $u_e \sim x^m$ arises in the potential flow around a wedge of included angle $\pi\beta$. These are the same corner-flow solutions that we saw in Section 3.10 can be

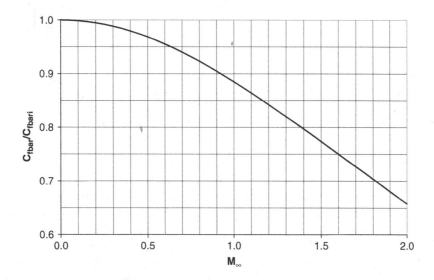

Figure 4.3.4   Effect of Mach number on adiabatic turbulent skin friction on a flat plate, Equation 4.3.7

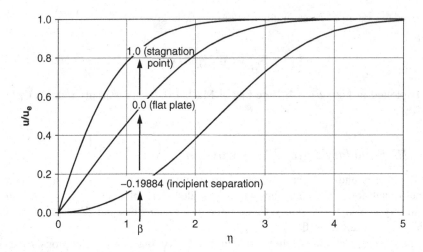

Figure 4.3.5 Velocity profiles for several laminar similarity solutions with outer-flow velocity distributions corresponding to the potential flow over wedges with included angles $\pi\beta$, as illustrated in Figure 4.3.6. η is the dimensionless normal coordinate $= (u_e/2\nu x)^{\frac{1}{2}}$

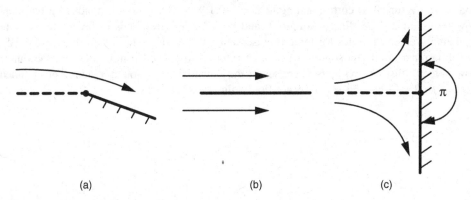

Figure 4.3.6 Illustration of the potential-flow wedge angles $\pi\beta$ corresponding to the laminar-boundary-layer similarity solutions. (a) Decelerating flow with negative included angle. (b) Flat-plate flow with zero included angle. (c) Accelerating flow in the special case of a 2D stagnation point flow ($\beta = 1.0$)

obtained by conformal mapping. Figure 4.3.6 illustrates how they work for the different types of flow. The decelerating flows correspond to wedges with negative included angles, or potential flow over a convex corner, as in Figure 4.3.6a. Flat-plate flow corresponds to a zero included angle, as in Figure 4.3.6b, and $\beta = 1$ corresponds to a 2D stagnation-point flow, as in Figure 4.3.6c, often referred to as Hiemenz flow. Note in Figure 3.10.2d that the edge velocity in the stagnation-point flow increases linearly away from the stagnation point, and the laminar-boundary-layer thickness is constant because the usual tendency of a boundary to grow is canceled by the strong favorable pressure gradient.

Of all of these similarity solutions, only two tend to appear in actual flowfields in something close to their ideal form: the flat-plate flow and the 2D stagnation-point flow. However, the other members of the laminar similarity family provide fairly good matches to local velocity profiles in more general laminar flows. As we discussed in Section 4.2.2, integral methods for predicting boundary-layer development often use assumptions about velocity-profile shapes. The laminar method developed by Drela and Giles (1986), for example, uses curve-fits based on the laminar similarity solutions.

In 2D incompressible turbulent flow, the velocity profiles themselves are never really similar, but the velocity-defect profiles, properly scaled, can be similar under the right conditions. Clauser (1954, 1956) found that if the $u_e(x)$ distribution is properly tailored, the velocity-defect profile in the form $(u_e - u)/u_\tau$ can be held constant, where $u_\tau = \sqrt{\tau_w/\rho}$ is the friction velocity, which we'll discuss further in Section 4.4.2. The defect profile is a function only of a turbulent pressure-gradient parameter,

$$\beta = \frac{\delta^*}{\tau_w} \frac{d\mathrm{p}}{d\mathrm{x}}, \tag{4.3.8}$$

not to be confused with the laminar similarity parameter of the same name. Note that the laminar and turbulent versions of β have opposite signs: laminar accelerating flows have positive β, while turbulent accelerating flows have negative β. The turbulent similarity relation has the form

$$\frac{(u_e - u)}{u_\tau} = f\left(\frac{\mathrm{y}}{\delta}, \beta\right). \tag{4.3.9}$$

Clauser (1954, 1956) coined the term *equilibrium turbulent boundary* layer for the special boundary-layer flows that obey the above relation. In principle, an equilibrium turbulent boundary layer is strictly similar only in the defect-profile sense of Equation 4.3.9, but because u_τ varies slowly with x, it will appear nearly similar in terms of the conventional velocity profile, provided the range of Reynolds number considered is not too wide.

Several researchers have succeeded in setting up experimental flows that were very close to "equilibrium" in the sense of Equation 4.3.8, ranging from strongly accelerating flows (negative β) to strongly decelerating (positive β). Flat-plate flow is a member of this family. The best-known of the non-flat-plate cases was published by Stratford (1959b), a turbulent boundary layer held in a constant state of incipient separation (nominally zero C_f), which we'll see in Section 7.4.3 has played a role in the theory of airfoil maximum lift.

Like the laminar similarity solutions, the family of equilibrium turbulent-boundary-layer velocity profiles has provided velocity-profile assumptions for integral methods of calculating boundary-layer development. The turbulent method by Drela and Giles (1986) is one example.

4.3.3 Axisymmetric Flow

Axisymmetric flow generally requires an axisymmetric body shape and zero angle of attack. Flow quantities by definition depend on only two coordinates (x, r), so that axisymmetric flow is 2D, mathematically speaking. What makes axisymmetric flow different from the usual planar 2D flow is that the body radius can vary with x, contracting or stretching the boundary layer circumferentially, as illustrated in Figure 4.3.7. The flux of momentum

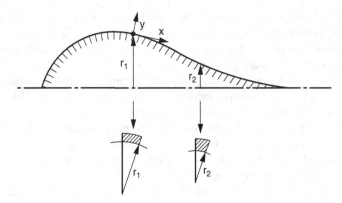

Figure 4.3.7 Illustration of the effect of a change in body radius on a boundary layer in axisymmetric flow

deficit in the boundary layer is thus distributed over a varying distance circumferentially, so that the radius variation joins with the skin friction and the pressure gradient as an additional factor that affects the evolution of the boundary-layer thickness along the body. When the radius grows, stretching the boundary layer circumferentially, the rate of boundary-layer growth is reduced or even reversed, and when the radius shrinks, the rate of boundary-layer growth is increased, compared with what it would be in a planar flow with the same pressure gradient.

These effects of radius variation are similar to the effects of convergence and divergence in 3D boundary layers that we discussed in Section 4.1.2, with the major difference that in more general 3D flows the effects are often enhanced by pressure-driven cross flow in the boundary layer. In axisymmetric flow, symmetry does not allow a pressure gradient in the cross-flow (circumferential) direction, so there is nothing to drive pressure-driven cross flow. Thus unless the body is spinning, producing shear-driven cross flow, there is no cross-flow in axisymmetric flow. In mathematical terms, the outer-flow streamlines are along longitude lines of the body surface. They are therefore geodesics of the body surface, defined by having zero intrinsic curvature, so that there is no tendency to produce pressure-driven cross flow, as we discussed in Section 4.2.1.

An axisymmetric body can either be simply connected, with no "holes," or it can be in the form of a ring with an open duct along the axis, as illustrated in Figure 4.3.8. If the leading "lip" of a ring-type body is blunt, the attachment of the flow to the lip will look locally like flow attachment to the leading edge of a 2D airfoil, and the boundary layer at the attachment will closely resemble the 2D similarity solution for $\beta = 1$ (Hiemenz flow, Section 4.3.2). If the nose of a simply connected body is blunt, flow attachment will be at a singular point of attachment, from which the flow spreads out radially, as in Figure 4.3.9. In the axisymmetric case, the boundary-layer equations for this flow can be transformed to the same ODE as for the similarity solution for 2D planar flow with $\beta = 1/2$, and the velocity profiles are therefore the same (see Schlichting, 1979). The boundary-layer thickness δ is constant in the neighborhood of the stagnation point, as it is in 2D planar flow.

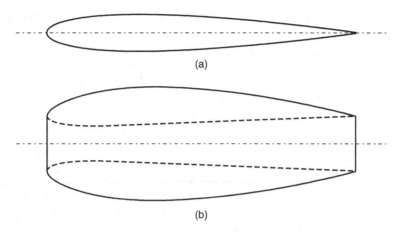

(a)

(b)

Figure 4.3.8 Illustration of two types of axisymmetric body. (a) Simply connected. (b) With open duct along axis

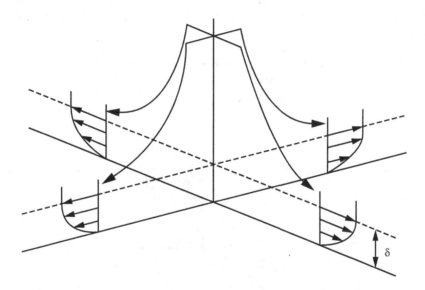

Figure 4.3.9 The boundary layer at a singular axisymmetric stagnation point of attachment

4.3.4 Plane-of-Symmetry and Attachment-Line Boundary Layers

Flows that have a nominal plane of symmetry are common in applications, as, for example, in the flow along the keel line of a ship hull at zero yaw or the flow along the vertical symmetry plane (crown and keel lines) of a fuselage at zero yaw. Another class of examples comes from the world of boundary-layer experiments. Many boundary-layer measurements have been made in special-purpose flow channels (air or water tunnels). The measurements are usually made in the boundary layer along the centerline of the channel floor or ceiling, and the channel sidewalls are usually parallel, with the intention of producing a 2D flow on the "test wall." But such setups usually produce significant unintended three dimensionality because of boundary-layer growth on the sidewalls and 3D flow effects in the corners.

Plane-of-symmetry flows generally differ from 2D planar flows in that the flow off of the plane of symmetry is either diverging away from the plane or converging toward it. In an axisymmetric flow, the flow along every longitude line of the body is a special case of plane-of-symmetry flow, with convergence or divergence if the body radius is changing with x. Why is axisymmetric flow a special case? Recall from Section 4.3.3 that an axisymmetric flow produces no pressure-driven cross flow. General plane-of-symmetry flows don't have this restriction: Off of the plane of symmetry, there can be pressure-driven cross flow. Whether the convergence or divergence involves cross flow or not, it can have a significant effect on the development of the boundary-layer velocity field, as we discussed in Section 4.3.3 in connection with axisymmetric flow.

The general situation in an idealized plane-of-symmetry flow is sketched in Figure 4.3.10. On the symmetry plane, the velocity profile is parallel to the plane and looks like a 2D boundary-layer profile. Off of the symmetry plane, divergence is shown, accompanied by pressure-driven cross-flow in which the divergence velocity component peaks inside the boundary layer with a larger value than at the edge.

The 3D boundary-layer equations are of course applicable to such flows, but we can take advantage of the symmetry to reduce the equations to a spatially 2D set that applies just on the symmetry plane itself. Starting with the 3D equations, the z-momentum equation degenerates to $0 = 0$ on the symmetry plane because the w velocity component is zero,

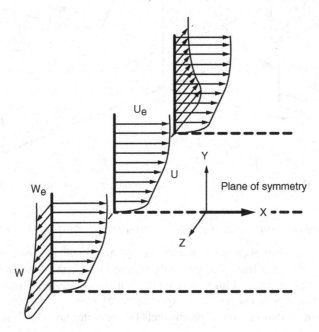

Figure 4.3.10   Sketch of a boundary-layer flow with convergence or divergence about a plane of symmetry

and $\partial/\partial z$ of all flow quantities other than w is zero. But differentiating the equation in the z direction turns it into a nondegenerate equation for the convergence/divergence profile $\partial w/\partial z$. For laminar constant-property flow in Cartesian coordinates, the resulting equations are:

- x momentum:

$$u\frac{\partial u}{\partial x} + v\frac{\partial u}{\partial y} = -\frac{1}{\rho}\frac{\partial p}{\partial x} + \frac{\mu}{\rho}\left(\frac{\partial^2 u}{\partial y^2}\right) \tag{4.3.10}$$

- z-differentiated momentum:

$$\left(\frac{\partial w}{\partial z}\right)^2 + u\frac{\partial}{\partial x}\left(\frac{\partial w}{\partial z}\right) + v\frac{\partial}{\partial y}\left(\frac{\partial w}{\partial z}\right) = -\frac{1}{\rho}\frac{\partial^2 p}{\partial z^2} + \frac{\mu}{\rho}\frac{\partial^2}{\partial y^2}\left(\frac{\partial w}{\partial z}\right) \tag{4.3.11}$$

- Continuity:

$$\frac{\partial u}{\partial x} + \frac{\partial v}{\partial y} + \frac{\partial w}{\partial z} = 0 \tag{4.3.12}$$

Note that the x-momentum equation is the same as the 2D boundary-layer momentum Equation 4.2.1 with no explicit appearance of $\partial w/\partial z$. The solution for the u velocity profile is thus affected by $\partial w/\partial z$ only through the continuity equation and its effect on the vertical velocity v.

Although the plane-of-symmetry x-momentum equation is the same as in 2D flow, the momentum-integral equation picks up an additional term representing the effects of convergence/divergence. Recall that the derivation of the 2D momentum-integral Equation 4.2.10 made use of the continuity equation. In plane-of-symmetry flow, $\partial w/\partial z$ appears in the continuity equation and thus shows up in the integrated momentum equation

$$\frac{C_f}{2} = \frac{d\theta}{dx} + (2+H)\frac{\theta}{u_e}\frac{du_e}{dx} + \frac{1}{u_e^2}\int_0^\infty (u_e - u)\frac{\partial w}{\partial z}\,dy \tag{4.3.13}$$

as the integral of $\partial w/\partial z$ weighted by the velocity defect $u_e - u$. This is consistent with the intuitive idea that the streamwise momentum balance is affected when momentum-deficient air is carried away by flow divergence. However, this intuitive interpretation applies directly only in the integrated sense of Equation 4.3.13. In the differential form of the momentum Equation 4.3.10, convergence or divergence is "felt" only "indirectly" through the $v\partial u/\partial y$ term.

Throughout the development of CFD codes for turbulent flows, the unintended three dimensionality (convergence/divergence) of nominally 2D boundary-layer experiments has been a problem for developers of turbulence models. Discrepancies between calculated and measured boundary-layer quantities are not always the fault solely of the turbulence model, but are also contributed to by convergence/divergence in the experimental flow. It takes only small amounts of convergence/divergence, usually undetectable by flow-direction measurements, to have a significant effect on the momentum thickness $\theta(x)$, for example. Mellor (1967) argued that fairer comparisons between his turbulence model and "2D" boundary-layer measurements could be made if he assumed that any discrepancy between calculated and measured $\theta(x)$ was due to convergence/divergence, and he "corrected" his calculations for it. To make the "correction," he solved the plane-of-symmetry equations instead of the 2D equations and set the code up to solve for the unknown $\partial w_e / \partial z$ boundary condition so as to match the measured $\theta(x)$. The turbulence model was then evaluated on how well it predicted $C_f(x)$, H(x), and detailed velocity profiles. One uncertainty in applying this calculation strategy is what to assume about the $\partial w / \partial z(y)$ profile at the initial x station, which can affect the solution over surprisingly long distances downstream.

One of the most important applications of plane-of-symmetry boundary-layer equations is to the band of flow attachment along the leading edge of a swept wing, usually called the *attachment line*. (In Section 5.2.2, we'll discuss how on a wing of finite span a single attachment line cannot be uniquely defined, but can for all practical purposes be pinned down very closely.) On most swept wings, the intrinsic curvature of the attachment line is small enough, and the boundary layer is thin enough, that cross flow on the attachment line itself is negligible, and the plane-of-symmetry equations apply. (See Section 4.2.1 for a definition of intrinsic curvature and a discussion of its relationship to cross flow.) In swept-wing attachment-line flow, the flow divergence is typically so strong that it keeps the boundary layer orders of magnitude thinner than it would become on a flat plate with the same streamwise distribution of edge velocity. We'll look further at the attachment-line boundary layer in our discussion of swept wings in Section 8.6.2.

4.3.5 Simplifying the Effects of Sweep and Taper in 3D

Boundary layers on swept wings, which we'll discuss in some detail in Section 8.6.2, are generally highly 3D in the sense of having strong pressure-driven cross-flow. In the most general cases, an adequate simulation of the flow requires solving the 3D boundary-layer equations. However, on well-designed swept wings, the isobars tend to line up along the constant-percent-chord lines, and simplified equations can then provide reasonably accurate predictions of boundary-layer development.

The simplest boundary-layer equations that include the effects of sweep are the infinite-span-swept-wing equations. We assume a boundary-layer coordinate system in which the two in-surface coordinates x_{sp} and x_{ch} are aligned parallel and perpendicular to the general sweep of the wing, as illustrated in Figure 4.3.11. We assume that all flow quantities are unchanging in the spanwise (x_{sp}) direction, as would be the case if the wing had a uniform

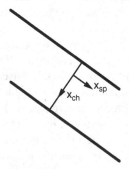

Figure 4.3.11 Coordinate system assumed in the infinite-span swept-wing boundary-layer equations

section and infinite span. When this assumption is applied to the outer inviscid flow instead of the boundary layer, we have simple sweep theory, which we'll discuss in Section 8.6.1.

Starting with the 3D boundary-layer equations in Cartesian coordinates and setting all $\partial/\partial x_{sp}$ terms to zero, we obtain a set of equations that still has three velocity components, but only two space coordinates. These are the infinite-span-swept-wing boundary-layer equations, sometimes referred to as the "2.5D boundary-layer equations." Because the spanwise coordinate x_{sp} no longer appears in the equations, the solution domain is only two-dimensional, along a single "cut" across the planform, in the perpendicular (x_{ch}) direction from leading edge to trailing edge. For purposes of numerical solution, the same upstream-to-downstream marching sequence can be used as in a 2D flow. Coding is simpler, and computing time is much shorter than for a full 3D calculation.

An important property of flows governed by these equations is that the influence that the flow in the spanwise (x_{sp}) direction has over the velocity profile in the chordwise (x_{ch}) direction is either loose or nonexistent, depending on the situation. The spanwise velocity u_{sp} doesn't appear in the chordwise momentum equation or in the continuity equation, so that the only way the spanwise flow can affect the chordwise flow is through the density and the viscosity (and eddy viscosity in the turbulent case). Thus for the spanwise flow to have any effect on the chordwise flow, the flow must be either compressible or turbulent. Thus in the case of laminar flow with constant properties, we have the *laminar independence principle:* The u_{ch} profile can be determined independently of the u_{sp} profile. This doesn't work in reverse, however. The u_{sp} profile depends strongly on the u_{ch} profile.

In turbulent and compressible flows, a rough kind of independence still tends to hold, in which the influence of the u_{sp} profile on the u_{ch} profile is weak. In Section 8.6.2, we'll look at CFD calculations that illustrate the typical magnitude of the deviation from independence in turbulent swept-wing flows.

The basic "2.5D flow" idea can be extended to wings with taper, with the help of an additional assumption. On a portion of a swept tapered wing, we can construct a coordinate system consisting of the constant-percent-chord lines ("spanlines") and either "chordwise" arcs constructed perpendicular to them, as shown in Figure 4.3.12a or conventional "rib" cuts, as in Figure 4.3.12b. Outer-flow quantities are assumed to be unchanging along the spanlines, but we can't assume boundary-layer quantities are. So we can't just set the $\partial/\partial x_{sp}$ terms to zero as we did in the case without taper. But we can make an assumption about

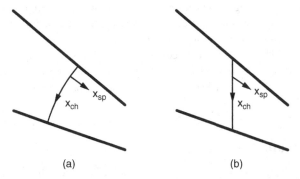

(a)   (b)

Figure 4.3.12 Coordinate systems assumed in the infinite-span swept-wing boundary-layer equations with taper. (a) A conical, orthogonal system. (b) A non-orthogonal system

them that will still give us an equation set that is spatially "2D," with no x_{sp} dependence that we have to solve for.

We assume a kind of conical similarity in which the distribution of the outer-flow velocities (both the spanwise and chordwise components) is the same along every chordwise arc, and thus the only thing that changes for the boundary layer as we move spanwise from one arc to another is the chord Reynolds number. We also assume that the velocity profiles are similar from one arc to another, and that only the boundary-layer thickness changes, in the same way that it would change with chord and Reynolds number in a flat-plate flow. This assumption will not be satisfied exactly in turbulent flow, and we thus incur some loss of physical fidelity.

To derive the equations for tapered-wing boundary layers, we start with the 3D boundary-layer equations. For the conical coordinate system of Figure 4.3.12a, we use the curvilinear orthogonal form, and for the rib-cut system of Figure 4.3.12b, they must be in nonorthogonal form. Next we introduce a scaled normal coordinate $\eta = y/\delta$. Now our assumption that the velocity profiles are similar from one arc to another is equivalent to assuming that boundary quantities are unchanging with x_{sp}, provided we follow them at constant η rather than constant y. For δ, we assume a flat-plate power law: $\delta \sim c_a^{0.5}$ for laminar flow and $\delta \sim c_a^{0.8}$ for turbulent flow, where c_a is the "chord" measured along the chordwise coordinate.

We'll skip the details, but when all is said and done, the infinite-span-swept-wing boundary-layer equations with taper have two types of terms that don't appear in the equations without taper:

1. Terms proportional to $\partial \delta / \partial x_{sp}$ that arise from the $\partial / \partial x_{sp}$ terms in the 3D equations, which we were able to set to zero when there was no taper. Using the power law for $\delta(c_a)$, we can express these in terms of $\partial c_a / x_{sp}$.
2. Terms related to the curvature of the "chordwise" arcs (Figure 4.3.12a) or the rate of change of the skew of the coordinates (the sweep of the spanlines in Figure 4.3.12b), which is proportional to the rate of taper $\partial c_a / \partial x_{sp}$.

So we've replaced the $\partial / \partial x_{sp}$ terms with terms containing $\partial c_a / \partial x_{sp}$. There are no other x_{sp}-derivative terms, and we can solve the equations along a single "chordwise" cut from leading edge to trailing edge, just as we could in the case without taper. Note that if the solution is obtained along a curved arc like those shown in Figure 4.3.12a, it can be transformed to apply to any other cut across the wing (a rib cut, for example), by use of the similarity assumption and the power law for $\delta(c_a)$.

Solutions to the 2.5D boundary-layer equations are almost as simple to compute as 2D solutions, but they can predict the effects of sweep and taper on the pressure-driven cross-flow with reasonable accuracy. In Section 8.6.2, we'll compare such predictions with those made with the full 3D equations.

4.4   Transition and Turbulence

In Section 3.7, we discussed in physical terms how turbulence affects the development of a flow and how the effects can be modeled in calculations (turbulence modeling). So far in this chapter we've taken a quick look at a turbulent mean-velocity profile (Figure 4.1.4b)

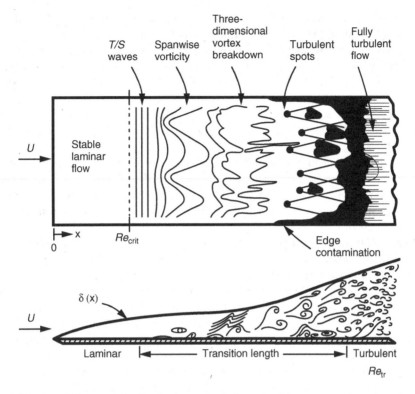

Figure 4.4.1 Possible stages in the laminar-to-turbulent transition process in a boundary layer. From White (1991). Used with permission of McGraw-Hill Companies

and we've seen how turbulent skin friction on a flat plate varies with Reynolds number (Figures 4.3.1 and 4.3.2). In this section, we'll look at how a laminar boundary layer transitions to turbulent, and then at the details of the turbulent boundary layer.

4.4.1 Boundary-Layer Transition

Laminar-to-turbulent transition in a boundary layer can involve a sequence of possible stages, as illustrated in Figure 4.4.1. Not all of these stages are always present, and discussions in the literature often refer to two prototypical paths to transition: (i) *Natural transition,* in which small disturbances grow larger through dynamic instabilities, eventually resulting in a breakdown into turbulence and (ii) *Bypass transition,* in which some large disturbance leads directly to turbulence, "bypassing" the process of instability growth and breakdown.

For the small-disturbance-growth phase of natural transition to be observable, the disturbance environment (freestream turbulence and acoustic levels) must be sufficiently quiet, and the surface must be smooth enough so that its interaction with the outer flow doesn't produce large disturbances. Of course, there are middle ranges of disturbance environments and surface quality in which the distinction between the two types of transition is blurred. And, of course, for transition to happen at all, conditions must be such that a turbulent boundary layer can be sustained downstream. This means that the Reynolds number must be high enough, and an extremely favorable pressure gradient cannot be present.

There is no quantitative theory for bypass transition, other than direct numerical simulation (DNS), which is almost never practical. Prediction of bypass transition is therefore practically always empirical. One application is the sizing of artificial boundary-layer tripping devices (*trips*), which are used in wind-tunnel testing to prevent excessive runs of laminar flow, given the low Reynolds numbers at which tests are usually run, but rarely used on actual flight vehicles. Empirical criteria for sizing trips can be found in Braslow and Knox (1958).

Predicting natural transition is important in the design of many types of flight vehicles. In Section 7.4.6, we'll look at laminar low-drag airfoils, and in Section 8.6.4, we'll consider laminar flow on swept wings. In such applications, theoretical calculations are now routinely used, but the calculations don't model the entire process, so that prediction must still rely heavily on empirical correlations.

The available theoretical framework for natural transition consists of *receptivity theory,* which models the process by which small environmental disturbances provide the initial amplitude for the growth of instabilities, *linear stability theory,* which predicts the relative rate of growth of the instabilities, once started, and *nonlinear stability theory,* or the theory of *secondary instabilities,* which extends the growth prediction into the nonlinear range. There is no quantitative theory for the final breakdown into turbulence.

Linear stability theory evolved in two major stages. In the earliest *inviscid stability theories,* realistic laminar-boundary-layer velocity profiles were assumed, but the flow was treated as inviscid for purposes of calculating the stability. The inviscid theory predicts that a laminar boundary layer is stable unless the velocity profile has an *inflection point* of the type produced by an adverse pressure gradient, as in Figure 4.1.5a. So flows with favorable pressure gradients are stable, flows with adverse pressure gradients are unstable, and flat-plate flow is neutrally stable, according to the inviscid theory. The later *viscous stability theory* predicts that viscosity stabilizes inviscidly unstable flows at very low Reynolds numbers, destabilizes inviscidly stable flows over a middle range of Reynolds numbers, and has a vanishing effect on stability at very high Reynolds numbers. The destabilizing effect of viscosity is crucial in the disturbance growth that leads to transition, so practical transition prediction has always relied on the viscous stability theory.

The e^n *method* was the first successful prediction method for natural transition, and elaborations on it are still the state of the art for practical prediction. In the e^n method, calculations are made only for the viscous linear-stability-theory phase of the disturbance growth, and transition is predicted to take place when the calculated amplitude ratio reaches a predetermined threshold value of e^n. In the early development of the method, the single value n = 9 was thought to be a good compromise for all flows. The resulting "e^9 method" was reasonably successful for 2D low-speed flows, but was much less so when applied later to more-complicated flows. The current state of the art is called *variable n-factor,* in which a variable value of n is used to account for the initial receptivity phase, and for the final non-linear growth and breakdown (see Crouch, 2008). Receptivity theory is not used directly for any calculations, but it does provide guidance in the determination of the empirical correlations that determine n. For example, receptivity theory suggests that the effects of surface roughness should be correlated with the roughness size normalized by boundary-layer thickness, not by a roughness Reynolds number. In the variable n-factor method, receptivity plays this important secondary role, while only the linear stability theory is used directly in the calculations.

In the viscous linear stability theory, small-amplitude velocity and pressure disturbances (u', v', w', p') are assumed to be superimposed on a base steady flow with a known velocity profile. This assumption is substituted into the NS equations, terms higher than first order in the disturbances are dropped, and terms containing only the base flow are subtracted out, given that the base flow is assumed to satisfy the equations. The resulting small-disturbance equations are homogeneous and are linear in the disturbances. At the usual kinds of boundaries (a solid wall with a no-slip condition or a freestream in which disturbances are assumed to die out far away), the boundary conditions on the disturbance velocities are also homogeneous. The problem is thus an *eigenvalue problem,* which will be satisfied by some sequence of *eigenfunctions.*

There are several different ways in which the disturbance equations have been simplified and used. The disturbance pressure is generally eliminated, reducing the number of equations. One of the steps is generally to assume that the disturbance velocities can be decomposed into a series of *normal modes* that are oscillatory in directions parallel to the surface and in time, leaving only the dependence on y to be determined, for example:

$$v' = v(y)e^{i(\alpha x + \beta z)}e^{-i\omega t}. \tag{4.4.1}$$

This reduces the problem to that of finding combinations of eigenvalues (spatial wave numbers α and β, and temporal frequency ω) for which physically meaningful solutions exist, and solving for corresponding eigenfunctions $v(y)$ that satisfy the equations. Assuming that growth takes place only in time (real α and β, complex ω) leads to the *temporal stability* problem, while assuming growth takes place in space (complex α and β, real ω) leads to the *spatial stability* problem. We can simplify the spatial stability problem further by lining the x direction up with the direction in which disturbance growth takes place, so that there is no growth in the z direction (β is real), thus reducing the number of parameters in the problem. Now ω and β (both real) are the only remaining free parameters that specify a particular disturbance mode. Once ω and β are specified, α is constrained by the eigenvalue problem, in that meaningful solutions exist only for particular values. Once α is determined, the real part defines the wavelength of the disturbance, and the imaginary part is the spatial growth rate.

Numerical solutions to the eigenvalue problem are used to determine α and thus the wavelength and the damping or growth rate of a particular disturbance. Because the theory is linear, the rate predicted is a relative rate, proportional to the current disturbance amplitude. Disturbances that are damped are referred to as stable, those that are amplified are referred to as unstable, and those on the boundary between the two are neutral. At a sufficiently low Reynolds number, all disturbances are stable. As Reynolds number increases, a *critical Reynolds number* is reached at which one particular disturbance crosses the neutral boundary from stable to unstable. As Reynolds number increases further, disturbances over a wider range of frequencies become unstable, and growth rates increase for the unstable disturbances.

For boundary layers, three basic types of potentially unstable modes (Figure 4.4.2) arise from the theory:

1. *Tollmien-Schlichting (TS) waves,* in which the velocity disturbances are "2D" in the direction perpendicular to the wave fronts. In 2D incompressible flat-plate flow, the fastest-growing modes propagate in the flow direction, while compressibility, pressure

gradients, or 3D effects can bring oblique modes into play. A streamwise mode in 2D flow is illustrated in Figure 4.4.2a.

2. *Cross-flow (CF) instabilities*, usually in the form of *stationary disturbance vortices*, as illustrated in Figure 4.4.2b, which can occur in boundary layers with moderately strong cross-flow velocities. As pure disturbances, CF vortices are counter-rotating, but when the background mean vorticity of the boundary layer is superimposed on them, they appear to be co-rotating. Traveling cross-flow disturbances can be significant, but only in wind-tunnel environments with high turbulence levels.

3. *Taylor-Goertler instability*, counter-rotating streamwise vortices, which can occur in boundary layers on concave surfaces, as illustrated in Figure 4.4.2c.

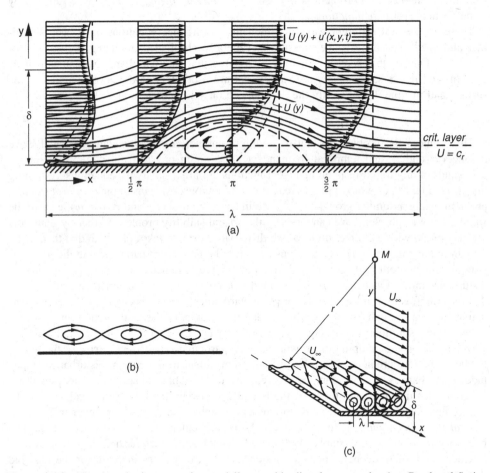

Figure 4.4.2 The three basic types of potentially unstable disturbance modes (see Reed and Saric, 1989). (a) *Tollmien-Schlichting waves* (2D disturbances). A case of streamwise propagation in a 2D flow is illustrated here in terms of velocity profiles and the streamline pattern. From White, (1991). Used with permission of McGraw-Hill Companies. (b) Cross-flow instability (stationary disturbance vortices, shown as a schematic streamline pattern in a plane perpendicular to the axes of the vortices). From Reed and Saric, (1989). Used with permission of Annual Review of Fluid Mechanics. (c) *Taylor-Goertler instability* (streamwise vortices on concave surfaces). From White, (1991). Used with permission of McGraw-Hill Companies

More than one mode may coexist in a given flow, and when their amplitudes are small, they grow independently of each other, according to the linear theory. Often, one or the other grows faster and leads to transition, as if the other mode hadn't existed. It is also conceivable that if two modes are of comparable strength when they reach the nonlinear range, that nonlinear interactions between them can play a role in transition. Figure 2.1c shows a smoke-flow visualization of the flow past a rotating body of revolution in which signs of both TS waves and vortices due to CF instability can be seen.

Real instability growth in an otherwise steady boundary layer is spatial, not temporal. But the spatial stability problem is harder to solve than the temporal problem, and in the early decades of stability research, only the temporal problem was dealt with. This turns out to have been forgivable, since the two problems lead to the same result at the neutral boundary, and early work concentrated only on determining the neutral boundaries for various boundary-layer velocity profiles. Later transition-prediction methods that depend on calculated growth rates generally use solutions to the spatial problem.

Another simplification is the assumption of parallel flow, in which the local variation of the boundary-layer velocity profiles in directions along the surface is ignored, and the base flow is assumed to depend only on y. Assuming 2D parallel flow and 2D disturbances, and eliminating u' from the equations, results in the classic ODE known as the Orr-Sommerfeld equation, which was the basis of much of the early work. Details are given by Schlichting (1979) and White (1991). More recently, methods have been devised for taking into account non-parallel-flow effects, which are found to increase the predicted instability. But practical transition prediction still often uses the parallel-flow assumption and accounts for the difference through calibration.

The stability or instability of a laminar boundary layer, and the growth rates when there are unstable disturbances, depend both on the Reynolds number and on the shape of the velocity profile. Stability is thus strongly dependent on the pressure gradient, as we saw above in our discussion of the inviscid and viscous stability theories, and can also be affected by distributed surface suction. In the viscous world, a favorable pressure gradient reduces or eliminates TS instability, but if the favorable gradient leads to stronger cross-flow, it increases CF instability (growth of stationary disturbance vortices). An adverse gradient usually results in such strong TS instability that CF instability is not the critical issue. Distributed surface suction changes velocity-profile shapes so as to reduce both TS and CF instabilities.

Figure 4.4.3 schematically illustrates how TS instability varies with Reynolds number and profile shape. Neutral stability boundaries are shown in terms of disturbance frequency versus displacement-thickness Reynolds number for favorable, zero (flat-plate) and adverse pressure gradients. Each neutral curve has a vertical tangent at its particular critical Reynolds number, below which all disturbances are stable. Above the critical Reynolds number, the range of unstable disturbance frequencies at first widens to both lower and higher frequencies. However, the high-frequency branch of each curve quickly levels off and comes down, and the high frequencies become stable again. The behavior of all three curves at high Reynolds numbers approaches the predictions of the inviscid stability theory. The favorable-gradient curve crosses the horizontal axis and disappears, reflecting inviscid stability. The flat-plate curve approaches the horizontal axis asymptotically, reflecting inviscid neutral stability. Only the adverse-pressure-gradient case retains a range of unstable disturbances at very high Reynolds numbers, in keeping with its inviscid instability.

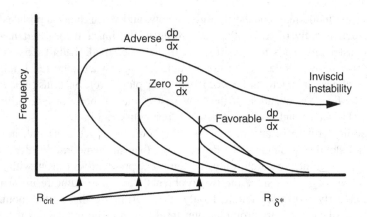

Figure 4.4.3   Schematic illustration of neutral stability boundaries for 2D (TS) disturbances in laminar boundary layers with favorable, zero, and adverse pressure gradients

   Disturbance growth of the kind predicted by linear stability theory is well established as part of the natural-transition process in boundary layers and planar 2D channels, but not in all "thin" viscous shear flows. Flows in circular pipes are unconditionally stable according to the theory, and yet there is still an upper limit to the Reynolds number at which laminar flow can exist. Why the theory doesn't work for pipe flow is not understood.

   Predictions of the stability theory for a flat-plate boundary layer (the neutral-stability boundary and the frequencies of amplified disturbances) were convincingly verified in a wind tunnel with very low turbulence levels at the U.S. Bureau of Standards in about 1940, but the results were not published to the international aerodynamics community until after World War II (Schubauer and Skramstad, 1947). The e^n transition-prediction method was first developed in the early 1950s (Smith and Gamberoni, 1956; Smith, 1981), and it was not until the 1970s that computers and solution algorithms progressed to the point that the spatial stability problem could be routinely solved for arbitrary pressure distributions.

   The e^n method involves marching in the direction in which disturbance growth is assumed to take place (the flow direction in 2D flows), and at each station solving the stability equations for a range of different modes to determine the growth rates. Then the growth rates are integrated in the growth direction to determine the amplitude ratio A/A_0 (the ratio of the disturbance amplitude to the initial amplitude at the station where the first instability appeared). The initial and final amplitudes themselves are not explicitly determined, only the ratio. This is usually expressed as $\ln(A/A_0)$, which is just n in the expression $A/A_0 = e^n$.

   The integration is carried out with differing degrees of rigor in different versions of the method. The rigorous choice is to do the integration for only one mode at a time as it evolves downstream, and to repeat the integration for a number of different modes covering the range of interest so as to determine at each station which one has produced the largest amplitude ratio. This generally means covering a two-parameter space, varying both β and ω, which is laborious. A shortcut is to repeat the integration only for multiple values of ω, using the value of β at each station that gives the most-rapid growth. The choice of β can be simple for incompressible flow because $\beta = 0$ (wave fronts perpendicular to the flow) usually leads to the fastest growth, but in transonic flow a range of β must generally be searched. Arnal and Casalis (2000) call this the "envelope method" because the resulting curve of n versus x is the envelope of multiple curves for different values of ω. A further shortcut is to use the value of ω that yields the maximum amplification rate at each local station as input to the integration. I would call this the "locally most amplified" method. For general flows, these shortcuts have no physical basis, but they save computing time because they don't require generating as many stability solutions. Of course such shortcuts produce higher n factors at transition than a rigorous integration would. While this is compensated for by calibration, the lower physical fidelity of the shortcut methods should presumably result in greater scatter in correlations with experiments.

Whether shortcuts are used or not, an e^n calculation determines a value of n at each station that is supposed to reflect the cumulative growth of the most-amplified disturbance. Transition is predicted to take place when n passes a threshold value determined by calibration against similar flows for which the transition location is known. The threshold value of n thus accounts empirically for the receptivity phase and the final nonlinear growth and breakdown. In principle, it can account for such factors in the initial-disturbance environment as freestream turbulence level and surface imperfections, but only if the calibration database includes flows in which these things were varied. In the variable-n-factor method, it is possible to base the calibration on results from more than one type of experiment. For example, the primary calibration would come from a flow similar to the target application, and data from other experiments could be used to adjust for other factors, such as surface roughness.

In 3D flows with cross-flow, TS and CF disturbances are treated as independent, in keeping with the linearity of the theory. There are several strategies for keeping track of the range of disturbance modes of both types (see Arnal and Casalis, 2000). The one that seems to produce the tightest correlations with experiments tracks separate n factors for TS and CF, and transition is predicted by whichever crosses its particular threshold first. Unless the surface is extremely (impractically) smooth, it is important to use a variable n-factor to account for the effects of surface roughness on the CF disturbances (see Crouch and Ng, 2000; Crouch, 2008). Data are available for flows with a wide variety of disturbance-growth progressions, leading to transition by TS growth in some cases and CF growth in others. For flows with low freestream turbulence and very smooth surfaces, typical threshold values are e^8 for TS transition and e^6 for CF. A formula relating n_{CF} to roughness height is given by Crouch and Ng (2000), and one relating n_{TS} to freestream turbulence level is given by Mack (1977).

Methods have been developed that carry out the e^n method by interpolating tables of precalculated growth rates instead of calculating solutions to the stability equations, saving orders of magnitude in computing time. For 2D flows with TS transition only, the ISES and MSES codes (Drela and Giles, 1987; Drela, 1993) use a two-parameter table-lookup scheme that returns growth rates as functions of the boundary-layer parameters R_θ and H. At each local station, the fastest-growing disturbance is used (as a function of ω), so that this is analogous to a "locally most amplified" method. A similar scheme that handles both TS and CF was developed by Crouch, Crouch, and Ng (2001). The TS growth rates used in this method are maximized locally with regard to β, but not with regard to ω, and multiple integrations are carried out for multiple values of ω. It is therefore analogous to an "envelope" method.

The final stage of nonlinear growth and breakdown is complex, and the details probably vary depending on the nature of the disturbances that feed it, and on local conditions such as pressure gradient and freestream turbulence. In a flat-plate boundary layer, Emmons spots like the one shown in Figure 2.1d form randomly and grow wider, merging together until the entire width of the surface is covered by turbulent flow.

4.4.2 Turbulent Boundary Layers

The turbulence in a fully turbulent boundary layer is inherently 3D and is random in space and time. But being random doesn't mean it's a featureless jumble of random swirls. Boundary-layer turbulence contains a variety of *coherent structures* over a range of length and time scales, from small structures that are characteristic of the region close to the wall to larger structures that span more of the full thickness of the boundary layer. Such structures form at essentially random locations over the surface and at random times, and even for structures of a particular type, there is wide variability in detailed evolution from one instance to another. But structures with distinctive features show up with surprising regularity. Many of their characteristics follow fixed statistical distributions.

Statistical distributions (probability distribution functions) for flow quantities within the structures are anchored in their vertical positions by the wall, and the vertical extent of the larger ones scales roughly with the boundary-layer thickness. In the horizontal directions there are probability distribution functions that move with the structures and reflect their typical shapes and sizes. And horizontal positions of structures relative to each other aren't completely random. There are two-point correlations that reflect "typical distances" and "typical cycle times," as well as "typical life spans" (paraphrased from Spalart, 2009, private communication).

Figure 2.1e shows flow visualizations of *large eddies* that appear in the outer part of the layer and illustrates how they differ in favorable and adverse pressure gradients. In these photos, the turbulence is made visible by smoke that was introduced through a slot in the wall upstream. The smoke has been transported away from the wall mostly by turbulent convection and has been diffused relatively little by molecular action, and it therefore tends to "mark" the turbulent, vortical regions. The regions empty of smoke are essentially nonturbulent and nearly irrotational, and we see that such regions can penetrate deep into the boundary layer. The boundary between turbulent and nonturbulent flow is highly convoluted but surprisingly sharp. As the structures move along with the general flow, the turbulence seen at fixed points in space in the outer part of the layer is intermittent in time. These large structures tend to scale with δ, the overall thickness of the layer.

The main smaller structures that appear in the inner part of the layer are associated with *streamwise vortices*. A pair of counter-rotating vortices is often linked in the form of a horseshoe, as shown in Figure 4.4.4. Where the flow between two vortices is downward toward the wall, higher speed fluid is carried toward the wall in what is called a *sweep*, and where the flow between two vortices is upward, lower speed fluid is carried away from the wall in a *burst*. The resulting streaky structure very close to the wall was shown in Figure 2.1f.

These turbulence structures have been studied intensively, both experimentally and by DNS. Comprehensive surveys are to be found in Robinson (1991) and Moin and Mahesh (1998). Surprisingly, this knowledge of organized structures has had very little effect on the practical prediction of turbulent boundary-layer flows. The reasons for this have to do with the nature of the turbulence modeling problem, which we've already discussed

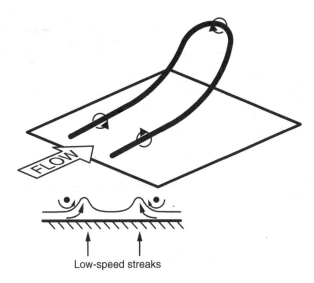

Low-speed streaks

Figure 4.4.4   Pair of streamwise vortices forming a horseshoe pair in the inner part of a turbulent boundary layer

in Section 3.7. In the RANS equations (Reynolds-averaged Navier-Stokes) governing the mean flowfield, the turbulent transport of momentum is represented by the *Reynolds stresses,* which are *local (single-point) averages* of fluctuating velocity components. Of course, the Reynolds stresses in the real world reflect contributions from the organized structures, but for modeling purposes, incorporating information about organized structures into single-point averages is difficult. As a result, such information that has found its way into the turbulence models has been of a very low-order variety, such as the idea that the motions that matter have length scales that decrease with decreasing distance from the wall, which was around long before we gained our current awareness of coherent structures. So as interesting as organized turbulence structures are, our detailed knowledge of them has not had much influence on engineering practice. Of more practical interest are the mean velocity profiles and the strategies that turbulence models have used for predicting them.

Typical mean velocity profiles in 2D turbulent boundary layers are illustrated in Figure 4.4.5. The most striking feature of turbulent profiles compared to laminar profiles is the very large velocity gradient at the wall. The highest gradient occupies a very thin region adjacent to the wall, too thin to be resolved on the scale of these plots. It is called the *viscous sublayer* because the shear stress there is effectively all viscous, with practically no contribution from Reynolds stress (see Section 3.7 for a discussion of Reynolds stress). But this lack of turbulent shear stress doesn't mean there is no turbulence in the sublayer. The flow in the sublayer is actually quite turbulent, with velocity fluctuations that are not small relative to the mean velocity. The Reynolds stress is nearly zero throughout a layer of finite thickness, not because the turbulence has disappeared, but because of the way the nearby wall constrains the turbulence, as we'll see next.

So the existence of the sublayer gives a distinctive appearance to turbulent mean velocity profiles. It also plays a vital role in the greater separation resistance of turbulent boundary layers relative to laminar, an issue we discussed in Section 4.1.4. There we noted that for a

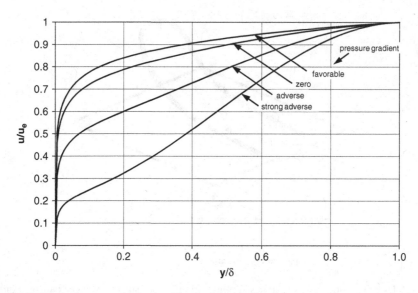

Figure 4.4.5 Typical mean-velocity profiles in 2D turbulent boundary layers in various pressure gradients

boundary layer to resist separation, fluid parcels near the wall must experience a favorable net viscous force associated with a positive shear-stress gradient $\partial\tau/\partial y$. In a constant-property laminar boundary layer, $\partial\tau/\partial y = \mu\partial^2 u/\partial y^2$, and the favorable viscous force thus requires positive $\partial^2 u/\partial y^2$. In the laminar case, the force is typically weak because μ is small. In a turbulent boundary layer, the rapid increase in the turbulent eddy viscosity outside the sublayer can provide a strong positive $\partial\tau/\partial y$ even though $\partial^2 u/\partial y^2$ is strongly negative there. One way to think of the turbulent boundary layer's greater resistance to separation is to imagine the strongly mixing part of the flow pulling the top of the sublayer along, keeping the velocity-profile slope and the skin friction positive.

In much of the earlier literature, the sublayer was called the "laminar sublayer," which wrongly implied that the flow in the sublayer is laminar. This unfortunate terminology has led to some misunderstandings, for example, in discussions of the effects of surface roughness, as we'll see in Section 6.1.8. It is clear from turbulence measurements and from flow visualization of the streaky structure in the sublayer in Figure 2.1f that the flow in the sublayer is anything but laminar. A closely related bit of misleading older terminology refers to the layer just outside the sublayer as the "buffer layer," implying that it somehow buffers "laminar" flow in the sublayer against disturbances from outside.

Many discussions of how the Reynolds stress decreases as the wall is approached use the term "damping" in one way or another. For example, Van Driest (1956) developed a successful and subsequently much-used formula for the eddy viscosity that includes a "damping factor" (see Section 3.7 for a discussion of the concept of eddy viscosity), and Mellor and Gibson (1966) speculated that the sublayer and part of the "buffer layer" constitute a region in which "all disturbances are damped." The term is not entirely inappropriate, in that the combination of viscosity and the no-slip condition at the wall does have a strong damping effect on the turbulence in this region, but it is potentially misleading because it suggests that the reduction of the Reynolds stress is due simply to a general reduction in turbulence level. The way that the Reynolds stress decreases as the sublayer is approached is actually more complicated than that.

The Reynolds stress of interest in a 2D boundary layer is $\rho\overline{u'v'}$ and thus dependent to the same degree on both v' and u' velocity fluctuations. DNS simulations by Spalart (1988) show that the rms u' level goes to zero linearly with y, pretty much across the entire sublayer. If v' did the same, we might expect that $\rho\overline{u'v'}$ would go to zero as y^2, and that there wouldn't be much of a seemingly finite sublayer. But rms v' goes to zero much faster (as y^2) than u', and $\rho\overline{u'v'}$ goes to zero as y^3. So the Reynolds stress disappears faster than rms u' in the sublayer and buffer layer because rms v' disappears faster.

So why does rms v' disappear faster than u' near the wall? Both are constrained to be zero at the wall, so the answer isn't just in the boundary conditions. At least part of the explanation has to do with the length scales of the turbulence structures and how viscosity affects motions with different scales. If l is the typical horizontal (x or z) distance over which instantaneous u', v', and w' vary, and given that u', v', and w' obey the continuity equation, we can make the following loose statement about how the magnitudes of u' and v' are related:

$$v' \sim y\frac{\partial v'}{\partial y} \sim y\left(\frac{\partial u'}{\partial x} + \frac{\partial w'}{\partial z}\right) \sim \frac{y}{l}(u' + w'). \qquad (4.4.2)$$

So for motions with length scales much larger than y, v' will tend to be much smaller than either u' or w', and there won't be much contribution to the Reynolds stress. The "active" motions (the motions that contribute to the Reynolds stress) tend to have length scales proportional to y, with v' of the same order as u'. But viscous damping has an increasing effect as l decreases, and as we enter the buffer layer the intensity of motions with length scale $l \sim y$ decreases. If we think of viscosity as imposing a lower cutoff on l (an oversimplification, but with a grain of truth to it), we would expect v' to decrease as y^2 while u' decreases as y, as was observed in the DNS simulations.

Outside the viscous sublayer, we find another distinctive feature of turbulent velocity profiles: the pronounced "corner," in which the second derivative of the velocity has a large negative value. In this region, the eddy viscosity increases rapidly with increasing y, and the velocity gradient diminishes accordingly, to the much lower levels typical of the outer part of the layer.

An interesting difference between turbulent and laminar profiles is in the appearance of inflection points in adverse pressure gradients. In our discussion in connection with Equation 4.2.7, we noted that in an adverse pressure gradient $\partial^2 u/\partial y^2$ must be positive at the wall and that a laminar profile must therefore have an inflection point somewhere away from the wall. The same applies to turbulent profiles, but the inflection isn't generally visible in profile plots like Figure 4.4.5. Deep in the sublayer, $\partial^2 u/\partial y^2$ must be positive, but just outside the sublayer it is strongly negative, and the location of the inflection point must be in the buffer layer, too close to the wall to be visible in ordinary profile plots. But inflection can also appear farther out in the profile. As indicated in Figure 4.4.5, a region of positive second derivative can appear in the outer portion of a 2D turbulent velocity profile if the pressure gradient is very strongly adverse or has been adverse over a long distance. Because the velocity profile tends to retain a pronounced negative second derivative both in the "corner" and near the edge of the boundary layer, a region of positive second derivative in the middle generally requires two inflections, as seen for the strongest adverse-pressure-gradient case in Figure 4.4.5.

Many of the available turbulence models can predict these distinctive features of turbulent velocity profiles, but before CFD became commonplace an understanding had already been developed based on physical reasoning, dimensional analysis, and simplified models. The physical reasoning starts with the idea that the boundary layer can be divided into an inner region and an outer region, and that in each region the velocity profile depends on the flow conditions in a distinctive way.

At the very bottom of a laminar boundary layer, the velocity profile is nearly linear and depends only on the wall shear stress and the viscosity, and not directly on the pressure gradient or on anything that happens in the outer part of the layer. In a turbulent boundary layer, the same dependence applies reasonably well to the wall region, which spans the viscous sublayer and the "corner" region of the velocity profile. To express this dependence in dimensionless form, we first define dimensionless *wall variables:*

$$u^+ = \frac{u}{u_\tau}, \tag{4.4.3}$$

$$y^+ = \frac{yu_\tau}{\nu}, \tag{4.4.4}$$

where

$$u_\tau = \left(\frac{\tau_w}{\rho}\right)^{1/2} \tag{4.4.5}$$

is called the *friction velocity,* and a unit of y^+ serves as a length scale often called a *wall unit.* The dependence we're assuming, that is, that the velocity profile depends only on the shear stress and the viscosity, then takes the form of a basic functional relationship called the *law of the wall:*

$$u^+ = f(y^+). \tag{4.4.6}$$

In the viscous sublayer, the velocity profile is nearly linear, as it effectively is at the bottom of a laminar boundary layer, and f takes an especially simple form:

$$u^+ = y^+. \tag{4.4.7}$$

Farther out in the wall region, turbulent shear stress comes to dominate relative to the viscous shear stress. But the velocity profile still depends on the viscosity because the presence of the viscous sublayer has a large effect on the profile shape well into the part of the region where the turbulent shear stress has taken over. Shortly, we'll determine what functional form f must take in the outer part of the inner region.

In the *outer region,* we assume that momentum transport is entirely turbulent and that the velocity profile has no direct dependence on viscosity. The velocity profile in the outer region goes to u_e at the outer edge of the boundary layer and must merge with the wall-region velocity profile somewhere in the neighborhood of the "corner" of the profile. The remaining physical arguments regarding the outer-region velocity profile then go as follows. The outer region is like a boundary layer with a "slip velocity" at the wall, the slip velocity just being the velocity provided by the wall-region profile. The processes (primarily large eddies) that determine the velocity profile in the outer region don't "know" how much slip velocity is being provided by the wall-region profile, only how much shear stress is being transmitted, which is assumed to at least scale with the wall shear. Because the flow in the outer region doesn't "know" what its velocity is relative to the wall, the velocity that matters is the *velocity defect*, $u_e - u$, and the outer region is therefore often called the *defect layer*.

The velocity-defect profile is assumed to depend on the overall thickness δ of the boundary layer, the wall shear stress, the pressure gradient, and upstream history, but *not* directly on the viscosity. Expressing this in dimensionless form, we have

$$\frac{u_e - u}{u_\tau} = F\left(\frac{y}{\delta}, \ldots\right). \tag{4.4.8}$$

In a flat-plate flow, F depends only on y/δ because there is no pressure gradient, and upstream history effects are negligible because the shape of the profile is changing slowly. Equation 4.4.8 is then known as the *velocity-defect law*, and data from many experiments on both smooth and rough walls closely define a single function F, provided R_θ is above about 6000. Then there is the special class of flows that we've already noted known as the *equilibrium turbulent boundary layers*, in which the pressure-gradient parameter β defined by Equation 4.3.8 is constant in x. In an equilibrium turbulent boundary layer, the function F is also constant in x, but with a different function for each value of β, as we discussed previously in connection with Equation 4.3.9. Equilibrium boundary layers were first described by Clauser (1954, 1956), and further theoretical issues were investigated by Mellor and Gibson (1966). For more general cases, F is not universal, but the important thing for our purposes is that the defect profile is a function of y/δ and not directly a function of viscosity.

So we have inner and outer regions in which the velocity profile has different kinds of functional dependence. The final physical argument in this analysis is that there is a significant *overlap region* in which both kinds of dependence (Equations 4.4.6 and 4.4.8 hold. In the overlap region, we then have

$$u^+ = f\left(\frac{\delta u_\tau}{\nu}\frac{y}{\delta}\right) = \frac{u_e}{u_\tau} - F\left(\frac{y}{\delta}, \ldots\right). \tag{4.4.9}$$

The velocity profile thus is given by a function f of y/δ *multiplied* by a constant, and at the same time by the function $F(y/\delta)$ with a constant *added* to it. As first pointed out by Millikan (1938) (see also White, 1991), the only possible functional form is logarithmic. The resulting *logarithmic law of the wall* for the overlap portion of the inner region is usually written

$$u^+ = \frac{1}{\kappa}\ln(y^+) + B, \tag{4.4.10}$$

where κ is called the von Karman constant and has been assigned values between about 0.38 and 0.44 depending on what experimental data are used. The constant B also varies depending on what value of κ was chosen, among other things. It is typically about 4.9 for a smooth wall and is smaller when the wall is rough, as we'll see in Section 6.1.8. The logarithmic overlap typically extends from $y^+ \approx 0$ out to about 0.15δ. At very high Reynolds numbers, the sublayer is very thin, and the log profile occupies nearly the whole region from the wall to 0.15δ, while at the lowest Reynolds numbers at which turbulent flow can be sustained, the log profile disappears.

Outside the viscous sublayer, the velocity profile blends smoothly from the linear sublayer profile to the log profile. Spalding (1961) devised a composite formula that fits the entire law-of-the-wall region:

$$y^+ = u^+ + e^{-\kappa B}\left[e^{-\kappa u^+} - 1 - \kappa u^+ - \frac{\left(\kappa u^+\right)^2}{2} - \frac{\left(\kappa u^+\right)^3}{6} \right],$$ (4.4.11)

which is illustrated in Figure 4.4.6. There are not many reliable sets of experimental data defining the profile in this region, but this formula fits the available data well (see White, 1991).

Why does it matter that a portion of the typical turbulent velocity profile is logarithmic? Well, there are at least two practical applications for this bit of knowledge. One is that surface roughness causes a shift in the logarithmic profile, and the shift provides a practical way of characterizing the effect on the skin friction, making it possible to take empirical roughness data from one flow situation and apply them to other flow situations, as we'll see in Section 6.1.8. Another is that the logarithmic profile provides a way to determine C_f from experimental measurements of the mean velocity profile. The viscous sublayer is typically so thin that measurements can't be made close enough to the wall to determine $\partial u/\partial y$ directly, but the logarithmic part of the profile can often be measured reasonably accurately. So in principle we should be able to determine the value of C_f that makes the measured profile fit the logarithmic law best. Clauser (1956) proposed one very effective way of doing this. First, recast Equation 4.4.10 in terms of C_f and u_e:

$$\frac{u}{u_e} = \frac{u_\tau}{u_e}\frac{1}{\kappa}\ln\left(\frac{yu_e}{v}\right) + \frac{u_\tau}{u_e}\frac{1}{\kappa}\ln\left(\frac{u_\tau}{u_e}\right) + \frac{u_\tau}{u_e}B,$$ (4.4.12)

a form of the log law in which increasing C_f increases both the slope and the intercept. Then all we have to do is to plot Equation 4.4.12 for an array of closely-spaced values of C_f and

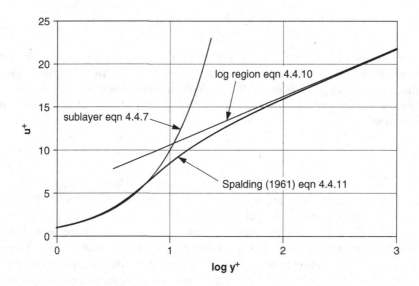

Figure 4.4.6   Velocity profile in the law-of-the-wall region of a turbulent boundary layer

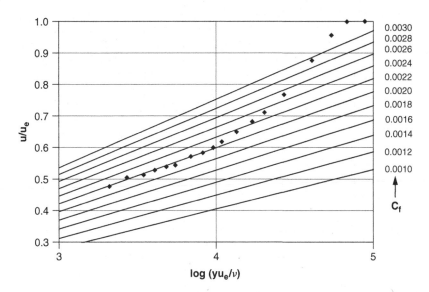

Figure 4.4.7   A "Clauser plot" of an experimental turbulent velocity profile compared with curves plotted from Equation 4.4.12, showing how an experimental estimate of C_f can be made ($C_f \approx 0.0020$ in this case). (Data from McLean, 1970; for a case about halfway between the onset of an adverse pressure gradient and separation)

plot the measured profile in terms of u/u_e versus $\log(y\,u_e/\nu)$ for comparison. Figure 4.4.7 illustrates how this works. The portion of the experimental profile that is logarithmic is easily identified, and the curve it lines up with best provides an estimate of C_f. This kind of plot is often called a *Clauser plot*.

The velocity-defect form in Equation 4.4.8 isn't very convenient for general flows because the function F isn't universal. Coles (1956) found another way of representing the outer-region velocity profile, in which a single universal function suffices. Instead of referencing the velocity in the outer region to u_e, he referenced it to the outward continuation of the inner-region profile, the law of the wall. In this formulation, the entire profile can be expressed in a single formula:

$$u^+ = f(y^+) + \frac{2\Pi}{\kappa} W\left(\frac{y}{\delta}\right), \tag{4.4.13}$$

where f is the law of the wall and can be expressed as a combination of Equations 4.4.7 and 4.4.10 or in terms of Equation 4.4.11, and the function W is well represented by

$$W\left(\frac{y}{\delta}\right) = \sin^2\left(\frac{\pi}{2}\frac{y}{\delta}\right). \tag{4.4.14}$$

The constant Π is determined by the difference between the law of the wall and the actual velocity u_e at $y = \delta$. The function W resembles the velocity profile in half of a turbulent wake and is therefore called the *wake function*, and Equation 4.4.13 is called the *law of the wake*. Equation 4.4.13 is compared with the same measured turbulent velocity profile used in Figure 4.4.7, first in terms of inner-region variables in Figure 4.4.8a, and then in terms of u/u_e versus y/δ in Figure 4.4.8b. This level of agreement with data is typical.

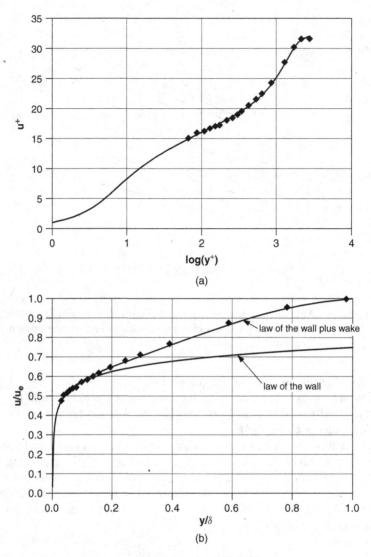

Figure 4.4.8 A typical experimental velocity profile compared with Coles' law of the wake. Same experimental data as in Figure 4.4.7. (a) In terms of wall variables. (b) Conventional velocity profile

The wake function W, shown in Figure 4.4.9, is "concave" (negative second derivative) in the bottom half and "convex" in the upper half, with an inflection point in the middle. Thus one component of the law-of-the-wake profile always has an inflection point, even when the pressure gradient isn't adverse. But the general convexity of the other component, the law of the wall, tends to keep the overall profile convex unless the adverse pressure gradient is strong or has been in force over a long distance. This can be seen in Figure 4.4.8b, in which the profile shown is in a moderately strong adverse gradient, at a station about half way from the onset of the pressure gradient to separation, and the profile is still convex, though just barely.

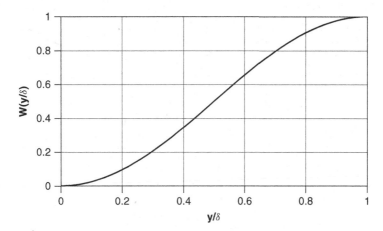

Figure 4.4.9 Wake function W(y/δ), defined by Equation 4.4.14, used in the law of the wake, Equation 4.4.13

The use of the word "wake" in the law of the wake is odd in a way. Coles (1956) noted the similarity of the function W(y/δ) to half of a wake velocity profile and proposed thinking of the outer part of a turbulent boundary layer as a half-wake. While a turbulent free wake and the outer region of a turbulent boundary layer are both regions of *velocity deficit* relative to u_e, the "wake component" $2\Pi \, W/\kappa$ of the law of the wake is, according to Equation 4.4.13, a *velocity excess* relative to the law of the wall. The entire law-of-the-wake velocity profile, including the law of the wall component, is often not at all wake-like in appearance, as, for example, in Figure 4.4.8b.

The skin friction is much larger for a turbulent boundary layer than for a corresponding laminar flow, as we saw for flat-plate flow in Figures 4.3.1–4.3.3. The difference between turbulent flow and laminar flow is even more dramatic when we look at it in terms of the eddy viscosity μ_t compared with the molecular viscosity μ. Away from the wall in a turbulent boundary layer, the mean-velocity gradient quickly becomes much smaller than it is at the wall, while the shear-stress level tends to remain high. This situation reflects levels of eddy viscosity that are orders of magnitude higher than the molecular viscosity. In Figure 4.4.10, we plot the ratio μ_t/μ for flat-plate boundary layers over a range of Reynolds numbers, according to the eddy-viscosity model of Mellor (1967). The μ_t/μ distribution in the inner region, plotted versus y^+, is independent of Reynolds number, as shown in part (a). It is the linear-looking part of the μ_t/μ distribution that produces the logarithmic part of the velocity profile. The outer region eddy viscosity is proportional to $R_{\delta*}$ as seen in part (b), a relationship first suggested by Clauser (1956). In some models, for example, Cebeci and Smith (1974), the eddy viscosity decreases toward the outer edge of the boundary layer to account for the intermittency of the turbulence. Mellor and Gibson (1966) claim this makes little difference in the calculated mean-velocity profiles.

In Section 4.1.2, we discussed the distinctive aspects of 3D boundary layers in general terms: shear- and pressure-driven cross flow, cross-flow velocity profiles, direction profiles, the nature of separation in 3D, and the effects of cross flow on the flow development and the displacement effect. An additional complication arises in turbulent 3D flows, and that is that the relationship between the mean-velocity gradient and the shear stress becomes more complex.

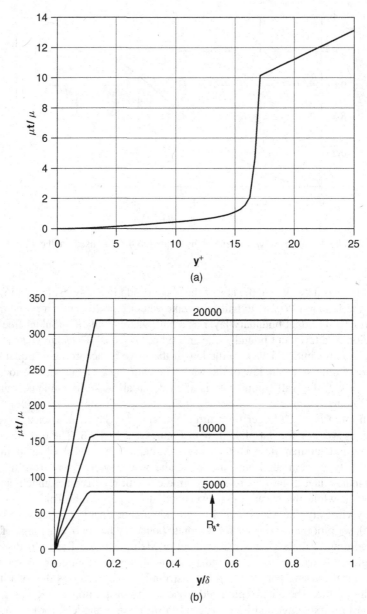

Figure 4.4.10 Levels of eddy viscosity given by Mellor and Gibson's (1966) model, for flat-plate boundary layers over a range of Reynolds numbers. (a) The inner region, in wall region variables, in which the variation is independent of Reynolds number. (b) Over the entire boundary-layer thickness, where the inner-region variation of (a) takes place too close to the wall to be visible

One way of looking at this relationship is to think of the shear stresses and the velocity gradients as vectors (τ_x, τ_z) and $(\partial u/\partial y, \partial w/\partial y)$ defining directions parallel to the surface. In 2D flows, these vectors are in the same direction, the direction of the entire flow, given by the vector $(u_e, 0)$. In 3D, the shear stress and the velocity gradient line up only in laminar flow or in the viscous sublayer of a turbulent boundary layer. In the turbulent part of a 3D boundary layer, they are generally skewed at an angle relative to each other, with the shear-stress direction often lagging behind when the velocity-gradient direction changes, as is the case in the example shown in Figure 4.4.11. Van den Berg (1988) compiled a list of possible physical mechanisms for the skewing, but what the mechanisms actually are in typical 3D boundary-layer flows has not been resolved.

The skewing means that the eddy viscosity has different values in the two directions, and it can have major effects on the development of the mean-velocity field. Many turbulence models ignore this, however, and assume that the eddy viscosity is isotropic (the same in all directions). We'll discuss this problem further in connection with the prediction of swept-wing flows in Section 8.6.2.

In most applications, we are not interested in the unsteadiness associated with a turbulent boundary layer, and we tend to measure or attempt to predict only the time-averaged (mean) flowfield. But the pressure fluctuations at the surface are sometimes important, as, for example, in the generation of boundary-layer noise. These tend to have broad-band spatial and temporal spectra reflecting the range of scales in the boundary-layer turbulence. In an attached boundary layer, the lowest spatial and temporal frequencies tend to be those associated with the passage of large structures such as those shown in Figure 2.1e. Boundary-layer separation from anywhere but a single, sharp trailing edge will give rise to lower frequencies associated with fluctuations of the instantaneous separation location and

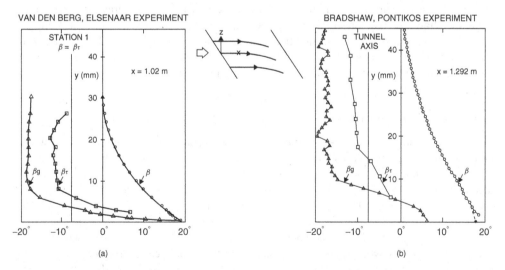

Figure 4.4.11 Examples of the skewing that can take place between the shear-stress direction β_τ and the velocity-gradient direction β_g compared to the outer-flow direction β in a 3D turbulent boundary layer in an adverse pressure gradient. See Bradshaw and Pontikos (1985). (a) Data from van den Berg, (1988). Published by AGARD. (b) Data from Bradshaw and Pontikos, (1985).

of the separated shear layer, and these fluctuations can be felt far upstream. Separation also increases the intensity of the surface pressure fluctuations, by something on the order of 10 dB, and reattachment of a turbulent shear layer increases them even more.

4.5 Control and Prevention of Flow Separation

As we noted in Section 4.1.4, delaying or preventing separation is important in many applications. In this section, we'll discuss some of the strategies that are used.

4.5.1 Body Shaping and Pressure Distribution

Separation from a smooth surface is generally related to a rapid or sustained rise in pressure, as we saw in Section 4.1.4. Thus a powerful means of delaying separation is through the shaping of the body so as to control the pressure rises. This usually means providing a generally *streamlined* shape, without sharp corners or large surface curvatures, and with a gradual closure at the back of the body to a sharp or almost sharp trailing edge or tail. If the closure is too abrupt, separation ahead of the tail will occur, as in Figure 4.1.2b or c. We'll consider the distinction between streamlined and bluff bodies further, as well as the underlying physics issues, in Sections 5.2, 6.1.3, 6.1.5, and 6.1.6. Even for a streamlined body, some fairly subtle details of the pressure distribution can be important. We'll consider this issue in some detail in connection with the maximum lift of airfoils in Sections 7.4.3 and 7.4.4.

Separation in 3D can form complicated flow patterns such as the *necklace vortex* separation that can occur in a wing-body junction, as shown in Figure 4.5.1a,b. A *fillet* or *strakelet* can eliminate the separation at the leading edge itself by easing the adverse pressure gradient and increasing the flow divergence ahead of the junction, as shown in Figure 4.5.1c. The change in the surface shear-stress pattern between (b) and (c) is dramatic. We should remember, however, that the surface shear-stress pattern tends to give an exaggerated impression of the change to the overall flowfield, as we pointed out in Section 4.1.2 and saw in Figure 4.1.11. Separations like the one in Figure 4.5.1b tend to be confined close to the body surface and to have very minor effects on the rest of the flow.

4.5.2 Vortex Generators

Generating streamwise vortices in the boundary layer upstream of where separation would otherwise occur can be an effective preventive measure. A *vortex generator* (VG) is usually a low-aspect-ratio vane with a planform ranging from delta to rectangular, attached perpendicular to the wall and placed at an angle of attack of about 15° to the local flow, so as to produce a strong vortex. The basic types are illustrated in Figure 4.5.2. A conventional VG is usually sized about as high as the boundary layer is thick ($h \approx \delta$), while a *subboundary-layer vortex generator* (SBVG) is much smaller ($h \ll \delta$). VGs are generally used in arrays and characterized by whether the vortices are all of the same sign (*co-rotating*) or of alternating signs (*counter-rotating*). In addition to vanes, various sorts of wedges, bumps, and scoops have been used to generate counter-rotating vortex pairs, but these devices are less common than vanes.

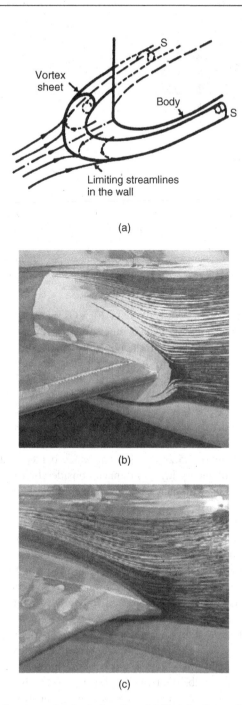

(a)

(b)

(c)

Figure 4.5.1   3D separation in a wing-body junction. (a) Sketch of a *necklace vortex* separation in a junction without a fillet. (b) Oil-flow photo of necklace vortex separation in a horizontal-tail junction. (c) Oil-flow photo of same junction as in (b) with *strakelet* installed

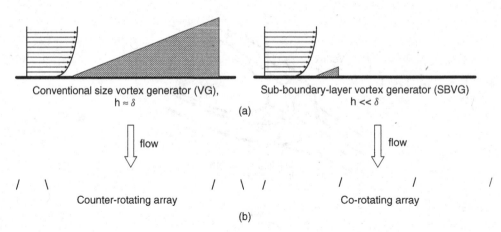

Figure 4.5.2 Types of vortex generators used for delaying or preventing separation. (a) Scaling of VG planform relative to boundary layer. (b) Arrangement of VGs on surface

A common explanation for the effectiveness of VGs is that they "enhance mixing." We've seen that a turbulent boundary layer is more effective than a laminar one at resisting separation, and it seems logical that introducing more mixing would make it even more effective. There is an element of truth to this idea, but it is an oversimplification, especially in the case of conventional large VGs. What a large VG does to the flow can be looked at either in terms of a large side force "pushing" the flow strongly to the side, or in terms of "induction" by a strong vortex. Either way, the result is a major rearrangement of the mean-velocity field, as illustrated in Figure 4.5.3 in terms of streamwise vorticity and velocity contours in a cross-plane downstream of the VGs.

Counter-rotating VGs (Figure 4.5.2a, left side) produce bands of strong flow divergence (Figure 4.5.3a), where the boundary layer is thinned considerably, which increases its resistance to separation, alternating with bands of flow convergence, in which the boundary layer is thickened, and resistance to separation is presumably reduced. As the vortices "migrate" laterally, the bands of divergence widen, and the bands of convergence narrow, until pairs of vortices converge on each other and begin to rise from the surface, as can be seen in Figure 4.5.3a. As the vortices rise, they lift some of the low-energy air away from the wall, so that it contributes less to the tendency to separate. If an array of counter-rotating VGs is placed insufficiently far upstream of where separation wants to occur, separation will occur in the convergence bands. If the array is placed too far upstream, the converged vortex pairs will rise too far and lose their effectiveness. The placement of an array, relative to its spacing, is thus a balancing act. And note that in counter-rotating arrays, the vanes are usually grouped in pairs, as shown in Figure 4.5.2c, such that vortex convergence and liftoff are delayed farther downstream than they would be with uniform spacing.

Co-rotating VGs (Figure 4.5.2b, right side) also produce bands of divergence (thinning) and convergence (thickening), as in Figure 4.5.3b, but not nearly as pronounced as those produced by counter-rotation.

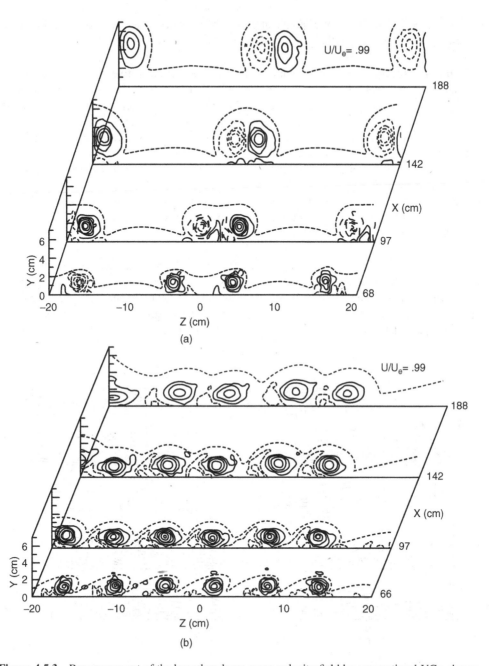

Figure 4.5.3 Rearrangement of the boundary-layer mean-velocity field by conventional VGs, shown in terms of streamwise vorticity contours and the $u/u_e = 0.99$ streamwise velocity contour at several streamwise stations. From Pauley and Eaton, 1988. Used with permission. (a) With counter rotating VGs. (b) With co-rotating VGs

Which type of array is more effective at preventing separation depends on several factors. Here is a brief summary of the pros and cons of the two types:

- **Counter-rotating**: A counter-rotating array can be more effective over short distances because of the strong divergence it can produce, but vortex merging limits the distance over which it remains effective. The effectiveness of a counter-rotating array can be seriously degraded if the vortices in a pair are not of equal strength (Pearcey, 1961), and counter-rotating arrays are thus not very tolerant of flow direction changes that unbalance the angles of attack of the vanes. And in 3D flows with strong cross-flow in the boundary layer, the two vortices in a pair behave differently, as explained below, which tends to favor a co-rotating array.
- **Co-rotating**: A co-rotating array will usually have lower nearfield effectiveness, but its effectiveness will persist farther downstream. And a co-rotating array is usually favored in cases with strong boundary-layer cross-flow.

Strong boundary-layer cross-flow has two main effects (Shizawa and Eaton, 1992): The vortices from VGs decay two to three times as fast as they would in a comparable 2D flow, and vortices of opposite rotation behave differently, as we mentioned above. When the main vortex from the VG has the same rotation as the background cross-flow vorticity in the outer part of the boundary layer, the vortex tends to diffuse more rapidly, as shown

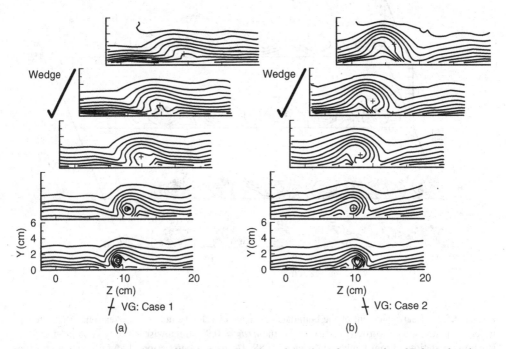

Figure 4.5.4 Illustration of the effects of the direction of rotation of a vortex in a 3D boundary layer with strong cross flow, shown in terms of streamwise velocity contours. From Shizawa and Eaton, 1992. Used with permission. (a) Vortex rotation from a VG that pushed the flow in the same direction as cross flow near wall. (b) Vortex rotation from a VG that pushed the flow in the direction opposing the cross flow near wall

in Figure 4.5.4a, than it does when the vortex rotates in the opposite direction, as in Figure 4.5.4b. So flow situations with strong boundary-layer cross-flow favor co-rotating arrays with rotation as in Figure 4.5.4b. An easy rule to remember is that the force exerted by the VG should push the flow in the direction opposite to the direction of the cross-flow. This is important in the application of VGs to swept wings. We'll discuss swept-wing boundary layers in some detail in Section 8.6.2.

One of the most common applications for large VGs is to delay shock-induced separation on wings with transonic flow. The physical issues involved were discussed by Pearcey (1961) and ESDU (1993) provides installation design guidelines. Large VGs have also been used on fuselage closures, engine-inlet and wind-tunnel diffusers, and windmill blades (McMasters, Crowder, and Robertson, 1985). In general, counter-rotating VGs have been favored in flows that are without strong cross flow and in which the flow direction can be depended on to remain the same, and co-rotating VGs have been favored on swept wings.

For modeling the effects of VGs in CFD calculations, it's tempting to treat the flow only in a spanwise-averaged sense and to modify the turbulence model so as to simulate the effects of enhanced mixing. But we saw in Figure 4.5.3 that a boundary layer downstream of conventional large VGs has a pronounced large-scale spanwise-periodic structure, the effects of which are not likely amenable to averaging. The most successful modeling of large VGs so far has been based on 3D RANS calculations that resolve the individual vortices (see Yu, Kao, and Bogue, 2000).

SBVGs, on the other hand, are much smaller than conventional VGs and have spacings that are not so large relative to δ, so that it is much more reasonable to think of their effects in spanwise-averaged terms. The method developed by researchers at QinetiQ, Ltd, based on experimental measurements by Ashill, Fulker, and Hackett (2001, 2002), follows this general strategy in simulating the effects of SBVGs in a 2D integral boundary-layer code.

4.5.3   Steady Tangential Blowing through a Slot

Blowing air through a tangential slot in the wall can be a very effective way to delay separation. The air can be introduced passively through an open slot in an airfoil, as in Figure 4.5.5a, or actively from a pressurized air source, as in Figure 4.5.5b. In either case, the result downstream of the slot is a velocity profile of the *wall-jet* type, with a velocity

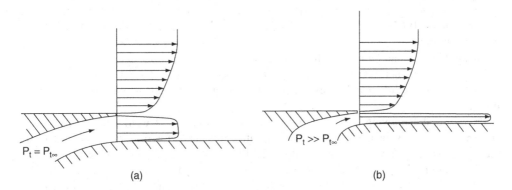

Figure 4.5.5 Tangential blowing slots and associated *wall-jet* velocity profiles. (a) Passive open slot. (b) Active slot with pressurized air supply

peak near the bottom of the layer. Note that in the passive case the maximum total pressure in the jet is the freestream total pressure, so that the maximum velocity at the jet exit is about the same as the inviscid-flow velocity at the edge of the boundary layer. In the active case, on the other hand, total pressures substantially higher than freestream are typically used, which results in higher jet velocities. Because of the typically higher momentum flux, active blowing tends to be implemented with a narrower slot than passive blowing.

Much of what is known about the performance of wall jets in preventing separation applies to the active case with a relatively narrow slot and a high jet velocity. In this situation, the slot height and jet volume flux required to control downstream flow separation tend to be relatively small, so that the details of the jet velocity profile don't matter much, and performance depends primarily on the integrated jet momentum flux. On the other hand, a passive wall jet, with its larger slot height and lower jet velocity, is not so simple and does not seem to perform in a way that correlates neatly with simple flow parameters. Understanding passive slot blowing requires taking a different view, looking at the problem in terms of the boundary layers and pressure distributions on the individual surfaces rather than in terms of a single jet. We'll discuss this approach to understanding the passive slot effect in Section 7.4.4.

Steady slot blowing is a "brute force" method of separation control in that it works by providing the fluid at the bottom of the boundary layer with the additional momentum it needs to overcome a longer run of adverse pressure gradient without separating.

The usual correlation parameter for the effectiveness of high-velocity wall jets in preventing separation on an airfoil of chord c is the momentum-flux coefficient:

$$C_\mu = \frac{\int\limits_{\text{slot}} \rho_j u_j^2 \, dy}{\rho_\infty u_\infty^2 c}. \tag{4.5.1}$$

The level of C_μ required depends on how far separation needs to be delayed and on other details of the particular application. To bring about a major change in separation behavior, for example, to achieve fully attached flow on a plain (unslotted) trailing-edge flap at a large deflection angle, might require $C_\mu \approx 0.02-0.06$, depending on the details. Performance of slot blowing correlates reasonably well with C_μ in the sense that for a given airfoil shape and flow condition, performance is insensitive to slot height and jet velocity, provided C_μ is kept constant. Lachman (1961) and Korbacher (1974) survey an extensive assortment of early investigations of the overall performance of steady blowing in delaying separation, and McGahan (1965) made detailed flowfield measurements in a series of separating flows with steady blowing.

Steady blowing can be highly effective in delaying or preventing separation, but it tends to be costly in an energy sense. For most of the practical applications where blowing might be considered, providing a C_μ level above 0.01 requires a large amount of compressor power.

Providing slot blowing can entail other practical problems as well. An effectively 2D slot interrupts the skin of the body, and an alternative path for possible structural loads must be provided. And if the required slot height is small, maintaining the height accurately can be a problem, especially if the jet supply pressure is high. In some cases, an attractive alternative is to replace the slot with a spanwise array of discrete nozzles discharging parallel to the surface. Only discrete holes through the skin are then required, and precise nozzle geometry is easy to maintain. Discrete-jet blowing has been found to be just as effective as slot blowing, provided that the nozzle array is placed far enough upstream so that the discrete jets have a chance to merge sufficiently to prevent separation where it needs to be prevented (McLean and Herring, 1973).

4.5.4 Active Unsteady Blowing

Active unsteady blowing is a much more recent development than steady blowing, and it involves entirely different physics. Instead of *preventing* separation by brute force addition of momentum to the bottom of the boundary layer, unsteady blowing *controls the effects* of separation by manipulation of the separated shear layer. Unsteady excitation provided by a pulsed blowing jet or by a small mechanical flapper (like a miniature version of the spoilers used on wing upper surfaces) excites instabilities in the separated shear layer, causing it to oscillate with a large amplitude. The time-averaged flowfield that results is intermediate between the separated flow that would occur without control and the fully attached flow that could be achieved with sufficiently intense steady blowing. On an airfoil or trailing-edge flap, for example, the increment in lift relative to the separated case is less than what could be achieved with steady blowing but can still be substantial.

So unsteady excitation can't achieve the performance levels that steady blowing can, but there are indications that unsteady excitation can achieve useful levels with much lower energy inputs than required for steady blowing. For example, Seifert *et al.* (1993) reported substantial increases in $C_{l\,max}$ for an airfoil with a plain flap deflected 30° with a time-averaged C_μ of 0.0006, roughly 2 orders of magnitude lower than steady blowing might have required. More recently, experiments and CFD simulations (the calculations by Shmilovich and Yadlin (2007), for example) have required much higher excitation levels, however. The excitation level that would actually be required to provide attractive performance for a real aircraft high-lift system is still highly uncertain.

There are two distinct approaches regarding the frequency of excitation. Earlier experiments on unsteady blowing (Seifert *et al.*, 1993) found a relatively low reduced frequency of excitation $fc/u_\infty \approx 1$ to be most effective. More recently, a high-frequency approach using frequencies an order of magnitude higher has been proposed (see Glezer, Amitay, and Honohan, 2003, for example).

4.5.5 Suction

Suction through slots or porous surfaces can be highly effective at delaying separation, essentially by removing the air with the lowest momentum at the bottom of the boundary layer and thus delaying the reversal of the velocity profile. Schlichting (1979) shows flow visualization images of slot suction delaying separation in a diverging channel and in the external flow around a circular cylinder. Thwaites (1958) shows experimental data for a circular cylinder on which porous-surface suction eliminated separation so that the theoretical potential-flow pressure distribution was nearly achieved. In Section 6.1.10, we'll look at another way of applying suction, in which the body shape is designed to localize all of the body's pressure recovery over a single suction slot, where a large fraction of the boundary layer arriving from upstream is removed. In practical applications, suction, like steady blowing, tends to be costly in terms of energy.

4.6 Heat Transfer and Compressibility

In our theoretical discussions so far in this chapter, we've generally assumed constant-property flow. Though the general momentum-transport processes and the development of the velocity field we've discussed so far wouldn't likely be changed qualitatively by variations in properties, variable fluid properties can have important quantitative effects on the flowfield, and there are situations in which the heat transfer and the wall temperature are of practical importance. Such effects can come into play through heating or cooling at the wall, or through the effects of compressibility, or combinations of these.

4.6.1 Heat Transfer, Compressibility, and the Boundary-Layer Temperature Field

When either heat transfer or compressibility causes a variation in temperature in the neighborhood of a wall, we have, by definition, a *temperature profile* and a *thermal boundary layer*. The development of the thermal boundary layer depends both on what goes on in the flowfield and on the *thermal boundary conditions* at the wall, where either the temperature, the heat flux, or a relationship between the two may be imposed by the thermal properties of the body and other aspects of the thermal environment.

As an illustration of the kinds of effects that can arise, consider a wing skin that is also part of a fuel tank. Fuel was pumped into the tank at a temperature close to the ambient temperature on the ground, and now the airplane is flying at high altitude, where the air is much colder. The fuel is being cooled by heat transfer that is driven by the temperature difference and regulated by the thermal conductivity of the fuel and the skin, and by the effective conductivity of the thermal boundary layer. The surface temperature adjusts itself to balance the transfer of heat from the tank with the transfer to the flow. The temperature difference that drives the heat transfer to the flow is the difference between the actual surface temperature and the *adiabatic wall temperature,* which is just the temperature the wall would reach in the absence of heat transfer, as, for example, if the wall were a perfect thermal insulator. At very low Mach numbers, the adiabatic wall temperature is just the static temperature of the flow, while at higher Mach numbers it reaches an elevated *adiabatic recovery temperature* because the kinetic energy of the flow is significant, and some of it is turned into heat in the boundary layer through viscous dissipation. We'll shortly consider a simplified theoretical estimate of the adiabatic wall temperature.

Other sources of heat flux (electromagnetic radiation, for example) can also affect surface temperature. The surface of a vehicle at rest in the sun can become quite hot to the touch, but the temperature will drop rapidly as flow speed increases. The net electromagnetic heat flux in this case is the difference between the absorbed part of the incoming solar flux (the part that isn't reflected), which is roughly fixed, and radiation from the body, mostly in the infrared. As the body is cooled by the flow, the net radiation into it increases somewhat because the outgoing infrared radiation is reduced, but the effective conductivity of the thermal boundary layer increases greatly with flow speed, so that the radiation overall becomes less and less of a player in the heat balance that determines the surface temperature.

Thermal boundary layers encompass wide ranges in the relative importance of the various effects. For example, if the Mach number is low, and heat transfer is driven by small temperature differences relative to the absolute temperature, the density and viscosity will be almost constant, and the flowfield will be affected very little by the heat transfer. In such cases, the heat is conducted and convected by the flow much as a passive contaminant would be. With larger temperature differences and/or Mach numbers, on the other hand, the changes in density and viscosity can have significant effects on the flowfield. In some flows, heat transfer dominates the thermal effects, while in others, heat transfer from the wall is insignificant, and compressibility is the only source of temperature differences. And many flows land between these two extremes.

4.6.2 The Thermal Energy Equation and the Prandtl Number

The thermal energy equation governs the temperature field and its interaction with the velocity and pressure fields (see Schlichting, 1979, for a detailed derivation). The equation expresses the balance between conduction and convection of thermal energy, and its conversion to and from kinetic energy by compression, expansion, and viscous dissipation. The equation gives rise to an important dimensionless material property, the *Prandtl number:*

$$\mathrm{Pr} = \mu c_p / k, \qquad (4.6.1)$$

where c_p is the specific heat at constant pressure, and k is the thermal conductivity. The Prandtl number appears in both the thermal conductivity and viscous dissipation terms in the energy equation (see Table 3.9.2) and measures the relative importance of viscous diffusivity and thermal diffusivity in the development of the velocity and temperature fields. Schlichting (1979) uses an order-of-magnitude analysis to arrive at a rough estimate of how the Prandtl number affects the relative thicknesses of the velocity and thermal boundary layers in laminar flow:

$$\delta_T / \delta \sim \mathrm{Pr}^{-1/2}. \qquad (4.6.2)$$

In gases, the Prandtl number is not far from 1, and the thermal boundary layer tends to be similar in thickness to the velocity boundary layer, as illustrated in Figure 4.6.1a. In liquids, the Prandtl number is typically orders of magnitude larger, and the thermal boundary layer tends to be much thinner than the velocity boundary layer, as illustrated in Figure 4.6.1b. Of course, in real situations the thickness of the thermal boundary layer also depends on the upstream history of the thermal boundary condition. For example, there can be situations in which a pronounced temperature difference is introduced at the wall well downstream of the start of the velocity boundary layer. The thermal boundary layer arising from such a change in boundary condition will be thinner than the velocity boundary layer until the temperature difference has had a chance to diffuse out into the flow, as illustrated in Figure 4.6.1c.

4.6.3 The Wall Temperature and Other Relations for an Adiabatic Wall

In many high-speed flow applications the wall boundary condition is very nearly adiabatic, and it is useful to be able to predict the wall temperature. Because the Prandtl number in gases is close to 1, we'll take as a starting point an *energy integral* derived independently by Busemann and Crocco (see White, 1991) for the special case of $\mathrm{Pr} = 1$. If we further assume a perfect gas and an adiabatic wall, the boundary-layer thermal energy equation can

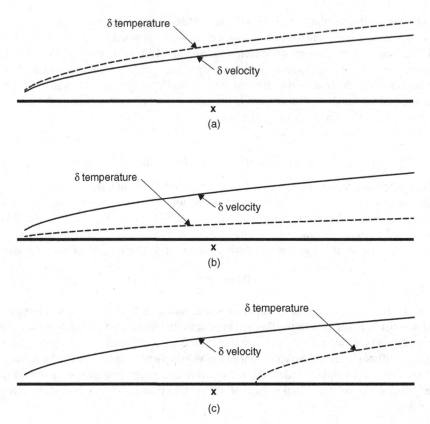

Figure 4.6.1 Schematic illustration of the thermal boundary layer in relation to the velocity boundary layer. (a) Gases (Pr about 0.7). (b) Liquids (Pr $\gg$ 1). (c) Gases (Pr near 1) after sudden change in wall temperature

be integrated, with the result that the total enthalpy is constant through the boundary layer:

$$H = h + u^2/2 = \text{constant}. \qquad (4.6.3)$$

Thus for Pr $= 1$, the adiabatic wall temperature, or the flow temperature at the wall, where $u = 0$, is the same as the stagnation temperature of the flow at the edge of the boundary layer, which in most external-flow situations will be the same as the far-field stagnation temperature.

But in gases the Prandtl number is a bit less than 1 (0.71 for air), and the adiabatic wall temperature is a bit less than the stagnation temperature at the boundary-layer edge. We define an adiabatic *recovery factor:*

$$r = (T_{aw} - T_e)/(T_{te} - T_e). \qquad (4.6.4)$$

We see that the Pr $= 1$ result of Equation 4.6.3 corresponds to r $= 1$. White (1991) presents approximate analyses indicating first that r should be very closely a function of Prandtl number only, and further that for laminar flow

$$r \approx \mathrm{Pr}^{1/2} \approx 0.84 \text{ for air,} \tag{4.6.5}$$

and for turbulent flow

$$r \approx \mathrm{Pr}^{1/3} \approx 0.89 \text{ for air.} \tag{4.6.6}$$

So we see that the adiabatic recovery temperature in air is much closer to the stagnation temperature than to the static temperature of the local outer flow, especially if the boundary layer is turbulent. These relationships should be good over the range of Prandtl number from 0.1 to 3.0, and provide a very convenient way to make rough estimates of the adiabatic wall temperature.

Consistent with the analyses that led to Equations 4.6.3–4.6.6, a reasonable approximation to the density and temperature profile for laminar or turbulent flow and an adiabatic wall is

$$\frac{\rho_e}{\rho} = \frac{T}{T_e} = 1 + \frac{\gamma - 1}{2} r M_e^2 \left(1 - \frac{u^2}{u_e^2} \right), \tag{4.6.7}$$

where r is given by Equation 4.6.5 for laminar flow or Equation 4.6.6 for turbulent flow. A typical temperature profile for turbulent adiabatic-wall conditions according to this equation is illustrated in Figure 4.6.2.

The increase in temperature through the boundary layer is accompanied by a reduction in density, which increases δ^* and reduces θ, so that the shape factor H is increased. Whitfield (1978) proposed a correlation for this effect for both laminar and turbulent flows up to $M_e = 3$, in terms of the kinematic shape factor H_K (H computed with the density ratio omitted from the integrals):

$$H = H_K (1 + 0.113 M_e^2) + 0.290 M_e^2. \tag{4.6.8}$$

A plot of this relationship in Figure 4.6.3 illustrates how a given velocity profile with a given H_K results in significantly higher values of H at high Mach numbers. The increase in temperature also tends to reduce C_f. An approximation for this effect in the case of a flat plate was seen in Equation 4.3.7 and Figure 4.3.4.

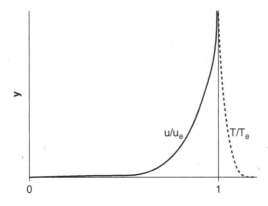

Figure 4.6.2 Velocity and temperature profiles for an adiabatic turbulent boundary layer in air at $M = 1$, according to Equation 4.6.7

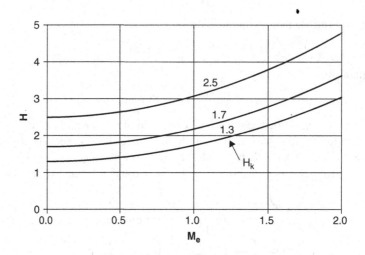

Figure 4.6.3 Effect of Mach number on the shape factor *H* for adiabatic-wall boundary layers, according to Equation 4.6.8

4.7   Effects of Surface Roughness

Distributed surface roughness can influence an aerodynamic flow in different ways depending on the situation: It can affect the location of laminar-to-turbulent transition, and it can alter the skin friction of a turbulent boundary layer and through combinations of these, it can alter the development of the boundary layer and the location of subsequent separation. In Section 4.4.1, we briefly discussed the effects of Reynolds number and roughness on transition, and in Section 6.1.6, we'll see how these can in turn affect the separation location and the pressure drag of spheres and cylinders. The maximum lift of an airfoil or wing also depends on separation, as we'll see in Section 7.4.3, so that roughness can affect maximum lift as well as drag. In Section 6.1.8, we'll look in some detail at how surface roughness can increase turbulent skin friction, which can influence total drag even when there is no change in transition or separation. This type of roughness effect is quite common in applications, caused, for example, by rough materials or paint, or by dirt or other contamination. Then in Section 6.2.2, we'll look at methods for using empirical data on roughness effects measured in simple flow situations for drag estimation in more-general flows. Riblets are an example of another sort of surface texture, one that would be applied purposely to reduce turbulent skin friction, and we'll discuss them in Section 6.3.2.

5

General Features of Flows around Bodies

In Chapter 4, we looked in considerable detail at the physics of the flow closest to the surface of a body, in the viscous boundary layer. In this chapter, we explore the generic characteristics of the entire flowfield in external flows around bodies, in preparation for more detailed discussion of specific aspects such as drag and lift in following chapters. While the basic physical laws (the NS equations: Navier-Stokes) govern all of the flow phenomena we'll look at, the equations by themselves won't help us much in our quest for intuitive understanding. As we saw in Chapters 2 and 3, there is just too much of a gap between the simplicity of the local physical laws and the potential complexity of the flow phenomena that they govern for us to be able to get much from the equations by mental exercise alone. Our discussion will therefore rely heavily on knowledge of phenomenology, that is, what actually happens in real flows. But we'll try to get beyond just describing what happens and understand why things happen the way they do. Some of the flow features we'll consider will seem at first to be trivial or obvious, but I'll ask you to bear with me. We'll find that explaining the "why" of even the most obvious flow features can be surprisingly tricky.

For purposes of this discussion, we'll put ourselves in the body-centered reference frame and assume the flow can be treated as steady. The unsteadiness in turbulent boundary layers and wakes will usually not be of direct interest to us because of the small length scales and short time scales involved. For most of the flow phenomena we'll look at, we'll be concerned with only the time-averaged effects of turbulence, which, as we saw in Section 3.7, are essentially the same as if there were local increases in the coefficients of viscosity and thermal conductivity. In terms of time-averaged effects, turbulence will have a qualitative effect on a flow only when it affects the pattern of boundary-layer separation, an issue we'll return to several times in the remainder of the book.

In Chapter 3, we discussed a general flow structure that we could deduce fairly directly from the local physics, that is, that a viscous boundary layer will develop close to the surface of a body, and the flow outside the boundary layer will tend to remain irrotational. Now, we'll try to understand features of global flowfields that are not easily inferred from the local equations.

Understanding Aerodynamics: Arguing from the Real Physics, First Edition. Doug McLean.
Images and Text: Copyright © 2013 Boeing. All Rights Reserved. Published 2013 by John Wiley & Sons, Ltd.

5.1   The Obstacle Effect

The main thing that characterizes all flows around bodies is that the body acts as an obstacle. It is obvious that flow cannot pass through a solid body unobstructed and must therefore move aside to flow around the obstacle. This general response is something we all take for granted, but the detailed mechanisms behind it are not so obvious.

Primitive intuition tends to imagine fluid particles flying toward the body without interacting with each other and hitting the front of the body unimpeded, like a hail of bullets, as illustrated in Figure 5.1.1. Even Newton developed theories of drag and lift based on such a model. Two major issues are not satisfactorily resolved by the "bullet model": (i) what happens to the particles after they collide with the body and (ii) what happens behind the body. With the bullet model, we might expect particles to just pile up on the front surface of the body or to bounce away from the body in all directions, never to interact with the body again. We would also expect to see a vacuum behind the body. Neither of these things happens in common aerodynamic flows. Of course, the root of the problem with the bullet model is in its assumption that fluid particles don't have random motion relative to the general flow direction and don't interact with each other. In the real world, the continuum behavior that results from frequent molecular collisions leads to a flow structure that begins to deviate from the bullet model well ahead of the nose of the body and differs drastically from it behind the body.

The real behavior of a continuum flow encountering an obstacle can be seen in nature, as shown in the sketch of flowing water by Leonardo da Vinci in Figure 5.1.2. What Leonardo's sketch shows, of course, is not the streamlines of the flow, but the pattern of waves on the water surface, but it is still clear that the flow ahead of the obstacle is reacting to the obstacle's presence, and the flow is filling the space behind the obstacle.

To maintain its integrity as a continuum and to obey the continuity equation, the flow must react ahead of the body and move aside. After it passes the body, in most cases, fluid from the main flow fills the space behind the body, though the flow path by which it gets there may be complicated, as we'll see later. In high-speed liquid flow, cavitation or ventilation may take place, leaving behind the body a cavity that is empty of liquid and filled only with gas, but this is a special case we'll not pursue further.

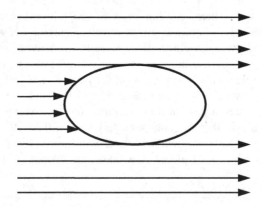

Figure 5.1.1   "Bullet model" for flow approaching a solid body

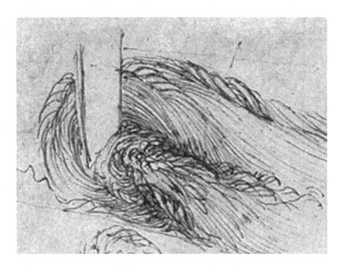

Figure 5.1.2   Water flowing around an obstacle as visualized by Leonardo da Vinci. Detail from a drawing circa 1510–1513, *A seated old man, and studies and notes on the movement of water.* Royal Collection Trust / © HM Queen Elizabeth II 2012, used with permission

A simple idealized case that illustrates the key features of the usual basic flow pattern is the 2D incompressible potential flow around a circular cylinder shown in Figure 5.1.3a–d.

First examine the streamline pattern in Figure 5.1.3a. The flow reacts ahead of the body by diverging away from the centerline to go around it, as we can imagine it must also be doing in Leonardo's sketch in Figure 5.1.2. After it passes the widest part of the body, the flow converges toward the centerline to close in again behind the body. The inviscid streamline pattern in Figure 5.1.3a exaggerates the suddenness of the reconvergence, or "closure" of the flow, but all flows eventually close behind a body, and we'll defer the details to later in the discussion.

Now examine the distributions of velocity magnitude and pressure coefficient C_p in Figure 5.1.3b, which can be shown as a single set of contours because the flow is assumed inviscid, Bernoulli's equation applies, and each value of the pressure corresponds to a single value of the velocity magnitude. Positive C_p corresponds to low velocity magnitude, and negative C_p corresponds to high velocity magnitude.

Keep in mind that the cause-and-effect relationship between the velocity and the pressure is circular, as I've stressed several times in the preceding chapters and will repeat several more. The region of high pressure at the front of the body coincides with a slowing of the flow and the start of the divergence of the flow around the body. The flow along the central streamline reaches stagnation pressure and zero velocity where it hits the nose of the body at the *stagnation point.*

The areas of low pressure at the "flanks" of the body speed the flow up again, actually overspeeding it by a factor of 2 in this example. The lateral pressure gradients in this region reverse the divergence away from the centerline and cause the flow to begin to converge toward the centerline so as to fill in behind the body. Note that the high velocity at the flanks coincides with pinching together of the streamlines in Figure 5.1.3a, in keeping with the 1D subsonic-flow relationship between velocity and streamtube area. Finally, the high

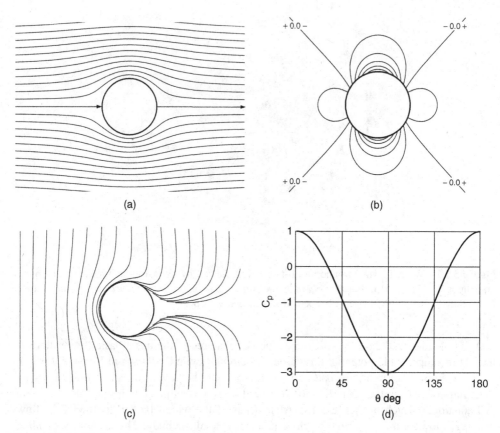

Figure 5.1.3 2D incompressible potential flow around a circular cylinder. (a) Streamlines. (b) Contours of constant pressure (Cp interval = 0.5) and constant velocity magnitude. (c) Timelines. (d) Surface pressure distribution

pressure at the back of the body coincides with a slowing of the flow and the stopping of the convergence.

Although the body imposes no hard constraint on the flow area, the wide part of the body acts as a sort of constriction, and the flow responds by speeding up to squeeze past. This streamtube-pinching effect is not a kinematic necessity, but a dynamic one. There are kinematically possible flow patterns in which there is no streamline pinching around the wide part of the body, like the pattern sketched in Figure 5.1.4. (In 2D, the pushing out of the streamlines around the body would have to extend unattenuated to infinite distances from the body.) Such a flow pattern is not forbidden by kinematic considerations or by conservation of mass, but it does not satisfy the momentum equation. One way to see this is to look at what happens in the flow crossing a vertical line above the crest of the body. All of the streamtubes crossing this line have the same area and therefore the same velocity and pressure. So the pressure is constant in the cross-stream direction, which is not consistent with the flow curvature.

We'll consider these issues again in connection with the streamtube-pinching explanation for the lift on an airfoil in Section 7.3.1.5.

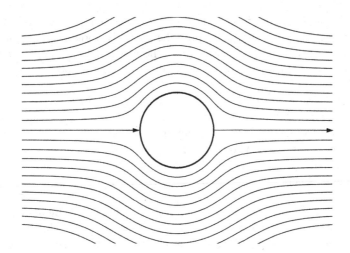

Figure 5.1.4   A kinematically possible flow pattern in which the pushing out of the streamlines around a 2D body extends unattenuated to infinite distances from the body, and there is no streamline pinching around the wide part of the body. Such a flow pattern is not forbidden by kinematic considerations or conservation of mass, but it does not satisfy the momentum equation

Adding timelines to the picture, as shown in Figure 5.1.3c, also provides interesting insights. The delay of the flow that passes through the low-velocity region around the front stagnation point is clearly seen. The high velocity around the flanks offsets the delay somewhat, but there is further delay at the rear of the body, and the final result downstream of the body is a net delay of the flow that passed near the body, and that delay is never recovered, even in this inviscid example.

The nonuniform delay means that parcels that pass close to the body experience a permanent displacement relative to their former neighbors and that there is thus a permanent deformation of the fluid. In a potential flow like this one, the deformation is a result solely of the passage of the flow through the pressure field surrounding the body, and as we saw in Section 3.8.2, such an interaction cannot introduce vorticity, or rotation. Thus each fluid parcel, step-by-step in its passage through the field, experiences purely irrotational deformations like that illustrated in Figure 3.3.5b. A global deformation pattern like that in Figure 5.1.3c is a result of an accumulation of such elemental local deformations. In a viscous flow, viscous forces in the boundary layer and wake contribute additional local deformations of the shearing type illustrated in Figure 3.3.5c.

A striking feature of the timeline pattern is the blank space behind the body, which has not yet been reached by any of the timelines. This void in the timeline pattern reflects the fact that the closer a fluid parcel is to the stagnation streamline, the longer the delay it experiences. In fact, on the stagnation streamline itself, the delay just to reach, or to leave, a stagnation point is logarithmically infinite. Thus timelines that start a finite distance upstream of the body never reach the front stagnation point, and never touch the body or any point on the aft stagnation streamline.

The corresponding pattern in the viscous flow at R ≈ 140 was shown in Figure 3.3.2c. The pattern is similar in front of the body, but separation from the flanks of the body, and the resulting viscous wake, prevent the timelines from closing in as much in the nearfield behind the body. Looking downstream in the real flow, it appears that the flow that was delayed passing close to the body is swallowed up by the unsteady vortex street in the wake, and the flow outside the wake doesn't appear to have been delayed on average. Denker (1996, web site) shows a nice illustration of the delay effect in connection with the flow around an airfoil.

The pressure distribution along the surface for the potential-flow case is shown in Figure 5.1.3d and is seen to be symmetrical front to back. This is consistent with the fact that this flow has zero pressure drag, as do all 2D potential flows, something we'll discuss further in Section 5.4. In flows with viscosity, the pressure rise at the back, often referred to as the aft "pressure recovery," is usually significantly reduced or completely absent, and the pressure drag is generally nonzero. Pressure drag in the viscous case is something we'll discuss further in Section 6.1.6.

Note that in examining flowfields like this it can be instructive to think of the interaction between the pressure and the velocity in two different ways. In what I would call the "integral" view, based on Bernoulli's equation, one makes use of the negative relationship between the pressure and the velocity magnitude. The "differential" view, based on the momentum equation, is more complicated but also more powerful because it can explain changes in flow direction as well as velocity magnitude. In this view, one looks at the pressure gradient and relates it to the flow acceleration in a vector sense. It is worthwhile to try taking both of these views in looking at the flowfield depicted in Figure 5.1.3.

Now we've seen the simplest general flow pattern associated with flow around an obstacle: high pressure, slowing down, and divergence as the flow approaches the front of the body; low pressure, speeding up, and pinching together of the streamlines as the flow squeezes past the widest part of the body; and slowing down again as the flow closes around the back. Around more-complicated body shapes, the pattern will become more complicated, but overall it will still have some of this general character.

Note the contrast between the continuum flow and the flow according to the "bullet model" with regard to how "much" of the flow comes into contact with the body. In the bullet model, all the particles in a wide swath hit the front of the body, while in the 2D continuum flow only the central streamline touches the body. This brings us to the next major aspect of continuum flows: how the flow attaches to the surface of the body and leaves it.

5.2   Basic Topology of Flow Attachment and Separation

An important aspect of the topology of flows around bodies has to do with what part of the flow contacts the body, what it does while in contact, and how it leaves the body. The part that touches the body is, of course, the flow at the bottom of the boundary layer, the physics of which we discussed in some detail in Chapter 4. The flow that occupies the bottom of the boundary layer must obviously have originated in some part of flowfield upstream and, after leaving the body, will end up in some part of the field downstream. We've already seen that the Newtonian, or "bullet," flow model is totally unrealistic in this regard. In this model, the entire streamtube "swept" by the cross section of the body collides with the forward-facing surfaces of the body, after which we have no realistic option for what happens to it. In a continuum flow, on the other hand, only an infinitely thin filament or slice of the flow coming from upstream can touch the body, something the mathematicians would call a "subset of zero measure."

We'll look at how this infinitesimal portion of the flow attaches to the body and separates from it, first in 2D, and then in 3D. Then we'll consider the topology of the flow patterns associated with it, both on the body surface and out in the field.

5.2.1 Attachment and Separation in 2D

In general, if the flow is to come into contact with the body, a first attachment of some sort and a final separation of some sort are both required. The simplest possible example of attachment and separation in 2D was shown in the flow illustrated in Figure 5.1.3, where the two streamtubes straddling the central streamline split apart and just graze the body. These streamtubes attach to the body at the front stagnation line and separate from it at the rear stagnation line. Once two streamtubes attach in this way in a 2D flow, it is impossible for another streamtube from upstream to touch the body because to do so it would have to cross one or the other of the streamtubes that are already attached.

The first attachment on a smooth surface will generally follow the simple pattern we just looked at in Figure 5.1.3. Separation, however, even in 2D flow, has a wider range of possibilities, determined by the physics of the viscous boundary layer on the surface upstream and by features of the body shape.

The separation pattern we see in Figure 5.2.1 is the simplest possible, in which the two attached streamtubes from upstream rejoin and separate from a single line. In part (a), the single separation line occurs on a smooth surface, a pattern seen only at very low Reynolds numbers, while in part (b), the separation line is anchored by a sharp edge at the rear of the body, as is typical of airfoil flows.

The other common pattern involves two separation lines that bound a "bubble" of recirculating flow that is isolated from the main flow and contains an additional line of attachment, as shown Figure 5.2.2. Topological consistency requires that the recirculating bubble must actually consist of two counter-rotating cells, as shown. In part (a), both separations are from the smooth surface; in part (b), one of the separations is anchored by a sharp edge; and in part (c), both are anchored by edges. Note that when a separation is not anchored by an edge, its location is sensitive to the flow environment upstream, especially the pressure distribution, as discussed in Section 4.1.4.

The flow patterns of Figures 5.2.1 and 5.2.2 illustrate the distinction between so-called "streamlined bodies" and "bluff bodies," as shown in Figure 5.2.3. When a body has a sharp or small-radius tail and is shaped so as to limit the rear separation to a single line as in

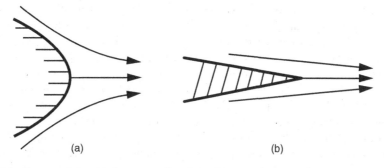

(a)   (b)

Figure 5.2.1 Pattern of the simplest separation from the rear of a 2D body. (a) Smooth surface. (b) Sharp edge

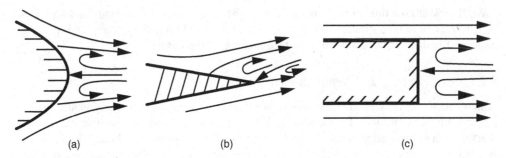

Figure 5.2.2   Patterns with two lines of separation from the rear of a 2D body. (a) Smooth surface.
(b) One sharp edge. (c) Two sharp edges

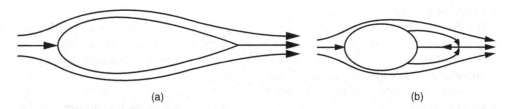

Figure 5.2.3   Streamlined and bluff bodies in 2D flow. (a) Streamlined. (b) Bluff

Figure 5.2.1b, or a practical approximation to it, it is often referred to as "streamlined."
A streamlined body at an unfavorable angle of attack may have a separation bubble of
significant size and thus fail to achieve "streamline" flow, as in Figure 5.2.2b. When the body
closes too abruptly to achieve streamline flow under any high-Reynolds-number condition,
as in Figure 5.2.2a,c, the flow is often referred to as "bluff-body flow." We'll discuss the
pressure drag associated with bluff-body flow in Section 6.1.6.

The above discussion raises an issue of terminology that can be confusing. In common
parlance, the term "separation" is often reserved for situations where the flow separates
"prematurely," before reaching the aft closure point of the body, and the flow patterns of
Figure 5.2.1, because they have no "separation" in this sense, would be referred to as "all
attached." But this use of the terminology ignores the fact that even if the flow doesn't
separate prematurely, it must leave the body somewhere, and it seems logical to refer to
that as "separation" as well. Topologically speaking, the only distinction between the simple
separation patterns of Figure 5.2.1 and patterns like Figure 5.2.2a or b with "premature"
separation is that the latter tend to divide the body surface into multiple regions and require
an additional attachment line, as well as a more complicated flow pattern behind the body.

In our discussion so far, we've really dealt with attachment and separation only in a
topological or geometric sense. We've identified attachment/separation streamlines that lie
in contact with the body over part of their length and are out in the field upstream and
downstream of the body. But it is also intuitively natural to want to think of attachment
and separation in terms of the Lagrangian time histories of fluid parcels that arrive at the
body, come into contact with it, flow along the surface, and then leave. Here we have to be
more careful. When attachment is at a stagnation point, as it is in the flow in Figure 5.1.3a,
infinitesimal fluid parcels on the attachment streamline itself spend infinite time in the

neighborhood of the stagnation point without actually arriving. Furthermore, because of the no-slip condition, fluid parcels on the surface itself can't really be said to be flowing along, and their "transit time" is also infinite. So to see a parcel arrive at the body and then leave, we must follow a parcel that passes at a finite distance to either side of the attachment streamline and doesn't actually touch the body, but only comes very close to it. This point will come up again when we discuss explanations of lift in Section 7.3.1.

5.2.2 Attachment and Separation in 3D

We've seen the simple patterns that attachment and separation can take on in 2D flows. If we look at cross sections of 3D flows, projecting the velocity vector into the plane of the cross section, the general 2D flow patterns we've already discussed will often appear, but the additional dimension makes possible other patterns as well. The velocity field in a cross-sectional cut is a 2D vector field, like a 2D flow, but it doesn't have to satisfy the 2D continuity equation. The flow has the additional dimension in which to move and in which it can vary, which opens up more possibilities for topological structure, even within a single cross section. And a single cross-sectional view will often fail to convey everything that's going on, so that to really understand a 3D flow we must look at the entire 3D field.

In 2D planar flows, a streamline should be thought of as a stream surface, and a streamtube should be thought of as a layer of flow extending to infinity in both directions spanwise. To understand 3D flows, on the other hand, we must keep track of discrete streamlines, and we should think of streamtubes as discrete tubes, not layers. With this in mind, let's look at flow attachment in 3D.

While in 2D planar flows an infinite stream surface attaches to the body at a stagnation line, in 3D only discrete streamlines can typically reach the body surface. Each of these attaches to the body at a *singular point of attachment*, which is most often a *nodal point of attachment* from which the flow spreads out in all directions along the surface. To "attract" an attachment streamline to attach at a nodal point, something in the local body shape must act, for flow purposes, like a locally pronounced "nose." A less common type of attachment point is a *saddle point of attachment*, of which we'll consider an example in our discussion of Figure 5.2.6b. In any case, common body shapes typically have only small numbers of attachment points, and often only one. All of the flow along the body surface can be thought of as emanating from these singular points of attachment, and the flow on any streamline other than an attachment streamline cannot contact the body, though it may come very close.

Although true flow attachment in 3D is nearly always limited to isolated singular points, there is often another kind of area on the body surface, generally an elongated band called an *attachment line*, where flow attachment also takes place, not in the strict sense, but for all practical purposes.

As an example of a body with attachment lines, consider the fuselage/swept-wing combination in Figure 5.2.4. For now, we'll think of the flow as inviscid and illustrate it schematically in terms the streamlines. Only one outer-flow streamline attaches to this body, at the blunt nose of the fuselage, in a nearly axisymmetric stagnation-point flow. Along the leading edge of each wing is an attachment line, where the flow is lined up along the leading edge. To either side of this narrow band of spanwise flow, the flow diverges rapidly to go over the upper and lower surfaces of the wing.

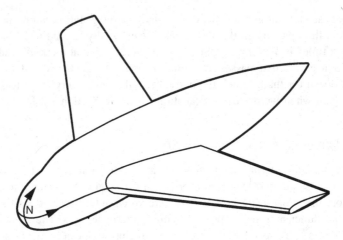

Figure 5.2.4 A fuselage/swept-wing combination with a singular point of attachment at the nose. The "attachment line" at the wing leading edge has the practical attributes of flow attachment but is not an attachment in the strict sense

An attachment line is generally characterized by strong divergence to both sides, and by flow that approaches the surface as if it were headed for a true attachment. But the flow approaching the surface along an attachment line of finite length never reaches the surface; though it often approaches it very closely. So an attachment line of finite length is an area of "effective attachment," not true attachment in the strict sense. An attachment line can be a location of true attachment only if it is infinitely long, as on the leading edge of the idealized swept wing of infinite span, which we'll discuss in Sections 8.6.1 and 8.6.2.

Let's look more closely at a leading-edge attachment-line flow. If we take a cross section of this flow perpendicular to the leading edge anywhere along the span of the wing, we see a flow pattern that is qualitatively the same as the flow around a 2D airfoil with attachment at a stagnation point near the leading edge, as illustrated in Figure 5.2.5. In terms of the velocity component in the plane of the cross section, the boundary layer development looks just like the stagnation-point boundary-layer flow we discussed in Section 4.3.2, and the flow at the edge of the boundary layer is being entrained into the boundary layer as if it were headed for a real attachment to the surface.

But this cross-sectional view doesn't tell the whole story. In a 3D flow, this is not a stagnation point, and there is spanwise flow along the leading edge at velocities that depend on the sweep and other details of the wing and fuselage shapes. And in the cross-sectional view shown in Figure 5.2.5, the transit time for a fluid parcel on the stagnation streamline to reach the surface at the stagnation point is infinite. So given the finite spanwise velocity in the 3D case, it would take an infinite distance along the span for a parcel entering the boundary layer to reach the surface. Thus if we look close enough to the surface in the boundary layer along the leading edge in the finite-span case in Figure 5.2.4, even at the tip, we'll find fluid that arrived there from the outer flow in the vicinity of the nodal point of attachment at the nose of the fuselage, not from anywhere along the wing. So in an ideal sense, the only flow attachment is at the fuselage nose. Of course, at any station along the span of the leading edge, we would have to look extremely close to the surface to find fluid that arrived there from the nose of the fuselage. A very short distance farther from the surface we would find fluid that was entrained into the boundary layer on the leading edge just a short distance upstream (inboard).

We used the term "attachment line" frequently in our discussion of 3D boundary-layer flows in Chapter 4, and we'll use it some more in connection with swept-wing leading-edge boundary layers in Section 8.6.2. This usage is justified because for all practical purposes flow attachment is taking place there, and it is only in the ideal sense discussed above that it is not a real flow attachment. And we use the term even though it is imprecise in another way: On a 3D wing, there is generally no single streamline that qualifies as uniquely defining the attachment line. Defining it as the line above which the flow diverges toward the upper surface and below which the flow diverges toward the lower surface doesn't pin it down exactly. On a wing of finite span, there will generally be a narrow but finite band of flow that satisfies this condition. So it would be more precise to call it an "attachment zone" or "attachment band."

Let's look at two further examples illustrating attachment in 3D. First, an isolated wing of finite span, with zero sweep at the leading edge is shown in Figure 5.2.6a. Like the case

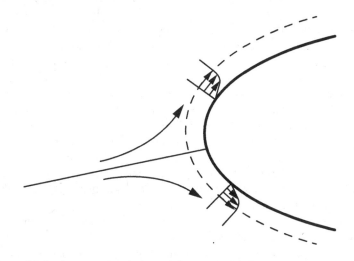

Figure 5.2.5 Cross-sectional view of flow at the leading edge of a swept wing

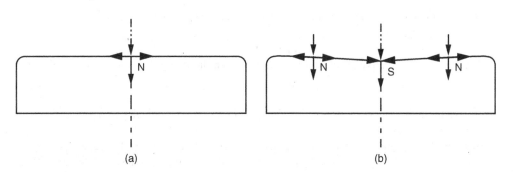

Figure 5.2.6 Further examples of attachment lines on leading edges of isolated wings, seen in plan view. (a) Zero leading-edge sweep. (b) Near-zero, slowly varying leading edge sweep

of the swept wing in Figure 5.2.4, this flow has an attachment line near the leading edge, with a nodal point of attachment where it intersects the symmetry plane. The nodal point on the symmetry plane is the only true attachment point, but the spanwise velocity elsewhere on the attachment line is so small that the flow in the neighborhood of a point anywhere on the attachment line is practically indistinguishable from the flow near a 2D stagnation point of attachment. For all practical purposes, any point on the attachment line qualifies as an attachment point like the nodal point at the symmetry plane.

Now consider another isolated wing, but with the sweep of the leading edge varying slightly from zero, as in Figure 5.2.6b. In this case, there are two nodal points of attachment outboard with spanwise flow diverging very slowly away from them. Between the nodal points, the spanwise flow along the attachment line converges toward another type of singular point, a *saddle point* of attachment. In this situation, we have three singular points of attachment, but every other point on the attachment line is practically indistinguishable from the singular points.

A further issue regarding attachment lines comes to light if we take any one of the bodies in Figures 5.2.4 and 5.2.6 and imagine morphing it gradually into a sphere. As the body becomes more sphere-like, the attachment point remains (or the multiple attachment points merge into one), and the attachment lines cease to exist as distinct entities. But there is no distinct threshold at which an attachment line disappears. We have no definite criterion for deciding when diverging flow constitutes an attachment line and when it doesn't. We can cite obvious qualitative differences between the flow around a sphere and the wing attachment-line flow in Figure 5.2.4: On the sphere, the flow divergence is axisymmetrically distributed (independent of the azimuthal angle), while the wing leading edge has a pronounced local maximum in the divergence. Another possibility would be to reverse the *region-of-origin* definition of 3D separation that we developed in Section 4.1.4 and define an attachment line as a near-discontinuity in the *region of destination* of the streamlines. Note that at typically very small distances to either side of a leading-edge attachment line the flow is headed to the upper and lower surfaces, which are widely separated destinations, practically speaking. But this is also a qualitative, not quantitative, criterion.

Let's summarize what we've learned about the concept of an attachment line. There is often a distinct elongated band on a body surface where flow attachment is taking place for all practical purposes, and the term "attachment line" is a useful shorthand for what is happening. But we've seen that the term is imprecise in three ways:

1. An attachment line of finite length is not an attachment in the strict sense, but only in a practical sense,
2. There is generally no single streamline that defines an attachment line, and
3. There is no quantitative criterion for deciding when diverging flow constitutes an attach-ment line.

We use the term "attachment line" routinely, though the theoretical ground underlying it isn't entirely firm.

So flow attachment to a body takes place through a combination of singular points of true attachment, and lines of near attachment called "attachment lines." Flow separation from a body is similar in that there are *singular points of separation* and *separation lines,* which are locations of separation for all practical purposes. But the similarity between attachment and separation extends only to these basic topological entities. The flow that separates from a body comes from the boundary layer and carries with it vorticity and a deficit in total-pressure, things not generally present in flow attachment. As a result, the flow downstream of separation is typically a very different animal from the flow upstream of attachment.

Let's look at some of the basic flow structures that can result from separations in 3D. In Figure 5.2.7a–c, we have cross sections of flow patterns that can occur on round-cross section bodies at angle of attack, based on observations reported by Han and Patel (1977) and Su, Tao, and Xu (1993). These start as free-vortex sheets emanating from open separation lines S_1 along the flanks of the body. Whether or not the sheets roll up significantly at their free edges depends on the strength of the vorticity in the sheets, which depends on the angle of attack, and on how far the sheets have evolved along the length of the body. Once rollup of the primary vortex sheets has occurred, secondary separations S_2 can appear, as in (b) and (c). Open separations from a smooth surface, as in these examples, can be unstable and move about with time, causing unsteady loads on the part of the body from which the vortex is shed or on structures downstream that may be impacted by the vortex, such as airplane tail surfaces.

The free vortex sheet shed from the side and trailing edges of a lifting wing generally begins to roll up immediately, as in Figure 5.2.7d. The further development of the vortex sheet from a lifting wing is a topic we'll consider in detail in Section 8.1.2, and we'll look at what happens to the vorticity far downstream in Section 8.5.5.

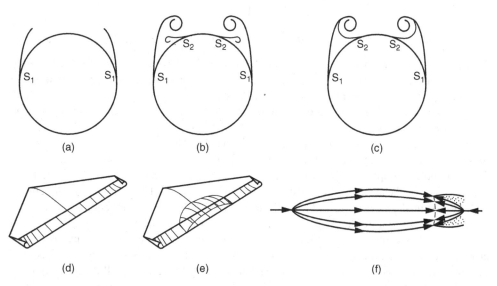

Figure 5.2.7 Flow structures associated with separation in 3D. (a) Free vortex sheets from open separation lines S1 along the flanks of a round cross-section body, either well forward on the body or at low angle of attack, where little or no rollup of the sheet has occurred. (b) Free vortex sheets as in (a), but farther along the body or at higher angle of attack, with some rollup of the sheets, and secondary separations S2. (c) Free vortex sheets as in (b), but where the primary and secondary sheets join to roll up into a single core. (d) Vortex sheet emanating from an open separation at the side and trailing edges of a wing. (e) Vortex sheet enclosing a closed separation bubble on the aft upper surface of a wing. (f) Vortex sheet enclosing a closed separation bubble at the tail of a body with abrupt closure (From Han and Patel, (1977))

A closed separation is generally a closed curve on the surface, and the vortex sheet emanating from it encloses a region of recirculating flow with low total-pressure and high levels of turbulence and lower-frequency unsteadiness. A closed separation line can form as separation moves forward from a wing trailing edge, as in Figure 5.2.7e, or can form at the tail of a body with abrupt closure, as in Figure 5.2.7f. The latter is the 3D counterpart to the 2D bluff-body separation of Figure 5.2.3b.

In many cases, as body angle of attack changes, for example, an open separation can change into a closed separation, and in other cases the two types of separation coexist. Thus the distinction between streamline behavior and bluff-body behavior isn't as simple in 3D as it was in 2D.

The flow that first attaches to a body is "clean" irrotational flow arriving from the farfield upstream. The flow that separates from the body, on the other hand, forms a vortical, usually turbulent shear layer. The body surface upstream of any separation can be thought of as being "wetted" by the outer flow, with nothing but a relatively thin attached boundary layer between the surface and effectively inviscid flow. The surface on both sides of an open separation line is wetted by outer flow in this way. A closed separation line, on the other hand, divides a surface upstream that is wetted by outer flow from a surface downstream that is wetted by the flow in the "separated region."

The patterns on the surface formed by the skin-friction lines, with their various isolated singular points and their "lines" of attachment and separation, obey the topological rules of critical-point theory, which is our next topic.

5.2.3   Streamline Topology on Surfaces and in Cross Sections

Much of our thinking about 3D flows is in terms of 2D views of the flowfield or of what goes on at the surface. Though we've seen that 2D views are limited in the information they can convey, they play a prominent role in our thinking because our graphics capabilities and our imaginations are better at dealing with things in 2D. We'll look at graphical visualization issues further in Section 10.7.

Two of the 2D views that we often take are of the skin-friction distribution on the surface of a body and of velocity vectors projected into cuts (imaginary surfaces) out in the flowfield. These are both examples of 2D vector fields defined on 2D surfaces. For purposes of studying the topology of these fields, we construct "streamlines" parallel to the vectors' directions, and we assign flow directionality, usually indicated graphically by directional arrows, but we can generally ignore the magnitudes of the vectors. For flowfields that are solutions to the NS equations, these "streamline fields" (really direction fields) are continuous except at isolated singularities, or *critical points,* at which the direction is undefined because the vector magnitude is zero. Continuity means that streamlines don't collide with adjacent streamlines. There are no singular lines of attachment or separation unless they happen to line up with a sharp edge of the body shape. Attachment and separation lines on smooth surfaces are not singular as in Figure 4.2.11b, but are indistinct as in Figure 4.2.11a.

Many interesting things can happen in a streamline field, but topologically speaking, the only things that distinguish a given field are the arrangement of its critical points and the character of its behavior on the boundary of the domain. The overlying topological structure of a field doesn't tell us everything about it, but it can be helpful to our understanding nonetheless. Being able to spot the critical points provides a useful way of thinking about a flow, and analyzing the critical points provides one way of testing the realizability of flow models. Knowledge of critical points can also be helpful in analyzing the oil-flow-visualization images often obtained in wind-tunnel testing, as we discussed in Section 4.1.2. Some of the most interesting things in a flowfield happen near the critical points, but because the surface shear stress is low there (remember that the shear stress is zero at the critical point itself), the oil tends to pile up and not resolve the details well. Understanding the characteristics of the various types of critical points can enable an observer to "fill in" details that are not always clearly visible.

Mathematical topology and critical-point theory (see Kaplan, 1958; Flegg, 1974) define certain rules that patterns of critical points must obey. Using these rules to analyze patterns in fluid flows is a relatively recent development (Perry and Fairlie, 1974; Hunt *et al.*, 1978). In the rest of this section, we'll look at a variety of situations that are of interest in studies of fluid flows, and the rules that apply. Much of the discussion follows that in Brune (1983).

The types of critical points are illustrated in Figure 5.2.8, from Brune. Of all of these types, we will usually have to consider only *nodes,* at which the flow either converges from all directions or diverges in all directions, and *saddles,* at which flow converges in one set of opposing directions and diverges in the other (pattern to the lower left of the origin in Figure 5.2.8). Critical lines, or degenerate critical points, tend to appear only in 2D situations (including axisymmetric). On 3D bodies, critical lines are all but impossible. Even in cases like those shown in Figure 5.2.6, there is no true degenerate line even though much of the attachment line is close to degenerate. Now let's look at some of the situations frequently encountered in 3D flows and at the topological rules that apply.

5.2.3.1 Skin-Friction Lines on a Complete Body Bounded by a Closed Surface

We'll start with the simplest case. In Figure 5.2.9, we have a simple wing-body combination and an isolated wing, each with its combination of nodes and saddles shown. These are both simply connected solid bodies, and their bounding surfaces are simply-connected closed surfaces, for which the topological rule is

$$\Sigma N - \Sigma S = 2, \qquad (5.2.1)$$

where $\sum N$ is the number of nodes and $\sum S$ is the number of saddles. We can see that the rule is satisfied in both cases in Figure 5.2.9.

Now consider a body that has holes through it. We're not talking here about holes just through the skin, but holes that go through the body, like ducts. The skin of the combined outer surface and ducts is still a single closed surface, just no longer simply connected. When holes are added, the flow patterns necessarily become more complicated, and the topological rule becomes

$$\Sigma N - \Sigma S = 2 - 2g, \qquad (5.2.2)$$

where g is the topological *genus* of the body. The genus is defined as the largest number of nonintersecting closed curves that can be laid out on the surface without dividing the surface into separate regions. The simple way to think of the genus is that it is the number of holes through the body. For example, the simple bodies of Figure 5.2.9 have $g = 0$, so that Equation 5.2.2 reduces to Equation 5.2.1. A torus, or donut, has $g = 1$.

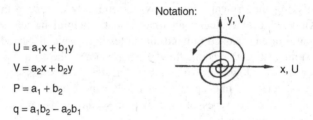

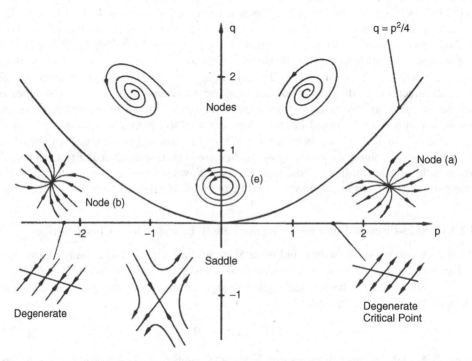

Figure 5.2.8 Types of critical points in a 2D direction field, where p and q are parameters that are used to categorize the form of the direction field around the critical point and are related to the "velocity" field in a coordinate system centered on the critical point, according to the diagram and equations at the top of the figure. (From Brune, 1983)

According to Equation 5.2.2, adding a hole through the body must add two more saddles than it does nodes. This is illustrated in Figure 5.2.10, which shows that the simplest possible effect of adding a duct through a fuselage-like body is to add two saddle points, one of attachment and one of separation. Other examples of opening holes in a body, such as deploying a slotted flap, which can open multiple holes between the main wing, the flap, and the support brackets, often add even more complexity.

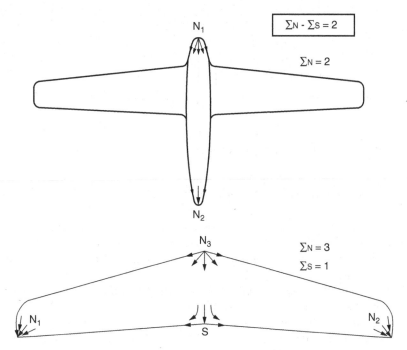

Figure 5.2.9 Examples of critical points on simply connected bodies bounded by closed surfaces. (From Brune, 1983)

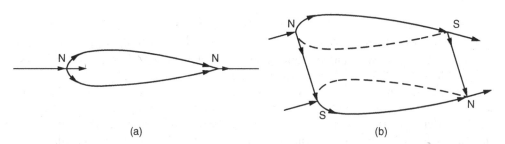

Figure 5.2.10 Simple example of adding a hole through a body and how it must add at least two saddle points. (a) Fuselage-like body with no hole, $g = 0$. (b) Fuselage-like body with duct, $g = 1$

5.2.3.2 Skin-Friction Lines on a Portion of a Body Surface

With more-complicated body shapes and flowfields, dealing with the entire surface as we did above can be difficult, and it is often more convenient to consider only a portion of the surface. The simplest option is to isolate a portion of the surface by drawing a closed curve around it in such a way that the curve doesn't pass through any critical point of the flow pattern. An example of this would be to take just a portion of the upper surface of

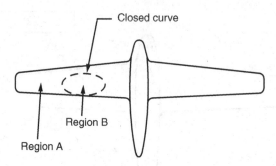

Figure 5.2.11 Isolating a portion of a surface, in this case part of the upper surface of the wing of a simple wing-body configuration (g = 0). (From Brune, 1983)

a wing, as in Figure 5.2.11. Such a portion is not a closed surface, so that before we can apply topological rules, we must determine the *index* of the surface.

The index of a surface is determined by what goes on around the closed contour at the boundary. It's a little like circulation, but instead of being the line integral of the "velocity" vector, it just counts the integral number of rotations the "velocity" vector goes through as we take one loop around the contour. This is illustrated in Figure 5.2.12a. When the boundary is circuited in the counterclockwise direction as shown, the index I is the number of counterclockwise rotations of the "velocity" vector. Examples of simple flow patterns and their indexes are shown in Figure 5.2.12b.

Once we know the index of a surface, based on the flow pattern at its boundary, the topological rule is

$$\Sigma N - \Sigma S = I. \tag{5.2.3}$$

The simplest example illustrating this rule would be to imagine that the surface portion in Figure 5.2.11 is covered by unidirectional flow, in which there would be no critical points, and for which we would have I = 0, trivially satisfying Equation 5.2.3.

Let's see how this works for more-complicated flow patterns. In Figure 5.2.13 are schematic sketches of skin-friction lines on a swept wing, showing a progression of features that have been observed in actual swept-wing flows. In each case, we've drawn the boundary near the leading edge, the tip, the wing-fuselage junction, and the sharp trailing edge.

In Figure 5.2.13a, we have a pattern typical of a low angle of attack, with no separation ahead of the trailing edge. This pattern is topologically the same as uniform flow, so it satisfies Equation 5.2.3 trivially. In (b), we've increased the angle of attack, and a closed separation line has moved forward from the outboard trailing edge, but there are no critical points, as before, and Equation 5.2.3 is still trivially satisfied.

The separated region in Figure 5.2.13b is bounded by a separation line, drawn in this sketch so as to be consistent with the *region-of-origin* definition of 3D separation from Section 4.1.4. This pattern illuminates some interesting features of separation on swept wings. Note that there are now two points on the trailing edge where the skin-friction lines run parallel to the trailing edge, and the skin friction perpendicular to the trailing edge is zero. Between these points, the skin friction perpendicular to the trailing edge is negative (denoted in the figure as "reverse ⊥ flow at TE"), and the skin-friction lines cross the trailing edge in the reverse-flow direction. But also note that this region of reversed perpendicular

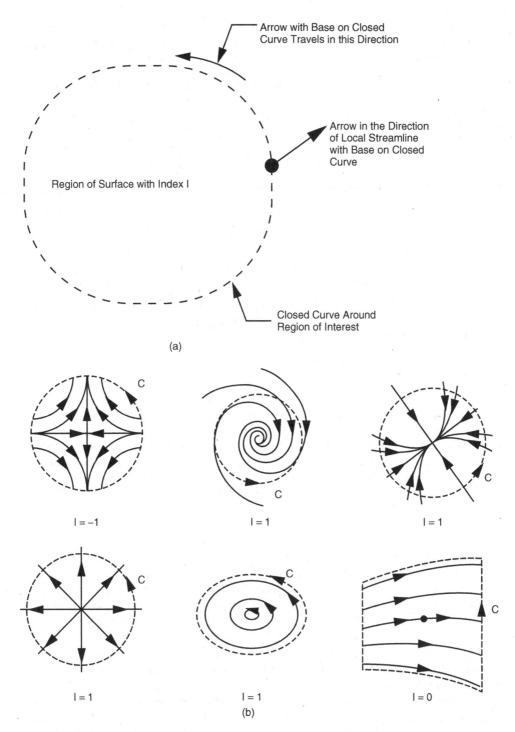

Figure 5.2.12 Illustrations of the index of a flow pattern on a bounded surface. (From Brune, 1983.)
(a) Determination of the index. (b) Examples from Kaplan (1958)

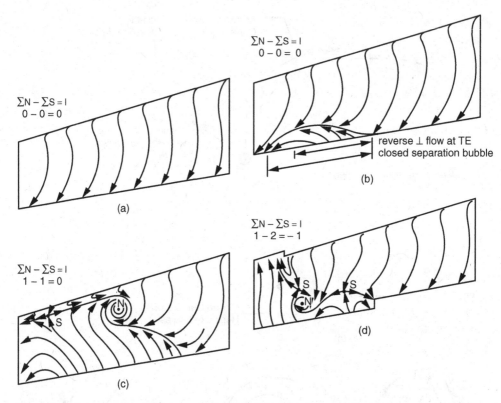

Figure 5.2.13   Examples of patterns of skin-friction lines on a portion of one surface of a swept wing. The boundary of the region has been drawn at the trailing edge, near the junction with a fuselage, the leading edge, and the tip. Schematic representation of progression that has been observed to occur with increasing angle of attack. (a) With no separation forward of the trailing edge. (b) Closed separation bubble appears along outboard trailing edge. (c) Outboard separation jumps forward to the leading edge, and a nodal point surrounded by spiral streamlines appears. (d) Pattern that can occur on the main part of a wing with high-lift devices deployed

skin friction at the trailing edge doesn't encompass the whole spanwise extent of the closed separation bubble. With a bubble of finite extent as sketched here, some outboard portion of the bubble will have positive perpendicular skin friction at the trailing edge. Zero or negative perpendicular skin friction at the trailing edge is an indicator of separation only over an inboard portion of the bubble. At the onset of separation, when the bubble first starts to move forward from the trailing edge, perpendicular skin friction would be zero at a single point on the trailing edge and positive everywhere else. Such a point of zero perpendicular skin friction would thus be an indicator of *incipient separation*. We'll discuss this further in Section 8.6.2.

In Figure 5.2.13c, we've increased the angle of attack further, and part of the closed separation line has moved up near the leading edge. Along the separation line, we now have a saddle point and a nodal separation with a spiral streamline pattern around it, and Equation 5.2.3 is still satisfied.

In Figure 5.2.13d, we show the main element of a wing that has part-span high-lift devices deployed, a leading-edge slat and a trailing-edge flap. The devices themselves are not shown in the drawing. These are slotted devices that enhance maximum lift, as we'll discuss in Section 7.4.4. Here the slat covers more of the span than the flap, which is typically the case. One of the patterns that can develop is shown here, a separation pattern with flow reversals at both the front and rear portions of our boundary, for which we have $I = -1$. The pattern satisfies Equation 5.2.3 with one node and two saddles.

5.2.3.3 Streamline Patterns Based on Velocities Projected into 2D Cuts of the Flowfield

Now let's look at examples of flow patterns in 2D cuts out in the field. First, we'll take the simplest case, a cut with no boundary (a cut that doesn't cut the body, and goes to infinity in two directions). This is not a closed surface, and even though it doesn't have a closed outer boundary, we must determine its index in order to establish the topological rule. Most external flows are close to uniform in the farfield, and if we follow any closed contour far from the body, we'll find $I = 0$, so that the rule is

$$\Sigma N - \Sigma S = 0. \tag{5.2.4}$$

The example in Figure 5.2.14, with one node and one saddle, satisfies this rule.

When a cut goes through a cross section of the body, we will still generally have $I = 0$ for the farfield flow behavior, but we can have nodes and saddles on the boundary at the body surface. The nodes and saddles at the solid surface are called *half-nodes* and *half-saddles*, which we'll designate by N' and S'. Examples of half-saddles can be seen in Figures 5.2.15 and 5.2.16. The number of critical points on the solid surface depends on how many separate

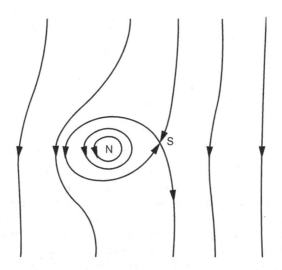

Figure 5.2.14   Combined node and saddle in an unbounded field. (From Brune, 1983)

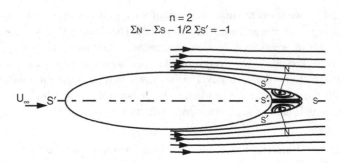

Figure 5.2.15 2D flow pattern from a Navier-Stokes solution for flow past an ellipse. (From Brune, 1983)

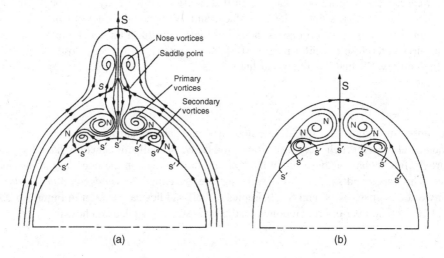

Figure 5.2.16 Postulated flow patterns in cuts through flows around blunt and sharp cones at angle of attack in supersonic flow, based on pitot-tube measurements. From Peake and Tobak, (1980), published by NASA. (a) Blunt cone (b) Sharp cone
Note: For both cones, a half-saddle point S′ also exists on windward ray. Sum of singular points satisfies (Equation 5.2.5)

pieces of the body are cut, and the rule is

$$\Sigma N + 1/2\Sigma N' - \Sigma S - 1/2\Sigma S' = 1 - n, \qquad (5.2.5)$$

where n is the connectivity of the 2D domain of the cut, which is just the number of separate pieces of the body that are cut, plus one. Thus in the common situation in which the body is cut only once, we have n = 2, and the RHS of Equation 5.2.5 is −1. Figure 5.2.15, from Brune (1983), shows the separation bubble that arises in a computational fluid dynamics (CFD) solution for the flow around an ellipse and how the whole pattern satisfies Equation 5.2.5 for n = 2. The separation points near the tail of the body are half-saddles (S′), and there are two recirculating cells (N) within the separation bubble.

In a 2D flow such as the one in Figure 5.2.15, a node will have closed streamlines, like node (e) in Figure 5.2.8, and is called a *center node*. A center node typically exists only

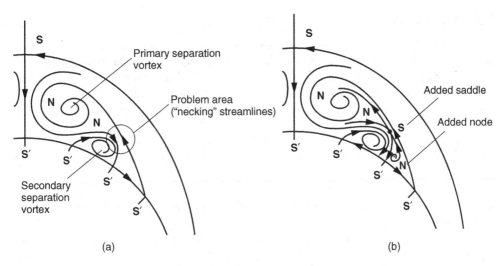

Figure 5.2.17 Detail from Figure 5.2.16b and a proposed modification to make it more physically consistent. (a) Circle added to the original pattern indicates "neck" flanked by flow in opposite directions. Pattern does not violate topological rules, but is physically questionable. The "neck" region is likely to contain a point of zero velocity. (b) Modified pattern with an additional saddle in the "neck" region and additional node to' satisfy the topological rule

in a 2D flow. It is an example of what is called a *transitional flow pattern* because it is on the boundary between two other types of patterns, the converging and diverging spiral nodes to the left and right of the vertical axis in Figure 5.2.8. It generally takes only a small change in the flow conditions, for example, the introduction of a small amount of three-dimensionality, to change a transitional flow pattern into one of the neighboring forms. 3D examples in which nodes take on the spiral form are shown in Figure 5.2.16, cuts through flows around blunt and sharp cones at angle of attack in supersonic flow. These examples also satisfy Equation 5.2.5 with n = 2.

The streamline patterns of Figure 5.2.16 were drawn based on actual flows, and they satisfy the topological rule. But there is a probable physical inconsistency in one of the details of the patterns, as drawn. Figure 5.2.17a shows a detail from one side of Figure 5.2.16b, with a circle superimposed to indicate the problem area: a narrow "neck" between two streamlines going in opposite directions. In effect, the secondary separation vortex with clockwise rotation has pinched a larger region of counter-clockwise rotation nearly into two parts: the primary separation vortex on the upper left, and a smaller region to the right, with the narrow "neck" in between. The "neck," with opposing velocities on opposite sides, must contain a region of very low velocity magnitude. Whether or not there is an actual point of zero velocity, the "neck" region is at least very nearly singular, and it would be clearer to show it as an actual saddle point, as indicated in the proposed modified pattern in Figure 5.2.17b. Topological consistency then requires adding a node, which would seem to fit best in the small circulating region on the right, as shown. This is an example in which physical intuition combined with knowledge of the topological rule helped clarify the interpretation of a flow pattern.

5.3   Wakes

In Section 5.2.2, we noted that the flow separating from a body carries vorticity and a deficit in total-pressure, and that downstream of the tail of the body this vortical flow becomes the *wake*. Wakes take on a wide variety of forms. The wake behind a lifting body in 3D carries strong streamwise vorticity and is called a *vortex wake*, something we'll discuss its own right in some detail in Sections 8.1.2 and 8.5.5. Wake flow not associated with lift in 3D is referred to as a *viscous wake*, not because the rest of the flow is not viscous, but because the wake is the part that has felt significant direct effects of viscosity. Vortex and viscous wakes are overlapping categories, and there is no rigorous way to assign a given wake to one or the other.

Wakes are practically always turbulent, and through turbulent diffusion they tend to spread out downstream of the body more rapidly than a laminar wake would. A viscous wake without significant streamwise vorticity spreads rapidly and becomes very diffuse, and the mean-velocity deficit decays to very low levels, as illustrated schematically in Figure 5.3.1. The cross-sectional outlines of wakes of different types evolve in distinctive ways. In 2D planar flows, wakes spread perpendicular to the general plane of the body, as shown in Figure 5.3.2a. In axisymmetric flows, wakes remain round as they spread, as in Figure 5.3.2b. A 3D wake with no streamwise vorticity would be expected to lose its distinctive shape and become round as it spreads, as in Figure 5.3.2c. But most 3D wakes

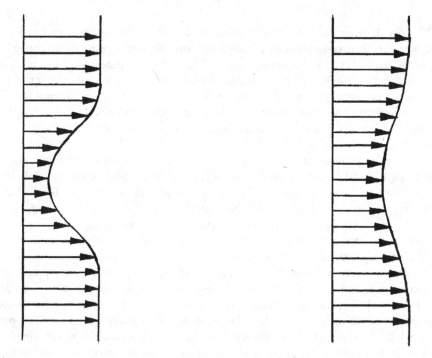

Figure 5.3.1   Schematic illustration of the spreading of a viscous wake and the associated decay of the mean-velocity deficit

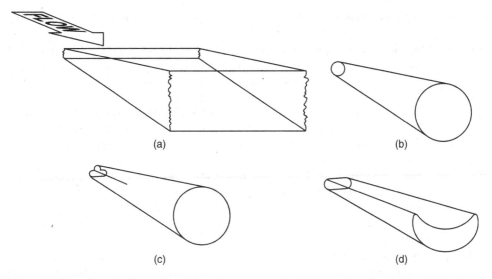

Figure 5.3.2 Schematic illustration of how the cross-sectional shapes of wakes evolve. (a) 2D planar wake. (b) Axisymmetric wake. (c) Initially 3D wake with no streamwise vorticity. (d) Initially 3D wake with streamwise vorticity and associated distortion

probably have enough streamwise vorticity that they become more distorted, not less, as they spread, as in Figure 5.3.2d.

Strong streamwise vorticity leads to complex changes in the cross-sectional shape of a wake as it flows downstream. The streamwise vorticity shed from the trailing edge of a lifting wing, for example, soon rolls up into distinct vortex cores. The vortex motion then inhibits turbulent mixing and slows the further spreading of the wake, so that the distinct vortex cores and the large velocity disturbances associated with them persist much farther downstream than disturbances do in a purely viscous wake. We'll look at this type of flow in detail in Sections 8.1.2 and 8.5.5.

The form a wake takes in the nearfield depends on whether the flow is of the bluff-body type or the streamlined type. Closed separation in the bluff-body case usually leads to a time-averaged (mean) flow characterized by multiple recirculating cells that persist into the near wake, as in Figures 3.3.1 and 5.2.15. At high Reynolds numbers, however, the wake flowfield is generally very unsteady. A bluff-body wake that is significantly thicker than the boundary layers that feed it is extremely unstable, which usually leads to periodic shedding of vortices that span nearly the whole thickness of the wake. As these large-scale structures form and move downstream, instantaneous flowfields look very different from the time average.

The classic example of this kind of flow is the wake behind a circular cylinder in 2D, in which a train of vortices of alternating sign (a von Karman vortex street) is shed into the wake. The vortex street is very effective at mixing, and the wake grows rapidly in thickness within the first few diameters downstream of the cylinder. The separation points on the flanks of the cylinder oscillate fore and aft with opposite phases, locked with the vortex shedding, but this "nodding" of the separations isn't essential to the shedding. Even if the separations were anchored in place, the separated wake would be unstable, and shedding would occur.

In bluff-body flows, patterns of vortex shedding tend to remain qualitatively similar over wide ranges of Reynolds number, at least in the nearfield. We looked at several views of the circular-cylinder flow at $R_d = 140$ in Figure 3.3.2, and Figure 5.3.3 shows views of it at $R_d = 50\,000$, from a DES calculation. At both of these Reynolds numbers, the separating boundary layers are laminar, but at the higher Reynolds number there is more small-scale turbulence superimposed on the shed vortex structures, as can be seen in Figure 5.3.3a, in terms of the instantaneous isovorticity surfaces. At the higher Reynolds number the small-scale mixing seems to be quickly smearing out the shed-vortex pattern, while at the lower Reynolds number the vortices look like they will remain coherent for many cycles.

The time-averaged streamline pattern in Figure 5.3.3b provides an interesting contrast. There is no hint of the periodic shedding, just a pair of recirculating cells that look qualitatively like the laminar flow in the range $20 < R_d < 60$, in which there is separation, but the flow is below the threshold of unsteadiness.

The pattern behind a body with true streamline flow is very different from the bluff-body pattern. The mean flow can actually leave a sharp trailing edge smoothly and without recirculation, as illustrated in Figure 4.1.2a. Still, there is turbulence of the usual boundary-layer variety upstream of the trailing edge. When the two boundary layers join and leave the trailing edge, the turbulence is no longer constrained by the solid surfaces, and the turbulence intensity and eddy viscosity increase. The resulting large shear stresses cause the mean-velocity profile to "fill in" rapidly, as shown in Figure 5.3.4, and with the reduction of the velocity defect comes a reduction and "necking down" of the displacement thickness, as indicated by the dashed lines. There can be vortex shedding in such a wake, but in cases with turbulent boundary layers and no significant separated wake, the shedding is typically very weak.

Except in a tiny region around the trailing edge itself, the mean velocity field in a 2D planar wake in a streamline flow obeys the boundary-layer equations, with the boundary

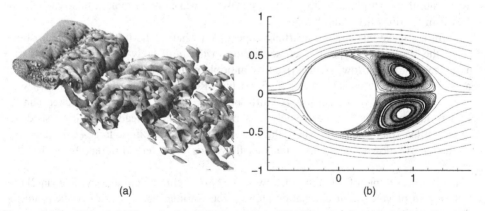

(a)   (b)

Figure 5.3.3 Wake behind a circular cylinder at $R = 50000$ from a Detached Eddy Simulation (DES) calculation. In spite of the appearance of rapid spreading of the unsteady wake, the mean-flow recirculating bubble is relatively compact. Rapid closure of the mean-flow bubble can be thought of as a result of strong mixing. From Travin *et al.* (2000). Used with permission of Springer Science + Business Media. (a) Unsteady isovorticity surfaces. (b) Mean-flow streamlines

conditions changed to reflect the fact that there is no solid surface with a no-slip condition. In viscous-inviscid interaction calculations for airfoils and wings, it is common practice to solve the boundary-layer equations for some distance downstream in the wake, to account for the effect on the pressure distribution of the necking down of the displacement thickness and to provide a better basis for calculating the viscous drag, as we'll see in Section 6.1.4. Because momentum transport in the wake is almost entirely turbulent, the development of the wake has relatively weak direct dependence on Reynolds number.

5.4   Integrated Forces: Lift and Drag

One of our main motivations for developing the discipline of aerodynamics is to know how to predict and control the integrated forces on bodies, and we resolve the forces in ways dictated by what is important in applications. For most purposes it is convenient to resolve the component perpendicular to the farfield flow, which we usually call *lift* even when it isn't oriented vertically upward. Lift is generally important for countering gravity and providing maneuvering forces for flight vehicles, and as part of the interaction that provides propulsion in sailing. The force component parallel to the farfield flow is generally a combination of *drag* (resistance) and *thrust* (propulsion). We'll look further at this distinction and the difficulties it entails in the introduction to Chapter 6.

We get the total aerodynamic force by integrating the distributed stresses exerted by the fluid over the surface of the body. It is generally convenient to resolve the surface stress into a normal component, which in flows that obey the NS equations is just the local hydrostatic pressure, and a shear component, which is just the viscous shear stress. Then the lift (in the **k** direction) is given by

$$L = \int_{S} (\mathbf{k} \bullet \overline{\overline{\tau}} \bullet \mathbf{n} - p\mathbf{k} \bullet \mathbf{n}) \; dS. \tag{5.4.1}$$

We'll look at the corresponding integral for the drag in Section 6.1.3.

In still air, the surface shear stresses are zero, and the normal stress is just the local atmospheric pressure. Under the influence of gravity, atmospheric pressure exerts a vertical hydrostatic buoyancy force on any body with volume. In airplane applications, this is small but not insignificant, and it is often accounted for in careful bookkeeping of airplane weight.

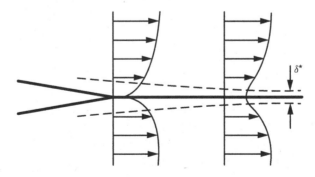

Figure 5.3.4   The filling in of the mean-velocity profile and the necking down of the displacement thickness in the near wake of a body in streamline flow

But gravity-induced buoyancy is usually ignored in dealing with aerodynamic forces, either from theory or from wind-tunnel measurements, which is tantamount to assuming that the contribution from buoyancy doesn't change between the static condition and a condition with flow.

In the integrated force on a closed body, it is only the variations in pressure that matter, not the background level. Under flow conditions, the pressure variations are typically of order one in terms of the usual coefficients normalized by free-stream dynamic pressure, much larger than the shear stresses, which tend to be of order 10^{-3}. As a result, the lift is dominated by pressure differences, with only a negligible contribution from shear. For drag, the relative contributions vary widely, depending on whether the body is bluff or streamlined, a distinction illustrated in Figure 5.2.3. The drag of a bluff body is dominated by pressure differences. A streamlined body also can cause substantial pressure disturbances, but because there are no large separated regions, the pressure distribution is close to what it would be in inviscid flow, and the pressure contribution to drag is small enough that the shear contribution dominates. But for bodies with significant thickness, even if they are streamlined, the pressure contribution to drag is never negligible like the shear contribution to lift. Even for well-designed streamlined bodies, the pressure contribution is rarely less than about 10% of the total drag. For airfoils, the pressure contribution to drag is generally larger than this, as we'll see in Section 7.4.2.

The pressure contribution to the force can be clearly interpreted as lift (Equation 5.4.1) or drag (Equation 6.1.1) only after it has been integrated over the entire surface of a closed body. If the pressure contribution is integrated over only a portion of a body, interpreting the result as "lift" or "drag" is problematic. We'll see the reasons for this when we consider the issue in detail as it applies to drag in Section 6.1.3.

Newton's third law requires that the force exerted by the flow on the body is reciprocated by an equal-and-opposite force exerted by the body on the flow. And the equations of motion require that this force must have manifestations in the flowfield. Generating lift requires setting some of the fluid in motion, as we'll discuss in great detail in Chapters 7 and 8. Drag requires either generating lift in 3D or the presence of the dissipative processes associated with viscosity, as we'll see in Chapter 6.

6

Drag and Propulsion

It is common to define an aerodynamic retarding force on a body moving through a fluid as *drag* and a propulsive force as *thrust*. This definition is unambiguous if we apply it to the flight-direction component of the total aerodynamic force on a complete body. But things are not so simple when we try to apply it to separate contributions to the total, as we often do. In practice we are often interested in vehicles made up of multiple parts and having active propulsion, with drag on some parts and thrust on others. Propulsive effort can be applied in a variety of forms: rotating propellers, flapping wings, complicated processes inside engines, and so on. The practical objective is generally to control the total flight-direction force on the vehicle, for example, to make it zero for steady, level flight or to make it a net thrust for climbing or accelerating flight. However, in assessing the aerodynamic performance of such a vehicle we have a natural inclination and some practical incentives to assess the drag-producing parts and thrust-producing parts separately. Thus we often divide the total flight-direction force into a drag on one portion of the vehicle's surface and a thrust on the other portion. This seems like a reasonable thing to do, and we've all seen the ubiquitous elementary force diagrams showing drag and thrust acting simultaneously on an airplane. But resolving partial contributions to the total drag or thrust in this way raises two potentially major difficulties. First, when a complete configuration is divided into portions, the integrated pressure force on each will often reflect spurious contributions that are not rightly viewed as either "thrust" or "drag," as we'll see in Section 6.1.3. And then even if we ignore this first difficulty, dividing the force in this way is ambiguous because choosing how to divide up the surface of the vehicle is always arbitrary to some extent.

There are some obvious tricks that one might think would allow us to determine separate drag and thrust for a vehicle designed to operate with active aerodynamic propulsion, but they don't work in general. For example, if we simply turned the engine off, wouldn't the total flight-direction force then be the drag? Well, it would be "drag" by definition, but its relevance to the situation with propulsion would be questionable. Parts of the propulsion system (e.g., stopped propeller blades) would now have drag that they wouldn't ordinarily have, and the flow around the whole configuration would be different.

Separating the effects of thrust and drag in the flowfield is problematic as well. In some parts of the field, we have *dissipation,* or *losses*, taking place, and in other parts, we have

Understanding Aerodynamics: Arguing from the Real Physics, First Edition. Doug McLean.
Images and Text: Copyright © 2013 Boeing. All Rights Reserved. Published 2013 by John Wiley & Sons, Ltd.

energy being added to the flow as a result of *propulsive effort* being expended. Dissipative effects tend to contribute to drag, and propulsive effort tends to contribute to thrust, but when both kinds of effects are present in the same flow, as they often are, these connections are not clean and simple. And if we look at the drag or thrust on just part of a body, the connections are often nonexistent. This is especially true of the pressure forces, and in Section 6.1.6 we'll address some misunderstandings that can arise in this regard. In most of our discussion of the basic physics of drag, we'll limit our attention to the passive case, without propulsion. Even then, to avoid errors, we must look beyond simply resolving the forces and come to understand drag in terms of the flow mechanisms responsible for it. Then in Section 6.1.10 we'll take a brief look at the basic physics issues associated with propulsion.

Drag, being parallel to the direction of a vehicle's motion through the local air mass, requires work to be done in the reference frame of the air mass. The energy to do this work can come from combinations of active propulsive effort, gravity, motions of the atmosphere, and the vehicle's kinetic energy. And of course when work is done, it has thermodynamic implications for the flowfield. These work and energy considerations can be tricky to deal with because they look different in different reference frames. We'll touch on this issue again in Section 6.1.3.

Our understanding of drag has a complicated history. Most early work in theoretical fluid mechanics assumed an inviscid fluid. For bodies without vortex shedding or shocks, inviscid theory predicts zero drag, in contradiction to all experiments, a result known as d'Alembert's paradox. Fluid dynamicists knew what the source of the problem was, but devising computationally manageable theories that could predict drag was more easily said than done. For drag in external flows, analytic solutions to the Navier-Stokes (NS) equations were feasible only for simple 2D bodies (cylinders and spheres) in the limit of low Reynolds number. Attempts to treat high-Reynolds-number bluff-body drag in an inviscid framework required empirical input and were not very successful. For streamlined bodies at high Reynolds numbers, a correct understanding finally arrived with Prandtl's boundary-layer theory in 1904. Accurate calculations of the total viscous drag, including the pressure contribution, didn't come until much later. We'll discuss the theoretical issues involved in Section 6.1.6 and look at the particular case of airfoils in Section 7.4.2.

6.1 Basic Physics and Flowfield Manifestations of Drag and Thrust

In this section, we discuss the overall physical process of drag production, looking first at the basic dissipative process internal to the fluid and then at its manifestations in various types of flowfields.

In flows of viscous fluids, when no active propulsive effort of any kind is exerted, a body must experience a *net retarding force* we call *drag*. It has several types of flowfield manifestations, but in general, an overall drag force is a thermodynamic necessity. The viscous shear stresses in the flow result in dissipation into heat, which, in keeping with the first law of thermodynamics, requires work to be done. To do work on the flow (in the reference frame fixed to the undisturbed air mass), the body must exert a net force on the air in the direction of the body's motion, and, in accordance with Newton's third law, the body experiences a force in the opposite direction.

First, let's review some background regarding viscosity and turbulence from previous sections.

6.1.1 Basic Physical Effects of Viscosity

In Section 3.6, we considered viscosity in physical terms and saw that it produces internal stresses in the fluid that alter the flowfield by transferring momentum, dissipating work into heat, and producing and diffusing vorticity. In our understanding of drag, the dissipation of work into heat is crucial because it is the ultimate result of all forms of drag. In Section 3.6, we saw that viscous stresses arise in the fluid only if the rate of deformation is nonzero, and in the discussion in connection with Figure 3.6.1 we noted that in a parcel-centered reference frame the stresses always respond to the rate of deformation in such a way that the work done on a parcel of fluid is positive and is dissipated irreversibly into heat. The no-slip condition at solid surfaces ensures that the fluid will undergo shearing deformation in the boundary layer and thus that shear stress (skin friction) will be transmitted to the body surface and that dissipation into heat will take place in the field.

6.1.2 The Role of Turbulence

In Section 3.7, we considered turbulence of the kind that appears in boundary layers and wakes and saw that its effects on the flow are qualitatively similar to those of viscosity. The transport of momentum by turbulent motions has, by and large, the same effects on the time-averaged flow as additional viscous stresses would and we refer to these apparent additional stresses as *turbulent stresses* or *Reynolds stresses*. The work done locally against the Reynolds stresses is transferred directly into the kinetic energy of the turbulent motions. This *turbulence production* is nearly always positive, like molecular viscous dissipation. Then the turbulent motions themselves contain local, unsteady velocity gradients, which produce molecular viscous stresses, which in turn dissipate the turbulence kinetic energy directly into heat (*turbulence dissipation*). Over most parts of a flow, production and dissipation are roughly in equilibrium, and it is as if the energy dissipated by turbulent stresses were being dissipated directly into heat. Turbulent stresses thus result in dissipation into heat, much as the time-averaged viscous stresses do, and their contribution to the production of drag is qualitatively similar.

Because the basic physical effects of turbulence and molecular viscosity are so similar, in all of the discussion that follows, terms such as "shear stress" and "dissipation" can be taken to refer to either the molecular or turbulent variety. Just remember that in the turbulent case the overall dissipation process involves the two steps of turbulence production and turbulence dissipation.

In Section 3.7, we saw that there are also important differences between turbulent and laminar flows. Outside the sublayer in a turbulent boundary layer, or in a wake, the apparent turbulent stresses are much larger than viscous stresses, and thus the direct contribution of turbulence to viscous drag is typically large. In most practical applications, however, turbulent flow over at least part of the surface is actually beneficial because of its greater resistance to flow separation, as we saw in Section 4.1.4.

6.1.3   Direct and Indirect Contributions to the Drag Force on the Body

The total drag of a body moving through a fluid is simply the flight-direction component of the stress imposed by the fluid on the body, integrated over the entire wetted surface:

$$D = \int_S (\mathbf{i} \bullet \overline{\overline{\tau}} \bullet \mathbf{n} - p\mathbf{i} \bullet \mathbf{n}) \, dS \qquad (6.1.1)$$

Here we've expressed the stress acting on each local element of the surface in terms of a component parallel to the local surface (the shear stress) and the pressure acting inward perpendicular to the surface. Further, we've resolved these two components of the stress into their components in the drag direction (opposite to the flight direction) by contracting them with the unit normal vector $\mathbf{n}$ outward from the body surface and the unit vector $\mathbf{i}$ in the drag direction. If we integrate the two parts of Equation 6.1.1 separately, the resulting forces are generally referred to as the skin-friction drag and the pressure drag. The skin-friction drag is entirely a result of viscous effects (viscosity and turbulence) in the boundary layers on the body's surfaces. The pressure drag is a result of a more complicated combination of flow mechanisms, including viscous effects, shocks, and the global effects of lift (induced drag). Given enough data defining the distribution of stresses on the surface, resolving the drag into a skin-friction part and a pressure part is straightforward, because it involves simply resolving a local stress vector into components. Dividing the drag into viscous drag, shock drag, and induced drag according to the mechanisms responsible isn't so simple.

Apart from such questions as to how the drag originates, which we'll get into later, the drag integration itself raises interesting issues because the two terms inside the integral in Equation 6.1.1 are so different in character.

The shear-stress term could be described as "wysiwyg" (what you see is what you get) in that when it is evaluated on a local part of the surface it is correctly seen as a local drag contribution that either adds to the total skin-friction drag or subtracts from it in a straightforward way, depending on its sign.

The pressure-drag term is fundamentally different and cannot be properly seen as a local drag contribution. There are reasons for this on two levels:

1. The local pressure-drag term depends on the overall pressure level in the environment, something that should have no direct bearing on drag. The integrated pressure drag depends only on the variation in pressure over the surface, and not directly on the overall pressure level. A mathematical statement of this that's even more general is that adding a constant to the pressure has no effect on the integrated pressure drag of a closed body because a constant pressure integrates to zero in Equation 6.1.1. Thus we can replace p in Equation 6.1.1 with $p - p_{ref}$, where p_{ref} is any constant reference pressure (it doesn't have to be the ambient pressure level), and the integrated pressure drag will be independent of the value of p_{ref}. The local pressure-drag term in its general $p - p_{ref}$ form cannot be properly seen as a local drag contribution because it depends on p_{ref}, which can be chosen arbitrarily. But couldn't we turn the pressure-drag term into a proper local drag contribution if we made the right particular choice of p_{ref} (the freestream pressure p_∞, say)? Well, no. There is no reason to expect p_∞ to be the right reference, and besides:

2. Much of the variation in pressure on the surface of a body is a largely inviscid response to the shape of the body and has nothing to do with drag. Thus there is no choice of p_{ref}

that can turn the pressure-drag term into a proper local drag contribution. The extreme example of this is a body in inviscid flow with no vortex shedding and therefore zero drag (d'Alembert's paradox). No matter what p_{ref} we choose, the pressure-drag term will generally have nonzero values except at isolated points, but integrated drag for a closed body will always be zero.

So we see that although the shear-stress term can be seen as a local drag contribution, the pressure-drag term makes sense as drag only after it's been integrated over the surface of a closed body.

This nonlocal character of the pressure drag makes life difficult when we want to assign a separate drag or thrust value to a portion of a complete body, which is usually of interest only when we are dealing with active propulsion. In the introduction to this chapter, we mentioned the idea of dividing the surface of a vehicle into parts and assessing drag or thrust separately on each, and we noted that the integrated pressure drag on a portion of a body will often reflect "spurious" contributions that aren't properly seen as either drag or thrust. There are two kinds of "spurious" pressure-drag forces that can appear on a portion of a body:

1. The pressure force on an "unopposed" segment of the surface of a body portion that is not closed. By "unopposed segment" I mean a forward-facing segment of surface for which there is no corresponding aft-facing segment directly behind it (in the flight direction), or an aft-facing surface segment without a corresponding forward-facing segment ahead of it (note that a closed body cannot have any such unopposed segment). The integrated pressure drag on an unopposed segment of a surface depends on p_{ref} and is thus "spurious." The only way to avoid this "spurious" force is to subdivide the body in such a way that no portion has any unopposed segment. An extreme example of how not to do it would be to divide a simple body into a forward portion that is completely forward facing and an aft portion that is completely aft facing. Each portion would be completely unopposed, and the integrated pressure drag on it would be completely "spurious."
2. The mutual buoyancy force that is generally present between bodies in proximity even if they're closed. Think of two spheres flying in formation, one behind the other, in potential flow. The forward one will experience a "thrust" force, and the rear one will experience an equal-and-opposite "drag," so that the total drag of the pair is zero (d'Alembert's paradox again). The forces on the individual spheres are parts of a passive inviscid interference effect, one that is also present in more-general cases with viscosity. It makes no sense to call such forces on portions of a body either "drag" or "thrust" because they make no net contribution to the total drag or thrust of the complete body, and they involve none of the flow mechanisms generally associated with drag (discussed below) or thrust (discussed in Section 6.1.10).

So dividing the surface of a body into parts raises questions about the meaning of the pressure "drag" or "thrust" on each, even if care has been taken to not to introduce unopposed surface segments. Now for the remainder of our discussion of drag in this section, we'll avoid this difficulty and assume a closed body that has not been subdivided.

Viscosity's contribution to the skin friction is obviously very direct, while its contribution to the pressure drag is more subtle. In 1910, Prandtl referred to this second contribution as the *form resistance*, attributing it to the presence of rotational flow in the portions of the field directly affected by viscosity (see Durand, 1967a, vol. I, p. 363). Even for a streamlined body with attached flow all the way to the back, Prandtl recognized that the rotational flow in the boundary layer could cause a pressure interaction contributing significantly to the total viscous drag. In Chapter 4, we interpreted this in terms of the displacement effect of the boundary layer, which can be relatively subtle if the boundary layer remains attached and can be much stronger if the boundary layer separates.

Setting in motion large volumes of air while producing lift in 3D, and dissipating energy in waves (shock waves in air or gravity waves in water), can also contribute to the total pressure drag. While all forms of drag ultimately end in viscous dissipation of work into heat, these two contributions are not conventionally considered to be part of the viscous drag. They both involve mechanisms by which energy is convected or propagated away from the body, and as a result, much of the dissipation associated with them takes place far from the body. Their nearfield manifestations are little influenced by viscosity. The involvement of viscosity in induced drag and wave drag is therefore even less direct than in the case of Prandtl's form resistance.

So three types of flow mechanisms, one that involves viscous effects in the nearfield and two that do not, contribute to the total pressure drag. But to what extent can we assign specific portions of the pressure drag felt by the body to the specific mechanisms in the flowfield? It turns out that we can do so only approximately, and only by appealing to theoretical idealizations. There is nothing about the distribution of the forces exerted on the surface that will tell us how much of the drag was caused by which flow mechanism. And looking at the flowfield doesn't yield a rigorous definition either. Because the different flow mechanisms overlap and interact, their effects do not add in a simple linear way to the total pressure drag, and an exact decomposition of the pressure drag into component parts is not possible. However, for practical purposes, it is possible to make an approximate decomposition, based on idealized theories regarding what goes on in the flowfield. For example, if the flow in the neighborhood of a shock is known, the shock's contribution to the drag can be estimated based on the Oswatitsch formula, as described in Section 6.1.4.8. Likewise, if the spanwise distribution of lift is known on the lifting surfaces, the induced drag can be estimated using Trefftz-plane theory, which is based on an idealized model of the flowfield associated with the given spanloading, valid in the limit of high aspect ratio, as we'll see in Section 8.3.4. Later in this section, we'll see a couple of different ways that the total viscous drag can be estimated, and if we subtract the skin-friction drag from that, we have an estimate of the viscous "part" of the pressure drag. We'll look at examples of how these ideas apply to simple 2D flows. In all of this, we must keep in mind that decomposing pressure drag into different "components," according to the flow mechanisms responsible, is an idealization. It is a useful one, however, and in practice, predictions of drag increments based on these idealized models have proved to be reasonably accurate.

6.1.4 Determining Drag from the Flowfield: Application of Conservation Laws

In this section, we look at the general problem of inferring the drag from information about the flowfield around the body. We consider some of the basic integral conservation laws and some relationships involving enstrophy (the local rate of dissipation into heat by the

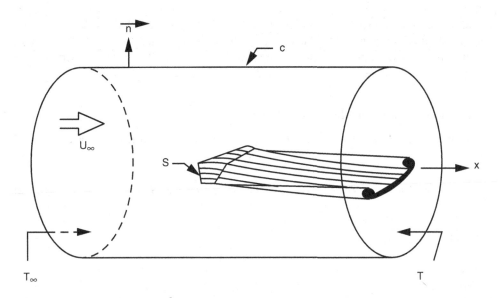

Figure 6.1.1   Control volume for application of conservation laws in body-centered reference frame

viscous stresses) and vorticity, with particular reference to what each one can contribute to a determination of the drag. Some of this discussion follows that found in Yates and Donaldson (1986).

6.1.4.1   The Body-Centered Control Volume

In looking at the local dissipation of work into heat in Section 3.6, we found it instructive to take the point of view of a fluid parcel moving through the field. Now, to consider the global problem of finding the force on a body, we take the reference frame in which the body is stationary and the fluid streams past it. For steady motion, the flow is then steady (turbulence has been time averaged), and we can drop the unsteady terms in all our conservation laws. To apply the integral conservation laws, we construct a control volume completely enclosing the body, as shown in Figure 6.1.1. For quantities that are conserved, such as momentum or total energy (thermal + kinetic + pressure work), we need only consider forces and fluxes at the bounding surfaces of the control volume, including the body surface. For other quantities, such as entropy, vorticity, or individual thermal or kinetic energy components, production or loss within the volume must also be considered.

6.1.4.2   Conservation of Energy

If no mass or heat is transferred through the body surface, and the body is mechanically passive with no moving parts, so that no work is exchanged between the body and the fluid in our body-centered reference frame, then the conservation of total energy reduces to the statement that total enthalpy is conserved at the outer boundary of the control volume (flux in equals flux out). In fact, for air, where the Prandtl number is close to unity, the total enthalpy is nearly uniform throughout the flowfield. Note also that turbulence kinetic energy plays no significant role in the total energy balance (see Cebeci and Bradshaw, 1984). Thus, because the body does no work on the fluid in this reference frame, the conservation of energy, by itself, tells us nothing about the drag.

6.1.4.3 Conservation of Momentum (General)

For a control volume enclosing our flying body as shown in Figure 6.1.1, integral momentum conservation gives

$$D = \int_{C+T+T_\infty} (-p\mathbf{i} \bullet \mathbf{n} - \rho u \mathbf{V} \bullet \mathbf{n}) \, dS \qquad (6.1.2)$$

Assumptions:

1. Conservation of momentum.
2. Steady flow.
3. No mass flux through body surface.
4. Viscous stress negligible on outer control surface.

Thus if we know the velocity, pressure, and density on a control surface enclosing the body at a moderate distance, we can, in principle, calculate the drag. The difficulty with this in practice is that the integrals over the outer control surface involve small differences between large contributions of opposite sign, both in the pressure and momentum flux terms. Whether we are dealing with measured or computed flowfield data, the momentum theorem by itself is not usually very useful without further assumptions.

6.1.4.4 Conservation of Momentum (Trefftz-Plane Representation)

One way out of this difficulty is to move the outer control surface so far away from the body that a linearized flow model is valid except in the vicinity of the viscous wake. When this is done, the velocity and pressure disturbances due to the body can be represented in terms of elementary singularities located at the body and along the wake. The flowfield around a body producing viscous drag generally has an apparent positive net source strength associated with it, and for our purposes, this is the only contribution that matters. Over the upstream and lateral portions of the outer control surface, the integrated effects of any higher-order singularities such as doublets or vortices can be made to vanish by taking the control surface far enough away. The farfield pressure and momentum-flux contributions of the source can be expressed analytically in terms of the effective source strength, which can be deduced from conservation of mass. The final result is that the drag can be expressed as an integral over the downstream face of the control volume (the Trefftz plane of induced-drag theory, which we'll cover in Section 8.3.4). The final result is

$$D = \int_T \left[(p_\infty - p) + \rho u \, (u_\infty - u) \right] \, dS \qquad (6.1.3)$$

Or, introducing the definition of *total head* (the same as the total pressure only in incompressible flow):

$$p_t = p + \frac{\rho v_\perp^2}{2} + \frac{\rho u^2}{2} \tag{6.1.4}$$

we have the alternative form

$$D = \int\limits_T \left[(p_{t\infty} - p_t) + \frac{\rho v_\perp^2}{2} - \frac{\rho (u_\infty - u)^2}{2} \right] dS \tag{6.1.5}$$

Assumptions:

1. Conservation of momentum.
2. Steady flow.
3. No mass flux through body surface.
4. Viscous stress negligible on outer control surface.
5. Subsonic flow in the farfield.
6. Farfield disturbances due only to simple singularities except near viscous wake.
7. Conservation of mass.

In that we now have only to integrate over a single downstream plane, these forms represent a distinct improvement over Equation 6.1.2. However, for a general, 3D, lifting body that sheds a vortex wake, the required integration is still over a very wide area, and in practice, these forms are of little use when we are dealing with numerical or experimental flowfield data. For our purposes, another shortcoming of these forms is that they do not help to distinguish the viscous drag from other drag components. Even the appearance of the cross-flow kinetic energy in the second term of Equation 6.1.5 does not help much. As we'll see in Section 8.3.4, in Trefftz-plane theories of induced drag, the cross-flow kinetic-energy integral is assumed to give the induced drag, and we might be led to assume that the remainder of Equation 6.1.5 would give the viscous drag. However, Equation 6.1.5 is valid only far downstream of the body, where in the real viscous world turbulent dissipation has already substantially reduced the cross-flow kinetic energy. In practice, even if the integral could be computed accurately, it would lead us to exaggerate the viscous component of the drag. To distinguish viscous drag from induced drag, we will need formulas applicable closer to the body, preferably with the integration limited to the viscous wake. This will require a different set of simplifying assumptions, as we will see next. But first we look at some special situations in which forms derivable from Equation 6.1.3 are useful.

6.1.4.5   Conservation of Momentum in Two-Dimensional and Axisymmetric Flows

A planar 2D or axisymmetric flow does not have the trailing vortices associated with lift in three dimensions. In this case, the pressure term in Equation 6.1.3 dies out much more rapidly with distance downstream of the body and can be treated approximately. If we put our downstream boundary sufficiently far from the body, the velocity outside the wake can be approximated by the sum of a uniform freestream and a potential-flow source and vortex

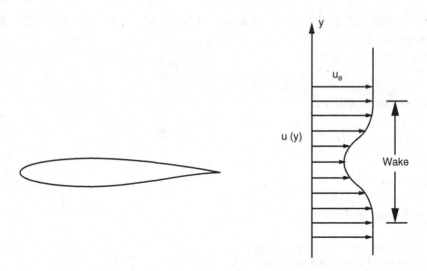

Figure 6.1.2   Airfoil profile drag determined by wake integration far downstream (Equation 6.1.7)

located at the body, and the pressure can be approximated by the corresponding linearized expression. The integrated pressure and momentum-flux contributions from the source and vortex can then be shown to cancel. All that remains is the second term in Equation 6.1.3, integrated only over the viscous wake as shown in Figure 6.1.2.

We then have the drag in terms of the momentum area of the wake:

$$D = \int_{wake} \rho u(u_\infty - u) \; dS \equiv \rho_\infty u_\infty^2 A_\theta. \tag{6.1.6}$$

In planar 2D flow, the momentum area per unit span is the familiar momentum thickness, and the profile drag coefficient is given by

$$C_d = \frac{2\theta}{c} \tag{6.1.7}$$

where θ is defined by Equation 4.2.11, with the integration including the entire wake.

Assumptions:

1. Conservation of momentum
2. Steady flow
3. No mass flux through body surface
4. Viscous stress negligible on outer control surface
5. Subsonic flow in the farfield
6. Except in the region within the viscous wake, farfield disturbances can be represented by a source located on or near the body
7. Conservation of mass
8. 2D or axisymmetric flow.

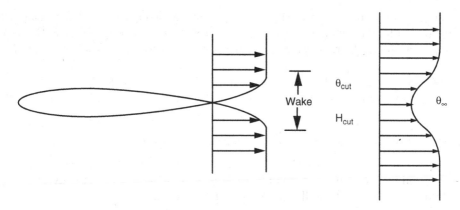

Figure 6.1.3   Airfoil profile drag determined by wake integration in nearfield using the Squire-Young formula (Equation 6.1.8)

In this situation, it is much easier to distinguish the viscous drag from other components than in the general 3D case. Under the above assumptions, the only other drag component that can be present is transonic shock drag associated with a bubble of supersonic flow in the near field, which we'll look at further in Section 7.4.8.

If we wish to infer the drag from a wake profile taken closer to the airfoil than Equation 6.1.7 requires, then further assumptions are required. So far, we have not had to make any assumptions about the flow in the nearfield, such as how the flow separates from the body. Now, we must limit our attention to flow that remains attached all the way to the back and only separates from a sharp trailing edge (in planar 2D flow) or a sharp tail (in axisymmetric flow). In the analysis put forward by Squire and Young (1938), the total (sum of upper and lower surfaces) momentum and displacement thicknesses of the boundary layers at the trailing edge are assumed to be known, and the momentum integral equation for the streamwise development of the wake is integrated approximately to give the wake momentum thickness θ_∞ far downstream, as illustrated in Figure 6.1.3.

Assuming the drag to be given as a function of θ_∞ by Equation 6.1.7, we now have a formula relating the drag to conditions in the wake close to the airfoil. The formula is ordinarily thought of as requiring trailing-edge data, but it should be equally valid for wake data taken at any station x_{cut} between the trailing edge and the farfield. As originally derived for incompressible flow, the formula is

$$C_d = \frac{2\theta(x_{cut})}{c} \left(\frac{u_e(x_{cut})}{u_\infty} \right)^{\frac{H(x_{cut})+5}{2}}, \tag{6.1.8}$$

where θ and H are defined the same way they were for a single boundary layer in Equations 4.2.10–4.2.12, just applied by integration across both the upper- and lower-surface boundary layers.

Assumptions:

1. Validity of Equation 6.1.7 far downstream
2. Incompressible flow

3. Integration of momentum-integral equation for wake, assuming a particular empirical relationship for the variation of the shape factor with edge velocity (linear ln(H) versus ln(u_e)). For this to be reasonable, there should be no reverse flow downstream of x_{cut}.

Corresponding formulas have also been derived for compressible flow (Cook, 1971), infinite-span swept wings (Young and Booth, 1950), and axisymmetric bodies (Young, 1939).

6.1.4.6 Formulas Relating Drag to Dissipation (Enstrophy)

Starting with the general Trefftz-plane drag formula, Equation 6.1.5, assuming incompressible flow, and invoking an expression for the generation of entropy by viscosity, Yates and Donaldson (1986) derive the following formula for the drag:

$$D = \frac{\mu}{u_\infty} \int_V |\vec{\omega}|^2 \, dV + \int_T \left[\frac{\rho v_\perp^2}{2} + \frac{\rho (u_\infty - u)^2}{2} + \left(1 - \frac{u}{u_\infty}\right)(p_{t\infty} - p_t) \right] dS. \quad (6.1.9)$$

Assumptions:

1. Conservation of momentum
2. Steady flow
3. No mass flux through body surface
4. Viscous stress negligible on outer control surface
5. Incompressible flow
6. Farfield disturbances due only to simple singularities except near viscous wake
7. Conservation of mass
8. Fourier heat conduction law
9. Conservation of entropy.

Note that the integrand in the volume integral (the enstrophy) makes a contribution only over the vortical part of the flowfield (boundary layer and viscous wake). Note also that all of the terms in the second integral (over the Trefftz plane) are of second order in disturbance quantities, so that the integral vanishes if we take the Trefftz plane far enough downstream. (For a lifting body with a trailing vortex system, the distance required could be very large.) In this limit, the volume integral approaches the total drag asymptotically, and we have

$$D = \frac{\mu}{u_\infty} \int_v |\vec{\omega}|^2 \, dV. \quad (6.1.10)$$

Assumptions:

1. Same as for Equation 6.1.9, plus
2. integral includes all significant dissipation.

In the next section, we apply this formula to two simple flowfields and find that it provides interesting insight into where in the flowfields the drag "originates."

6.1.4.7   Inferring Drag from Vorticity and Total Pressure at a Single Downstream Plane

A theoretical framework originally developed by Maskell (1972) allows both the induced drag and the viscous drag to be inferred from the velocity (vorticity) and total-pressure fields sampled only in the viscous wake at a single cross-plane behind the body. The approach was later extended by Brune and Hallstaff (1985) and applied to experimental data specially obtained for the purpose. This type of analysis could also be applied to computed flowfields, but in that case flowfield data are more easily available, and there would be less incentive to limit the input data to a single cross-plane. The physical assumptions involved in using a single cross-plane almost certainly limit the configurations for which accurate results can be obtained, but these limits have not been fully determined. For example, given data only in a single cross-plane, the x-component of the vorticity is the only one determined, and it is assumed that the induced drag can be computed based only on the x-component. This assumption is likely to be better in the farfield than in the nearfield, which implies that the data plane should be taken far from the body. However, to minimize the effects of dissipation, which tend to reduce the apparent induced drag and inflate the viscous drag, the inclination has been to place the data plane close to the configuration, where the assumptions of the method appear to be quite limiting. In spite of these questions, results for configurations with high-aspect-ratio wings have been good.

6.1.4.8   Shock Drag and the Oswatitsch Formula

In Section 3.11, we discussed the major physical properties of shocks. Whether a shock is normal to the flow or oblique, it doesn't generate any significant shear stresses, but passage of air through a shock is a dissipative process that reduces total pressure and increases entropy. Thus when a shock appears in the flowfield around a body, the dissipation in the shock itself contributes to the drag through the pressure term in Equation 6.1.1. A shock can also interact with the boundary layer on the body surface, causing a thickening of the boundary layer and an increase in the viscous form drag, an effect we'll consider in more detail in Section 7.4.8 in connection with transonic airfoils. Here we'll concentrate on the more direct contribution to drag of the shock dissipation itself.

By invoking conservation of momentum for steady flow through a control volume surrounding a body, Oswatitsch (1956) derived the following formula for the drag:

$$D = \frac{T_\infty}{V_\infty} \iint_{Shock} (S - S_\infty) \rho \mathbf{V} \bullet \mathbf{n} ds. \qquad (6.1.11)$$

Assumptions:

1. Conservation of momentum.
2. Steady flow.
3. No mass or energy flux through body surface.
4. Viscous stress negligible on outer control surface.
5. Farfield disturbances accounted for only to first order.

Tognaccini (2003) derived a higher order version of this formula with terms to second order in $(S - S_\infty)$. Note that Equation 6.1.11 represents the shock drag only if the shock is the only source of entropy increase in the flow. This requirement is not satisfied in real viscous flows because of entropy generation in the boundary layers and viscous wake. Still, this equation is often used for estimates of the contribution of the shock to the drag. In calculations of this kind, the control volume is taken to surround only the shock itself, so that only the entropy increase through the shock itself is integrated. This is somewhat justified, given that the integrand represents the flux of entropy excess, and in upstream and downstream regions where no entropy is being generated, the flux is constant. But it ignores the fact that in real flows, the streamtube that passes through the shock will eventually be entrained into the boundary layer and/or the wake, where further entropy will be generated. Use of the formula in this way ignores nonlinear interactions between the entropy-generation mechanisms in the shock and the viscous layers.

In Section 3.11, we also discussed how a shock can be represented in a computational fluid dynamics (CFD) solution in potential flow, in which the flow is treated as isentropic and irrotational. In such representations of shocks, there can be no entropy change, so Equation 6.1.11 cannot be used directly. The fictitious momentum jump that is generated at a shock in such flow solutions is sometimes used as an estimate of the shock drag, but it is significantly in error, as discussed by Steger and Baldwin (1971). The best alternative is to interrogate the calculated flow conditions upstream of the shock (the perpendicular Mach number), to use the normal-shock relations to calculate what the entropy jump would be, and then to use that as the input to Equation 6.1.11.

6.1.5 Examples of Flowfield Manifestations of Drag in Simple 2D Flows

Here we apply the principles developed above to a thin flat plate and a 2D airfoil to illustrate the manifestations of drag in flowfields with and without pressure gradients. For these two examples, we have carried out CFD calculations and plotted several quantities relevant to the drag to show how they vary with distance along the flow. The calculations are for a chord Reynolds number of 10^7 and are based on boundary-layer equations (specifically the computer code developed by McLean, (1977)), which should be valid except for some flow details near the trailing edges. The quantities plotted are

$$\text{Skin} - \text{friction coefficient}: \quad C_f \equiv \frac{\tau_w}{\frac{1}{2}\rho u_\infty^2}, \tag{6.1.12}$$

$$\text{Cumulative skin friction}: \quad I_{CF} \equiv \int_0^x C_f \, dx', \tag{6.1.13}$$

$$\text{Drag Equation 6.1.7}: \quad \frac{2\theta}{c} \equiv \frac{2}{c\rho_\infty u_\infty^2} \int_0^\infty \rho u(u_e - u) \, dy, \tag{6.1.14}$$

$$\text{Cumulative enstrophy}: \quad I_{ENS} \equiv \frac{2}{\rho_\infty u_\infty^3 c} \int_0^x \int_0^\infty \mu_{eff} \left(\frac{\partial u}{\partial y}\right)^2 dy \, dx'. \tag{6.1.15}$$

The quantities $2\theta/c$ and I_{ENS} are both valid farfield expressions for the drag, according to Equations 6.1.7 and 6.1.10, respectively, so both are expected to approach the total drag asymptotically at large distances downstream.

6.1.5.1 Thin Flat Plate

For discussing viscous drag in external flows, this is the simplest case. As illustrated in Figure 6.1.4, boundary layers form on both sides of the plate, laminar at first, followed by transition to turbulence. At the trailing edge, the boundary layers merge to form a wake that continues downstream. Because the plate surface has no projection in the x direction, the pressure contribution to the drag in Equation 6.1.1 is zero, and the drag on the plate is entirely skin friction. Thus I_{CF} at the trailing edge is equal to the total drag coefficient. Also, because we have assumed no pressure disturbance in our calculation, I_{CF} and $2\,\theta/c$ are equal all along the plate and into the wake. So for this very special case, the Trefftz-plane

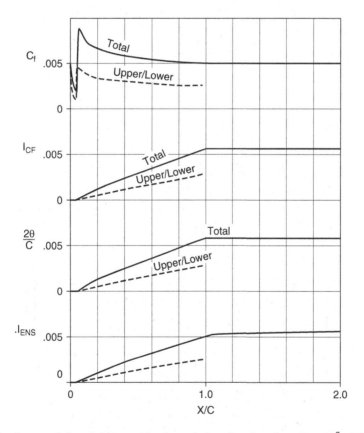

Figure 6.1.4 Computed flow-field quantities for turbulent flat-plate flow, $R = 10^7$. Calculations by P.P. Sullivan using method of McLean (1977)

drag Equation 6.1.3 is applicable at any station in the flow, not just in the farfield. The enstrophy integral I_{ENS} shows that dissipation has accounted for about 92% of the drag by the time the trailing edge is reached and that the drag is essentially completely accounted for about 50% chord downstream of the trailing edge.

6.1.5.2   Two-Dimensional Airfoil

The assumed pressure distribution and computed flowfield results for a hypothetical 2D lifting airfoil are shown in Figure 6.1.5.

The differences between this and the flat plate, resulting from the strong pressure disturbance of the airfoil, are immediately apparent. The sum of I_{CF} at the trailing edge is only 70% of the total drag; the remainder is the form drag. (Remember that the form drag is a viscous effect, since 2D, incompressible potential flow would have zero drag.)

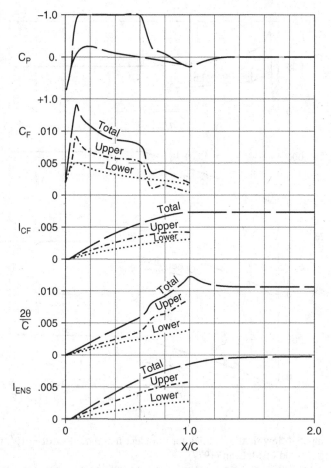

Figure 6.1.5 Computed flow-field quantities for turbulent flow on a hypothetical lifting airfoil, $R = 10^7$. Calculations by P.P. Sullivan using method of McLean (1977)

I_{ENS} at the trailing edge accounts for about 90% of the drag, slightly less than the corresponding 92% for the flat plate. Again, we need go only 50% chord downstream of the trailing edge before dissipation has accounted for essentially all of the drag.[1]

For a symmetrical, nonlifting airfoil or a slender axisymmetric body, the pressure gradient effects would generally be weaker, and we would expect the results to be intermediate between those of the lifting airfoil and the flat plate.

6.1.6   Pressure Drag of Streamlined and Bluff Bodies

We've discussed three flowfield mechanisms that contribute to pressure drag:

1. Viscous effects in the nearfield, either the relatively subtle displacement effect of an attached boundary layer or the much stronger effect of a separated boundary layer;
2. The global effects of lift in 3D (induced drag); and
3. Energy dissipation in waves, either shocks or gravity waves.

The first mechanism involves immediate dissipation into heat in the viscous layers near the body surface and just downstream. The last two are largely "inviscid" in their nearfield manifestations, but they too eventually produce dissipation into heat.

There is a common tendency to think of pressure drag in terms of the local surface pressures on parts of the surface and to think that the pressures can be manipulated by design of the surface shapes, in an essentially inviscid way, so as to influence the pressure drag. In this way of thinking, there is a simple criterion for assessing design changes that affect the pressure distribution: Low pressures on forward-facing surfaces and high pressures on aft-facing surfaces lead to low drag, and vice versa. This could lead us to think, for example, that it would be favorable to place an engine inlet so that it imposes higher-than-ambient pressures on an adjacent aft-facing surface. Another example might be to think that it would be favorable to place a blister fairing in an area on the surface where the pressure gradient is positive (increasing pressure in the flow direction) so that the volume of the fairing experiences a favorable "buoyancy" effect.

But this kind of thinking is mistaken more often than not. A simple counterexample shows why it is unreliable in general. Consider a body in an inviscid potential flow with no shocks and no shedding of vorticity. The pressure drag in this case must be zero, the result known as d'Alembert's paradox, as we saw in Section 5.4. The erroneous line of thinking says we ought to be able to put a blister fairing on a part of this body where the pressure gradient is positive, and thereby reduce the drag. But because we're starting from zero drag, the drag would have to become negative, which is impossible. So in global terms, we see that the addition of a blister fairing in this simple case cannot reduce the pressure drag, nor increase it, for that matter. Looking at it locally, in terms of the local pressures on the surface, the blister fairing may indeed produce a favorable incremental buoyancy effect on its own surface, but the pressures elsewhere on the body must change in response to the addition of the fairing so as to cancel that effect and make the total pressure drag zero

---

[1] Note that I_{ENS} and $2\theta/c$ should agree far downstream, but that in this case, the discrepancy at $x/c = 2$ is larger than it was for the flat plate in Figure 6.1.4. Presumably this is due to cumulative integration error, which could be reduced by grid refinement. The trailing-edge region was very likely the major contributor to the error buildup, since the calculations were in the direct mode, and the trailing-edge singularity was simply glossed over.

again. Such pressure changes on the rest of a body will often be subtle and difficult to see on a pressure plot, but they can act over large areas of the surface and cannot be ignored.

So in a simple inviscid case with no shocks or vortex shedding, changes to the pressure distribution cannot affect the pressure drag. In more general situations in which the pressure drag is nonzero, the same mechanism can still come into play, in which a change to the pressure distribution on one part of the body is offset by changes diffusely distributed over the rest of the body, with the result that there is little or no change to the total pressure drag. The key point to remember is that a change to the pressure distribution on the surface cannot change the pressure drag unless one or more of the pressure-drag-producing mechanisms in the flowfield is affected. So for purposes of understanding how a change in the surface pressures might affect the pressure drag, it can be very misleading to think just about the surface pressures themselves. One must consider how the change might affect either the viscous dissipation in the field, the lift distribution, or the strength of any shocks. It is good to be in the habit of asking "where in the field is the drag being 'produced,' and how is this change likely to affect that process?" Even if the result of asking that question is uncertainty, that answer is preferable to being misled by oversimplified reasoning.

In Sections 5.2 and 5.3, we discussed the distinction between streamlined bodies with attached flow all the way to a reasonably sharp "tail," and bluff bodies with large regions of separated flow at the back. For streamlined bodies, skin-friction drag dominates. In the 2D lifting-airfoil example above, we saw that skin-friction drag accounted for roughly 70% of the total, with pressure drag due to boundary-layer displacement accounting for roughly 30%, a breakdown that can vary depending on the thickness ratio, the lift, and the detailed pressure distribution. Pressure drag tends to be a smaller fraction in axisymmetric flow than in 2D planar flow. The pressure-drag fraction for a commercial-airplane fuselage, for example, is typically less than 10%. Because the skin friction is such a dominant part of the drag for a streamlined body, for purposes of drag estimation it is often convenient to relate the total drag to an idealized skin-friction drag through a *form factor,* which we'll discuss in Sections 6.2.1 and 7.4.2.

As we saw in Section 5.2, there is a range of possible nonstreamline flow patterns in which separation occurs other than at a sharp tail, either because an otherwise streamlined body is at an angle of attack or because the body shape isn't amenable to streamline flow. Separations of the 3D open type (Sections 4.1.4 and 5.2.2) don't by themselves form recirculating "deadwater" wakes, but they do result in the shedding of vortex sheets that tend to roll up into streamwise vortex cores. Examples of this type of separation are seen on the highly upswept aft fuselages of military transport airplanes and along the sides of the sloping roofs of some hatchback cars. The pressure drag that results from such separations can be quite high and is similar to the lift-induced drag of wings, which we'll discuss in detail in Sections 8.1 and 8.3.

Classic bluff-body flow is characterized by separation of the closed type that produces a substantial recirculating wake. For a body with separation of this type, skin friction tends to be only a small fraction of the total drag. Pressure drag dominates because of the strong displacement effect of the large separated region, and the frontal area of the body becomes a more relevant reference quantity than the wetted area or the skin friction. The pressure drag of a bluff body depends on the thickness of the separated wake, which depends on both the cross-sectional area of the body and the location of separation. When separation takes place at sharp edges in the pattern we discussed in connection with Figure 5.2.2c, the

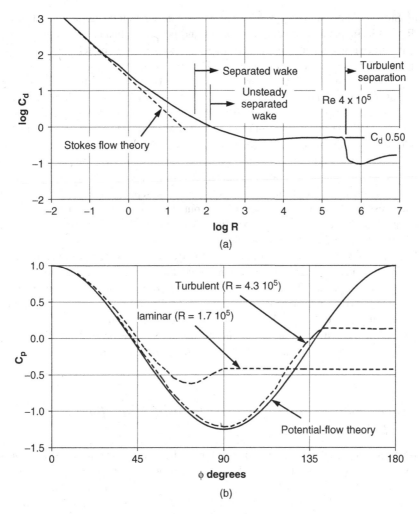

Figure 6.1.6 Drag behavior of a smooth sphere over a wide range of Reynolds number in incompressible flow. (a) Drag versus R curve defined by data from numerous experiments plotted by Hoerner (1965). (b) Surface pressure distributions for high Reynolds numbers with laminar and turbulent separation, plotted versus angle from the front stagnation point

separation location is independent of Reynolds number and surface condition, and for all but very low Reynolds numbers, the drag coefficient will tend to be nearly independent of Reynolds number.

On bodies with smooth contours such that separation is not fixed by sharp edges, the flow pattern and the drag vary more strongly with Reynolds number and can be influenced by surface roughness. For example, Figure 6.1.6a shows the drag coefficient of a smooth sphere over a wide range of Reynolds number. At very low Reynolds numbers, it follows the Stokes-flow solution (creeping flow, in which inertial effects are negligible) for which $C_d \sim R_d^{-1}$. Above $R_d \approx 1$, inertial effects become significant, and the drag diverges upward from the Stokes solution. At $R_d \approx 50$ there is a small separated region behind the sphere, and at $R_d \approx 130$ the separated region is longer and becomes unsteady. The variation of C_d with R_d is surprisingly smooth in spite of these changes in flow pattern.

At higher Reynolds numbers (up to about 4×10^5), the boundary layer remains laminar to the separation point. Laminar separation occurs well forward, only about 80° from the front stagnation point, leading to a large separated wake with considerable unsteadiness and a roughly constant drag coefficient of about 0.5 over the range of Reynolds number from 1×10^3 to 4×10^5. Above 4×10^5, the boundary layer transitions to turbulent upstream of separation, the separation point moves back to about 140°, and the drag coefficient drops abruptly to about 0.1. Figure 6.1.6b shows the surface pressure distributions for these cases with laminar and turbulent separation, compared with potential-flow theory. In both cases, separation reduces the level of the suction peak ahead of separation and produces a leveling off of the pressure downstream of separation, but the effects are much more dramatic in the laminar case, due to the earlier separation: Not only is the separated region wider, the pressure downstream of separation is lower.

For bodies that exhibit this kind of behavior, the Reynolds number at which the separation switches from laminar to turbulent is often called the *critical Reynolds number*, and the abrupt drop in the drag coefficient is called the *drag crisis*. A golf ball is one such body, and its dimpled surface is a purposeful adaptation to the Reynolds-number range in which it flies. Over the range of speed achievable by even the strongest golfer, the Reynolds number is below the critical value for a smooth sphere. The purpose of the dimpled surface is to provoke early transition and thereby move the drag coefficient toward a lower post-crisis level.

6.1.7   Questionable Drag Categories: Parasite Drag, Base Drag, and Slot Drag

The term "parasite drag" is often used to refer to the entire portion of an airplane's drag other than induced drag, and has even found its way into the dictionary (Websters, 1976). "Parasite drag" thus has a workable definition, as long as we're willing to have it depend on a theoretical value for the induced drag. (Recall from Section 6.1.3 that induced drag doesn't have a rigorous definition apart from the idealized theory.) But there is a semantic problem in that the term imputes a particular character to the drag that isn't logically justifiable. In general, "parasite" refers to something that is supported by something else and gives nothing in return (paraphrased from Websters, 1976). The viscous drag of the wings and fuselage of an airplane is a necessary byproduct of producing lift, and housing structure and payload. These functions can't be said to "give nothing in return," so calling their byproduct a "parasite" makes no sense.

If a bluff body or 2D airfoil-like shape is "squared off" in back so that separation takes place at two corners as illustrated in Figure 5.2.2c, the "additional" drag due to the squared-off base is often called "base drag." The problem with the term "base drag," however, is that it has no defensible definition. Here are the three logically possible candidates I know of, along with the reasons why they don't work:

1. Define base drag as the portion of the actual drag that is somehow caused by the presence of the blunt base.

Problem: While it's true that a body with a squared-off base tends to have higher drag than one that doesn't, there is no good way to isolate the portion of the drag that can be "blamed" on the base. Whether we assess the drag in terms of the stresses on the body surface (Equation 6.1.1) or the dissipation in the field (Equation 6.1.9), there is no rigorous way to divide either the surface or the volume to decompose the drag.

2. Define base drag relative to the drag of a comparable sharp-tailed body.

   Problem: There is no unique way to define the comparable body.

3. Define base drag as the deficit in the static pressure on the base (which does tend to be nearly constant) relative to ambient pressure, times the area of the base.

   Problem: There is no reason why ambient pressure is the relevant reference, as we noted in Section 6.1.3. In Section 6.1.6, we discussed how misleading it can be to try to infer differences in drag from local static pressures, and this situation is no exception. That this definition doesn't work is evident from two counterexamples. We'll see in Section 7.4.1 that a streamlined airfoil typically has a region of positive C_p (pressure higher than ambient) near the trailing edge. We could truncate the airfoil within that region and have positive C_p on the base, and the current definition would then assess negative base drag, which makes no sense. The example of a sphere is also instructive. A sphere doesn't have a squared-off back end, but it does have a large separated region in the back. Note that in the case of turbulent separation in Figure 6.1.6b, the surface C_p in the separated region is positive. But it certainly doesn't make sense to say that a sphere has negative base drag.

So it seems that defining base drag in a rigorous quantitative way is hopeless. And without a workable definition, "base drag" seems questionable, even just as a concept.

Hoerner (1965) in his section III.8 uses the term "base drag" but doesn't provide an explicit definition. He presents experimental values of "base drag" in his Figures 37 and 39 but doesn't specify whether they were determined by comparison with sharp-tailed baseline bodies (candidate definition 2 above) or by measurements of base pressure (candidate definition 3 above). In his Figure 38, he plots total drag for a series of bodies that includes a sharp-tailed baseline, which suggests that he is thinking in terms of candidate definition 2. Whatever the definition, he proposes that "base drag" is determined by a "jet-pump mechanism," which is supposed to work as follows (paraphrased):

> The "jet" is the outer flow, "placed like a tube around the space behind the base," and it "mixes with the dead air and tries to pump it away," thus reducing the base pressure. The separated boundary layer has an "insulating" effect that reduces the effectiveness of the jet pump. The longer the forebody, the thicker will be the boundary layer that separates from the edge of the base and the less effective the jet-pump mechanism will be in reducing the base pressure. The "base drag" thus correlates with a parameter related to the fineness ratio of the forebody and decreases with increasing fineness ratio.

The data presented by Hoerner support the correlation reasonably well, but I propose that it is not for the reason that he supposes. The problem with Hoerner's explanation is that it ignores the dominant role of the outer inviscid flow in determining the pressures on the body. The separated shear layer that forms the outer boundary of the near wake can support very little pressure difference across it, so the base pressure must be essentially the same as the static pressure in the inviscid flow outside the boundary layer where it separates from the edge of the base. This static pressure is a result of the external inviscid flow, as influenced by the effective displacement shape of the forebody and the separated wake downstream. The thickness of the boundary layer at separation influences the effective shape of the wake to some extent, but the static pressure at separation, and thus on the base, is much more strongly driven by the overall shape and fineness ratio of the body, through the inviscid flow. So the correlation of the "base drag" with forebody fineness ratio is much more of an inviscid-flow effect than a result of any "jet-pump mechanism." The data appear to be just as consistent with this explanation as with Hoerner's.

We've seen why base drag is a questionable concept. In a similar vein, airfoils with high-lift slots, which we'll discuss in Section 7.4.4, tend to have higher drag than airfoils without slots, and the excess is sometimes referred to as "slot drag." This terminology has the same problem: There is no good way to isolate the portion of the drag that can be "blamed" on the slot. The slot doesn't represent a mechanism that can produce pressure drag that is separate from the viscous drag. In a 2D inviscid, shock-free flow, a slotted airfoil has zero drag, just like any other airfoil. In the real world, a slot imposes strong pressure gradients and curvature effects on the boundary layers, as well as separation and reattachment if the slot has a "cove" (see Section 7.4.4, especially Figure 7.4.18). These are things that tend to increase the total viscous dissipation in the field, but they do not make slot flows in any way unique, and they do not define a lower limit on the amount of "additional" drag a slot must cause.

6.1.8   Effects of Distributed Surface Roughness on Turbulent Skin Friction

In Section 4.7, we briefly listed the various ways surface roughness can affect an aerodynamic flow. In this section, we look at how surface roughness can increase turbulent skin friction, which can influence total drag even when there is no change in transition or separation. We'll look at basic physical explanations of how the skin-friction increase comes about, the associated scaling issues, and empirical data for different types of surfaces. Then, in Section 6.2.2, we'll look at methods for using this kind of information in drag estimation.

In turbulent flow, surface roughness elements that are sufficiently large are known to cause significant increases in area-averaged skin friction. It seems intuitively obvious that they should, but explaining how this happens is trickier than one might expect. The roughness elements are like small bluff bodies immersed in the highly nonuniform and unsteady flow at the bottom of the boundary layer, and they are often so closely spaced that the flow around each element is strongly affected by neighboring elements. The physical explanations offered in most of the classical sources tend to be incomplete at best and sometimes contradictory (see Bradshaw, 2000). But before we address explanations, let's look at how these skin-friction increases have actually been observed to behave, first in the idealized situation of uniform roughness along the entire length of a flow and then locally in more-general flow situations.

The effects of two different types of roughness on the average skin friction coefficient $\overline{C_f}$ of a turbulent flat-plate boundary layer are plotted in Figure 6.1.7. The solid roughness curves are for *standard sand-grain roughness,* which consists of a single layer of densely packed grains of uniform diameter k_s and was the focus of many early investigations of roughness effects both in pipes and boundary layers. Curves are shown for three different grain sizes relative the length of the plate. These are idealized curves based on the

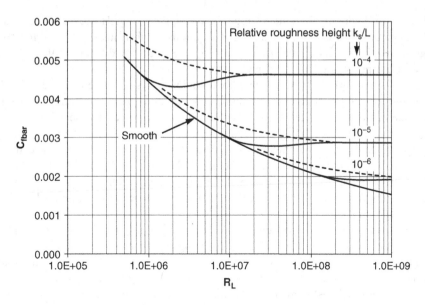

Figure 6.1.7 Average turbulent skin friction coefficient for a flat plate of length L in a smooth condition and uniformly covered with roughness of two different types: *standard sand-grain roughness* (densely packed sand grains of uniform diameter k), shown by the solid curves, and roughness typical of *commercially rough* pipes, shown by the dashed curves

pipe-flow measurements of Nikuradse (1933), see Figure 20.21 in Schlichting (1979), for example), converted to flat-plate $\overline{C_f}$ by the "equivalent-flat-plate method" that we'll discuss in Section 6.2.2. Starting at a low Reynolds number, each curve follows the smooth-surface curve (nominally the same as the turbulent curve in Figure 4.3.1) up to a *critical Reynolds number* and then diverges from it and eventually levels off. Before $\overline{C_f}$ levels off, it dips to a minimum about 5% below the final value. This general progression has led to the definition of three regimes: a *hydraulically smooth* regime before the divergence, a *fully rough* regime in which $\overline{C_f}$ is independent of Reynolds number, and a *transitionally rough* regime in between. As pointed out by Bradshaw (2000), it is unfortunate that this "transition" terminology came to be applied to roughness in turbulent flow, as it risks confusion with laminar-to-turbulent transition, which is unrelated. But it is frequently used in the literature, and for lack of anything better, we'll continue to use it.

Standard sand-grain roughness is a highly artificial kind of surface, and surfaces more representative of those usually found in real-world applications don't generally show the same kind of behavior in the transitional regime. A well-known example of a real-world surface is documented in the measurements by Colebrook (1939) in *commercially rough* pipes, which served as part of the basis for the famous *Moody chart* for resistance in pipe flow (Moody, 1944). The dashed curves in Figure 6.1.7 are flat-plate roughness curves derived in the same way as the sand-grain curves (solid), but based on Colebrook's pipe data instead of Nikuradse's. In Colebrook's curve-fits, the drag of a rough surface approaches the smooth-surface curve only asymptotically with decreasing Reynolds number, and there is no critical Reynolds number below which the effect of roughness is taken to be zero, so that there is no hydraulically smooth regime, strictly speaking. However, there is a fully rough regime in which $\overline{C_f}$ is independent of Reynolds number, just as there is for standard sand-grain roughness. $\overline{C_f}$ in the transitional regime does not dip below the fully-rough value, as it did in the standard-sand-grain case.

For a real-world surface, there is generally no unique physical scale length corresponding to the grain diameter that we used for uniform sand grains. As a substitute for defining a length scale based on roughness geometry, Schlichting proposed the concept of *equivalent sand-grain roughness* (see Bradshaw, 2000), in which the surface is characterized by the roughness height k_s of the standard sand-grain roughness that would produce the same skin-friction coefficient in the fully-rough regime. Obviously, knowing the equivalent sand-grain-roughness height for a surface provides a reliable indication of its drag behavior only in the fully rough regime, because different kinds of surfaces have different behaviors in the transitional regime.

The relationship between the equivalent sand-grain height k_s and the actual physical height h of the roughness varies greatly, depending on the nature of the surface. Even if the height of the roughness elements is uniform, the drag depends strongly on the shape of the elements and on the density of the surface coverage. Waigh and Kind (1998) correlated data for uniform roughness elements of various shapes (e.g., spheres, cylinders, rectangular blocks) and with a wide range of coverage densities and found that the ratio k_s/h varied from 0.1 to almost 30 depending on shape and area coverage.

When the height of the roughness elements is not uniform, there is no single height h to characterize the surface, so an average measure such as R_a (the average of the absolute value of the deviation from the mean surface) or k_{rms} (the rms deviation) is often used, usually measured in practice by a stylus profilometer. We should expect the ratio k_s/R_a or k_s/k_{rms} to vary depending on the nature of the surface, but not over as wide a range as k_s/h does for roughness of uniform height. Much of the variation in the k_s/h values correlated by Waigh and Kind was due to the wide range of surface coverage. When coverage is sparse, the drag penalty is reduced, but so are R_a and k_{rms}, so that using R_a or k_{rms} instead of h partly compensates for this effect. But the ratio k_s/R_a or k_s/k_{rms} still varies depending on the nature of the surface.

Basic roughness geometry is another factor that must be kept in mind when using R_a or k_{rms}. These average measures are typically considerably smaller than the maximum element size (diameter). This is because maximum deviations are, of course, larger than average deviations, and diameter is larger by another factor of 2 because it is a peak-to-peak measure rather than a deviation from the mean. A careless error that sometimes arises is to forget this and to take R_a or k_{rms} to be interchangeable with k_s. This can lead to a serious underestimation of the roughness effect, because R_a and k_{rms} are typically almost an order of magnitude smaller than k_s. A simple geometric argument shows that this is true even for standard sand-grain roughness. If we ran a stylus profilometer over a standard sand-grain surface, the stylus would rarely penetrate into the crevices between grains to more than about a half grain diameter below the peaks, because of the relatively blunt shape of the stylus, and the resulting R_a or k_{rms} would be a small fraction of the grain diameter.

In Figure 6.1.7, we looked at the global effect of roughness in the idealized case of a flat plate with roughness that is uniform over the whole surface. In flow situations that are more general, with roughness distributions that may be nonuniform, we must deal with the effects of roughness locally. Unless the characteristics of the roughness change too suddenly as we

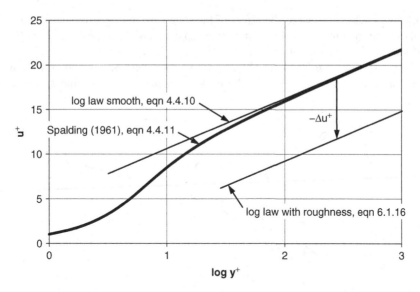

Figure 6.1.8   Shift, due to roughness, in the logarithmic portion of a turbulent wall-region velocity profile

move along the surface, the local effect is closely determined by the relationship between the roughness geometry and the mean velocity profile in the *wall region* of the boundary layer, which we discussed in Section 4.4.2. As we saw there, for a smooth surface, this wall-region velocity profile is very close to being a function only of the viscosity and the shear stress at the surface. The velocity profile expressed in *wall variables* is then universal, and the functional relationship $u^+ = \text{fcn}(y^+)$ is known as the *law of the wall*, which applies to the viscous sublayer and the logarithmic region, as illustrated in Figure 4.4.6. The effect of roughness on this local velocity profile is to shift the logarithmic portion of the velocity profile downward by a constant Δu^+ as shown in Figure 6.1.8.

The size of the shift depends on the roughness Reynolds number k^+, with a functional relationship $\Delta u^+(k^+)$ that depends on the particular roughness geometry. The logarithmic velocity profile is then given by:

$$u^+ = \frac{1}{\kappa} \ln y^+ + B - \Delta u^+(k^+), \tag{6.1.16}$$

where κ and B are the same as they were for a smooth surface. Roughness thus adds one parameter, k^+, and one function, $\Delta u^+(k^+)$, to the problem.

The functional relationship $\Delta u^+(k^+)$ is a fixed characteristic of the particular roughness geometry, and the value of Δu^+ in a particular situation is the dominant factor determining the local increase in skin friction coefficient, ΔC_f. In general, the larger the shift Δu^+, the larger ΔC_f will be. But because C_f is generally normalized by a velocity outside the wall region (U_e or U_∞ usually), ΔC_f also depends to some extent on the global flow conditions. To estimate the rough-surface C_f in a given flow situation, the first thing we need to know is the function $\Delta u^+(k^+)$ for the surface in question. Then we need to use that information to estimate ΔC_f. The function $\Delta u^+(k^+)$ can be incorporated into a turbulence model for use in detailed CFD calculations, or the effect of the roughness can be estimated for a flat-plate flow that is assumed to be similar to the target application. In Section 6.2.2, we'll discuss both of these approaches in more detail.

The shift functions $\Delta u^+(k^+)$ for three types of roughness are shown in Figure 6.1.9a, including standard-sand-grain and Colebrook's commercially rough pipe surfaces, whose

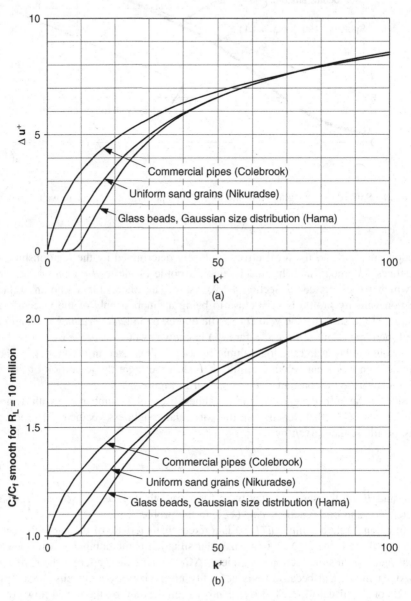

Figure 6.1.9 Local behavior of three types of distributed roughness. (a) Log-law shifts. (b) Relative change in local C_f

global flat-plate behavior as a function of Reynolds number was shown in Figure 6.1.7. What these shifts mean in terms of $\Delta C_f/C_f$ as a function of k^+ depends on the global flow situation. $\Delta C_f/C_f$ for the particular case of a flat-plate boundary layer at a length Reynolds number of 10^7, is shown in Figure 6.1.9b. Looking first at $\Delta u^+(k^+)$ for the standard sand-grain roughness, we see that the divergence from smooth-surface behavior takes place at about $k^+ = 5$. The fully rough regime starts at about $k^+ = 70$, after which $\Delta u^+(k^+)$ increases in a straight line. The length scales k for the other roughness types are the equivalent sand-grain roughness heights, defined so that the curves match the standard-sand-grain curve in the fully rough regime.

The question of whether the drag penalty for standard-sand-grain roughness is actually zero up to $k^+ = 5$ has not been settled. The experimental data on which the curve-fits are based show considerable scatter. Bradshaw (2000) has pointed out that a curve-fit in which there is no finite hydraulically smooth regime, for example, $\Delta u^+ \approx 0.02(k^+)^2$ for $k^+ < 7$, would fit the data just as well as the curve in Figure 6.1.9a does. The C_f increase at $k^+ = 5$ might be a couple of percentage instead of zero, but is still small compared with about 20% for commercial pipes.

The Δu^+ and ΔC_f curves of Figure 6.1.9 make it even more clear what a wide range of behavior different types of surfaces can display in the transitionally rough regime and why the concept of equivalent sand-grain roughness doesn't apply there. To show that the standard-sand-grain and commercially rough pipe curves don't bracket the range of possible behavior, I included Hama's (1954) curve for glass beads with a Gaussian size distribution, which is seen to lie far outside the range defined by the other two types. The nonuniform element size might lead one to expect behavior intermediate between uniform sand and commercial pipe roughness, but instead this surface shows behavior that is roughly twice as far from commercial pipes as the standard-sand-grain surfaces are, and Hama's curve-fit defines an even wider hydraulically smooth regime. I have seen no convincing explanation for this in the literature. The message we should take from Figure 6.1.9 is one of caution regarding roughness in the transitional regime. In the neighborhood of $k^+ = 10$, the C_f increase can be anywhere from near 0 to about 30%. This high level of uncertainty is unfortunate, as the transitional regime is the range of the greatest practical importance in aeronautical applications. There is much less uncertainty in the fully rough regime, but roughness in that range is very rarely tolerated in applications because C_f is nearly double its smooth-surface value at the start of the regime and goes up from there.

Now that we've seen what the behavior looks like, how do we explain it physically? The specifics, such as the details of the different roughness curves in the transitional regime, seem to be beyond the reach of simple qualitative explanations. But we should at least be able to explain the general trend: Whether or not there is a finite hydraulically smooth regime in which the effect is zero, the drag increment eventually "takes off" and increases mono-tonically for larger roughness Reynolds numbers. Explanations for this that have appeared in the classical sources are often fuzzy and in some cases obviously incorrect.

In one category of explanation, a major distinction is drawn between the sublayer and the rest of the boundary layer outside it. It is assumed to be obvious that roughness elements that are submerged in the sublayer should have little or no effect and that the farther the elements protrude outside the sublayer, the more drag they should cause. But a detailed argument as to why this should be so is not offered. Some versions are clearly under the influence of an erroneous interpretation of a "laminar sublayer," while in others it is ambiguous as to how the sublayer is being interpreted. Here is a sampling:

- **Hoerner (1965):** "there is always a comparatively thin sublayer within which stable laminar flow is prevailing. As long as the protuberances of a rough surface are deeply enough submerged within the sublayer, the surface is, therefore, hydrodynamically equal to a smooth one."
- **Schlichting (1979):** "It is clear that roughness will cause no increase in resistance in cases where the protrusions are so small (or the boundary layer is so thick) that they are all contained within the laminar sublayer."
- **White (1991):** "even a small roughness will break up the viscous sublayer and greatly increase the wall friction."

Hoerner seems to have been misled by the terminology into thinking the flow in the sublayer is actually "laminar" and presumably steady ("stable"). We saw in Figure 2.1f that the flow in the sublayer is far from laminar, and we've noted that the only thing "laminar" about the sublayer is that the Reynolds stress is effectively zero due to the effect of the wall in constraining the turbulent motions that are large enough not to be damped much by viscosity (see Section 4.4.2 and Equation 4.4.2). Schlichting doesn't explicitly refer to "laminar flow" in the sublayer, but it is implied that there is something different about the sublayer. White uses the more modern "viscous" to designate the sublayer, but he doesn't explain what he means by "break up." If the sublayer isn't laminar, what is there about it that is prone to being broken up? In any case, whether these explanations are based on correct interpretations of the sublayer or not, they don't provide much of a physical explanation as to why the skin friction increases when the roughness elements protrude outside the sublayer.

Another category of explanation hinges on the type of flow around the roughness elements and its dependence on the local Reynolds number. The general line of argument is that the skin friction begins to increase when the flow around individual roughness elements leaves the very low Reynolds number regime and begins to display a separated or unsteady wake. Smith and Kaups (1968) make an argument of this type. First, they draw an analogy between the roughness elements and isolated bluff bodies in free air. They show that in the part of the boundary layer where the velocity profile is linear, the local roughness Reynolds number is related to k^+ by $R_k = ku_k/\nu = (k^+)^2$, where $u_k = u$ at $y = k$. Thus the critical roughness height $k^+ = 5$ corresponds to $R_k = 25$, "at which isolated bluff bodies begin to develop a substantial and perhaps unstable wake." They also point out that 25 is the minimum R_k "that first affects transition in laminar flows," though this presumably has no bearing on the turbulent-skin-friction issue. So the onset of the shedding of unsteady wakes from the roughness elements seems to correspond to the onset of the skin-friction increase in a turbulent boundary layer. But then Smith and Kaups acknowledge that the analogy with bluff bodies in free air doesn't seem to apply to the behavior of the drag curves. In the range of Reynolds number where unsteady wake shedding begins, the drag coefficient of a sphere, for example, does nothing remarkable, and in fact, it is decreasing smoothly with Reynolds number, not increasing (see Figure 6.1.6). In view of this, Smith and Kaups conclude

> Hence the drag increment of roughness does not relate directly to the drag of bluff bodies at low Re in a free stream. Instead the drag begins to be significant when a substantial wake region develops or when this wake region becomes unstable.

They leave it at that and don't explain how the onset of unsteady wake shedding is supposed to produce a change in the drag curve for a rough-wall boundary layer, while it doesn't for isolated bluff bodies.

In addition to the fact that it doesn't really explain the skin-friction increase, this argument has an inconsistency that Smith and Kaups overlooked. The argument for the correspondence between $k^+ = 5$ and $R_k = 25$ makes sense in the context of an isolated roughness element exposed to a linear velocity profile that has not been disturbed by nearby roughness elements. Yet the $k^+ = 5$ threshold is derived from the skin-friction curve for standard-sand-grain roughness, in which the grains are tightly packed together on the surface. For this type of surface, most of each grain is shielded by adjacent grains, and the average deviation of the surface contour from the mean is much less than the grain diameter k. Thus the effective Reynolds number of the protuberances is much less than 25 at the $k^+ = 5$ threshold where the drag curve deviates from that of a smooth surface. In view of this, it seems likely that there is no unsteady wake shedding involved in the onset of the skin-friction increase for standard-sand-grain roughness.

So we see that the two main explanations that have been offered don't explain the skin-friction increase in a satisfying way. Is there a better way to explain it? I think there is. But we should expect to be able to explain it only in loose, general terms. After all, we're dealing with complicated three-dimensional roughness shapes immersed in the complex turbulent flow at the bottom of a boundary layer.

I propose that the general trend of increasing drag with increasing k^+ is related to the height of the roughness elements relative to the thickness of the viscous sublayer and that it has nothing to do with changes in the wakes of the roughness elements. Two key features of the sublayer play roles: the shape of the mean velocity profile, and the distribution of Reynolds shear stress. The explanation is conveniently divided in two parts. To summarize before we dive into the details:

1. When roughness elements are submerged in the sublayer, where the mean-velocity profile is approximately linear, their effects needn't increase the area-averaged skin friction, because the boundary layer can "adjust" to their presence with nothing more than a shift in its effective origin. This should be true regardless of what flow regime the roughness elements find themselves in, for example, whether the wakes are separated or not, or unsteady or not. A shift in effective origin is consistent with the equations of motion, as we'll find below, and it can prevent an increase in skin friction by keeping the average velocity of the flow around the roughness elements low.

2. When the roughness height is increased, and the elements are no longer sufficiently submerged in the sublayer, a shift in the effective origin can no longer prevent an increase in skin friction. Taller roughness elements would require a larger shift in effective origin, and a thicker layer of low-velocity fluid would have to be maintained to prevent an increase in skin friction. But turbulent mixing prevents the maintenance of sufficiently low velocities over a thicker region. Recall from our discussion of Equation 4.4.2 that effective turbulent mixing (as embodied in the Reynolds stress) is damped by viscosity near the wall, but that the damping is able to keep the Reynolds stress near zero only over a thin region (the sublayer). So when roughness elements reach a certain size, sufficiently low velocity cannot be maintained across their whole depth, and turbulent mixing will raise the average velocity around them, increasing the average skin friction. This explanation

is supported by the success of the "discrete element" approach to predicting roughness drag (see Taylor, Coleman, and Hodge, 1984). In this approach, the drag of roughness elements is assumed to be affected by the turbulence physics (the local Reynolds stress) in just the way I've proposed here, through the shape of the mean-velocity profile.

The idea of a shift in effective origin for small roughness is supported by a physical argument that starts with an idealized Couette flow. Consider the parallel flow between two flat plates, the lower one stationary and the upper one moving to the right as illustrated in Figure 6.1.10. For the smooth-surface case in part (a), this is the same flow situation we considered in our discussion of the basic effects of viscosity in Section 3.6 (Figure 3.6.2a). The shear stress is constant across the gap between the plates, and we'll assume the flow to be laminar, so that the velocity profile will be linear, and the Reynolds stress will be zero, as in the sublayer. In part (b), we have added roughness to the lower plate, and now the flow near the rough surface is nonuniform on the scale of the roughness elements. Some short distance above the tops of the elements, however, the flow becomes practically uniform in both horizontal directions, and we can adjust the speed of the upper plate so that the velocity gradient $\partial u/\partial y$ in the uniform part of the flow is the same as it was in the smooth-surface case in part (a). The shear stress across the part of the gap not occupied by roughness elements is now the same as in the smooth-surface case, and the shear forces per unit area on the two plates are the same as well. So if the velocity gradient a short distance above the tops of the roughness elements is the same as it was in the smooth case, the skin friction will also be the same, and there will be no drag penalty due to roughness. The only difference will be a shift in the apparent origin of the mean velocity profile, defined by extrapolation of the velocity profile to zero, as indicated by the dashed line in part (b). And note that we didn't have to assume anything about the type of flow that surrounds the roughness elements, other than that it didn't result in transition to turbulence across the entire gap in this idealized laminar example.

Now consider the same comparison between a smooth surface and a rough surface, but in a boundary layer, which can be either laminar or turbulent. The situation is illustrated figuratively in Figure 6.1.11, with the smooth case shown in part (a). In the rough case in part (b), we assume that the roughness elements are well submerged in the region where

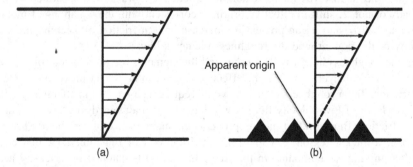

(a)   (b)

Figure 6.1.10 Laminar Couette flow like that of Figure 3.6.2a, illustrating the probable effect of surface roughness in a flow with a linear mean-velocity profile. (a) Smooth wall. (b) Rough wall with the same velocity gradient far above the roughness elements as in (a), but with a shift in the apparent origin of the velocity profile

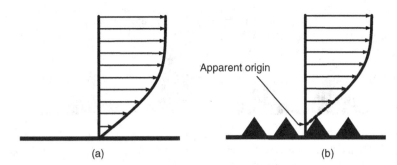

Figure 6.1.11   Boundary-layer velocity profiles illustrating the probable effect of roughness elements that are submerged within the region where the profile is linear. (a) Smooth wall. (b) Rough wall case in which the velocity profile well above the tops of the roughness elements is the same as it was in the smooth-wall case, with a small shift in the apparent origin. The size of the roughness elements is exaggerated for clarity

the velocity profile is nearly linear, and where the Reynolds stress is negligible (the viscous sublayer in the turbulent case). Then we know from the Couette-flow argument above that the relationship between the skin friction and the velocity gradient at some distance above the tops of the roughness elements is unchanged by the roughness. And then *if* the velocity profile above the tops of the roughness is a shifted version of the profile in the smooth case, the skin friction in the two cases will be the same. In the Couette-flow case, we could force the velocity profile to be a shifted version of the smooth wall profile by adjusting the speed of the upper wall. But why should such a simple shift apply in the case of a boundary layer? For a plausible answer, consider the equations of motion.

The displacement thickness of the boundary layer would be affected by such a shift, and the interaction with the outer flow could therefore change the pressure distribution imposed on the boundary layer, but in most external flow situations this will be a very small effect. (In a pipe or duct flow, the effect wouldn't be so small because such flows are sensitive to small changes in the effective area of the duct.) What is more important for our purposes is that the integrand in the definition of the momentum thickness (Equation 4.2.11) vanishes when the velocity goes to zero, so that the shift in origin would not affect the momentum thickness to a significant degree. Thus a simple shift of the velocity profile, with no change in either the skin friction or the momentum thickness, is consistent with the rough-wall flow satisfying the same integrated momentum balance as the smooth-wall flow, as expressed by Equation 4.2.10. Furthermore, in the region well above the tops of the roughness, the shifted velocity profile of Figure 6.1.11b should satisfy the differential equations of motion in the same way the smooth-wall flowfield of Figure 6.1.11a does, provided the roughness doesn't directly affect the Reynolds stress. The conclusion from this arm-waving line of argument is that roughness elements sufficiently submerged within the linear part of the velocity profile should cause no significant increase in skin friction, and only a small shift in the apparent origin of the boundary layer.

But when roughness elements are not sufficiently submerged in the viscous sublayer, a shift in origin can no longer prevent an increase in skin friction, as we argued in item (2) above. We also noted that this explanation is supported by the "discrete element" approach to predicting roughness drag in Taylor, Coleman, and Hodge (1984). In this

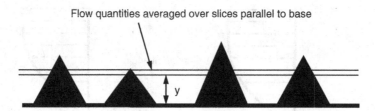

Figure 6.1.12   In the "discrete-element" approach to predicting roughness drag by Taylor, Coleman, and Hodge (1984), the flow domain into which roughness elements protrude is divided into thin slices parallel to the nominal base of the rough surface

approach, the flow domain that has roughness elements protruding into it is divided into thin slices parallel to the nominal base of the rough surface, as illustrated in Figure 6.1.12. In each slice, the flow is treated as 2D in plan view, and the effects of the roughness elements protruding through the slice are accounted for in terms of flow quantities averaged over the planform area of the slice. Based on the geometry of the protrusions, an average element diameter and the average flow area blocked by protrusions are calculated. The average flow velocity in the slice and the average mass flow are related through a reduced cross-sectional area that accounts for this blockage. In each slice, the average drag of the roughness elements is calculated from the local average dynamic pressure and a drag coefficient that depends on the Reynolds number based on the local average velocity and the average element "diameter" at that value of y. Nothing magic is assumed regarding how this drag coefficient varies with Reynolds number. The turbulent shear stress exerted across interfaces between slices is determined by a conventional mixing-length model for smooth-wall turbulent boundary layers, which is not modified to account for roughness.

When all is said and done, the theory predicts that C_f increases when roughness elements protrude outside the sublayer, and agreement with experiments for various rough surfaces is fairly good. The "complicated" drag curve of a given rough surface is thus found to be predicted fairly well by simple local physics and geometry, and detailed bookkeeping of the effects of the nonlinear mean-velocity profile.

So the success of the "discrete element" approach seems to support the proposal that a shift in the effective origin of the boundary layer prevents a significant increase in skin friction when roughness is sufficiently small, but not when roughness is too large to be sufficiently submerged in the sublayer. And the fact that predicting even the gross features of the C_f-versus-k^+ curves requires detailed bookkeeping calculations offers an excuse as to why our "simple" qualitative argument can explain only the existence of a smooth regime and the sign of the trend above that, but not the shapes of the curves.

6.1.9   Interference Drag

Aerodynamic interference between aircraft components can alter the total drag compared with what would be estimated for the components in isolation. Drag increases are much more common than drag reductions ("favorable interference"). Whatever the sign, such increments are often referred to as *interference drag*. This terminology incurs the same problem we encountered above in our discussion of "base drag" and "slot drag": there is no rigorous way to define the baseline for comparison. Separate airplane components, by themselves, don't generally form closed bodies that can be realistically "flown" in isolation for a baseline drag evaluation. And even if they could, the flow around them in isolation wouldn't be representative of a reasonable baseline condition, because the components were designed to fly in conjunction. Thus interference drag is not a "component" of drag with a physically rigorous definition. But it does make sense in the context of drag estimation, where we can accept a looser definition of the baseline in terms of "what would be estimated" (usually by handbook methods) for the drag contributions of the separate airplane components, before any accounting for interference.

Taking this looser definition, interference drag can comprise contributions from induced drag, shock drag, and viscous drag. Changes to induced drag and shock drag are the only interference mechanisms that look effectively "inviscid" in the near field. The mutual inviscid buoyancy effect that occurs when two bodies with volume are "flown" in proximity is not properly considered interference drag, as we concluded in Section 6.1.3. In a shock-free inviscid flow, such effects would always add up to zero for the total configuration. Again, as we saw above in connection with pressure drag, the total drag can change only if one or more of the drag-producing mechanisms in the field is affected. Now let's look specifically at the viscous contribution to interference drag.

For a classical aircraft configured according to the Cayley ideal (see Küchemann, 1978) with relatively high-aspect-ratio wings and distinct fuselage and tail surfaces, much of the viscous drag is generated in boundary layer and wake flows that are generally similar to the simple flows discussed in Section 6.1.5. Over large portions of the fuselage surface, the boundary layer behaves qualitatively like that on an axisymmetric body. Even if the wings are swept, the flow over the airfoil sections is nearly 2D in the sense that flow quantities change relatively slowly in the spanwise direction. (With sweep, the wing boundary layer is 3D in the sense that substantial cross-flow generally develops, but 2D in the sense of slow spanwise change. We discussed this type of behavior in theoretical terms in Section 4.3.5, and we'll look at the behavior on wings in more detail in Section 8.6.2.) The generation of viscous drag in these flows follows the qualitative pattern of the airfoil flow in Figure 6.1.5, in that pressure drag is a relatively small fraction of the viscous drag, more of the dissipation takes place in the boundary layer than in the wake, and the dissipation in the wake is essentially complete within a relatively short distance.

Aerodynamic interference between aircraft components generally increases the viscous drag over what the components would experience if the simple flow patterns described above prevailed. The viscous part of this interference results mainly from strong three-dimensionality and local 3D separations that occur at tips and junctions. Figure 4.5.1a schematically illustrates the "necklace vortex" type of separation that is common in wing-root junctions. Even a junction with a strakelet or fillet that does not produce a distinct separation, as shown in Figure 4.5.1c, still induces strong three-dimensionality in the fuselage boundary layer, which probably still produces a drag increase.

The drag penalties associated with such viscous interference effects are difficult to isolate and assess. Three-dimensional NS calculations can reproduce such flows as necklace vortices at least qualitatively, but the fidelity of turbulence models is not good enough that we should expect accurate quantitative assessments of the associated drag penalties. Assessment of these effects therefore often relies on testing or on empirical correlations, which are necessarily rather crude. The most extensive collection of correlations is in Hoerner (1965).

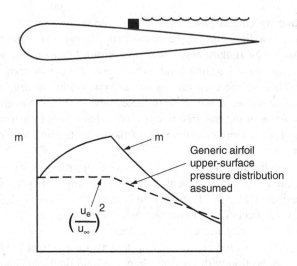

Figure 6.1.13 Example from Nash and Bradshaw (1967) of an excrescence on an airfoil, showing how the excrescence-drag magnification factor m can differ from a simple dynamic-pressure correction.

Another source of viscous drag is surface excrescences and roughness. The total drag associated with such items is not limited to the forces on the items themselves, but is usually increased by the effects of aerodynamic interference with the rest of the configuration. The flowfields around these items produce changes in the skin-friction on the surrounding surface that can be either positive or negative, but the net change in the pressure drag of the rest of the configuration is generally positive. The resulting "magnification" effect on the drag due to an item is usually larger than what would be expected just because the item might be immersed in a part of the flowfield where the local velocity is greater than that of the freestream.

One theoretical model for these effects is that of Nash and Bradshaw (1967), and the example from their paper is shown in Figure 6.1.13. The theory is 2D and is based on some very crude simplifying assumptions. In the general 3D case, estimates of the magnification effect based on this 2D theory should be regarded as very uncertain.

Another model for this kind of effect, really more of a rule of thumb, is based on the idea of work done, and is even more crude than Nash-Bradshaw, but is not restricted to 2D. If the local flow field surrounding the excrescence has an identifiable local "freestream" velocity different from the global freestream, the rate of work done against the part of the total drag that we can blame on the excrescence should be close to the local drag force on the excrescence (in the local flow direction) times the local flow velocity, not the freestream velocity. This suggests a magnification factor of $(q_{local}/q_\infty)^{1.5}$ for the total drag. The difference between the force on the excrescence (this time resolved in the global drag direction) and the magnified drag increment would presumably be mostly pressure drag on the larger body, due to the source effect of the viscous wake of the excrescence. An idealized example that supports this rule is shown in Figure 13 of Drela (2009).

We've looked at two crude models for a magnification effect on the drag of an excrescence, Nash-Bradshaw and a local-$q^{1.5}$ model. Both of these models suggest that the total drag increment does not depend on the direction of the local flow in which the excrescence is immersed, only the local velocity magnitude. For example, an excrescence located on an airfoil leading edge where the local flow direction is vertical should produce the same total drag increment as if the same excrescence were placed somewhere else on the airfoil where the local velocity magnitude is the same. Thus even if the local drag force on the excrescence is perpendicular to the global drag direction, the excrescence will still cause a drag penalty.

6.1.10 Some Basic Physics of Propulsion

As we noted in the introduction to this chapter, the purpose of propulsion is generally to control the total flight-direction force on a vehicle, for example, to make it zero for steady, level flight or to make it a net propulsive force for climbing or accelerating flight. Thus although propulsion systems often affect other components of the aerodynamic force to some extent, their effect on the flight-direction component is usually the largest one and the one we're most interested in. In this section, we'll touch on only the most basic physics of propulsion and its effects in the external flowfield. A more detailed discussion of the different types of engines and their internal flows is beyond the scope of this book.

The total aerodynamic force in the flight-direction affects the momentum balance in the flowfield in the same way whether propulsion is present or not, and the analysis that we applied to drag at the beginning of Section 6.1.4 also applies to situations with propulsion, with one caveat that is usually minor. The exhaust gasses of most propulsion systems include some mass of fuel that was not part of the flow from upstream, but this is usually small compared to the mass of the flow affected by the propulsor. Thus Equations 6.1.2 and 6.1.3 apply to situations with propulsion, with only a small error due to the neglect of fuel mass addition. When we applied the theory to pure drag in Section 6.1.4, we saw that drag implies a wake with a net deficit in momentum flux and/or pressure. In cases with propulsion, we'll see a combination of viscous wake and *propulsion plume* that will have a combined net deficit or excess in momentum-flux and/or pressure, depending on how much propulsion is being applied. For example, when the total flight-direction force is zero, as required for steady level flight, the net deficit would be zero.

Whether the wake and propulsion plume are separately distinguishable depends on the how close to the vehicle we look and on the degree to which the propulsion system is integrated with the rest of the configuration. And there is a spectrum of possibilities for the degree of propulsion integration, with the two extremes illustrated in Figure 6.1.14. At one end of the spectrum is the Cayley concept of a classical aircraft (see Küchemann, 1978), which assumes that the propulsion system is distinct from the rest of the configuration. The aircraft wake and the propulsion plume are distinct in the near field, and this distinctness provides one basis for determining separate drag and thrust. At the other end of the spectrum we have the Ackeret ideal, in which propulsion is integrated with the configuration to such an extent that the propulsion system energizes only the air that has experienced losses in the boundary layer, and no viscous wake or propulsion plume remains to produce further dissipation. Assigning separate thrust and drag in this case is problematic. The Ackeret ideal makes no provision for making up the induced losses associated with lift in 3D, which we'll discuss in Section 8.3.

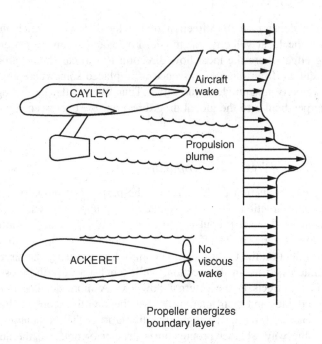

Figure 6.1.14   The Cayley and Ackeret ideals of propulsion

On the spectrum between these two ideals, real aircraft fall much closer to the Cayley than to the Ackeret. However, there is always some interference between the propulsion system and the rest of the aircraft, such as the interference effects of propeller slipstreams, the complicated mixture of propulsion effects and drag effects on the surfaces of bodies or nacelles near jet exhausts (the "boat-tail drag" associated with narrowing of a body ahead of a jet nozzle, for example), and the interaction of propulsion inlets with the rest of the flowfield.

Much of the theory of propulsion deals with propulsors as isolated thrust producers, as they would be in the Cayley ideal. When an isolated propulsor produces thrust, according to Equation 6.1.3 its exhaust plume must have a net excess of momentum flux and/or pressure. In a simple propulsion plume without circumferential velocities, the pressure far downstream must decay to freestream pressure, as we assumed in our analyses of simple viscous drag in Section 6.1.4. In this case, only an excess of velocity remains, as in the Cayley "propulsion plume" sketched in Figure 6.1.14. When a propulsion plume retains significant circumferential velocities, as in the case of a single-rotation propeller, the plume will have a pressure deficit consistent with centrifugal effects, and for a given thrust this must be accompanied by a correspondingly larger excess in streamwise velocity. Note that this is not the same issue as the energy loss due to "swirl" that we'll discuss later in connection with induced efficiency.

To produce a plume with an excess in momentum flux, the propulsor must add energy to the flow through some combination of mechanical work and heat addition. The possible combinations range from work only (a propeller with no exhaust gas added to the flow) to heat only (a ram jet). The most practical modern propulsors are open propellers and turbofans, both of which use rotating airfoil-shaped blades to transfer mechanical work to and from the flow.

If the situation is nominally steady in the long term and we ignore or time-average fluctuations associated with the passage of propulsor blades, we can distinguish a *propulsion streamtube* in which the flow is acted on directly by the propulsor. For a propeller, it is the streamtube that passes through the disc swept by the propeller blades, and for a ducted engine such as a turbojet or turbofan, it is the streamtube that enters the inlet. Defining a propulsion streamtube identifies the part of the field that is most directly affected by the propulsor, and in which the most important changes in total-pressure and total-temperature take place. When work is done on the fluid in the propulsion streamtube, it increases both the total-pressure and the total-temperature of the stream. Heat addition further increases the total-temperature but reduces total-pressure (see Shapiro, 1953, Section 7.2). Unsteady fluctuations (due to blade passage) that extend outside the propulsion streamtube can affect total-pressure, but these effects are usually small and are ignored in the simpler theories. Downstream of the propulsor, there is generally turbulent mixing between the propulsion streamtube and the external flow, redistributing total-pressure and total temperature, but these effects are also usually neglected in the theories.

To delve into the physics issues further, let's consider the open propeller, the simplest propulsor in terms of basic physics, with no duct and no heat or mass addition to deal with.

The simplest and oldest theories for propellers ignore the details of the propeller itself and use control-volume analysis to predict average velocities in the propulsion streamtube and to make rough estimates of propeller performance trends, such as how efficiency varies with the thrust loading coefficient (defined shortly in Equation 6.1.18). The flowfield assumed in these sorts of theories is illustrated in Figure 6.1.15, with the propulsion streamtube running from left to right. The flow is assumed to be axisymmetric, steady, inviscid, and incompressible. The discrete effects of passing propeller blades are ignored, and nothing is assumed about the shape of the propeller required to produce the integrated effects. The theories can be thought of as representing either the time average of a real flow, or the flow in the limit as the number of blades goes to infinity and their chord goes to zero. They incur significant error by neglecting discrete blade effects, and of course they cannot predict differences due to different numbers of blades.

The propeller itself is represented by an idealized *actuator disc* across which the pressure increases as a result of the work done by the propeller, and the tangential velocity can change to account for circumferential velocity (swirl) in the downstream flow. No change in axial velocity occurs across the disc, as required by mass conservation in incompressible flow. The pressure jump across the disc is assumed uniform for simplicity. The axial velocity just upstream and downstream of the disc is also assumed uniform, though we'll see later that this isn't realistic. Because the jumps in axial velocity and axial momentum flux are zero, momentum conservation requires that the integrated pressure jump equal the thrust. The assumption of axisymmetric flow means that there is no mechanism for pressure forces to produce changes in circumferential velocity and thus dictates that circumferential velocities must be zero upstream of the actuator disc. The integrated flux of moment of momentum must therefore be zero upstream of the disc. Changes in circumferential velocity and in the flux of moment of momentum can take place only as jumps across the disc. The integrated flux of moment of momentum downstream of the disc must equal the torque applied by the propeller, and continuing downstream it must remain constant.

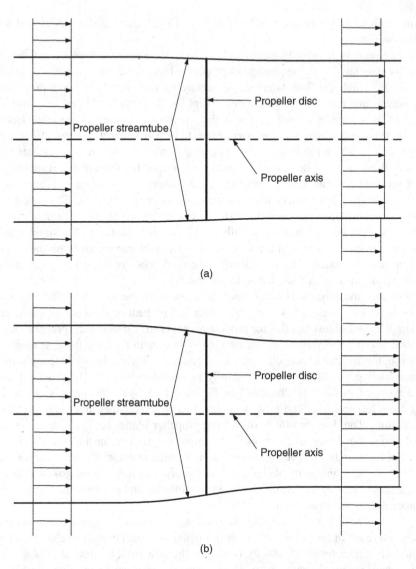

Figure 6.1.15   Illustrations of the idealized propulsion streamtube and propeller actuator disc assumed in actuator-disc theories for propellers. The contraction of the propulsion streamtube and the excess axial velocity downstream are indicated approximately to their correct scales for the indicated thrust loadings. (a) Moderate thrust loading, $\tau = 0.2$, typical of cruising flight. (b) High thrust loading, $\tau = 1.2$, typical of climbing flight

For purposes of physical understanding, actuator-disc theory is useful because it indicates the general time-averaged shape that the propulsion streamtube must have, to satisfy both mass and momentum conservation. For the idealized propeller flowfield we're considering, the axial momentum balance requires axial velocity higher than freestream at the propeller disc and even higher still in the propulsion streamtube downstream. The propulsion streamtube must therefore contract as illustrated in Figure 6.1.15. The theory indicates that half the change in axial velocity occurs upstream of the disc and half downstream, with the streamtube area decreasing inversely with the velocity. The degree of streamtube contraction and the level of excess velocity downstream both increase with increasing thrust loading and are shown approximately to scale in Figure 6.1.15. Part (a) corresponds to a moderate thrust loading typical of a single-engine light plane cruising, and part (b) to a high loading typical of climbing.

The velocity field we've assumed requires the bounding surface of the propulsion stream-tube downstream of the actuator disc to be a vortex sheet across which there is a jump in the velocity vector. Associated with the jump in longitudinal velocity is a circumferential vorticity component, and associated with the jump in circumferential velocity is a longitudinal vorticity component. The vortex lines in the bounding sheet are thus helical, and the vorticity in the sheet is uniformly distributed circumferentially. The real distribution of vorticity produced by the propeller blades is of course complicated and unsteady.

An early version of actuator-disc theory that doesn't account for swirl is credited to Rankine in 1865 and was motivated more by nautical than aeronautical applications. Later elaborations on the theory account crudely for swirl and for the integrated effects of the viscous drag of the propeller blades, as described by von Mises (1959).

Discussions in the literature (including that in von Mises, 1959) generally fail to mention a problem with simple actuator-disc models, and it has to do with the axial velocity where the bounding vortex sheet abuts the edge of the actuator disc. We've assumed no jump in axial velocity across the disc, but the jump in total-pressure across the disc will generally dictate a jump in axial velocity across the vortex sheet that abuts the edge of the disc. Spalart (2003) has proposed that a spiral singularity in the shape of the vortex sheet at the edge of the disc might solve the problem. This is not a settled issue.

Moving beyond the idealized actuator-disc models, a detailed discussion of the physics must include the effects of the discrete, rotating blades. As the propeller moves forward and rotates about its axis, its blade tips follow helical paths as illustrated in Figure 6.1.16. The ratio between the distance traveled in one revolution and the diameter is called the *advance ratio*, $J = V/nd_p$, where V is the forward velocity through the air mass, n is the rotations per unit time, and d_p is the propeller diameter. Blade sections inboard of the tips move along helices with the same pitch but smaller radii, thus moving at steeper angles relative to the circumferential direction. In the reference frame of a blade section, the flow approaches the section from a direction determined by a combination of the section's helical motion and the disturbance velocities produced by the propeller. Preferably, the blades are twisted to account for all of this, so as to keep each section at a favorable angle of attack to the local relative flow and to produce a favorable distribution of loading along the blade. We'll see later how one theoretical estimate of the "optimum" load distribution is determined.

An intuitive starting point for understanding how a propeller works is to think of it as a *screw*. This is an image that has been with us for a long time: In the early days an aeronautical propeller was called an *airscrew*, and a nautical propeller is still sometimes referred to as a screw. Imagine the propeller as a short section of a machine screw being turned by some applied torque and thus moving itself along a threaded hole and applying thrust against something that resists its axial motion. (Or, instead of a threaded hole and a machine screw, imagine the helical tip path of Figure 6.1.16 as a rigid, stationary rail along which the propeller tip slides. And the rail analogy needn't be limited to the propeller

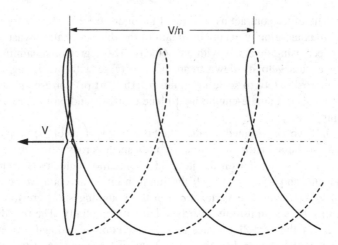

Figure 6.1.16 A propeller moving axially through the air mass at velocity V, rotating at n revolutions per unit time. The propeller tips trace helical paths as shown. The ratio between the distance traveled in one revolution and the diameter is known as the *advance ratio*, $J = V/nd_p$

tip. We can just as well imagine any section along the span of the blade sliding along its corresponding helical rail.) Further, imagine that there is no mechanical friction. Without friction losses, and with the propeller reacting against a stationary thread or rail, the input work done in a given time by the turning torque and the output thrust work done against the axial resistance would be equal, and the propeller efficiency (thrust work out/shaft work in) would be 100%. Of course in a fluid, especially a viscous fluid, propeller efficiency must be less than 100%, and for a given thrust, forward speed, and rotation speed, the torque required is always greater than it would be for the idealized screw.

We can think of the required additional torque as comprising two parts, one to make up for *viscous loss,* and one to make up for *induced loss.* The viscous loss is due to the viscous drag of the blade sections, the same thing as the profile drag of an airfoil, which we'll discuss in Section 7.4.2. In the rail analogy, mechanical friction between the blade and the rail would be analogous to sectional viscous drag. The induced loss results from the propeller's setting air in motion, both axially and circumferentially, and thus leaving kinetic energy behind. To model the induced loss in our mechanical rail analogy, we would have to allow the rail to move (or "yield") in the directions of both the thrust force and the torque, and thus to absorb some of the work done by the turning torque. Holding the same rotation speed and forward speed would then require a steeper helix angle, and greater torque would be required to make up for the loss and deliver the same thrust. Aerodynamically speaking, the induced loss is analogous to the induced drag of a wing of finite span, which we'll discuss in Section 8.3. Note that this decomposition of the additional torque into viscous and induced parts is not exact and can only be estimated with idealized models, the same situation we pointed out with regard to drag in Section 6.1.3.

A propeller blade acts like a lifting wing operating in a non-uniform flowfield in which the relative local velocity vector varies along the span. A schematic vector diagram of the situation seen by a typical section of the blade at radius r from the propeller axis is

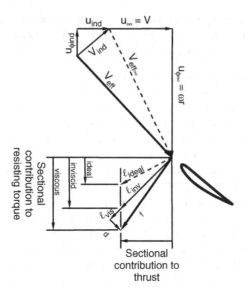

Figure 6.1.17 Schematic diagram of the relative flow velocities and forces seen by a propeller blade section at radius r from the propeller axis. The section is moving from right to left with flight velocity V and upward with circumferential velocity ωr resulting from the angular velocity $\omega = 2\pi n$ of the propeller. The view is radially inward along the axis of the blade and is consistent with right-hand rotation of the propeller, as in Figure 6.1.16. The upper portion of the diagram shows the components of the local relative flow velocity $\mathbf{V}_{\text{eff}}$ seen by the section, and the lower portion shows three versions of the resulting force components on the section for a fixed net thrust contribution

shown in Figure 6.1.17. The upper portion of the diagram shows the components of the relative flow velocity seen by the section. First, the forward velocity V of the propeller and the circumferential velocity ωr of the blade section combine to produce an effective local "undisturbed freestream" velocity vector $\mathbf{V}_{\text{eff}\infty}$. Next we take into account the local inviscid disturbance velocity produced by the propeller at the location of the section, which we call $\mathbf{V}_{\text{ind}}$ because it is often referred to as the "induced" velocity. When $\mathbf{V}_{\text{ind}}$ is added to $\mathbf{V}_{\text{eff}\infty}$, the resultant is the total effective local flow velocity vector $\mathbf{V}_{\text{eff}}$. The effect of $\mathbf{V}_{\text{ind}}$ is generally to increase the axial velocity and decrease the circumferential velocity seen by the section.

The lower portion of Figure 6.1.17 illustrates three versions of the resulting force components on the section, with and without the effect of $\mathbf{V}_{\text{ind}}$, and with and without the effect of viscous drag **d.** First, note that the local sectional lift is defined as the component of the sectional force perpendicular to the local relative flow direction (in contrast with the global definition we introduced in Section 5.4, where total lift was the component of the total aerodynamic force perpendicular to the global freestream). Note further that the axial component of the local sectional lift contributes to the thrust, and the axial component of the sectional drag subtracts from thrust. For this comparison, the lift components are varied so as to hold the net sectional contribution to thrust constant, at the value indicated by the horizontal arrow at the bottom of the diagram. The circumferential components of both lift and drag contribute to the torque that retards the propeller's rotation. The net circumferential components contributing to the torque are indicated by the downward arrows at the lower left. The three versions of the force vector are:

"Ideal:"	In the idealized situation with no "induced" velocity (which would apply in the limit of zero load on the propeller), the lift vector l_{ideal} is perpendicular to the effective local "undisturbed freestream" velocity vector $V_{eff\infty}$. This situation is analogous to having the section slide along a stationary frictionless rail, and with no mechanism for any loss, all of the work done against the resisting torque shows up as thrust work at an efficiency of 100%.
"Inviscid:"	The inviscid "induced" velocity V_{ind} is now included, and the resulting lift vector l is now perpendicular to the local effective flow velocity vector V_{eff}. Note that holding the sectional contribution to thrust constant requires the magnitude of l_{inv} to be larger than that of l_{ideal}. Compared to the "ideal" case, the resisting torque has increased, and even in this case with zero viscous drag the efficiency is less than 100%.
"Viscous:"	When both the inviscid "induced" velocity and a viscous-drag component d are included, the resultant is the total sectional force vector f. Note that to maintain a constant net contribution to thrust, the lift magnitude has had to increase further, that is, the magnitude of l_{vis} is larger than that of l_{inv}. This would produce a larger V_{ind}, and to be strictly correct, l_{vis} and l_{inv} should be shown as having slightly different directions. But in practical situations, in which d tends to be small, this is a small effect, so it is ignored in the diagram. Compared to the "inviscid" case, viscous drag has increased the resisting torque, further reducing efficiency.

Note that holding the net sectional thrust constant in these comparisons simplified the vector diagram a bit, but was not essential. Instead, we could have held the magnitude of the lift vector constant, for example. In that case, including the effects of V_{ind} and d would have increased the required torque less but decreased the thrust. The directions of the resulting force vectors would be the same, as would our conclusions regarding efficiency. With the diagram in its present form with net thrust held constant, the longer vertical arrows on the lower left for the "inviscid" and "viscous" cases represent the additional torque required for a given thrust, relative to that required to turn the "ideal" screw. But keep in mind the caveat above, that this decomposition of the additional torque into an "induced" part and a viscous part is an idealization. In the view illustrated in Figure 6.1.17, this would be reflected in the fact that in a real flow the "induced" velocity V_{ind} cannot be defined exactly, but only through the use of an idealized inviscid model.

The flow produced by the propeller is quite three-dimensional and unsteady. The flow around and downstream of each blade has much in common with the flow around a 3D wing, which we'll discuss in Section 8.1.1-3. Propeller blades and 3D wings both shed vortex sheets from their trailing edges. In both cases the sheets start out streaming back in the local flow direction, and as they progress downstream they deform and roll up at the outer edges to form "tip vortices." In the case of a propeller, the vortex sheets and the tip vortices follow roughly helical paths, but relative to the helical paths of the propeller tips in Figure 6.1.16, the tip vortices generally migrate aft faster and migrate inward, consistent with the increased axial velocity of the flow and the general contraction of the propulsion streamtube illustrated in Figure 6.1.15.

The torque exerted by the propeller introduces angular momentum into the downstream flow in the form of circumferential velocities that tend to be everywhere in the same direction as the propeller rotation. This circumferential motion is loosely referred to as "swirl," a term that can be misleading because it is easily confused with the axial vorticity. Although the swirl tends to have the same sign everywhere, it is easy to show that axial vorticity of both signs must always be present. If we ignore the radial velocity, which is usually small in propeller flows, the axial vorticity is given by

$$\omega_x = \frac{1}{r}\frac{\partial}{\partial r}(ru_\varphi), \qquad (6.1.17)$$

where u_φ is the circumferential velocity. Moving from the axis outward, u_φ must first increase from zero and then return to zero outside the slipstream, and ω_x must have opposite signs in the regions of increase and decrease. And this is consistent with an even stronger statement about the axial vorticity, which is that the integrated axial vorticity over a any constant-x surface that cuts the entire propulsion streamtube must be zero. This can be established by applying Stokes's theorem to any closed contour that surrounds the entire propulsion streamtube, combined with a capping surface bounded by the closed contour, where the capping surface bulges out in front (upstream) of the propeller and cuts none of the downstream propulsion streamtube. The flow everywhere on such a capping surface is irrotational, and the integrated vorticity flux through it must therefore be zero, so that the circulation around any closed contour outside the propulsion streamtube is also zero. This point is not always correctly understood. There have been depictions incorrectly showing propeller streamtubes as having net axial vortex strength that "induces" circumferential velocities at locations outside the propeller streamtube (Loth and Loth, 1984, for example). Actually, we've just shown that this can't happen. Zero net axial vorticity in the propeller streamtube means that there can be no such effect extending outside the propeller streamtube. Further, it means that in the idealized case of a time-averaged, axisymmetric propeller flow there can be nonzero swirl velocities only inside the propulsion streamtube, and only in the part downstream of the propeller.

The fact that the propeller streamtube carries zero integrated axial vorticity means that the blade tip vortices must be offset by vorticity of opposite sign inside the streamtube. For efficient propeller designs, the vorticity feeding the tip vortices is shed close to the blade tips, and the offsetting vorticity of opposite sign is shed from the inboard portions of the blades. For an idealized isolated propeller with no fuselage or nacelle behind it, the vorticity shed inboard gathers downstream into a vortex core similar to the tip vortex behind a 3D wing that we'll discuss in Section 8.1.2. In real propeller installations, this inboard vorticity forms a kind of shell streaming back along the outer surface of the fuselage or nacelle.

Between the inboard and outboard regions of shedding, there is usually a significant expanse of the blades' radius from which relatively little vorticity of either sign is shed. There is thus an annular region in the propeller streamtube in which there is little vorticity and the flow is similar to that around a potential vortex, with the circumferential velocity u_φ going roughly as 1/r.

We've seen that a propeller flowfield is complex, with contorted, moving vortex sheets and tip vortices, and complicated unsteady distributions of axial and circumferential velocity. But despite the complexity, we can use a simple dimensional argument to predict how propeller performance should scale, and idealized models to predict the major performance trends. The dimensional argument is as follows. At low Mach numbers and high Reynolds numbers, the dependence on Mach number and Reynolds number should be weak. For a propeller of a given shape and operating at a given advance ratio, the flowfield in dimensionless terms is effectively determined, regardless of the geometric scale or the velocity level, and the pressure disturbance anywhere on a blade element will scale closely with V^2. Then the following dimensionless performance measures should depend almost exclusively on propeller shape and advance ratio:

$$\text{Thrust loading: } \tau = T/qs_p, \tag{6.1.18}$$

$$\text{Torque coefficient: } \mu = M/qs_p\, d_p, \tag{6.1.19}$$

$$\text{Shaft power loading: } \sigma = P/qVs_p, \tag{6.1.20}$$

$$\text{Efficiency: thrust power out/shaft power in} = \eta = TV/P = \tau/\sigma, \tag{6.1.21}$$

where s_p is the propeller disc area. Actually predicting values for these coefficients requires calculating the detailed forces on the propeller blades, which can be done to various levels of fidelity under the general rubric of *blade-element theory*.

The problem blade-element theory deals with is similar to that of predicting the spanwise distribution of lift (the "spanload") of a 3D wing, which we'll discuss in Section 8.2. Just as in the case of a wing, the load on a section of a propeller blade influences the flow direction seen everywhere along all the blades and thus influences the loading at all blade sections. In the simplest blade-element theory, this influence is ignored, and blade section loadings are calculated based on undisturbed local conditions, an approximation valid in the limit of light loadings (see von Mises, 1959). Then there are more-accurate potential-flow theories for wings and propellers, in which this nonlocal influence is inferred from the vorticity distribution in the wake, that is, the velocities "induced" by the wake vorticity are calculated using the Biot-Savart law. In such theories, the vortex sheets are usually assumed not to distort and roll up at their edges, which for a propeller means that a helical screw shape is assumed for the vortex sheets. Prandtl (in Prandtl and Betz, 1927) derived approximate expressions for the velocities "induced" by such sheets, which have been used in both analysis methods (calculating the loading on blades of a given shape) and design methods (calculating the twist of the blades so as to produce a given loading). Betz (also in Prandtl and Betz, 1927) showed that the loading that produces the minimum induced loss for a given thrust, diameter, and advance ratio produces a velocity field that ideally should not distort the assumed helical-screw-shaped vortex sheets (i.e., a velocity field compatible with "rigid" but not stationary vortex sheets), and Goldstein (1929) derived exact potential-flow solutions for these cases, for propellers with two blades or four blades. Goldstein's solutions define "optimum" load distributions that have been used for the design of propellers for minimum induced loss (see Drela, 2006), which should come close to maximizing efficiency overall. Although the potential-flow solutions on which these designs are based are exact, the physical model is not, because it assumes an idealized configuration of helical vortex sheets of fixed pitch, ignoring the overall contraction of the propulsion streamtube and the sheets' distortion and rollup into tip vortices.

Now we'll use numerical calculations to explore how propeller efficiency varies with thrust loading and advance ratio for propellers designed to meet Goldstein's minimum-induced-loss blade loadings. The QMIL code (the design version of the QPROP analysis code, see Drela, 2006) takes the number of blades, thrust loading, advance ratio, and propeller blade airfoil characteristics as inputs, and designs the propeller to meet an approximate version of the appropriate Goldstein blade loading. Design outputs are the distributions of blade chord and pitch angle. Flow solution outputs include distributions of forces on the blades, radial distributions of "induced" velocities, both at the blade locations and circumferentially averaged, total torque and power, and induced and total efficiencies. Blade section viscous drag is taken into account in the calculations or not, as desired. For the plots presented below, cases were calculated for several dense arrays of (τ, J) pairs, and curves of constant efficiency and maximum efficiency were interpolated from the resulting database. Keep in mind that every point on any of these curves represents a different minimum-induced-loss propeller operating at its design condition. These are not performance curves for fixed propellers.

First, let's look at induced efficiency η_i for two-bladed propellers designed and operating with no viscous drag on the blades. Curves of constant η_i are shown on a J-versus-τ plot in Figure 6.1.18. Two basic trends are evident. First, induced efficiency decreases with increasing thrust loading. Second, induced efficiency decreases with increasing advance ratio (and the associated lower rotation speed and higher torque). Both trends are consistent with a simple momentum-versus-energy interpretation. The "payoff" (thrust) is proportional to the excess momentum imparted, while part of the "cost" (power required) is related to the kinetic energy left behind. A low thrust loading imparts a lower excess velocity and thus lower kinetic energy for a given momentum, and yields higher induced efficiency. Likewise, a low advance ratio corresponds to higher rotation speed for a given forward

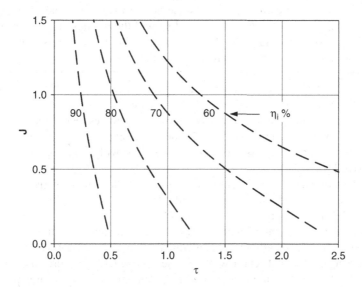

Figure 6.1.18 Curves of constant induced efficiency η_i in terms of J versus τ for two-bladed propellers with zero viscous drag on the blades. Calculated by the QMIL code (Drela, 2006)

speed and diameter, which delivers the required power with lower torque, thus imparting less swirl, leaving behind less kinetic energy, and giving higher induced efficiency. In these inviscid calculations, maximum induced efficiency for a given induced efficiency. In τ comes in the limit of zero advance ratio, as rotation speed goes to infinity, and torque goes to zero. Obviously this part of the trend won't hold up for blades with viscous drag, as we'll see next.

We saw in Figure 6.1.17 how viscous drag increases the resisting torque when thrust is held constant, reducing efficiency. Now, before we look at the results of detailed calculations, let's think about how we would expect the effect of viscous drag to vary with advance ratio J. Two limiting cases are informative:

Low J: A typical blade element advances along a helix with a shallow angle relative
             to the circumferential direction. The sectional lift (l in Figure 6.1.17)
             lines up closely with the thrust direction and contributes nearly 100% to
             thrust. This minimizes the blade lift forces required for a given thrust,
             and provided the blade chord is sized so that the section operates near its
             maximum lift/drag ratio, this also minimizes the viscous drag forces. But
             for a given distance of forward travel by the propeller, the distance
             traveled by the section is large, so that even the relatively small viscous
             drag forces produce a large energy loss and a large loss of efficiency.

High J: A typical blade element advances along a helix with a steep angle (nearly
             90°) relative to the circumferential direction. The sectional lift (l in
             Figure 6.1.17) lines up close to the circumferential direction and
             contributes little to thrust. If the helix angle is too close to 90°, the
             viscous drag will more than cancel the small thrust contribution from the
             lift, and net positive thrust will be impossible. Even if we back off on the
             helix angle enough to get some net positive thrust, the lift required for a
             given thrust will be large, and so will the viscous drag. Efficiency will be
             low even though the distance traveled by the section is only slightly
             longer than the forward travel of the propeller.

So with the inclusion of viscous drag we expect efficiency to drop off strongly for either extremely low or extremely high J. Between these two extremes there should be a favorable range where we avoid the high loss due to either excessive distance traveled or excessive lift required. It's not that the losses due to viscous drag will be negligible, just that we'll avoid excessive losses.

Now let's look at what detailed calculations predict for the effects of viscous drag. Recall that for each calculated operating point QMIL designs a propeller to a particular radial distribution of lift along the blades, that is, the load distribution given by the approximated Goldstein solution for minimum induced loss for that operating condition. When viscous drag is to be included, an arbitrary radial distribution of sectional viscous drag versus sectional lift can be specified. On a well designed propeller, the chord distribution would be tailored so that each section operates near its maximum sectional lift/drag ratio. Thus for generic calculations a constant sectional lift/drag ratio is not a bad assumption, and for these calculations, a constant ratio of 50 was enforced everywhere along the blades. The resulting

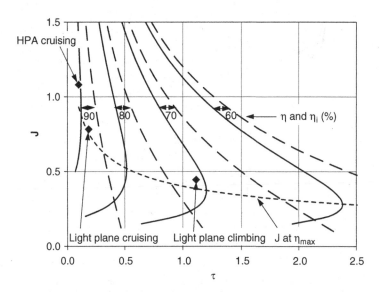

Figure 6.1.19 Curves of constant overall efficiency η (solid) in terms of J versus τ for two-bladed propellers, calculated by QMIL (Drela, 2006) for a blade-section lift/drag ratio of 50. The corresponding constant-η_i curves (long dashed) are included and are slightly different from those in Figure 6.1.18 because blade viscous drag constitutes a negative contribution to the thrust, so that for a given total τ the inviscid part of the thrust must be larger, and the induced losses are greater than in the inviscid cases

curves of constant overall η for two-bladed propellers are shown on a J-versus-τ plot in Figure 6.1.19. The corresponding η_i curves are included for comparison. They are slightly different from those in Figure 6.1.18 because the blade viscous drag constitutes a negative contribution to the thrust, so that for a given total τ the inviscid part of the thrust must be larger, and the induced losses are greater than in the inviscid cases. First, note that η is everywhere lower than η_i, as it should be, given that an additional loss mechanism is included. Note also that the efficiency drops off dramatically for low values of J, as we expected above. The maximum efficiency for a given τ now occurs at nonzero advance ratio. Over the practical range of J shown in Figure 6.1.19, we see no sign of the dramatic drop-off in efficiency that we expect for very high values of J.

Figure 6.1.20 summarizes the maximum-efficiency trends, showing η_{max} and the corresponding J at η_{max} as functions of τ. This makes clear the relationship between higher induced velocity and reduced efficiency as thrust loading increases. It also shows how the optimum advance ratio decreases as thrust loading increases. Going to four blades instead of two increases efficiency slightly and increases J at η_{max}. It makes intuitive sense that having more blades decreases the losses due to discrete blade effects and would favor slower rotation speed for a given forward speed, that is, higher advance ratio.

For a practical perspective, the operating points for some propellers in real applications are indicated in Figure 6.1.19. The points for a light plane cruising and climbing are the same ones for which the ideal propulsion streamtubes were sketched in Figure 6.1.15. The Human-Powered-Aircraft (HPA) point shows the result of seeking very high efficiency, that

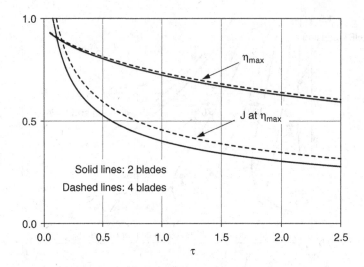

Figure 6.1.20 Trends for the point of maximum efficiency as functions of τ, for two-bladed and four-bladed propellers, calculated by QMIL (Drela, 2006) for a blade-section lift/drag ratio of 50

is, a very low thrust loading and high advance ratio. The light plane accepts more of a penalty in efficiency to avoid having to use a larger-diameter, slower-turning propeller, which would require a longer landing gear and a gearbox to reduce speed between the engine and propeller.

Figure 6.1.21 shows the induced axial and circumferential velocities, at the blade locations and circumferentially averaged, calculated for $\tau = 1.0$, $J = 0.5$, which is near the operating

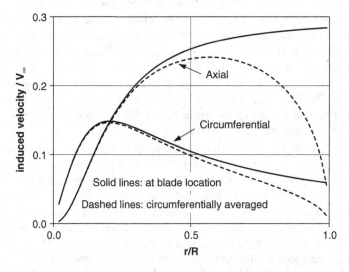

Figure 6.1.21 Radial distributions of induced velocities for a two-bladed propeller designed and operating at $\tau = 1.0$, $J = 0.5$, calculated by QMIL (Drela, 2006) for a blade-section lift/drag ratio of 50

condition for a lightplane climbing as shown in Figure 6.1.19. It is clear that the axial velocity is not constant as is assumed in the simple actuator-disc theories. The averaged circumferential velocity goes roughly as $1/r$ from $r/R = 0.3$ to 0.8, which is consistent with the flowfield we described earlier in which most of the vorticity shed from the blades is shed in an inner region near the axis and an annular outer region.

For flight at high-subsonic and supersonic speeds, turbojets and turbofans have been strongly preferred over open propellers. The reasons for this involve several aspects of how speed affects the different propulsion systems.

First, consider the factors affecting energy consumption by an open propeller. For a given flight mission, practical airframes tend to have optimum lift-to-drag ratios that fall within a narrow range regardless of the target speed, and propellers tend to have practical optimum efficiencies that also fall within a narrow range. Thus the amount of shaft work required to drive an airplane of a given weight over a given distance is roughly fixed. Furthermore, piston and turbo-shaft engines tend to burn a roughly fixed amount of fuel per unit of work done, regardless of the target rate (power level) for which they're sized. So far, this indicates that the amount of fuel required for a given mission should be relatively insensitive to the target speed up to the point where compressibility effects begin to reduce propeller efficiency. And there are other considerations that come into play, such as the fact that higher speeds favor higher cruise altitudes, and the effects of altitude on engine efficiency and on the power available from an engine of a given size and weight. The upshot is that the use of conventional open propellers has mostly been limited to cruise Mach numbers below 0.60.

Turbojets and turbofans allow operation at higher Mach numbers because their inlets slow the flow down before it encounters the rotating blades. But these engines don't just allow operation at higher speeds, they actually favor it because of the way their fuel consumption varies with speed. They don't share the open propeller's characteristic of requiring a roughly fixed amount of fuel per unit of thrust work. Instead, they favor speed (or are penalized for low speed), a bit like a pure rocket propulsion system. A pure rocket tends to require a fixed amount of fuel per unit thrust per unit time, regardless of speed, so that the fuel required for a given thrust over a given distance decreases with increasing speed, as $1/V$. Because turbojets and turbofans are air-breathing, they don't favor speed quite as much a pure rocket. The fuel consumption for a given thrust over a given distance varies approximately as $V^{-0.9}$ for the turbojet and as $V^{-0.5}$ for the modern high-bypass-ratio turbofan. Thus even the turbofan favors flight at high subsonic Mach numbers, where airfoils should be designed specifically for transonic flow, as we'll discuss in Section 7.4.8, and wings are usually swept, as we'll discuss in Section 8.6.

Even a turbofan tends to operate at relatively high thrust loadings, in the neighborhood of 0.6 in cruise, compared with 0.2 for the open propeller of a light plane in Figure 6.1.19. The other major difference in the flowfield is that while an open propeller producing positive thrust always speeds the flow up ahead of it, a ducted engine can slow the flow down before it enters the fan. Part of this deceleration takes place ahead of the inlet, and part inside. The result is that the pressure ahead of the fan is higher than freestream, a situation that is possible only because of the presence of the inlet duct. To support the high pressure inside, the inlet duct carries a load radially outward, effectively acting as a ring-shaped wing lifting outward. Deceleration of the flow ahead of the fan is essentially taking place in the large-scale flowfield of this ring-wing. We'll discuss airfoil and wing flowfields in detail in Sections 7.3.4 and 8.11.

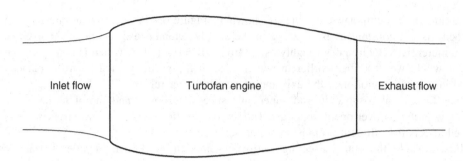

Inlet flow   Turbofan engine   Exhaust flow

Figure 6.1.22   Sketch showing the general character of the inlet and exhaust streamtubes of a turbofan engine operating at a thrust loading typical of cruise. The flow entering the engine slows down as it approaches the inlet, and the inlet streamtube grows larger. At higher thrust loadings typical of takeoff and climb, the opposite happens, more like the propeller streamtubes of Figure 6.1.15

So the flow ahead of a turbofan in cruise actually slows down ahead of the fan: And while the propulsion streamtube ahead of an open propeller always contracts, the "captured streamtube" of a turbofan in cruise starts off smaller than the inlet and grows larger approaching the fan, as sketched in Figure 6.1.22. However, at the low forward speeds and higher thrust loadings typical of takeoff and initial climb, the captured streamtube of a turbofan starts off larger than the inlet and contracts, more like that of an open propeller.

The actuator-disc approach can be extended to compressible flow and to include heat addition. For modeling ducted engines, the duct must generally be included. Actuator-disc modeling of ducted engines is used mainly in CFD applications, as we'll discuss in Section 10.4.6.

This concludes our treatment of isolated propulsors. Now let's return to the fully integrated ideal. Propulsion according to the Ackeret ideal as in Figure 6.1.14 can also be called *boundary-layer propulsion*, in which the boundary-layer air leaving the tail or trailing edge is reenergized to freestream total head, so that the dissipation that would otherwise take place in the wake (at least the part associated with the viscous drag) would be prevented. The airfoil example in Figure 6.1.5 indicates that the potential power saving might be equivalent to eliminating on the order of 10% of the viscous drag. Propulsion systems in some nautical applications (ships, submarines, and torpedoes) take at least some advantage of this effect. But the concept has never found an application in aeronautical practice because it would require distribution of propulsion energy from an engine to points throughout the airframe (all along the wing trailing edge, for example), and the resulting internal losses and extra weight and complexity would more than offset the benefit.

There is one vehicle configuration where such penalties would be minimal, and that is a body of revolution, as might be used for a lighter-than-air (LTA) airship or an underwater vehicle. Goldschmied (1982) developed a design for such a body that carried the concept of boundary-layer propulsion a step further, combining boundary-layer propulsion with boundary-layer control by suction to prevent separation. The body was designed to produce a pressure distribution similar to that of the Griffith-type laminar-flow airfoils (see Thwaites, 1958), with the pressure recovery concentrated over a single suction slot where most of the boundary-layer air is removed. There is no adverse pressure gradient on any of the external surface of the body, as there would have to be on the aft portion

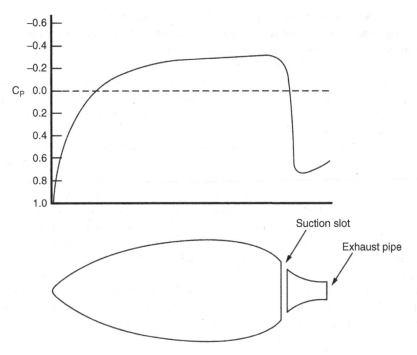

Figure 6.1.23 Shape and incompressible potential-flow pressure distribution of a self-propelled body of revolution using a combination of boundary-layer control and boundary-layer propulsion. (Based on Goldschmied, 1982)

of a conventional body designed to achieve attached flow passively. The body shape and incompressible potential-flow pressure distribution are sketched in Figure 6.1.23.

The suction air is pressurized by a fan and expelled through a central exhaust pipe at the tail. Because the boundary layer on the tailcone develops at very low velocity and in a favorable pressure gradient, its momentum deficit when it reaches the tail (the exhaust-nozzle exit) is small, and the exhaust air needs very little momentum excess to make up for it. Thus if the tail pipe is sized properly, the body achieves self-propulsion with an exhaust total head only slightly above that of the freestream, and, in principle, the body comes very close to achieving the ideal of Ackeret propulsion. Goldschmied's experimental results verified that attached flow was achieved at a suction flow rate consistent with the criterion that all boundary-layer air with total pressure lower than the static pressure on the downstream attachment surface must be sucked away (Taylor's criterion, see Thwaites, 1958). They also seemed to indicate that self-propulsion was achieved at the predicted low exhaust total pressure, but there was an unresolved discrepancy between the drag determined by the force balance and that determined by the wake surveys.

6.2 Drag Estimation

Predictions of viscous drag range from crude, preliminary estimates to predictions based on very detailed (and often laborious) calculations, usually combined with experimental test results. What level of prediction is appropriate depends of course on the purpose. In the initial stages of preliminary design, estimates based on empirical correlations are often sufficient. In the final stages of development, very accurate predictions can be needed, for example, to support performance guarantees. At this stage, a great deal of effort is usually justified, including very careful model testing and extrapolation to full scale, making use of the best CFD tools available. Between these extremes, particularly with regard to detailed design and design optimization, lies a gray area where the right balance between empiricism, testing, and computation is difficult to define.

CFD is playing an ever-increasing role in the detailed design process, and with good reason. In general, CFD is reasonably successful at predicting the pressure distribution on an aircraft, at least near the cruise condition where the flow tends to be well-behaved. CFD-based inverse design and optimization methods now make it easy to design shapes that produce good aerodynamic performance, and this has speeded up the design process considerably. However, the evaluation of the drag of a configuration, especially the viscous drag, is currently one of CFD's weaker suits (along with the prediction of separated-flow phenomena). While CFD drag predictions can provide a useful indication and are improving rapidly, a good deal of caution is still in order. For example, current turbulence models shouldn't be relied on to provide very accurate predictions of swept-wing flow, as we'll see in Section 8.6.2. For some time to come, testing, and the experience of the designer, will continue to play a role in drag evaluation during detailed design.

6.2.1 Empirical Correlations

6.2.1.1 Flat-Plate Skin Friction

The simpler a flow is, the more likely it is that it can be represented by an empirical correlation involving only a few dimensionless parameters. For viscous drag prediction, the simplest flows are those in which the drag is dominated by laminar or turbulent skin friction, and pressure gradients are mild enough that flat-plate flow is a good model. For slender fuselage-like bodies or thin flying surfaces at very low angles of attack, flat-plate skin friction can provide a very reasonable viscous-drag estimate. In Section 4.3.1, we discussed flat-plate boundary layers in some detail. For purposes of drag estimation, we use the average skin friction coefficient given by Equation 4.3.1 for laminar flow and Equation 4.3.2 for turbulent flow. These curves are plotted in Figure 4.3.1, along with curves for flows that transition from laminar to turbulent at a range of locations. For a flow that is either all laminar or mostly turbulent, only the wetted area and length Reynolds number need be known. For flows with transition, an accurate estimate is difficult, because the drag depends strongly on transition location, as is clear in Figure 4.3.1, and the transition location is sensitive to pressure gradients, surface imperfections, and freestream turbulence (see Section 4.4.1). For high subsonic speeds and above, an adjustment for Mach number should be applied, such as the one given for turbulent flow by Equation 4.3.7 and illustrated in Figure 4.3.4.

6.2.1.2 Airfoils with Significant form Drag

Moving up the complexity ladder, we come to effectively 2D wing sections with significant thickness and/or lift in subsonic flow. Of course, if there is flow separation, the drag can increase dramatically, and no simple correlation can be expected to work. Even if the flow remains attached all the way to the trailing edge, the significant pressure gradients associated with thickness and lift require correction to the flat-plate skin friction. First, the average skin friction is different from that on a flat plate, and second, the form drag must be added. These two effects are generally lumped together into an empirical *form factor* applied to the corresponding flat-plate skin-friction drag. In Section 7.4.2, we'll look at how the form factor varies with lift for a typical subsonic airfoil. Correlations relating form factors to airfoil geometric parameters can be found in several sources (see Hoerner, 1965, and the ESDU, 1990, data units). Such simple correlations for profile drag are risky, as illustrated by the following example. An airfoil with a Stratford recovery pressure distribution (zero skin friction throughout the pressure-rise region; see Smith, 1975) can have a very low skin-friction drag, which would be at least partly compensated for by a relatively high form drag. The distinction between this and a more conventional airfoil would be missed by any simple correlation scheme based on airfoil geometry.

6.2.1.3 Full Configurations

Preliminary-design methods for estimating the drag of an entire configuration make use of the above ideas for estimating the drag contributions of the major components (i.e., flat-plate skin friction corrected by an appropriate form-drag factor) and then add induced drag, interference drag, and the other viscous-drag contributions such as junctions, excrescences, and so on. Probably the most wide ranging compilation of empirical data on which to base estimates of all these contributions is Hoerner (1965). More recent data and theoretical derivations can be found in the ESDU (1990) data units.

6.2.1.4 Excrescence Drag

The drag contribution of small excrescence items such as fastener heads, steps, and gaps depends strongly on how deeply the object is immersed in the boundary layer. In the correlation of these effects, the height of the object can be scaled in two distinctly different ways, that is, relative to boundary-layer wall variables or boundary-layer thickness. It is important to use a correlation that is scaled in the way that is appropriate to the situation at hand, as illustrated in Figure 6.2.1.

If the object occupies a small fraction of the boundary layer thickness, the height should be expressed in terms of wall variables $h^+ = h u_\tau / v$, because the velocity profile in the near-wall region correlates best with these variables. If the object protrudes higher than about 10–20% of δ, it is more appropriate to use a correlation in terms of y/δ. Once a suitable correlation has been found in the literature, it should be assumed only to give the local effect of the excrescence. To estimate the effect on total drag, a correction for the "magnification" effect of pressure gradients should be applied, such as the one by Nash and Bradshaw (1967), or the local-$q^{1.5}$ model, both discussed in Section 6.1.9.

6.2.2 Effects of Surface Roughness on Turbulent Skin Friction

In Section 6.1.8, we looked at the behavior of the turbulent skin friction for various types of rough surfaces and discussed physical explanations for the skin-friction increase. In this

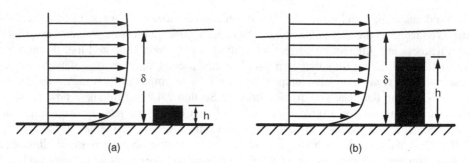

Figure 6.2.1 Types of scaling appropriate for excrescence items, depending on height relative to boundary-layer thickness. (a) h $\lesssim$ 0.2δ: wall-region scaling. (b) 0.2δ $\lesssim$ h: h/δ scaling

section, we'll look at how empirical data from one situation can be applied in other situations for purposes of drag estimation.

6.2.2.1 Scaling of Roughness Effects

While it is often important to be able to assess the drag impact of deviations from a smooth surface condition, it is rarely practical to measure the effects in place in the actual application. Instead, we must usually rely on measurements made in other flow situations, usually in ground-based test facilities. Most such measurements are made in simple flow situations, such as a turbulent boundary layer on a flat plate in a wind tunnel or towing tank, a pipe or channel flow, or flow between rotating concentric cylinders. In such experiments, skin-friction levels with and without roughness may be measured directly or inferred from pressure and/or flowfield measurements. In any case, the use of the resulting increments for drag estimates in flow situations different from that of the basic experiment requires some analysis. Below we look at the theoretical basis for two different ways of doing the required analysis, one using detailed CFD calculations and the other based on an equivalent flat plate. In Section 6.1.8, we looked at curve-fits based on some of the available empirical data for several types of roughness, and in Section 6.3.2, we'll look at typical data for riblets.

We wish to estimate the drag increment due to a particular surface texture exposed to a turbulent boundary layer in the flow environment of some practical application. We have available experimental measurements of the skin friction of the surface in question, preferably taken in some simple flow situation such a flat-plate flow, pipe flow, or rotating-cylinder test rig. Getting accurate increments from the experiment generally requires that measurements were also made for a reference smooth surface. To use such experimental results for estimation purposes, we must first put the measured smooth-to-rough increments in a form that is universally applicable and then convert them into increments for the application in question. In this section, we describe the theoretical basis for two practical ways of doing the required analysis.

The effect of a given rough surface on the flow depends on the flow conditions, particularly the Reynolds number. What complicates the situation, and the basic reason that analysis is required, is that the drag increment due to roughness depends on both the local roughness Reynolds number and the global Reynolds number of the flow, and that we can't generally match both of them between a generic experiment and a specific application. The best we can usually hope for is to match the roughness Reynolds number, since the drag increment has the strongest dependence on it, and to adjust by analysis for the unavoidable mismatch in the global Reynolds number and for differences in other variables such as Mach number. As we saw in Section 6.1.8, the local roughness Reynolds number is defined by $k^+ = ku_\tau / \nu$, where k is a length scale characterizing the size of the roughness elements, and u_τ is the "friction velocity," which as we saw in Section 4.4.2 is the velocity scale of choice for the "wall region" of a turbulent boundary layer. The global Reynolds number can be characterized in different ways depending on the situation, for example, in terms of R_x, R_θ, or $R_{\delta*}$.

The approaches outlined here are applicable to compressible flows, but we assume that the flow around the roughness elements themselves is not locally "compressible." The local density and temperature in the wall region may be significantly different from freestream, but the local Mach number applicable to the roughness, say $M^+ \equiv u_\tau/a$, is assumed to be small, so that the single parameter k^+ suffices to correlate the local roughness effect. This assumption is probably valid for most applications with small roughness levels in the transonic and low supersonic regimes. For stronger compressibility effects, we would presumably reach a point where something like M^+ should be included as a correlation parameter in addition to k^+.

It is clear that for the results of a generic experiment to be useful, the measurements must cover a sufficiently wide range of k^+ to include what will be encountered in the application. One way to do this is to run the experimental rig at a fixed overall Reynolds number and vary the size of the roughness, while keeping the character of the roughness the same. But because it is difficult to accurately reproduce a given surface texture at more than one scale, this is very rarely attempted for surfaces other than standard sand grains. Instead, the usual way is to use a single rough surface and to vary k^+ by varying the overall Reynolds number in the test rig. In either case, the results of such an experiment can be expressed fairly directly as a smooth-to-rough increment ΔC_f as a function of k^+, where each value of k^+ corresponds to a particular value of the overall Reynolds number, say $R_{\delta*}$. What we need now is a way to adjust ΔC_f for the fact that at a given k^+, $R_{\delta*}$, and possibly other flow variables in the application will generally be different from what they were in the experiment. To start, we recall from Section 6.1.8 that the one universal local effect of roughness is to shift the logarithmic part of the wall-region velocity profile by an amount Δu^+ that depends on k^+, as expressed in Equation 6.1.16.

The variation of Δu^+ as a function of k^+, that is, the log-law-shift function $\Delta u^+ (k^+)$, is a universal characteristic of the particular roughness shape in question, and within the limitations of the Law-of-the-Wall approximation, is independent of the type of flow (e.g., boundary layer, duct, pipe) as long as the Law of the Wall applies to it. What is most useful to us is that $\Delta u^+ (k^+)$ is also independent of the overall Reynolds number of the flow and that it can, with some calculation, be related to ΔC_f.

We will discuss two ways of relating $\Delta u^+ (k^+)$ to ΔC_f: a "turbulence-modeling approach" for detailed CFD calculations, and an "equivalent-flat-plate approach" for preliminary estimates.

6.2.2.2 The Turbulence-Modeling Approach

In this approach, we modify a turbulence model so that when it is used in CFD calculations, the flow solutions match the known $\Delta u^+ (k^+)$ for the roughness in question. We then use the modified turbulence model to calculate the flow in the experiment or target application with and without roughness and thus to predict ΔC_f. The advantage of this method is physical fidelity. It takes into account whatever level of flow complexity can be handled by the CFD code and provides as accurate an estimate of ΔC_f as the turbulence model is capable of. Furthermore, the total drag calculated by the CFD code contains an estimate of how ΔC_f affects the viscous pressure drag of the body through the change in displacement thickness. The disadvantage of this approach is that it requires modifying a CFD code and running detailed CFD calculations.

Turbulence models that allow accurate calculation of the mean flowfield in the near-wall region of a turbulent flow generally have an explicit dependence on the distance y from the wall, usually expressed as y^+. This functional dependence on y^+ is typically tailored so that numerical solutions using the model match experimental velocity profiles in the Law-of-the-Wall region, including the viscous sublayer and the logarithmic region. The algebraic eddy-viscosity model of Mellor (1967) as used in the 3D boundary-layer code by the author (McLean, 1977) is one example of such a model.

A CFD solution for a turbulent flow can be made to simulate the effects of surface roughness through a simple modification to the turbulence-modeling part of the calculation. The modeling equations themselves are not changed. Only the value of y^+ input to the near-wall part of the model is modified by a shift Δy^+ added to the geometric y^+. When a Δy^+ shift is applied, the resulting flow solution no longer realistically represents the flow at the very bottom of the near-wall region, where part of the volume in the real flow domain is occupied by roughness elements. However, the logarithmic part of the calculated velocity profile experiences a shift Δu^+ like the shift in a real flow with roughness, and the skin friction increases by the right amount. With the right input for Δy^+, the shift Δu^+ can be made to match the Δu^+ that a real flow would have at the same k^+. When that is done, the calculated ΔC_f, relative to a baseline calculation for a smooth wall, should closely match ΔC_f for the corresponding real flows.

Thus what is needed is a way to determine the right Δy^+ to input to the turbulence model so that flow solutions match the Δu^+ (k^+) behavior for the particular type of rough surface in question. This is made easier by the fact that the near-wall turbulence model determines a unique relationship between Δy^+ and Δu^+, which in practice can be determined numerically by a series of flow solutions with different constant input values of Δy^+. When both the $\Delta u^+(\Delta y^+)$ behavior of the turbulence model and the desired Δu^+ (k^+) behavior for the roughness in question are known, it is just a matter of interpolation to define a function Δy^+ (k^+) to be used to determine the input Δy^+. In the author's 3D boundary-layer code, Δy^+ (k^+) functions for several different types of roughness are included in the code in the form of numerical tables. For a calculation of a flow over a rough surface, the user specifies which of the roughness functions is to be used and inputs the roughness size k as a function of location on the surface (it needn't be constant). Every time the eddy-viscosity subroutine is called in the course of such a calculation, the friction velocity u_τ is known from the previous iteration of the flow solution, k^+ is calculated based on the local k, and the corresponding value of Δy^+ is determined by numerical table lookup.

The $\Delta u^+(\Delta y^+)$ behavior of Mellor's eddy-viscosity model is shown in Figure 6.2.2. The resulting Δy^+ (k^+) curves corresponding to the Δu^+ (k^+) roughness curves of Figure 6.1.9a are plotted in Figure 6.2.3. Note that over the entire transitional roughness regime ($k^+ > 70$)

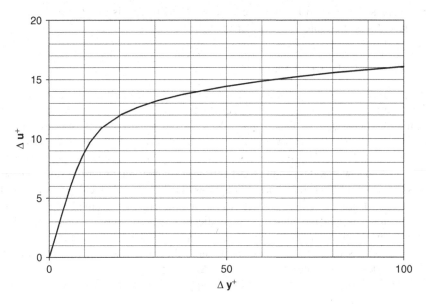

Figure 6.2.2 Illustration of how the logarithmic wall-region velocity profile, as calculated with a typical near-wall turbulence model, responds to an input Δy^+. This $\Delta u^+(\Delta y^+)$ response is used to determine $\Delta y^+(k^+)$ curves like those shown in Figure 6.2.3 that can be used to predict roughness effects. These calculations used the algebraic eddy-viscosity model by Mellor (1967)

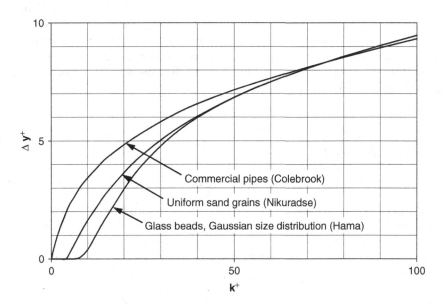

Figure 6.2.3 Illustration of $\Delta y^+(k^+)$ curves that can be used in a near-wall turbulence model to simulate the effects of roughness. These curves are the result of combining the $\Delta u^+(k^+)$ roughness curves of Figure 6.1.9a with the $\Delta u^+(\Delta y^+)$ response of the turbulence model in Figure 6.2.2

the Δu^+ values in Figure 6.1.9 never exceed 8, so that Δy^+ is nearly the same as Δu^+, and Figure 6.2.3 is therefore very similar to Figure 6.1.9.

6.2.2.3 The Equivalent-Flat-Plate Approach

In this approach, we define an equivalent-flat-plate flow to represent the experiment or target application, and then calculate the development of the equivalent-flat-plate flow with and without roughness, thus determining ΔC_f. In these calculations, we use an explicit relation between Δu^+ (k^+) and C_f that is valid locally for a flat-plate boundary layer, in combination with the von Karman momentum integral equation and an empirical smooth-wall skin-friction law that defines the local skin-friction coefficient for the smooth case. The roughness size k and even the type of roughness can vary with location along the plate. The advantage of this method is that it does not require running detailed CFD calculations. The disadvantage is that the accuracy of predicted C_f increments is compromised by the neglect of flow features in the application that are different from a flat-plate flow, for example, the effects of pressure gradients or three dimensionality. A further disadvantage is that it provides no estimate of the effect of ΔC_f on pressure drag.

For flat-plate flow, we have the following approximate relation between C_f and Δu^+ (see Hama, 1954, for a derivation):

$$\Delta u^+ = \left(\frac{2}{C_{f(smooth)}}\right)^{1/2} - \left(\frac{2}{C_{f(rough)}}\right)^{1/2}, \tag{6.2.1}$$

which holds not at the same x station in the flows over the rough and smooth walls, but at stations where the two flows have the same $R_{\delta*}$. Because the shape factor H of a flat-plate velocity profile varies slowly with Reynolds number and is only weakly affected by roughness, stations with the same $R_{\delta*}$ also have approximately the same R_θ, and in practical calculations, we will find it more convenient to assume that Equation 6.2.1 applies at constant R_θ. Our other two basic equations are the von Karman momentum integral equation specialized for a flat plate:

$$\frac{d\theta}{dx} = \frac{dR_\theta}{dR_x} = \frac{C_f}{2}, \tag{6.2.2}$$

and a smooth-wall skin-friction law:

$$C_{f(smooth)}(R_{\theta(smooth)}, M_\infty) = C_{fi(smooth)}(R_{\theta(smooth)})(1 + 0.13M_\infty^2)^{-1}, \tag{6.2.3}$$

where $C_{fi(smooth)}$ on the right-hand side is not an analytic expression, but instead has to be evaluated by interpolation from a numerical table derived to be consistent with the expression for local turbulent skin-friction as a function of R_x, Equation 4.3.5.

For purposes of calculating the smooth equivalent-flat-plate flow, Equation 6.2.2 is integrated as follows:

$$R_{\theta(smooth)}(Re_x) = R_{\theta(smooth)0} + \int_{Re_{x0}}^{Re_x} \frac{C_{f(smooth)}(R_{\theta(smooth)})}{2} dRe_x. \tag{6.2.4}$$

For purposes of calculating the rough equivalent-flat-plate flow, Equation 6.2.1 can be solved for $C_{f(rough)}$, and Equation 6.2.2 can be integrated as follows:

$$C_{f(rough)} = 2\left\{\left(\frac{2}{C_{f(smooth)}(R_{\theta(rough)})}\right)^{1/2} - \Delta u^+(s^+)\right\}^2 \tag{6.2.5}$$

$$R_{\theta(rough)}(R_x) = R_{\theta(rough)o} + \int_{Re_{xo}}^{Re_x} \frac{C_{f(rough)}(R_{\theta(rough)}, s^+)}{2} d\,Re_x. \tag{6.2.6}$$

The numerical table to be used for Equation 6.2.3 and for the value of $C_{f(smooth)}$ on the right-hand side of Equation 6.2.5 is generated by numerical integration of Equation 6.2.2, with C_f from Equation 4.3.5. Note that $C_{f(smooth)}$ on the right-hand side of Equation 6.2.5 is evaluated at $R_{\theta(rough)}$ in keeping with the discussion above regarding how Equation 6.2.1 should be applied.

The numerical calculations required to implement this method can be programmed in a spreadsheet. A fixed array of points in R_x is set up, and the solutions for both a smooth wall and rough wall are advanced step by step by trapezoidal integration. Note that Equations 6.2.3 and 6.2.4 for the smooth case and Equations 6.2.5 and 6.2.6 for the rough case represent circular references, that is, $C_{f(smooth)}$ from Equation 6.2.3 is used on the right-hand side of Equation 6.2.4 but also depends on the result (left-hand side) of Equation 6.2.4. In a spreadsheet, these circular references can be resolved by automatic iteration.

Figure 6.2.4 shows an example of such a calculation, for an all-turbulent flat-plate flow at $R_L = 10^7$ with roughness only over a portion of the length ($0.2 < x/L < 0.4$). At the start of the rough region, the local C_f increases by about 30%. At the end of the rough section,

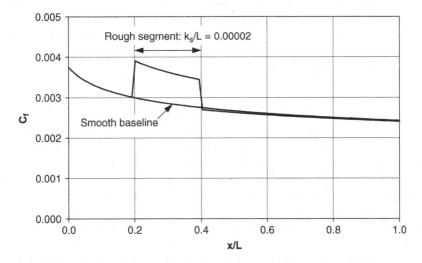

Figure 6.2.4   Local turbulent C_f for a flat plate at $R_L = 10^7$ with roughness only over a portion of the length ($0.2 < x/L < 0.4$), calculated by the equivalent-flat-plate method implemented in a spreadsheet

C_f decreases to a level slightly below that of the smooth baseline because the roughness thickened the boundary layer more than in the smooth case. In terms of integrated skin friction, this "undershoot" effect offsets only a small fraction of the drag penalty of the roughness.

6.2.3 CFD Prediction of Drag

As we saw in Section 6.1, drag involves subtle higher-order phenomena in the flowfield. Thus it is more difficult to compute accurately than the other force components, and it places higher demands on the accuracy of the numerical method and on the adequacy of the grid. We'll defer further discussion of these issues to Chapter 10.

6.3 Drag Reduction

Efforts to reduce aerodynamic drag can be aimed at the lift-induced drag, the shock drag, the skin-friction drag, or the viscous pressure drag. Reducing one component of the drag often entails compromise with the other drag components or with nonaerodynamic considerations. In this section, we'll concentrate mainly on the skin-friction drag, but first let's put all the components in perspective:

- **Lift-induced drag**: In Sections 8.3 and 8.4, we'll discuss the induced drag and how reducing it generally requires increasing the effective span and thus the structural weight of wings, and must therefore be subject to compromise.
- **Shock drag**: The wing is often the dominant contributor to the shock drag of an airplane in the transonic regime, and for high-aspect-ratio wings, the shock drag is largely attributable to the sectional behavior of the airfoils. The shock drag of a transonic airfoil depends on how well the airfoil was designed. Assuming good design, shock drag can generally be reduced by thinning the airfoil, which tends to increase the structural weight of the wing, or by increasing the chord, which can increase viscous drag. We'll look at these airfoil-design issues in more detail in Section 7.4.8.
- **Viscous drag**: The dominant issues in viscous-drag reduction depend on whether the body is bluff or streamlined.

In Section 6.1.6, a bluff body was defined as one that fails to produce attached streamline flow at the back end and instead has a large separated wake. We saw that because of the large region of separated flow, the drag of such a body is relatively high and is mostly pressure drag. Still, such bodies can benefit from careful contouring forward of the bluff back end, so as to avoid early separation. This is why the front ends of some cars and trucks have become more slender and rounded as fuel mileage has become more important.

The viscous drag of a streamlined body is mostly skin friction, but there is always some pressure drag due to the displacement effect of the attached boundary layer. Reducing the wetted area by shortening the body can reduce the skin-friction drag, but it tends to increase the pressure drag. Generally speaking, a major goal of good design of the body shape is to optimize the trade between skin-friction drag and pressure drag, subject to all of the nonaerodynamic considerations that apply. We'll see how this applies to airfoil design in particular in Section 7.4. Finally, there are various ways of reducing the skin friction directly, which is where we'll concentrate the rest of this section.

6.3.1 Reducing Drag by Maintaining a Run of Laminar Flow

In Figures 4.3.1 and 4.3.2, we saw that laminar skin friction is much lower than turbulent skin friction and that delaying the transition of the boundary layer from laminar to turbulent should therefore be a powerful means of drag reduction. To delay transition, we must generally do something to delay the onset and reduce the growth of the instabilities that beset the laminar boundary layer. In Section 4.4.1, we discussed the associated theory and saw that instability growth is strongly affected by the pressure gradient and can also be influenced by distributed surface suction.

There are three main approaches to implementing laminar flow for low drag:

- **Natural laminar flow (NLF)**: Instabilities are controlled through the pressure distribution alone.
- **Laminar flow control (LFC)**: Instabilities are controlled through a combination of the pressure distribution and distributed suction.
- **Hybrid laminar flow control (HLFC)**: A specialized technology for swept wings, in which LFC is applied only in the leading-edge region of a swept wing, and NLF persists for some further distance aft.

In practice, the distributed suction required for LFC or HLFC is usually approximated by suction through closely spaced narrow slots or small round holes.

The application of NLF to drag reduction on 2D airfoils is discussed in Section 7.4.6, and laminar flow on swept wings is discussed further in Section 8.6.2. State-of-the-art design for laminar flow generally requires doing calculations of instability growth rates and integrated growth factors, using the theory discussed in Section 4.4.1. Implementing laminar flow in high-Reynolds-number applications generally requires very smooth, wave-free surfaces, which can be difficult to produce and maintain.

6.3.2 Reduction of Turbulent Skin Friction

Like death and taxes, turbulent skin friction on major parts of a vehicle's surface is practically unavoidable. Both through the direct sheer force at the wall and it's indirect effect on pressure drag, it accounts for a substantial fraction of the drag of most airplanes and water-born vessels. In Section 6.1.8, we saw that surface roughness can increase turbulent C_f, so that simply making the surface smooth enough to avoid this penalty is the first option that should always be considered. Beyond that, to reduce C_f below the smooth-surface level, many schemes have been proposed and tested, but most either don't work or are impractical for one reason or another. Because being proposed and rejected doesn't prevent their being proposed again and again, it's good to be familiar with the things that have already been rejected and the reasons why they didn't work. To this end, I've compiled the following list (Table 6.3.1) of turbulent-skin-friction reduction concepts, with brief explanations of their purported mechanisms and their actual workability:

Table 6.3.1 Turbulent skin-friction reduction concepts

Passive turbulent skin-friction reduction

Slick surfaces or coatings	Intuitively, it seems like a slick surface ought to allow the air to just slide along and thus greatly reduce C_f. But in reality, such surfaces have the same no-slip condition that other surfaces do. Coatings that smooth the surface can sometimes delay laminar-to-turbulent transition and reduce average skin friction as a result, but to reduce turbulent skin friction locally, a surface would have to allow significant molecular "slip" at the surface, and no material has ever shown this capability at ordinary atmospheric conditions.
Wavy surfaces	2D surface waves of low amplitude can reduce average skin friction, but the reduction is more than offset by the additional pressure drag caused by the waves (see Lin *et al.*, 1983).
Compliant surfaces	Surfaces that deflect in response to the surface-pressure fluctuations of the turbulent boundary layer have never worked in ordinary turbulent air flows (see Hough, 1980).
Outer-layer manipulators or large-eddy-breakup devices (LEBUs)	These are thin plates or airfoils positioned in the outer part of the boundary layer with the intent of interfering with the large eddies. In experiments at low Reynolds numbers, these have produced drag reductions ranging from none to substantial, but at Reynolds numbers typical of full-scale flight, they have never demonstrated a net drag reduction (Anders, 1989).
Riblets	A pattern of tiny, alternating ridges, and grooves aligned longitudinally (approximately in the direction of the flow), these have been demonstrated to work in many different experiments and applications, but the reduction is modest, a maximum reduction in turbulent skin friction of about 6.5% in practice at high Reynolds numbers. Because these actually work, they are discussed further below
Shark-skin surfaces	The tiny scales (dermal denticles) that make up the outer surface of shark skin typically have longitudinal ridges that look like riblets (3–9 ridges on each scale) and seem to be scaled to the flow conditions in the same way riblets would be (see Bechert and Bartenwerfer, 1989). These have not been investigated extensively, but they probably are no better than riblets and would be harder to produce.
"Negative dispersion" surfaces (Gao and Chow, 1992)	These were a mosaic-like pattern of diamond-shaped protuberances that were predicted theoretically to extract turbulence kinetic energy and return it to the mean flow. Initial experimental results looked promising but have not been replicated.
Chevrons with random phases (Sirovich and Karlsson, 1997)	This concept involved spanwise rows of small chevron-shaped protuberances with random phases between rows and was derived from theoretical calculations. Experiments were not conclusive.

(*continued overleaf*)

Table 6.3.1   (*continued*)

Active turbulent skin-friction reduction	
Upstream slot injection or distributed injection through a porous surface (microblowing, as described by Huang, 1998)	These injection schemes can reduce skin friction, but the reduction is more than offset by the ram-drag penalty of taking the injection air on board.
Electromagnetic tiles (Nosenchuck and Brown, 1993)	Oscillating electric currents and magnetic fields introduced through a mosaic of surface electrodes applies spatially and temporally periodic body forces to the fluid that reduce turbulence production. This scheme works only in fluids with much higher electrical conductivity than ordinary air.
MEMS (micro-electro-mechanical systems) actuators	Arrays of tiny actuators on the surface that operate on the turbulence locally, these will probably never be practical for applications at full-scale Reynolds numbers. The scheme would require very large numbers of very small actuators and sensors, covering large areas, and they would be easily damaged or contaminated by dirt or ice.

Of this entire list, only riblets have been shown to have a chance of working in practice, so let's discuss them further.

6.3.2.1   Riblets

Riblets are a pattern of tiny, alternating ridges, and grooves aligned longitudinally (approximately in the direction of the flow). They reduce turbulent C_f by a modest amount by inhibiting the lateral turbulence motions near the bottom of the boundary layer, primarily the motions associated with the near-wall streamwise vortices, which are associated with the streaks in the sublayer (see Figure 2.1f). Inhibiting the near-wall vortices slightly reduces the overall production of turbulence in the boundary layer.

The effect scales with turbulent-boundary-layer wall variables, just like the effects of surface roughness in the transitional regime that we looked at in Sections 6.1.8 and 6.2.2. The C_f reduction correlates with s^+, the ridge-to-ridge spacing in wall units, and typical behavior is shown in Figure 6.3.1. In this example, C_f is reduced over a range of s^+ up to about 23, a range that varies somewhat depending on the particular riblet profile. The peak C_f reduction is around $s^+ \approx 14$, which also varies somewhat with the profile. The behavior of the effect in the range of low s^+ below about 5 is not well established by the data, just as we saw in the case of surface roughness in Section 6.1.8. At high s^+, C_f increases above the smooth-surface value, much as it would with surface roughness. The optimum spacing of about 14 wall units means that riblets must typically be quite small, just as the permissible roughness of a surface is small. We'll look at a specific sizing example next.

The general shape of the curve in Figure 6.3.1, with a C_f reduction over a midrange of s^+ and a C_f increase at high s^+, reflects a competition, as s^+ increases, between increasing

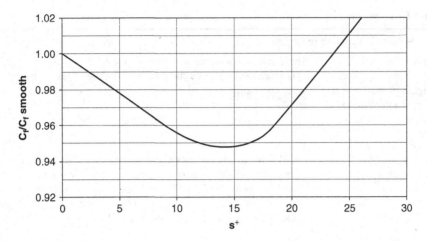

Figure 6.3.1 Typical turbulent-C_f reduction by riblets as a function of the dimensionless spacing s^+

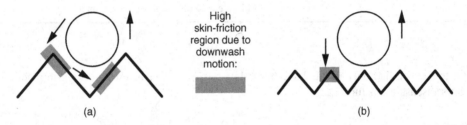

(a)   (b)

Figure 6.3.2 Schematic diagram of the mechanisms for drag increase and drag reduction by riblets. From Choi, Moin, and Kim, (1993); used with permission of *Journal of Fluid Mechanics*. (a) $s^+ \approx 40$ (b) $s^+ \approx 20$

effectiveness of the ridges in inhibiting lateral motions, and increasing penetration of high-velocity fluid into the grooves. How this competition plays out seems to be strongly related to how the dominant scale of the sublayer vortices in the instantaneous flowfield varies relative to s as s^+ increases. Based on direct numerical simulation (DNS) calculations of the turbulent flowfield, Choi, Moin, and Kim (1993) point out that the vortices' average diameter is $d^+ \approx 30$ and that they tend on average to be centered at $y^+ \approx 20$. They present a schematic diagram, reproduced here in Figure 6.3.2, indicating that at high s^+, the vortices are able to settle farther into the grooves and to bring high-velocity fluid close to the groove walls, thus increasing the average shear stress. This also seems to be reflected in the time-averaged flow in an interesting way. Suzuki and Kasagi (1993) found, based on flowfield measurements, that there is a time-averaged downward flow into the middle of the groove coupled with upward flow along the groove walls. The pattern is very weak in the drag-reducing range of s^+, but it becomes much stronger in the drag-increasing range, as shown in Figure 6.3.3.

The drag-reduction effect of riblets seems to be fairly forgiving of misalignment of the ridges in yaw. There have not been many measurements of the yaw effect, so there is

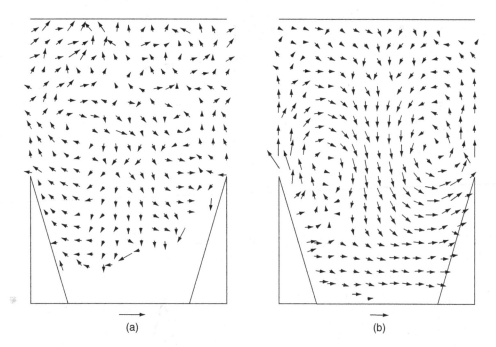

(a)   (b)

Figure 6.3.3 Time-averaged flow patterns in a riblet groove. From Suzuki and Kasagi, (1993). Used with permission. Reference vector at the bottom of each plot represents $0.005u_e$. (a) $s^+ = 15$ (b) $s^+ = 31$

no well-established curve. A few degrees of misalignment seems to make little difference. The drag reduction is roughly half gone at 15°, and completely gone at about 30°. With this level of tolerance, groove alignment is not a major practical issue in most applications.

The effects of profile shape on the C_f reduction are known almost entirely through experimental investigations, with the most extensive single collection to be found in Bechert *et al.* (1997). It has been found that sharp and narrow ridges are favored. For ridges with a triangular cross section and a depth ratio h/s = 0.5, the dependence on the included angle at the ridge is as shown in Figure 6.3.4. The C_f reduction is maximum for a 0° -included-angle "razor blade" ridge and decreases to less than half the maximum at 60°. A profile with a 30° included angle, as shown in Figure 6.3.5, gives up a little performance but is a more practical compromise than the razor blade. The depth of the grooves seems to be less important than the shape of the ridges, as long as the depth is about half the spacing or greater.

The effects of riblets on the near-wall turbulence have been studied using DNS calculations, but only at relatively low Reynolds numbers (see Choi, Moin, and Kim, 1993, for example). Besides DNS, there is no quantitative theory that deals with these effects at the level of the actual turbulence physics and explains how the C_f reduction varies as a function of the riblet profile. The closest thing we have to a theory is the concept of *protrusion height,* which is defined as follows. Consider the riblet profile to be immersed at the bottom of a theoretical laminar shear flow at very low Reynolds number (Stokes flow). Far above the tops of the ridges, the flow is uniform in horizontal planes, and the vertical

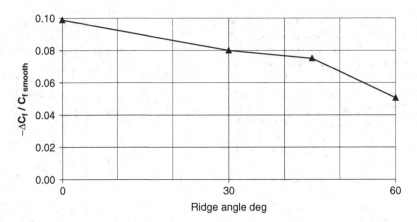

Figure 6.3.4 C_f reduction of riblets with triangular-cross-section ridges as a function of the ridge included angle. (Data from Bechert *et al.*, 1997)

Figure 6.3.5 Riblet profile with flat valleys and ridges with 30° included angle

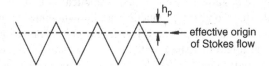

Figure 6.3.6 The *protrusion height* is defined as the vertical distance between the riblet ridge and the effective origin of an idealized Stokes flow (the point of zero velocity extrapolated from the uniform velocity gradient far above the riblets)

gradient of the velocity is uniform. Extrapolating the linear velocity profile to zero defines an effective origin that is always found to be between the bottoms of the grooves and the tops of the ridges. The protrusion height h_p is the distance that the ridges protrude above this effective origin, as shown in Figure 6.3.6. The idea of protrusion height has been used in two different ways. Bechert and Bartenwerfer (1989) correlated turbulent C_f reduction with the protrusion height calculated for flow parallel to the ridges (*streamwise protrusion height*). Lucini, Manzo, and Pozzi (1991) noted that the *cross-flow protrusion height* is always greater than the streamwise protrusion height and used the difference between the two to correlate turbulent C_f reduction. The applicability of these theories to the actual

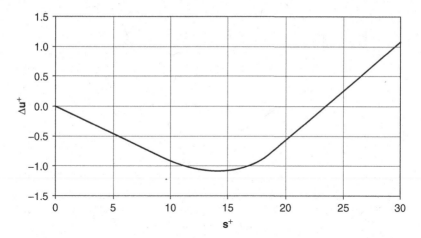

Figure 6.3.7 $\Delta u^+(s^+)$ curve derived from experimental data for the riblet profile of Figure 6.3.5. Analogous to the $\Delta u^+(k^+)$ roughness curves of Figure 6.1.9a. Can also be used as a $\Delta y^+(s^+)$ curve for simulating riblet effects in calculations with a turbulence model. It is thus analogous to the curves for simulating surface roughness in Figure 6.2.3

physics is questionable, but both correlate with the observed qualitative trends. For example, they both predict that sharp ridges are best and that C_f-reduction effectiveness increases with decreasing included angle at the ridges.

Choosing the appropriate ridge spacing and estimating the drag reduction for a particular riblet application can be done to varying levels of accuracy. A rough estimate can come from a plot like Figure 6.3.1, if one is available for the riblet profile in question, combined with an estimate of the average s^+ for the surface, based on the dimensional s for the riblets and the average turbulent C_f from Equation 4.3.2. But $C_f/C_{fsmooth}$ for a riblet surface varies with Reynolds number and other aspects of the flow conditions, and more accurate estimates require using the same kinds of calculations outlined for distributed roughness in Section 6.2.2. For many applications, the equivalent-flat-plate approach can be adequate, but detailed CFD calculations with a turbulence model modified to simulate the particular riblet profile can provide more accurate accounting for complications such as pressure gradients and 3D flow.

When experimental data are available for the riblet profile in question, they are usually expressed in terms of $\Delta C_f/C_{fsmooth}$, as a function of the Reynolds number that was varied in the test facility. With appropriate calculations applied to the situation in the test facility, $\Delta C_f/C_{fsmooth}$ (R) can be converted to Δu^+ (s^+). This can be used directly in the equivalent-flat-plate approach as described in Section 6.2.2. Figure 6.3.7 shows the Δu^+ (s^+) curve derived in this way from experimental data for the riblet profile of Figure 6.3.5.

In Section 6.2.2, we described how Δu^+ (s^+) data for a rough surface can also be converted to a Δy^+ (s^+) table for inclusion in a turbulence model for CFD calculations. In the case of riblets, such a conversion isn't generally required because their Δu^+ values are small and are generally in the linear part of the $\Delta u^+(\Delta y^+)$ response curve for near-wall turbulence models, where $\Delta u^+ = \Delta y^+$, as can be seen in Figure 6.2.2. The Δu^+ (s^+) curve of Figure 6.3.7 can therefore also be used directly, without conversion, as the Δy^+ (s^+)

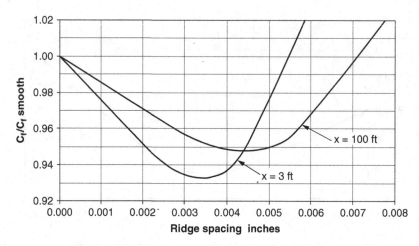

Figure 6.3.8 Turbulent-C_f reduction by riblet profile of Figure 6.3.5, as a function of spacing, calculated for a flat plate at Mach 0.85 and a Reynolds number of 1.7 million per foot

curve for inclusion in a turbulence model. We also noted in Section 6.2.2 that using detailed CFD calculations to predict ΔC_f due to roughness has the advantage that the indirect effect on pressure drag is also predicted. The same advantage applies to using CFD to predict the effects of riblets. In applications of riblets to bodies with pressure gradients, the effect on pressure drag is typically a significant portion of the total drag reduction.

The fact that the riblet effect scales with turbulent-boundary-layer wall variables makes the use of riblets in applications surprisingly simple. First, turbulent C_f tends to vary slowly along the length of a surface, and s^+ varies as $C_f^{0.5}$, which varies even more slowly. Thus it turns out that a single riblet spacing can be used along the length of a long aerodynamic surface, and the average C_f reduction can be nearly equal to the maximum that could be achieved with a variable spacing. To illustrate how this works, Figure 6.3.8 shows the C_f reduction as a function of riblet spacing at 3 and 100 ft from the leading edge of a flat plate at a Reynolds number of 1.7 million per foot, typical of flight at Mach 0.85 at 35 000 ft. In this example, a spacing of between 0.003 and 0.004 in. works well over the whole length. This example also gives an indication of how much $\Delta C_f/C_{fsmooth}$ varies with length Reynolds number, and thus illustrates the inaccuracy that can be incurred by using a single $C_f/C_{fsmooth}$ curve like the one in Figure 6.3.1 to make drag estimates.

The fact that optimum riblet spacings tend to be so small raises a host of practical issues. Producing sharp ridges on such a small scale is difficult, and whatever material the ridges are made of, durability is a problem. So far, the favored approach is to mold the profile into a plastic or elastomeric appliqué that can then be applied to the surface. But covering large areas of an airplane's surface with "tape" poses many problems, and this technology has not yet found its way into routine service on airplanes, though it may in the future.

7

Lift and Airfoils in 2D at Subsonic Speeds

In a way, lift is the most visible of the aerodynamic forces. We see heavier-than-air animals and machines flying through the air every day, and something has to be holding them up there. Of course, we can predict the existence of aerodynamic lift mathematically by solving equations of motion for the flow around the lifting object. The accuracy of such predictions depends on the level of fidelity of the equations we choose to solve and varies with the type of lifting-surface shape and with the flow situation. Some types of lifting flow are easier to predict accurately than others. In principle, however, if we had the computing power available to carry out a direct numerical simulation (DNS) solution of the Navier-Stokes (NS) equations for the flow, we would be able to predict any lifting flow, 2D or 3D, with high accuracy. So in one sense, the physics of lift is perfectly understood: Lift happens because the flow obeys the NS equations with a no-slip condition on solid surfaces.

On the other hand, physical explanations of lift, without math, pose a more difficult problem. Practically everyone, the nontechnical person included, has heard at least one nonmathematical explanation of how an airfoil produces lift when air flows past it. Such explanations fall into several general categories, with many variations. Unfortunately, most of them are either incomplete or wrong in one way or another. And some give up at one point or another and resort to math. This situation is a consequence of the general difficulty of explaining things physically in fluid mechanics, a problem we've touched on several times in the preceding chapters.

In the real world, lifting flows are never precisely two dimensional, even when we try to make them so, as we do in so-called "2D" wind-tunnel testing. It seems like it should be possible to produce precisely 2D airfoil flow in a wind tunnel: Just mount a 2D airfoil model so that it spans the space between parallel tunnel sidewalls. But in reality 2D flow is practically impossible to achieve because of viscous effects on the tunnel sidewalls and in the junctions between the model and the sidewalls, and results of "2D" wind-tunnel testing are always questionable to some extent. On many flight vehicles, however, wings are of high enough aspect ratio that the local flow at stations over most of the span behaves at least qualitatively like the 2D ideal. Thus exploring the physics and doing some of our design work in the ideal 2D world makes sense. Even trying to simulate 2D flow in the wind tunnel can be useful in spite of the generally imperfect results. In this chapter, we'll concentrate on nominally 2D flow because it's simpler, and we can learn a lot that is generally applicable.

In this chapter, we'll start with the easy part: the mathematical prediction of lift through solutions of equations of motion, and the closely related (and still mathematical) explanations of lift in terms of circulation and vorticity. Then we'll look at physical explanations of lift in 2D, starting with a discussion of the strengths and weaknesses of many of the explanations already in circulation. With these cautionary tales in mind, we'll try to develop our own "best we can do" physical explanation of lift in 2D. Then we'll discuss some of the major physics and design aspects of airfoils. In Chapter 8, we'll extend the discussion from 2D to 3D.

7.1 Mathematical Prediction of Lift in 2D

The earliest attempt that we know of to predict lift mathematically was Newton's theory for the lift of an inclined flat plate. Unfortunately, Newton got it wrong, but that is forgivable, given that his theory predates any rational theory of continuum fluid dynamics by more than a 100 years. Newton assumed a "bullet" model for the flow, as shown in Figure 7.1.1, in which particles do not interact with each other before they strike the forward-facing surface of the body, and no particle strikes the aft-facing surface, a model whose shortcomings we've already discussed in Section 5.1.

Assuming that the particles that strike the plate transfer the perpendicular component of their momentum to the plate and retain the parallel component, Newton predicted that lift is proportional to the square of the angle of attack, for small angles. The square relationship arises because the theory predicts that lift is proportional to both the number of particles intercepted per unit time and the momentum absorbed from each particle and that both increase with angle of attack. Later we'll see that in continuum lifting flows, we get a linear relationship (ideally) between lift and angle of attack, largely because the angle of attack has very little effect on the amount of fluid per unit time that the lifting surface effectively interacts with. So the square relationship is simply wrong, and for reasonable angles of attack, the magnitude of the lift predicted by Newton's model is much too small and was cited even into the late 1800s as a reason why heavier-than-air flying machines would never be possible.

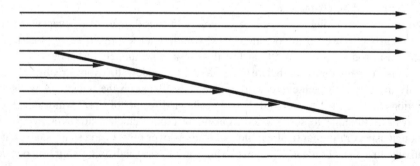

Figure 7.1.1   Newton's model for the lifting flow around an inclined flat plate (an application of the "bullet model" of Figure 5.1.1)

To get the right answer, we must model the continuum behavior of the fluid correctly. I've already claimed that a DNS solution of the full unsteady NS equations does this and should, in principle, predict any lifting flow with high accuracy. This claim is untested, however, because such a calculation has yet to be carried out for any Reynolds number typical of an aeronautical application. For routine calculations with current computers, we must scale our ambitions back to solutions of the *Reynolds-averaged Navier-Stokes* (RANS) equations with turbulence modeling (see Section 3.7). At this level of fidelity, we can predict lift with reasonable accuracy, provided that the airfoil or wing is nicely shaped, has a reasonably sharp trailing edge, and is at a low enough angle of attack that the flow stays attached to both surfaces at least to within a short distance of the trailing edge. Fortunately, most airfoils that are of practical interest meet these requirements, and the attached-flow regime covers much of the flight envelope of most flight vehicles, so that this predictive capability is highly useful.

As the angle of attack is increased, however, and boundary-layer separation moves forward on one surface, the accuracy of RANS predictions generally deteriorates, primarily because turbulence models are less accurate for separated flows than they are for attached boundary layers. As the angle of attack increases further, the separated region becomes larger yet, the real flow becomes increasingly unsteady, and most numerical schemes for the steady RANS equations will at some point fail to converge to a steady solution. Obtaining realistic solutions for flows with large separated-flow regions requires solving for at least the large-scale unsteady motions associated with the separation. It is not enough simply to solve the unsteady Reynolds-averaged Navier-Stokes (URANS) equations. The calculation procedure must somehow enforce a proper distinction between the large-scale motions that can be computed and the small-scale turbulence that must still be modeled. Large-eddy simulation (LES) and detached-eddy simulation (DES) are approaches that are under exploratory development, and we'll look further at them in Chapter 10, but they are not yet in routine use.

At the next step down in fidelity from steady RANS solutions with turbulence modeling, we find the various viscous-inviscid coupling schemes in which the flow is divided into an outer inviscid region and an inner boundary layer. With the right kinds of coupling and solution schemes, such methods can predict much of the same class of flows that can be predicted by RANS, including flows with small separated regions, and within the range where they can produce converged solutions, their accuracy can be comparable to that of RANS solutions. The widely used ISES and MSES codes (Drela and Giles, 1986; Drela, 1993) are examples of this class of methods. A disadvantage of coupled methods is that they tend not to handle geometric complexity as well as RANS, especially in 3D.

Even viscous-inviscid coupling solutions require extensive computation and have been practical only since the early 1970s. Before that, flowfield prediction for airfoils was restricted to *inviscid flow*. The loss of fidelity incurred by the neglect of viscous effects depends on the type of airfoil and the flow conditions. At low Reynolds numbers, boundary layers are thick and usually have major effects on the pressure distribution. In the transonic regime, the pressure distribution is extremely sensitive to the effective shape of the airfoil and is therefore sensitive to the displacement effect of the boundary layer. But for airfoils with sharp trailing edges in the attached-flow regime, at moderate-to-high Reynolds numbers, and outside the transonic regime, predictions of lift and pressure distributions from inviscid-flow theory can be adequate for many purposes. In Section 7.4.1, we'll look at this issue further, using computational examples to assess the magnitude of viscous effects on lift and pressure distributions in incompressible flow at moderately high Reynolds numbers.

The easiest way to obtain solutions for inviscid flow is through *potential-flow theory*, which for much of the twentieth century provided our only means for predicting the lift and pressure distributions of airfoils in 2D flow. As simple and powerful as potential-flow theory is, however, applying it to lifting flows gives rise to a couple of interesting mathematical difficulties. The first difficulty is in representing a lifting flow at all in terms of a velocity potential in the manner defined by $\mathbf{V} = \nabla \phi$ (see Section 3.10 for an introductory discussion of potential-flow theory). Two requirements must be satisfied. First, if a potential function is to represent a continuous velocity field, the potential must be continuous and have continuous first derivatives. Second, the Kutta-Joukowski theorem, which we'll discuss in detail in Section 7.2, tells us that we must have circulation around the airfoil if we are to have lift. These requirements conflict: A single potential function that is continuous throughout the domain surrounding the airfoil cannot represent a flow that has nonzero circulation. This follows from the definition of the velocity as the gradient of the potential. The line integral of the velocity on a contour from any point A to any other point B must be equal to the change in the value of the potential between the two points. If there is nonzero circulation on a closed contour from point A back to point A, then the potential must have two different values at point A. Thus if we are to be able to represent flows with nonzero circulation, we must relax the requirement for continuity of the potential by defining a *branch cut* from some point on the surface of the airfoil to infinity, as shown in Figure 7.1.2, across which a jump in the value of the potential, but not the first derivatives, can take place. Note that the jump in potential across the cut is equal to the circulation around the airfoil and must therefore be the same everywhere along the cut. Note also that for any given velocity field with circulation, it doesn't matter where we put the branch cut. The only requirement is that there be a cut.

Even after we've introduced a branch cut to allow for flows with circulation, another difficulty remains. The potential jump across the branch cut is now a free parameter in the problem that our potential equation and boundary conditions are not sufficient to determine. A solution exists for any value of the potential jump and the circulation. Thus the lift, which is the main thing we seek to predict, is indeterminate. Figure 7.1.3 illustrates this problem,

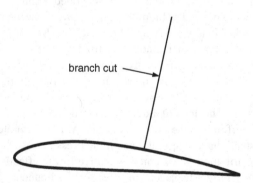

Figure 7.1.2 Illustration of a branch cut to allow a velocity potential to represent a flow with circulation around an airfoil

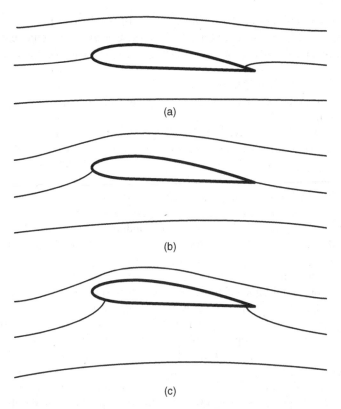

(a)

(b)

(c)

Figure 7.1.3   Sketches of potential-flow streamline patterns around the same airfoil shape with different values of circulation. (a) Zero circulation; zero lift. (b) Circulation such that flow leaves trailing edge smoothly; some lift. (c) More circulation; higher lift than (b)

showing streamline patterns for the same airfoil shape with different values of circulation, all of which satisfy the potential equation and the boundary conditions. Kutta (Durand, 1967a) proposed a resolution to this indeterminacy, observing that the only physically reasonable solution is the one in which the flow leaves the trailing edge smoothly, as in Figure 7.1.3b. This criterion for choosing the preferred value of the circulation is known as the *Kutta condition*. Note that it resolves the non-uniqueness of the potential-flow solution only if the airfoil has a single sharp trailing edge.

Over the many years since Kutta introduced it, there has been much discussion of what the Kutta condition really means, physically speaking. Some critics have seen it as a mathematical artificiality and argued that the need for it demonstrates the physical inadequacy of potential-flow theory. The holders of this harsh view seem to be in a minority, however. The more prevalent view is that when a physical theory leaves out some physical effect, it is reasonable to adopt a rule or adjustment if it can be shown to effectively compensate for what was left out. In the case of potential-flow theory, what was left out is the combination of viscosity and the no-slip condition. It is often observed that viscous flow cannot generally negotiate a sharp corner, as in Figure 7.1.3a,c, without separating, and that the presence of viscosity is therefore the reason that the flow pattern of Figure 7.1.3b is the only physically reasonable one. In this view, the Kutta condition is seen as a perfectly reasonable proxy for one of the major effects of viscosity; that is, that viscous flows tend to separate from sharp edges.

The compensation that the Kutta condition provides is not perfect, but it is quite effective. It resolves the indeterminacy of the lift and enables potential-flow theory to predict something close to the right lift-versus-alpha curve and airfoil pressure distributions in the attached-flow regime, but it leaves the displacement effect of the boundary layer unaccounted for (see Section 4.1.3 for a discussion of the displacement effect in general, and Section 7.4.1 for examples of the effects of displacement on the lift curve and pressure distribution of an airfoil).

Of course, an "edge" doesn't have to be perfectly sharp to provoke viscous-flow separation, and in the real world separation will still anchor itself near the trailing "edge" of an airfoil even if there is a considerable degree of rounding. Thus even if the trailing edge is rounded, something like the Kutta condition still applies, albeit with some uncertainty as to where it applies.

So the Kutta condition is reasonably seen as accounting for the major effect of viscosity. A logical extension of this line of thinking is that lift would not exist without viscosity. It can be shown, based on starting the flow from rest, that in the absence of viscosity, the nonlifting flow pattern of Figure 7.1.3a is the one that would occur (see Gentry, 2006). There is some experimental support for this conclusion, though it is not extensive. Fluids without viscosity exist only in the form of superfluids such as liquid helium, and these have been produced in the laboratory only in small volumes in which the usual kinds of aerodynamics experiments can't be done. There has been one experiment in which a tiny propeller was suspended on a slender fiber in a flow of superfluid liquid helium (Donnelly, 1967). There was no detectable torsional deflection of the fiber, indicating that there was no lift on the propeller's blades. I don't know how definitive this result is. Given the small scale of the experiment, it's unlikely that the trailing edges were very sharp on the scale of the blade chord.

A possible alternative mechanism for the Kutta condition in gas flows is compressibility. Compressible flow around a sufficiently sharp trailing edge would expand to a vacuum condition, with no plausible way for the flow to reach the stagnation point on the other surface. Thus even in the absence of viscosity, we might still see the Kutta condition obeyed through the effects of compressibility, a possibility we consider in more detail in Section 9.1.5.

In the early years of our discipline, even solutions for incompressible potential flow around general airfoil shapes required too much computation to be practical. As we noted in Section 3.10, analytic conformal mapping provides a way to generate solutions for fairly "complicated" shapes with practically no computing. By applying this method, Joukowski (Durand, 1967a) was able to define some reasonably practical-looking airfoil shapes and predict their pressure distributions. Another simplification, developed by Munk around 1920 (Durand, 1967a), is the linearized inviscid theory, applicable in the limit of small thickness and angle of attack. We'll consider the linear theory further in Section 7.4.1, where we find that it provides interesting insights into the effects of shape and angle of attack on an airfoil's pressure distribution.

7.2   Lift in Terms of Circulation and Bound Vorticity

A major relationship that can be derived from momentum conservation and kinematic considerations (circulation and vorticity theorems) is that lift is always accompanied by circulation and vorticity. In Section 7.1, we saw how this relationship must be accounted for in potential-flow solutions for 2D airfoil flows. It is also a key ingredient in quantitative theories of lift and induced drag in 3D, as we'll see in Sections 8.2 and 8.3. And it plays a role in a particular kind of explanation of lift in general, as we'll see below, though not the kind of physical explanation we'll be seeking in Section 7.3.

The most basic statement of this relationship is the *Kutta-Joukowski theorem* (Joukowski, 1906). As originally derived, it applies to lift in 2D, but it also applies to stations along the span of a 3D wing in the limit of high aspect ratio. The theorem states that the lift per unit span on a 2D airfoil of any shape in a steady, inviscid, irrotational flow is proportional to the circulation Γ on a closed contour enclosing the airfoil:

$$l = \rho U_\infty \Gamma, \tag{7.2.1}$$

where ρ and U_∞ are the density and velocity at infinity. The contour on which Γ is defined can be any closed contour enclosing the airfoil, because the flow everywhere outside the airfoil surface is assumed irrotational, and Stokes's theorem (Equation 3.3.4) then requires that Γ must have the same value regardless of what closed contour is chosen. In a real, viscous flow, the theorem should still apply with high accuracy, provided the contour is outside the boundary layer and doesn't cut the viscous wake too close to the trailing edge.

Now, given that lift requires circulation around the airfoil, Stokes's theorem also requires that any closed contour enclosing the airfoil must have a net spanwise vorticity flux passing through it, equal to the circulation. This vorticity is referred to as *bound vorticity* because it is associated with the airfoil itself and is not convected away from the airfoil by the flow. Consider four ways that the bound vorticity can be thought to reside with the airfoil, in decreasing order of realism:

1. In a real, viscous flow with a physical boundary layer on the surface of the airfoil, the bound vorticity is the actual physical vorticity in the boundary layer, as illustrated in Figure 7.2.1a. The bound vorticity is therefore distributed over some short distance off the surface into the flowfield. In a steady boundary-layer flow, vorticity is replaced by convection from upstream and diffusion from above or below as fast as it is convected away by the flow, so that it is as if the vorticity were stationary, and we can still think of it as bound vorticity (see Section 4.2.4 for a discussion of the vorticity "budget" of a boundary layer).
2. In an inviscid flowfield like we assumed in the derivation of Equation 7.2.1, the bound vorticity cannot reside anywhere in the interior of the field because the flow is irrotational. The vorticity must therefore reside on the airfoil surface, as shown in Figure 7.2.1b. It takes the form of an ideal vortex sheet, which is a sheet of zero thickness in which the vorticity is infinite, but the vorticity flux (strength) per unit width of sheet is finite, a concept we discussed in Section 3.3.7. The strength of the sheet is equal locally to the magnitude of the potential-flow velocity at the surface, which is consistent with having the interior of the airfoil filled with fluid at rest and with the idea that a vortex sheet defines a velocity jump, as we saw in Section 3.3.8.

3. In either of options (1) or (2) above, the vortex sheets on the upper and lower surfaces consist of vorticity of opposite signs, and their effects are partially offsetting. What really counts out in the field is the sum of the upper and lower surface contributions, or the *net vorticity* at each longitudinal station along the chord, which in some models of the flow is taken to reside in a vortex sheet on a *camber line* between the upper and lower surfaces, as illustrated in Figure 7.2.1c. This picture simplifies things considerably and is usually what is meant when the term "bound vorticity" is used. We'll find that thinking just of the net vorticity makes it easier to think about what happens in the flowfield around a 3D wing, which we'll take up in Section 8.1.2. The picture also applies naturally to an airfoil with zero thickness, in which case the vorticity strength is equal to the magnitude of the velocity difference (jump) between the upper- and lower-surface flows.
4. In some highly simplified models that are used when only the farfield effect is important, the bound vorticity is represented by a single vortex, usually placed at the quarter chord, as shown in Figure 7.2.1d.

Of course, the viscous wake downstream of the trailing edge in 2D flow also contains vorticity, but of opposite signs in the upper and lower portions of the wake. At any longitudinal station along the wake, the net vorticity (integrated over the thickness of the wake) is thus a difference between contributions of opposite sign. The near wake of a lifting airfoil usually has some curvature and therefore a pressure difference and a velocity jump across it, both usually small. This means that the net vorticity convected past the trailing edge is small but not exactly zero. As the wake straightens out downstream, the net vorticity in it must go to zero. For this to happen, any net vorticity that was there in the near field must be destroyed by viscous diffusion between the vorticity of opposite signs in the upper and lower portions of the wake. Then practically no net vorticity is convected out of the near field by the wake. The global situation is as if all the vorticity resided close to the surface (the bound vorticity), and there were no other vorticity in the field, just as it was for the inviscid-flow case.

The usual derivation of Equation 7.2.1 invokes momentum conservation applied to a control volume with its inner boundary at the airfoil surface and its outer boundary far

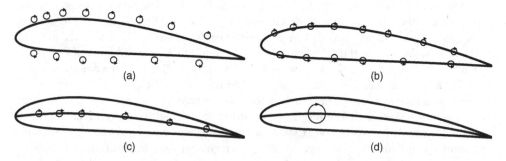

(a)

(b)

(c)

(d)

Figure 7.2.1 Illustrations of the bound vorticity associated with a lifting airfoil, represented in different ways, in order of decreasing physical fidelity. (a) As the vorticity in the physical boundary layer of a real viscous flow. (b) As an ideal vortex sheet on the surface itself, in an inviscid flow. (c) As the net vorticity of the upper and lower surfaces, located on a camber line between the surfaces. (d) As a single vortex at the quarter-chord location

enough away that the velocity field can be approximated by the combination of the uniform freestream and a single vortex of strength Γ located on or near the airfoil, representing the integrated bound vorticity on the airfoil surface. The original derivation assumed incompressible flow, but it is sufficient to assume subsonic flow and small disturbances in the farfield.

So conservation of momentum in control-volume form, combined with Stokes' theorem, tells us that lift must be accompanied by circulation and bound vorticity. Note, however, that Equation 7.2.1 is just a relationship between the lift and the circulation, and that it doesn't predict what the lift or circulation will be for a particular airfoil shape and flow condition. To predict the lift, we must still solve for the flowfield to some level of fidelity, as discussed in Section 7.1. And, of course, Equation 7.2.1 doesn't explain what causes the circulation.

7.2.1   The Classical Argument for the Origin of the Bound Vorticity

An argument to explain the origin of the circulation and thus the existence of lift is given by Prandtl and Tietjens (1934) and by Batchelor (1967). It follows the airfoil through the process of a start from rest and relies more on observation of experimental flow visualizations and on circulation and vortex theorems than on direct physical reasoning. The argument can be paraphrased as follows:

> With the air initially at rest relative to the airfoil, the circulation around any closed contour enclosing the airfoil is zero. The air is then suddenly put into "uniform translatory motion" (motion that is uniform in the limit in the farfield). One of Kelvin's theorems states that when a nonviscous fluid is suddenly put into uniform motion, the circulation around any closed contour is unchanged, so that the circulation around the airfoil is still zero. Immediately after this sudden start, the velocity field looks like the steady potential-flow solution without circulation, as in Figure 7.1.3a. At this point, the same argument comes into play that was used to support the Kutta condition that determines the lift in steady potential flow: The flow around the sharp trailing edge has infinite velocity, and now, because air actually has a small amount of viscosity, the flow separates smoothly from the trailing edge. Prandtl and Tietjens state that as a result of the separation, a "surface of discontinuity," or vortex sheet, begins to shed from the trailing edge. According to one of Helmholtz's theorems, this vorticity must be convected with the flow, and it will therefore be carried downstream. This sheet rolls up into what is called the *starting vortex*. Prandtl and Tietjens don't explain why the separation from the trailing edge leads to the shedding of a vortex, but appeal instead to flow-visualization photos that show the vortex forming. The photos also show that the formation of the starting vortex is accompanied by the formation of a circulatory flow around the airfoil in the direction opposite to that of the starting vortex. Prandtl and Tietjens point out that this is consistent with another one of Kelvin's theorems, which requires that the total circulation around the airfoil and the starting vortex continues to be zero, and that the individual circulations around them must therefore be equal and opposite. Finally, the flow around the airfoil settles into a steady state with nonzero circulation, and the existence of lift follows from the Kutta-Joukowski theorem.

Note that this scenario is more of a description of the process by which circulation is established than it is an explanation. To the extent that it is an explanation, it is more logical and mathematical than physical, as all the crucial steps along the way are justified by mathematically derived theorems: Kelvin, Helmholtz, and Kutta-Joukowski. The theorems are correctly applied, so the logical inferences are correct. But they do amount more to logical inferences than to physical explanations.

A further shortcoming of this argument as an explanation of lift in general is that it concentrates on a particular time sequence by which lift can be established (an impulsive start from rest at a fixed angle of attack) and "explains" only one aspect of the final steady state (the circulation). One might wrongly infer from the Prandtl-and-Tietjens scenario that the final steady-state lift depends on the details of the initial time history. But the final steady state is nearly always a unique "stable attractor" that is the end point of any one of an infinity of time histories (impulsive starts, gradual starts, varying angle of attack, etc.). Thus the final steady state is in a sense more fundamental than any of the possible starting sequences, and a more satisfying explanation would deal directly with it and show us why it does what it does. And this would require establishing more details than just the circulation.

The Prandtl-and-Tietjens scenario falls short of being a real physical explanation in other ways as well. First, it assumes the existence of the starting vortex and supports it only by appealing to experimental flow visualizations. Just how or why the vortex forms is never explained. Then it strays into questionable cause and effect when it says that "Owing to the formation of the starting vortex, the velocity field is changed . . . " This implies that vorticity is somehow a cause of the velocities that occur elsewhere, which reflects an incorrect interpretation of the Biot-Savart law, as we saw in Section 3.3.9. Otherwise, the Prandtl-and-Tietjens scenario, as presented, neglects to assign any cause-and-effect relationships or at least fails to make a distinction between logical inference and physical cause.

The general order in which the argument is presented leads the reader to infer that lift is a result of the formation of the starting vortex. The overall logical inference is true, that if a starting vortex was shed, then there must be lift on the airfoil. But this particular inference works just as well in the opposite direction: If there is lift on the airfoil, then a starting vortex must have been shed. The general direction of physical cause and effect is such that it would be closer to the truth to say that the lift force is the prime mover, and the starting vortex is a result of it, not a cause.

Batchelor's version of the explanation also refers to experimental flow visualizations, but it dwells in greater detail on the initial phase of the formation of the starting vortex, and it makes a more convincing argument that the shedding of vorticity is a necessary consequence of viscosity. Overall, however, it also implies that the shedding of the starting vortex somehow causes the establishment of the circulation on the airfoil.

A close analogy to the relationship between lift and the starting vortex is the relationship between a bear walking through the woods after a fresh snowfall, and the paw prints he leaves behind in the snow. The presence of the prints allows the logical inference that a bear has passed by since the snowfall. But this works in the opposite direction as well: Knowing that a bear has passed through soft snow allows the equally logical inference that prints must have been left. The logical inference can work either way, but the physical cause-and-effect relationship is clearly one-way: The passage of the bear caused the prints, but the prints did not cause the passage of the bear. Like a paw print in the snow, the starting vortex is a mostly passive trace left behind by other physical events. It is not a cause of those events.

In "The Origins of Lift," Gentry (2006) recounts essentially the same scenario as Prandtl and Tietjens, but he goes on to describe some key features of the airfoil flowfield, such as the upwash ahead of the airfoil, the higher velocity over the upper surface, and the downwash behind the airfoil, as being caused by the "circulatory flow." Actually, these features are just parts of the circulatory flow pattern, and it isn't logical to say they are caused by it. Assigning causation to "circulatory flow" in this way is also closely related to the idea that vorticity at one location can "cause" velocities elsewhere, which is incorrect, as we saw in Section 3.3.9.

7.3 Physical Explanations of Lift in 2D

It's easy to explain how a rocket works, but explaining how a wing works takes a rocket scientist.
                                                                                    – Philippe Spalart

In this section, we seek to explain lift in 2D flows, in the context of continuum fluid mechanics, but without appealing to mathematics, both to further our own physical understanding and to have a satisfactory explanation that we can share with nonexperts. We've already alluded to what a difficult task this is, and because it is so difficult we'll end up devoting a lot of attention to it. Not only do we have to deal with the complexity of the physics, but we will also have to address an extensive collection of nontechnical lore that this topic has generated over the years, most of which is deeply flawed. Because of this, we'll end up spending more time on the background than on the explanation itself. But all this effort is justified by the general importance of the issue. An explanation for what causes lift is probably the single most important thing most laymen want to know about aerodynamics.

First, we'll look at some of the general characteristics of many of the explanations that are already in circulation (no pun intended) and discuss their strengths and weaknesses. With that as background, we'll set down our "requirements" for a more satisfying explanation, and then we'll proceed to compile our "best" explanation. We'll divide this into a basic part that can be shared with a nontechnical audience, followed by some additional technical details.

7.3.1 Past Explanations and their Strengths and Weaknesses

The purpose of this section is not to provide an exhaustive recounting of the explanations that have been offered in the past, but to look at some of the general lines of argument they share and to assess their strengths and weaknesses.

7.3.1.1 A General Observation on the Nature of the Problem

I propose that an underlying reason most of the explanations we'll discuss below fall short is that they set out to do more than is logically possible, given the nature of continuum fluid mechanics. In Section 3.5, we discussed how predicting what will happen in a fluid flow requires solving the equations of motion, something we can't generally do in our heads with sufficient precision to choose the "right" solution from the many kinematically possible flow patterns. Yet, in most of the proposed physical explanations of lift, there is an implied assumption that a simple linear argument starting from a few basic principles can both predict and explain the existence of lift and that no prior knowledge of the characteristics of the flowfield is required. We'll discuss this issue further when we set down our "desired attributes of a more satisfactory explanation" in Section 7.3.2.

7.3.1.2 The Flowfield-First Fallacy

A line of thinking that characterizes many of the physical explanations for lift that have been proposed is something I call the "flowfield-first" fallacy. It shows up most seriously in Bernoulli-based explanations, but it creeps into other explanations as well. The general line of argument is first to determine, by some argument or other, what the flow around the airfoil does and then to deduce that the flow exerts a lift force on the airfoil. There is no mention of whether the lift force influences what the flow does. It is not always explicit that the flow "causes" the force, but causation is implied by the way the inferences run. What is implied is one-way causation of the kind that I argued in Section 3.5 is incompatible with continuum fluid mechanics, in which cause-and-effect relationships tend to be reciprocal. Any argument that claims to establish what a flowfield does, without reference to the forces that it exchanges with its environment, must be at least incomplete. Note that the mathematical predictions of lift that we discussed in Section 7.1 don't share this failing because, in solving equations of motion for the continuum behavior of the fluid, they determine the flowfield and the force together, taking into account the proper circular cause-and-effect relationships.

So the flowfield-first approach is faulty at the outset, in assuming that we can determine key features of the flowfield that will lead to a lift force, and that we can make the determination without knowing that the lift force is there. The Bernoulli-based explanations are the most serious offenders in propagating this fallacy, but they tend to have other faults as well, as we'll see.

7.3.1.3 Bernoulli-Based Explanations

The key flow feature that Bernoulli-based explanations appeal to is a region of high velocity that forms over the upper surface of the airfoil, which is then said to imply, or even cause, low pressure on the upper surface, as a consequence of Bernoulli's principle. Causation is not always explicitly stated, but it is implied. We'll see later that lower pressure and higher velocity over the upper surface are indeed necessary for lift. But the implication that the high velocity causes the low pressure amounts to the same kind of one-way causation that we just saw in connection with the flowfield-first fallacy, applied now at another level of detail in the flowfield, and it is wrong for the same reason. Again, one-way causation is simply not compatible with the physics. In this case, the high velocity and the low pressure are indeed related, but it is wrong to explain the causation as running only in one direction. And this error effectively dooms any explanation built along these lines. Because the low pressure is seen only as a result of the high velocity, and not as part of the cause, it is impossible to explain correctly how the high velocity got there in the first place. Most attempts to explain the high velocity without appealing to the pressure follow either of two main approaches, both unsatisfactory.

7.3.1.4 Longer Path and Equal Transit Time

This is an argument that is widespread in explanations aimed at the layman. In this approach, it is assumed that the upper surface of the airfoil is more convex than the lower surface,

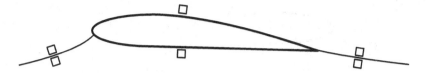

Figure 7.3.1 Fluid parcels splitting at the leading edge of an airfoil and rejoining at the trailing edge according to the erroneous equal-transit-time assumption

which is often true but not always, and that the path the air must follow around the upper surface is therefore longer than the path around the lower surface. It is further assumed that fluid parcels that are split apart at the leading edge to traverse the upper and lower surfaces must rejoin at the trailing edge as shown in Figure 7.3.1. Thus fluid parcels negotiating both paths must do so in equal transit times, and we conclude that the velocity over the upper surface must be higher than that over the lower surface.

This explanation is wrong for several reasons that we'll address next, but first we should note that relative path length doesn't work well as an indicator of how much lift an airfoil can produce. First, lift can be produced with zero difference in path length. For an airfoil consisting only of a camber line (no thickness), there is an *ideal angle of attack* at which the flow attaches smoothly to the leading edge, and the upper- and lower-surface flows see the same path length. For example, a camber line consisting of a segment of a circular arc produces lift at its ideal angle of attack of 0°. Then, even among airfoils that do have a difference in path length, the difference tends to be no more than a small percentage. This is an order of magnitude too small for the path-length explanation to account for the lift that airfoils actually produce, as Craig (1997) has pointed out. Airfoils under real lifting conditions can produce much lower pressures and much higher velocities over the upper surface than any reasonable path-length difference can account for. So where does the path-length explanation go wrong?

First, the assumption of equal transit time is wrong. There is no reason why fluid parcels that split at the leading edge must rejoin at the trailing edge. But to say more about relative transit times, we need to observe one careful distinction regarding what fluid parcels we are going to follow to measure the transit time. In real viscous flows, we assume that the no-slip condition holds, with zero velocity at the solid surface, so that transit times for parcels passing close to the surface approach infinity. Even in ideal inviscid flows there is a similar problem. In practically all 2D airfoil flows, the initial attachment of the flow is at a stagnation point at which the velocity is zero and away from which the velocity initially increases linearly, and again, transit times for parcels passing close to the surface approach infinity. This is an example of the infinite delay effect for blunt-nosed obstacles in general that we discussed in connection with Figure 5.1.3c. To get around this difficulty, we must measure the transit time for parcels that start their journeys at some arbitrary distance above or below the stagnation streamline upstream, and in the case of viscous flow, the distance should be chosen large enough so that the parcels remain outside the boundary layers on the upper and lower surfaces. When we do this, we find that under lifting conditions, parcels that traverse the upper surface make the trip in less time and get to the trailing edge before the corresponding parcels that traverse the lower surface. Parcels that started close together near the attachment streamline ahead of the airfoil end up permanently displaced from each other after they pass the trailing edge, as shown in Figure 7.3.2.

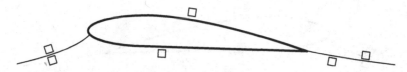

Figure 7.3.2   Fluid parcels splitting at the leading edge of an airfoil and ending up displaced from each other downstream in an ideal lifting flow that satisfies the Kutta condition

The equal-transit-time explanation is problematic on another level as well, and that is that it is not a real physical explanation. Simply saying that something must go faster than something else to get somewhere at the same time does not explain *how* it goes faster, for example, by identifying the physical force that accelerates it to a higher velocity.

7.3.1.5   Hump, Half-Venturi, or Streamtube Pinching

As usually presented (see Anderson, 2008, Section 5.19, for example), this approach also assumes that the upper surface is more convex than the lower surface. The argument then assumes a general flow pattern along the lines of what we discussed in Section 5.1, what I called the "obstacle effect." The upper surface acts as a kind of "hump" or larger obstacle to the flow than does the lower surface, with the result that the streamlines over the upper surface are pinched together more than those over the lower surface. Then, as a result of streamtube mass-flux conservation, the velocity is higher. As an alternative to the "hump," an analogy is sometimes drawn between the upper surface of the airfoil and the inner surface of a Venturi tube, but this is essentially the same argument.

This explanation is better than the longer-path-length explanation in that the velocity difference that it might account for is not so limited. But as usually presented, it has two major flaws:

1. It doesn't really explain *how* streamtube pinching comes about at all, let alone why it is greater over the upper surface than over the lower surface. Streamtube pinching is not a kinematic necessity, as we saw in Section 5.1 in the discussion of Figure 5.1.4, so a dynamical explanation for it is needed. Anderson's (2008) version offers only that the flow somehow "senses the upper portion of the airfoil as an obstruction" and pinches down to go around it. I assume this is not meant to say that a fluid flow actually has some kind of remote-sensing capability, just that in pinching down the flow behaves as if it were sensing the presence of the airfoil. Still, this leaves unexplained what physical principle is at work in the pinching-down response. Really explaining streamtube pinching requires getting into the details of the dynamics, as we did in our discussion of the obstacle effect in Section 5.1.
2. Appealing to mass-flux conservation (continuity) isn't very satisfying as a physical explanation for higher velocity. Conservation of mass is a fundamental physical principle, but at the flowfield level it is really more of a kinematic constraint than a dynamical relation (see Section 3.4.1). Really understanding why something speeds up requires looking at the forces.

Some versions of this explanation argue that the airfoil needn't be more convex on the upper surface and that a positive angle of attack is sufficient to cause the leading edge to act as a hump and produce high velocity over the upper surface near the leading edge. The explanation by Eastlake (2002) is in this category. This version of the hump argument has some appeal in that it provides a basis for the variation of lift with angle of attack, but it still has the major fault of not providing a satisfactory explanation for how the streamtube pinching and the high velocity happen. We can also counter this argument with the observation that it doesn't rule out the zero-lift flow pattern of Figure 7.1.3a, in which the trailing edge also acts as a hump, producing high velocity on the lower surface near the trailing edge.

7.3.1.6   Confusion Regarding Low Pressures

Confused thinking about what low pressure actually means, physically speaking, seems to be widespread, and the confusion isn't limited to explanations of airfoil lift. We all tend to think intuitively of lower-than-ambient pressure as something that can exert a pull on surfaces that it touches. For example, Shevell (1989) uses a tornado as an example in his discussion of the low pressure in the core of a vortex, and states that "The low pressure is the force with which a tornado removes the roof from a house." This of course can't be true in a literal mechanical sense. Even the lowest pressure reachable in the core of a tornado must still be positive in an absolute sense, and therefore cannot by itself be the force that lifts a roof. If a roof is lifted, it is lifted by the pressure beneath it. The low-pressure air above the roof plays its part by not pushing down as hard as it ordinarily would. The low pressure doesn't directly provide the force that lifts the roof.

The idea that low pressure can exert a pull finds its way into some airfoil explanations that discuss the low pressure on the upper surface. It is mostly explanations aimed at popular audiences that are guilty of this, and it is often seen in both the graphic illustrations and the words.

In the graphics accompanying many explanations, the pressure distribution around the airfoil surface is plotted as an array of vectors (arrows) normal to the surface. The trouble is that the vectors' magnitudes are made proportional to the local pressure difference, relative to ambient, so that high pressure is shown as arrows pointing toward the surface, and low pressure is shown as arrows pointing away from the surface, as seen in Figure 7.3.3a. One could argue that this convention is perfectly fine for conveying quantitative information and that it shouldn't cause confusion for a technically literate audience, provided the basis is explicitly spelled out. However, the quantitative data could be conveyed more easily and clearly in a conventional Cartesian plot. In representing the pressure as arrows with directions, the intent is clearly not just to convey data, but also to provide a physical feel for how the pressure acts as a force. In this regard, an illustration like Figure 7.3.3a is misleading, at least to a nontechnical reader, because it gives the impression that the air in regions of low pressure is pulling on the surface.

To provide the physical feel that Figure 7.3.3a seeks to provide, but to do it without misleading, is problematic. We could make the lengths of the arrows proportional to the absolute pressure, as in Figure 7.3.3b. In this case, we have had to assume the equivalent of a relatively high subsonic Mach number, so that the pressure differences are large enough

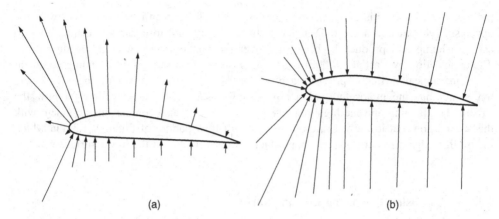

Figure 7.3.3 Airfoil pressure distributions represented graphically as vectors. (a) Arrows proportional to the pressure difference $p - p_\infty$. (b) Arrows proportional to the absolute pressure

to stand out relative to the magnitudes. And this is the only drawback to this type of presentation: It exaggerates pressure differences compared to what would actually occur, say, in a general-aviation application at low Mach numbers where the pressure differences would be too small to see clearly.

The words that accompany an illustration like Figure 7.3.3a are often as misleading as the illustration. The word "suction" is often used to refer to the pressures that are lower than ambient. As a technical usage, there is nothing wrong with this, as long as everyone understands that it just means "lower pressure than ambient." But for a nontechnical reader, it is potentially confusing because it tends to evoke an image of the air pulling on the surface, like the arrows in Figure 7.3.3a.

Another idea that is often put forward is that the upper surface produces most of the lift, because the negative pressure differences on the upper surface, relative to ambient, are on average larger in magnitude than the positive pressure differences on the lower surface. Of course, if we look at it in terms of the actual physical forces, it is more correct to say that the air pushing against the lower surface is responsible for all of the lift, and then some, and that the air over the upper surface helps out by not pushing down as hard as the lower-surface air pushes up. But the view in terms of differences from ambient also has some merit. After all, the absolute ambient pressure would be there even if the airfoil were at rest, and it is the differences from ambient that result from the dynamics of the flow and are responsible for the lift. And it is true that the negative differences on the upper surface are generally larger than the positive differences on the lower surface.

It is often implied that this is a general characteristic of the production of lift, which is only partly true. According to the linearized theory we'll discuss in Section 7.4.1, lift results from the angle of attack and camber shape only, and airfoil thickness makes no contribution, to first order. For a cambered-plate airfoil (zero thickness) in the linear limit, lift is characterized by equal-and-opposite differences on the upper and lower surfaces. So for a very thin airfoil at a low lift level, the pressure differences associated with the lift are almost equal-and-opposite. But in most practical situations, both thickness and non-linear effects tend to reduce the pressure over most of the chord on both surfaces of the airfoil, so that the negative differences on the upper surface are actually larger than the positive differences on the lower surface.

7.3.1.7 Momentum-Based Explanations and the Coanda Effect

In momentum-based explanations, it is generally argued that the airfoil produces a flowfield in which some of the air is "deflected" downward and thus has downward momentum imparted to it. To acquire downward momentum, the air must have a downward force exerted on it by the airfoil, and thus, by Newton's third law, the airfoil must have an upward force exerted on it by the air. At the highest level, this general approach avoids the flowfield-first fallacy that mars most of the Bernoulli-based explanations. At least the mutuality of the force exchange between the airfoil and the air is explicitly acknowledged. But most explanations of this type fall short of providing a complete explanation of how the airfoil accomplishes the downward deflection of the stream.

Some momentum-based explanations emphasize that it is not just the lower surface of the airfoil that deflects the flow, and that the flow pattern over the upper surface also contributes strongly to the overall downward deflection. This general assertion is correct, but it is often followed by incorrect reasoning as to how the upper-surface flow does what it does. For example, Anderson and Eberhardt (2001) and Craig (1997) invoke the *Coanda effect* as the reason that the flow is able to follow the curved upper surface. This is problematic on more than one level:

1. Applying the term "Coanda effect" to an airfoil flow is inaccurate and therefore confusing. The Coanda effect usually refers to the tendency of a powered jet flow (in which the jet has higher total pressure than the surrounding fluid) to attach to an adjacent solid surface and to follow the contour of the surface. Although the attached boundary layer on the surface of an airfoil is a shear layer, it is not the same as a powered jet, and is not usually considered an example of the Coanda effect. As we'll see below, there is a limited way in which something like the Coanda effect can be construed as playing a role in airfoil flows, but it is a bit of a stretch.
2. The Coanda effect is erroneously seen as implying that viscosity plays a direct role in the ability of a flow to follow a curved surface. Anderson and Eberhardt assert that viscous forces in the boundary layer tend to make the flow turn toward the surface, specifically, as they put it, that the "differences in speed in adjacent layers cause *shear forces,* which cause the flow of the fluid to want to bend in the direction of the slower layer." Actually, there is no basis in the physics for any direct relationship between shear forces and the tendency of the flow to follow a curved path.

First, we'll look at what the term "Coanda effect" properly encompasses, and then we'll consider what "flow attachment" or "boundary-layer attachment" in ordinary aerodynamic flows really involves and how it differs from the Coanda effect.

The phenomenon that Coanda himself investigated, for which others later coined the term "Coanda effect," was limited to powered jet flows in which the jet is of the same phase (gas or liquid) as the surrounding fluid and has higher total pressure. A relatively thin 2D or annular jet (a jet in the form of a sheet) has a tendency to bend and attach itself to an adjacent solid surface, and to follow the surface even if the surface has strong convex streamwise curvature. This tendency is a result of the jet's entrainment of surrounding fluid and of the requirements imposed on the flow pattern by the continuity equation.

The effect is illustrated in Figure 7.3.4. In (a), we see an isolated 2D jet flowing into otherwise quiescent air. Whether the flow issuing from the nozzle is turbulent or not, at high Reynolds numbers the jet downstream will be turbulent, and the important thing for our purposes is that a turbulent jet strongly entrains fluid from the surroundings as it spreads downstream. Outside of the turbulent jet itself, the fluid flows toward the jet, thus feeding the entrainment. The velocities required to feed the entrainment are not large, but they are important, as we'll see. In (b), we've introduced a solid surface adjacent to the jet, with the leading edge of the surface close to the edge of the jet, but with the rest of the surface curving increasingly away from the jet. Because we've placed the surface in a region that in the case of the isolated jet was nearly quiescent air, we might naively expect it not to have much effect on the high-velocity flow in the jet, and for the jet to continue to flow straight, as it did in (a). But blockage of the flow feeding the entrainment makes it impossible for the flow pattern of (b) to be sustained. The air between the jet and the surface is entrained faster than it is replaced from the surroundings, and the flow quickly switches to the pattern in (c), where the jet bends to flow along the curved surface. The pressure field simultaneously adjusts (Remember circular cause and effect between velocity and pressure!) so that a pressure gradient normal to the local flow direction balances the centrifugal force associated with the curvature of the flow. So we see that even in a jet flow that exhibits the Coanda effect, the curvature of the flow is not a direct result of viscous forces (or their turbulent counterparts), but an indirect one.

In ordinary aerodynamic flows without powered jets, *flow attachment* has very little in common with the powered-jet effect that Coanda investigated and is really just the *absence of boundary-layer separation.* In Section 4.1.4, we saw that boundary-layer separation from a smooth surface in a 2D flow generally requires rising pressure (an adverse pressure gradient) to stagnate the low-velocity fluid. Counteracting the effect of an adverse pressure gradient is the favorable viscous force by which the higher velocity fluid farther from the surface drags the low-velocity fluid along. Boundary-layer separation thus involves a tug-of-war between the adverse pressure gradient and an opposing viscous force. At any given station along a surface subjected to an adverse pressure gradient, the pressure gradient will generally be winning the tug-of war locally, slowing the fluid near the wall, and reducing the velocity slope at the wall. How far the boundary layer perseveres into the adverse pressure gradient before it separates depends on the *rate* at which the pressure gradient wins and the velocity slope at the wall decreases. Until the slope of the velocity profile at the wall is brought to zero, the boundary layer remains attached, just like the corresponding inviscid flow would under the same conditions. Separation occurs only when the adverse pressure gradient has acted over a long enough distance to produce reversal of the velocity profile. Thus the role of viscous forces in maintaining boundary-layer attachment is to reduce the deceleration caused by the pressure gradient and to help the low-velocity fluid at the bottom of the boundary layer keep moving, and the viscous forces are needed only in situations where the pressure gradient is adverse. Viscous forces have nothing direct to do with causing the flow to turn and follow a curved surface.

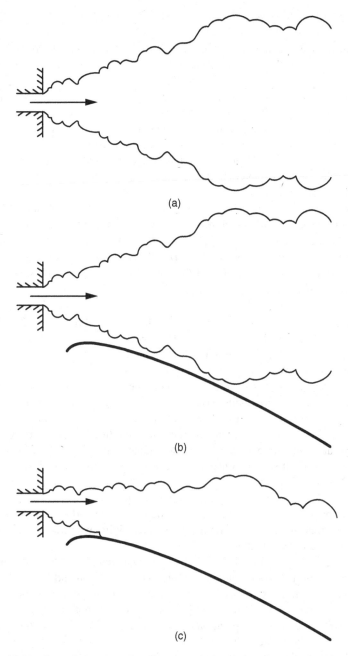

(a)

(b)

(c)

Figure 7.3.4 Illustration of the Coanda effect. (a) Isolated free jet in otherwise quiescent air. (b) Curved surface added adjacent to the jet, but not altering the flow. (c) Jet attached to the curved surface

There are two things that very likely have contributed to confusion regarding this issue:

1. The boundary layer on the surface of the airfoil is effectively a sheet of vorticity. In Section 3.3.5, we saw that one of the kinematic properties of vorticity is that fluid parcels have a solid-body-rotation component to their motion, with angular velocity $\omega/2$. If we picture an airfoil oriented so that the flow is from left to right, fluid parcels in the upper-surface boundary layer will have a solid-body-rotation component to their motion, in the clockwise direction. It seems intuitively natural that fluid parcels that are rotating will tend to follow curved paths (curved downward in this case). This is a pre-Newtonian sort of intuition, however, that has no basis in the physics (the meaning of "pre-Newtonian" is discussed at the end of Chapter 2). There is no connection between fluid-parcel rotation and curved paths in the flow.
2. There is an indirect association between surface curvature and the need for viscous effects in the maintenance of flow attachment. Convex surface curvature is often, though not always, associated with an adverse pressure gradient, in which case favorable viscous forces are needed to prevent separation. But the viscous forces prevent separation by dragging fluid along in the direction of the local flow, not by directly contributing to the turning of the flow.

If viscous forces make no direct contribution to the turning of the flow when the surface is curved, what actually causes the flow to turn? The answer to this question lies in the interplay between the velocity field and the pressure field, which works in the same way whether the fluid is viscous or not. When a flow turns to follow a curved surface, it is able to do so because the pressure field adjusts so as to provide the force needed to accelerate the fluid toward the center of curvature. Thus the centrifugal force generated when the flow follows a curved path is countered by a pressure gradient perpendicular, or *normal,* to the local flow direction. The normal pressure gradient and the flow curvature have a reciprocal relationship in which they cause and support each other simultaneously. This is just another aspect of the circular cause-and-effect relationship between the pressure and velocity fields that we discussed in Section 3.5 and that will figure heavily in the physical explanation for lift that we'll develop in Section 7.3.3. Note that the normal pressure gradient is perpendicular to the streamwise pressure gradient that we considered earlier and that it is the streamwise pressure gradient that plays the important role in determining whether the viscous boundary layer separates or remains attached.

So viscous forces play no direct role in ordinary flow attachment, contrary to Anderson and Eberhardt's explanation. A good counterexample to the Anderson and Eberhardt argument is the flow around a rotating circular cylinder with the freestream perpendicular to the cylinder axis. This is an example of a flow in which the tangential motion of the surface, due to rotation, affects the location of separation. In this case, the flow follows the curved surface farther around the side of the cylinder where the surface is moving with the flow, and it separates earlier from the side on which the surface is moving against the flow. But if we apply the Anderson and Eberhardt argument to this flow, it predicts the opposite of what is observed. On the side where the surface is moving with the flow, the viscous stresses are reduced or even reversed, and the ability of the flow to follow the curved surface would be reduced, according to their argument, but in fact it is enhanced. And vice versa for the other side of the cylinder. The observed effects on both sides are consistent with ordinary

Figure 7.3.5   Candidate flow patterns around an airfoil at a moderate angle of attack. (a) Separation ahead of the trailing edge. (b) Attached flow to the trailing edge

boundary-layer theory, which correctly accounts for the effects of pressure gradient and surface motion.

We've looked in some detail at what the Coanda effect actually is, as it is usually understood to apply to a "powered" jet flow, and why it is not needed for ordinary flow attachment. But might there still be a way that it can properly be said to apply to airfoil flows? Perhaps, but it is at most a very limited way. Figure 7.3.5 shows two candidate flow patterns around an airfoil at a moderate angle of attack.

In Figure 7.3.5a, the flow separates ahead of the trailing edge and doesn't reattach to the surface, and in Figure 7.3.5b, the flow remains attached all the way to the trailing edge. At a low enough angle of attack, where the attached-flow pattern of (b) is the correct one, we could argue that what rules out the alternative separated-flow pattern of (a) is the same entrainment mechanism that is responsible for the Coanda effect. The separated shear layer of (a) entrains fluid faster than it can be replaced from downstream of the trailing edge. So for airfoils with attached flow to the trailing edge, we could say that it is something like the Coanda effect that rules out a separated-flow pattern. But this is not the same as saying that the Coanda effect is needed for the attached-flow pattern to be possible. The fact is that we needn't appeal to any viscous-flow mechanism, boundary layer, Coanda, or otherwise, to explain how flows can follow curved surfaces. Even ideal inviscid "flows" represented by solutions to the potential equation have no trouble following curved surfaces. And in real viscous flows, the natural tendency of the boundary layer is to remain attached unless it is provoked to separate by an adverse pressure gradient that is too strong, as we saw in Section 4.1.4.

7.3.1.8   The Water-Faucet Demonstration

In some explanations that refer to the Coanda effect (Anderson and Eberhardt, 2001, for example), a "simple experiment" is cited as a demonstration. If a curved object, like a spoon or a water glass, is held tangent to a small vertical stream of water from a faucet, the stream will deviate from its original vertical path and follow the curved surface, as shown in Figure 7.3.6a. This effect is obviously not a result of the same turbulent-jet entrainment that is responsible for the regular Coanda effect. Instead, it seems to be due to molecular attraction between the liquid and the solid and to the fact that the stream resists being torn apart, because of the surface tension at the water-air interface. These forces should be independent of the speed of the flow. Thus the amount of deflection they can produce should decrease as the speed of the stream increases, and, indeed, this seems to be the case. You can demonstrate to yourself that if the stream is sufficiently slow, the entire stream will follow the curvature of the glass, but if the stream is too fast, it will break up, and much of it will leave the surface of the glass without being deflected much, as shown in

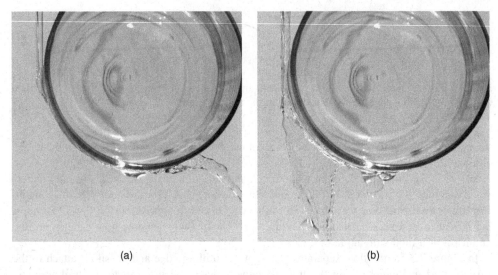

(a)   (b)

Figure 7.3.6 Water-faucet experiment that is sometimes proposed as a demonstration of the "Coanda effect" but actually demonstrates molecular attraction and surface tension (Photos by the author). (a) The slowest coherent stream that an ordinary garden nozzle can produce enters vertically downward on the left, adheres to the surface of a drinking glass, and is deflected through more than 90°. (b) A somewhat faster stream from same nozzle breaks up and is not deflected nearly as much

Figure 7.3.6b. In any case, such water-stream demonstrations are not relevant to explaining how aerodynamic flows follow convex surfaces.

7.3.1.9 A Momentum-Based Argument in Which Flow Turning Comes First

A variation on the momentum-based argument is put forward by Weltner and Ingelman-Sundberg (2000) on their web site. It is noteworthy because it explicitly rejects the one-way causation from velocity to pressure that is common in Bernoulli-based explanations, but then implies one-way causation in the other direction. Their argument goes as follows, (paraphrased):

> It is wrong to argue that the high flow speed over the upper surface of an airfoil "causes" the low pressure there because the pressure difference whose existence we're trying to justify must have been there in the first place to accelerate the flow to higher speed. So where does the pressure difference come from? It arises because the airfoil deflects the flow, or causes it to change direction. So the change in flow direction causes the reduction in pressure, which in turn causes, or at least implies, the increase in flow speed.

This argument claims, in effect, that it is not correct to invoke a longitudinal acceleration (a change in speed) as the sole reason for a pressure change, but in the case of an airfoil flow, it is permissible to invoke the normal acceleration (a change in flow direction). The implied justification is that the primary effect of the airfoil surface is to force the flow to change direction and that it is therefore logical for the normal acceleration to precede the pressure change in the chain of cause and effect.

This idea has considerable intuitive appeal, but it is not entirely correct. The problem is that the interaction of most of the flow with the solid surface is not as direct as this argument implies. Only one vanishingly thin streamtube (the stagnation streamline) actually comes into contact with the airfoil surface, and the normal accelerations of all other streamtubes happen out in the field, just like the longitudinal accelerations do. For most fluid parcels, there is no direct interaction with the airfoil surface, only with adjacent parcels, and in this situation, there is no basis in the physics for making a distinction between the normal and longitudinal components of the acceleration. They are both just accelerations, and neither one has a one-way causal link to the pressure. The original argument correctly states that a change in flow speed requires a pressure difference. We can say the same thing about a change in flow direction: The only thing that can cause a change in the velocity vector is a pressure gradient. Thus for the normal acceleration to happen, the normal pressure gradient must already be there. And then if we incorrectly limit ourselves to one-way causation, we leave unanswered the question of what causes the pressure gradient. A correct explanation must acknowledge circular causation between the pressure and velocity fields.

7.3.1.10 The Bernoulli-versus-Momentum "Controversy"

Despite their flaws, both the Bernoulli approach and the momentum approach manage to get at parts of the truth. As we saw in Section 5.4, it is the pressure that transmits the lift force to any lifting body. And lift is invariably accompanied by low pressure on the upper surface and must therefore also be accompanied by elevated velocity outside the boundary layer, consistent with the Bernoulli explanation. Generating lift also requires deflecting the flow downward and imparting downward momentum, as in the momentum explanation. The Bernoulli and momentum explanations just appeal to different aspects of the same global phenomenon. The Bernoulli explanation is based on the near-surface flow and the surface pressure, while the momentum explanation is based on a manifestation of lift that extends into the far-field. They are not contradictory. However, having two different explanations that both seem "right" has led to some heated arguments, especially on some web sites that offer explanations of lift.

For example, there has been a tendency for advocates of either the Bernoulli approach or the momentum approach to argue that one is correct and the other is wrong, leading to a Bernoulli-versus-momentum (or Bernoulli-versus-Newton) "controversy," as seen, for example, in much of the discussion accompanying the Wikipedia article on "Aerodynamic lift." An alternative put forward by some other commentators is that both approaches are correct, but that they represent two separate mechanisms and two separate types of lift. The explanation on the ALSTAR instructional web site (Florida International University) is in this camp. According to this view, a flat plate at an angle of attack produces only "reaction lift," and an airfoil with camber produces "Bernoulli lift" at zero angle of attack and adds "reaction lift" as angle of attack increases. The Bernoulli-versus-momentum controversy and the Bernoulli-lift/reaction-lift distinction are false, of course. Lift is a phenomenon that always involves a seamless combination of pressure difference and momentum transfer.

Another argument that is often made, as in several successive versions of the Wikipedia article "Aerodynamic Lift," is that lift can always be explained either in terms of pressure or in terms of momentum and that the two explanations are somehow "equivalent." This "either/or" approach also misses the mark. It's true that lift can be accounted for quantitatively either by the integrated pressure difference or by the momentum transfer. But such accounting doesn't constitute a satisfying physical explanation for how it all happens. A complete explanation must address all of the necessary aspects of the flowfield, and neither the Bernoulli nor the momentum approach is complete in this sense. As we'll see in Sections 7.3.3 and 7.3.4, a complete explanation must address both the velocity magnitude (Bernoulli) and direction (downward turning) because the flow we're trying to explain is not one-dimensional, but two-dimensional.

7.3.1.11 An Explanation Based on Flow Curvature

Babinsky (2003) begins by explaining the relationship between streamline curvature and the pressure gradient in the cross-stream (normal) direction, and he does so without any incorrect implication that the cause-and-effect relationship is one way. He then presents graphic illustrations of streamline patterns in flows around airfoils, showing that under lifting conditions the streamlines above and below the airfoil are generally curved downward. The pressure gradient (low pressure to high) is thus upward, and given that the pressure far above and below the airfoil must be close to ambient, the pressure on the airfoil upper surface must be low, and the pressure on the lower surface must be high, so that there is lift. Babinsky further shows that if the airfoil is thick, the high pressure on the lower surface may not materialize, but the pressure difference between the upper and lower surfaces can still provide lift. This explanation is correct as far as it goes, but it is incomplete in that it doesn't explain how the pressure gradients in the streamwise direction are sustained.

7.3.1.12 Lanchester's Explanation

An interesting early explanation was put forward by Lanchester (1907) and is paraphrased in the historical sketch by Giacomelli and Pistolesi, in Durand (1967). Lanchester first imagines a horizontal 2D flat plate (finite chord, infinite span) moving vertically downward, in effect a 2D planar parachute. There is, of course, a lift force on the plate, with low pressure above the plate and high pressure below, and air both above and below the plate is dragged downward. Because there can be "no permanent change of density or accumulation of matter in the lower strata of the atmosphere," the downward current above and below the plate must be accompanied by upward currents around the edges of the plate, driven by the difference in pressure. This is the same general flow pattern that we discussed in Section 5.1 in connection with the "obstacle effect." Lanchester imagined this motion to be associated with "a field of force established around the plane when the load was first applied: a field of force everywhere defined by the acceleration of the air particles."

Lanchester then considers adding a horizontal motion to the vertical motion, so that the plate becomes a 2D "glider" descending along a sloped path. He deduces the resulting velocity field not by superposition of velocities, but by arguing that fluid particles passing through the same "field of force" that was there in the case of pure vertical motion

"will receive an upward acceleration as they approach the aerofoil, and will have an upward velocity as they encounter its leading edge. While passing instead under or over the aerofoil, the field of force is in the opposite direction, viz. [that is,] downward, and thus the upward motion is converted into a downward motion. Then, after the passage of the aerofoil, the air is again in an upwardly directed field, and the downward velocity imparted by the aerofoil is absorbed."

Note that although the velocity field with only vertical motion was symmetrical fore and aft, with upwash off both the leading and trailing edges, the field in the presence of forward motion is asymmetric, with downwash behind the trailing edge rather than upwash. Lanchester deduced from this flowfield picture that a slightly cambered airfoil, with the trailing edge turned down to enhance the downwash and the leading turned down to meet the oncoming upwash, should be superior to an airfoil with zero camber. Lanchester concludes that the vertical velocity of the air particles far ahead of and far behind the airfoil is zero, by the same argument as before, that a nonzero velocity would result in an accumulation of matter. He also concludes that there is no "continual transmission of energy to the air" and that the drag associated with such a 2D lifting motion is zero, ignoring skin friction.

Lanchester's explanation is intriguing for several reasons. The first is that it was developed so early in the history of the field, before the Kutta-Joukowski theorem was known and before potential-flow solutions for flows around airfoils had been derived. Further, and to its credit, Lanchester's explanation acknowledges a mutual relationship, rather than one-way cause and effect, between the velocity and the pressure (the "field of force"), and it provides an essentially correct account of some key features of 2D lifting flow: upwash ahead of the airfoil, and downwash behind, decaying to zero in the farfield. It does not explicitly address how lift varies with angle of attack, but it could easily be extended to do that. One weakness is that the 2D-parachute flowfield is not a very good model for the flow in the presence of forward motion. Of course, a steady parachute-type flow, with a massive separated wake above the plate, would be a terrible model. Lanchester avoided the problem of a separated wake by stipulating that we take the "field of force" to be that which existed at the moment when the load was first applied (the pressure field immediately after an impulsive start). Even so, this leads to a pressure field that is wrong in one important detail: The pressure field for a flat plate in vertical motion alone is symmetrical fore and aft, while the pressure field associated with forward motion at an angle of attack has a marked fore-and-aft asymmetry, as we'll see later. So Lanchester's explanation avoids some of the shortcomings of the other explanations that we've seen, but his flow model is inaccurate in some details.

7.3.1.13 An Unusual Argument for Downward Turning in the Nearfield

Hoffren (2001) argues that the streamline coming off the trailing edge of an airfoil at an angle of attack is directed downward and asymptotically levels off at a level below that of the trailing edge. Then, using essentially the same "no accumulation" argument that Lanchester (1907) used, he argues that the stagnation streamline approaching the leading edge must have started at the same low level. This means that the flow generally rises ahead of the airfoil and descends behind it and must therefore experience downward turning in the neighborhood of the airfoil. This kind of argument establishes that if the flow leaves the airfoil at a downward angle, there is a kinematic necessity for downward turning. But it doesn't establish a physical mechanism for the downward turning.

7.3.1.14 Two Other Recent Explanations

Craig (1997), in *Stop Abusing Bernoulli – How Airplanes Really Fly,* a book that has been popular in the airplane home-builder community, proposes an interesting wrinkle on the conventional momentum-based explanation, but unfortunately ends up generating more confusion than enlightenment. He draws attention to the upwash ahead of the leading edge of an airfoil and outboard of the wingtips of a finite wing, invoking the same reason for it that Lanchester did, that is, that to avoid an accumulation of matter, the downwash above and below the lifting surface must be accompanied by upwash elsewhere. For this upwash, he coins the term "recirculation," apparently to reflect the idea that the air involved in the upwash motion was previously part of the downwash. But this idea correctly reflects the actual flowfield only far downstream of a wing of finite span, where the flow actually circulates around each member of a pair of trailing vortices. (We'll discuss the trailing vortex wake of a finite wing in some detail in Section 8.1.2.)

While the air immediately around the airfoil is part of a flow with *circulation,* as we saw in Section 7.2, none of it remains in the vicinity of the airfoil long enough to "recirculate" in any reasonable sense of the word, so the terminology is confusing at best. Furthering the confusion, Craig's Figure 2.5 shows a perspective view of a wing with arrows indicating that air from below the lower surface circulates forward, then upward in front of the leading edge, and then back, ending up above the upper surface. This picture might be correct if we isolated the circulatory part of the velocity field and regarded the arrows as *streamlines* of that part of the field, but Craig doesn't stipulate that that is what his Figure 2.5 represents.

If we interpret the arrows as *particle paths,* the picture is simply wrong. (Ordinarily in steady flows, particle paths and streamlines coincide, but not in this case, because the circulatory streamlines were constructed based on only a part of the velocity field.) In any case, the fact is that the air involved in the upwash ahead of the leading edge (and directly outboard of the tips of a finite wing) has not yet been involved in any of the downwash. Nevertheless, Craig refers to the upwash ahead of the airfoil as "recirculation" and assigns to it a major role in the production of lift. The accompanying discussion of transfers of kinetic energy from downwash to upwash and of lift being a "regenerative" process is confusing, and it is difficult to assess its physical correctness or lack of it.

A very clear description of the flowfield around an airfoil is given in *See How It Flies,* a web site for pilots, produced by a physicist (Denker, 1996). Many of the details of the flowfield and the pressure field around an airfoil are described and illustrated through excellent diagrams and animations. Cause-and-effect relationships at the local level, such as the relationship between pressure differences and flow curvature, and pressure differences and flow speed are described. At the global level, circulation and the Kutta-Joukowski theorem are discussed. Denker's presentation provides the reader with a wealth of information and physical interpretation, all to a high technical standard. Its only shortcoming, to my mind, is that it doesn't quite manage to show how all the pieces fit together, or to establish a complete web of cause-and-effect relationships at the global level.

7.3.2   Desired Attributes of a More Satisfactory Explanation

Every explanation we've looked at so far has been flawed in one way or another. We would of course like our explanation to avoid such flaws. But beyond that, what other attributes would we like it to have? Remember that we seek a physical explanation that does not require mathematics. In this regard, observing that we can predict lift by solving the NS equations is not a satisfactory explanation. Some, for example, Shevell (1989), have concluded that circulation and the Kutta-Joukowski theorem provide the only satisfactory explanation, but this is really just math as well. Besides, even if we were to accept the Kutta-Joukowski theorem as part of our explanation, it would just shift our problem from that of explaining where the lift comes from to that of explaining where the circulation comes from.

We acknowledge at the outset that it is not possible, without math, to predict the existence of lift, without knowing a priori some things about the flow that produces it. Providing ironclad "proof" that lift must exist is too much to expect, because that would require quantitative precision that is not available without computation or at least a simplified quantitative theory. Our objective is therefore to *explain,* not to *predict* or *prove.* Even though we've thus limited our objective, we still want our explanation to be as complete as possible, consistent with the correct physics. By "complete," we mean the following:

1. Though it is fair game to assume a priori some knowledge of *what* the flow does, we should explain in a satisfying way *how* the flow does it.
2. Our explanation should be based on a logical application of physical principles and leave no gaps.
3. Not only should the logical inferences be correct, but the direct physical cause-and-effect relationships should be made clear.
4. The explanation should make clear not just how lift is produced, but also how the magnitude of the lift can be "controlled," that is, how it varies with angle of attack and airspeed.

Although we have strong justification for relying on prior knowledge of what the flow does, we still tend to think of it as "cheating" somehow. We therefore have an inclination to try to define the minimum prior knowledge that needs to be assumed, and to try to deduce everything else from first principles. There is a strong appeal to the idea of an explanation that is maximally "self-sufficient" in this sense. But would such self-sufficiency really serve a pedagogical purpose? I think not. It seems to me that it would be better to assume the level of prior knowledge that makes for the clearest explanation, and that there is no disadvantage in assuming more than the minimum, provided it is all explained in the end.

So completeness and clarity are high on our list of objectives. There are also some faults we want to avoid:

1. Misrepresenting the phenomenon as being simpler than it really is. Given the subtlety of the phenomenon we're trying the explain, oversimplification is a great temptation, and it is a key weakness in many of the explanations we discussed in the previous section. Von Karman is reported to have said that in explaining things to a nontechnical audience, a Plausible Falsehood is preferable to the Difficult Truth (see Sears, 1994). I don't share this pessimistic view.
2. One-way causation of the kind we know is inconsistent with the physics, for example, "Lift is due to circulation" or "The pressure difference is caused by a velocity difference."
3. Assuming something that isn't subsequently explained, for example, "Circulation means there is lift," but not explaining where the circulation comes from.
4. Dependence on things that we know are not needed, for example, a sharp trailing edge or a curved upper surface.
5. "Naming" as a substitute for explaining, as, for example, in saying that a jet flow follows a curved surface because of the Coanda effect, where "Coanda effect" is just a name for the tendency of jet flows to follow curved surfaces.

An explanation that meets these requirements is not going to seem as tidy as most of the explanations we discussed in the previous section. However, as much as we'd like our explanation to be short and simple, we should resist the temptation to oversimplify and should aim for a level of detail and fidelity that does justice to the physics and meets our "completeness" requirement. A satisfactory explanation must resolve the Bernoulli-versus-momentum "controversy" and show how both pressure differences and momentum transfer are necessary parts of the picture. And it must blend these two seemingly independent aspects of the phenomenon into a coherent whole.

Ease of sharing favors an explanation that can be conveyed in relatively simple words, with no more graphic aid than can easily be sketched on a napkin. A drawback to simple graphics, however, is that the levels of detail and physical fidelity are limited. One way to increase the level of detail and to ensure fidelity is to introduce graphics with real quantitative accuracy. Going further in this direction, digital animation could probably be used to provide a level of "feel" for the physics that would be difficult to get any other way, for example, by illustrating time histories of forces and velocities as fluid parcels move through the field. But either of these high-tech options would reduce "portability." My choice here is to try to achieve a satisfactory basic explanation without leaving the "talking/ sketching" realm.

Because our intent is to share our basic explanation with a "public" that will not have read the background in this book, and because we'd like our explanation to have some staying power, it behooves us to give some thought to its "fitness for survival" in the marketplace of ideas. Practically everyone we share our explanation with will have heard other explanations before ours and will hear more in the future. Superior correctness on the part of our explanation can help, but is not likely to be sufficient, because the general public receives little real-world feedback regarding the correctness of its understanding of aerodynamics. Dawkins (1976) drew an analogy between the survival of ideas in a human population and the survival of genes in a biological gene pool, and proposed that Darwin's insights on natural selection apply in both universes. He coined the word *meme* to represent units or groups of ideas that can be passed from brain to brain and whose survival depends on their ability to "reproduce" (spread) and on their resistance to being displaced by other ideas. The survival fitness of our explanation as a meme will be greatly enhanced if, in addition to providing superior correctness, it inoculates its hosts (the audience) against the errors they are likely to encounter in other explanations. To do that, our explanation should discuss some of the most common errors and explicitly call attention to what is needed to avoid them.

The basic explanation that follows is intended to stand alone and therefore repeats in simplified form some of the background in other parts of the book and some of the critiques of other explanations in Section 7.3.1. It comes as close to achieving the desired attributes stated above as I have been able to manage. It is longer than other explanations, but I have not found any way to shorten it significantly without compromising either completeness or correctness.

7.3.3 A Basic Explanation of Lift on an Airfoil, Accessible to a Nontechnical Audience

When a relatively thin, flat *lifting surface* such as a wing, a sailboat sail, or a shark's fin moves through air or water, it can produce a force perpendicular to its direction of motion. This force is called *lift*, whether or not it is in an upward direction.

Mathematical theories of lift have been agreed on by the experts since the early twentieth century, but there has been a long history of disagreement on how to explain lift in simple physical terms, without math. Over the last 100 years or so, many different nonmathematical explanations have appeared in books, popular magazines, pilot-training materials, museum exhibits, and so forth. These explanations follow a variety of approaches, but they generally try to make lift simpler than it really is, explaining too little about the flow and leaving important parts of the phenomenon unexplained. We'll consider the common examples below under "How the main popular explanations are incomplete." And some explanations resort to erroneous arguments to explain features of the flow. These are discussed below under "Popular misconceptions." Naturally the existence of numerous different explanations has been a source of confusion and controversy.

The following is a nonmathematical explanation of lift that attempts to resolve the controversies and to be scientifically complete and correct. It is necessarily more complicated and longer than earlier popular explanations.

7.3.3.1 Airfoil Shape and Angle of Attack

The cross-sectional shape of a lifting surface, as illustrated in Figure 7.3.7, is called an *airfoil*. How much lift a lifting surface produces depends on the shape of the airfoil, on the *angle of attack* at which it approaches the oncoming flow, and on flow speed and density. A positive angle of attack means that the *leading edge* (front) of the airfoil is positioned higher than the *trailing edge* (back), relative to the direction of the oncoming flow, as in Figure 7.3.7.

Almost any shape, as long as it is not too thick, will work as an airfoil and produce some lift when the angle of attack is in the right range. The main types of airfoil shapes are shown in Figure 7.3.8. A simple *flat plate* shape (a) is used on many toy gliders. The Wright brothers used a *curved-plate* airfoil shape (b) for the wings of their early airplanes because they found that adding curvature, or *camber,* increased lift. A toy glider will also fly much

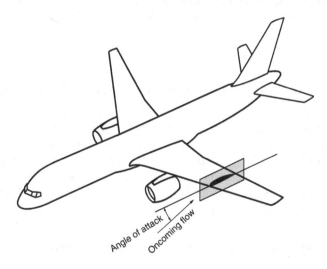

Figure 7.3.7 The cross-sectional shape of a wing is called an airfoil, and the angle at which oncoming air approaches it is called the angle of attack

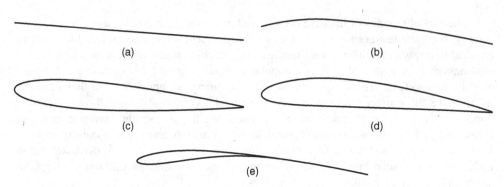

Figure 7.3.8 Types of airfoil shapes. (a) Flat plate. (b) Curved (cambered) plate. (c) Symmetrical streamlined shape. (d) Cambered streamlined shape. (e) Bird's wing

better with a curved-plate airfoil than with a flat-plate airfoil. Sailboat sails also generally take on the curved-plate form. Finally, larger model airplanes and full-size airplanes use *streamlined* airfoil shapes that are rounded at the *leading edge* and sharp at the *trailing edge,* (c) and (d). Aerobatic airplanes often use the symmetrical type (c), which works equally well right-side-up or inverted, while most other airplanes use streamlined airfoils with camber (d). Streamlined airfoils, especially those with camber, can produce more lift with less drag than other shapes, and they enable an airplane to fly over a wider range of speeds. Birds' wings tend to have a combination of the curved-plate and streamlined shapes (e).

7.3.3.2 The Airfoil Reference Frame

The phenomenon of lift generation is the same regardless of what reference frame we view it from. We could watch the airfoil move through the air, but everything is easier to understand if we imagine ourselves moving along with the airfoil, so that the airfoil appears to us to be standing still and the fluid appears to flow past. And for simplicity we'll refer to the "upper" and "lower" surfaces of the airfoil assuming that the wing is positioned horizontally and is lifting upward.

7.3.3.3 Lift Involves Action and Reaction (Newton's Third Law)

An intuitive way to imagine that lift is possible is to think of the airfoil shape and angle of attack as working together so that the airfoil pushes downward on the fluid as it flows past. The fluid must then push back with an equal and opposite (upward) force, which is the lift. This is an example of *Newton's third law,* that every action has an equal and opposite reaction. Thus lift is an interaction in which the airfoil and the fluid exchange equal and opposite forces. The aspects of the interaction that need to be explained further have to do with how the moving fluid actually pushes back.

7.3.3.4 Lift Is Felt as a Pressure Difference on the Airfoil Surfaces

A fluid always exerts *pressure,* which just means that it always pushes against itself and against any surface that it touches. When the fluid around an airfoil is at rest, the pressure is practically the same everywhere, so that it pushes upward on the lower surface and downward on the upper surface equally, and there is no lift. When the fluid is moving and there is lift, the fluid exerts the lift force directly on the airfoil's surfaces as a difference in pressure: higher pressure on the lower surface than on the upper surface. Under lifting conditions, the average pressure on the lower surface is usually higher than ambient unless the airfoil is very thick, and the average pressure on the upper surface is always lower than ambient.

How large is this pressure difference, typically? In the atmosphere at sea level, the ambient pressure is about 2100 pounds per square foot. Compared with this, it doesn't take much of a pressure difference to provide a practical amount of lift. Even the heaviest airplane requires no more than about 150 pounds of lift per square foot of wing area. So the pressure difference that an airfoil must produce to support any airplane is much less than the "background" atmospheric pressure, and even when a wing is lifting, the pressure on the upper surface pushes down almost as hard, in absolute terms, as the pressure on the lower surface pushes up. The difference is what counts, and even a relatively small pressure difference, spread over a large enough area, can lift a 747.

7.3.3.5   Lift Involves Force and Acceleration (Newton's Second Law)

Explaining how the flow maintains the pressure difference described above requires looking at the forces exerted on the air and the resulting accelerations of the air, not just at the surface of the airfoil, but in an extended region around the airfoil. In the explanation below, we first identify the major features of the flow that are essential to maintaining the pressure difference, and then we consider how the whole combination satisfies Newton's second law.

The outline of the explanation is as follows:

- The fluid flows as if it were a continuous material that deforms to follow the contours of the airfoil.
- The airfoil affects the direction and speed of the flow within a deep swath above and below the airfoil in what is called a *velocity field*. Flow above and below the airfoil is deflected downward. Flow above the airfoil always speeds up, and flow below usually slows down.
- The airfoil affects the pressure over a wide area in what is called a *pressure field*. When lift is produced, a diffuse cloud of low pressure always forms above the airfoil, and a diffuse cloud of high pressure usually forms below. Where these clouds touch the airfoil they constitute the pressure difference that exerts the lift on the airfoil.
- The velocity field and the pressure field support each other in a reciprocal cause-and-effect relationship, an interaction in accordance with *Newton's second law of motion*.

Let's look at each part in more detail.

7.3.3.6   The Fluid Flows as if It were a Continuous Material that Deforms to Follow the Contours of the Airfoil

Fluids such as air and water consist of huge numbers of individual molecules that move randomly in all directions, even when the fluid appears to be at rest. In water, molecules are

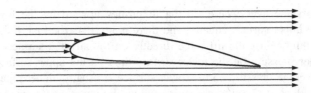

Figure 7.3.9 If molecules had no random motion and never interacted with each other, they would fly directly into the forward-facing parts of the airfoil and not touch the aft-facing parts

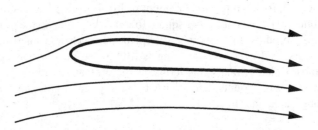

Figure 7.3.10 Because of the random motion and many collisions among its molecules, a fluid actually flows like a continuous material, deforming and changing course to flow around the airfoil, in what is called a *flowfield*

in constant contact with their neighbors as they move. In air, they are not in constant contact, but they collide frequently and travel only very short distances between collisions. If the molecules didn't have this random motion and never interacted with each other, they would fly directly into the forward-facing parts of the airfoil and not touch the aft-facing parts, as in Figure 7.3.9. Instead, because of the random motion and the frequent contact between molecules, the fluid flows as if it were a continuous material. It deforms and changes course to flow around the airfoil, and it fills all of the space around the airfoil and touches all of its surfaces, as illustrated in Figure 7.3.10.

7.3.3.7   The Airfoil Affects the Direction and Speed of the Flow within a Deep Swath above and below the Airfoil in What Is Called a Velocity Field. Flow above and below the Airfoil Is Deflected Downward. Flow above the Airfoil Always Speeds Up and Flow below Usually Slows Down

Because the fluid deforms continuously as it flows, changes in direction are gradual, and the speed and direction of the flow vary over a wide area around the airfoil. A spread-out pattern of variations like this is often referred to as a *flowfield* or *velocity field*.

The airfoil's solid surface forces the flow very close to it to follow the direction of the airfoil contour, with the result that the speed and direction of the flow are affected over a wide area. And the flow is thus affected by both the airfoil shape and angle of attack.

When an airfoil produces lift, it deflects the fluid flow downward, as indicated by the streamlines sloping downward to the right in Figure 7.3.10 and by the downward-turning arrows in Figure 7.3.11. To produce the downward turning, the aft portion of the airfoil

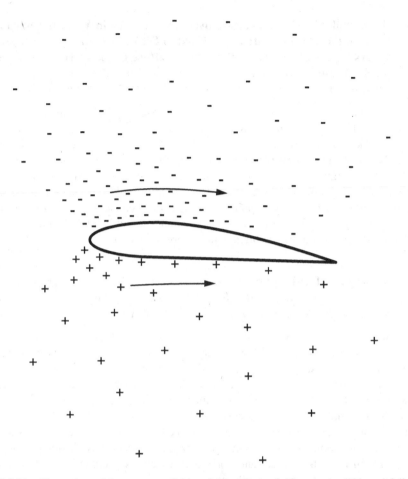

Figure 7.3.11 Illustration of the *pressure field* and the *velocity field* around a lifting airfoil. Minus signs indicate pressure lower than ambient, and plus signs indicate pressure higher than ambient. The tighter the spacing of the symbols, the larger the pressure difference

surfaces must have a predominantly downward slope. Thus to produce lift, the airfoil must have either camber or a positive angle of attack, or a combination of the two.

It seems obvious that downward-sloping airfoil contours should force the flow very close to the surface to be deflected downward, but it is important to understand that it is not just the flow close to the surface that is affected, and that the downward turning is spread over a deep swath of flow above and below the airfoil. The amount of downward turning is greatest close to the airfoil and dies away gradually far above and below.

In addition to the downward turning, the flow above the upper surface is always speeded up, as indicated by the longer arrow above the airfoil in Figure 7.3.11, and the flow below the airfoil is usually slowed down, as indicated by the shorter arrow. Like the downward turning described above, this effect is also strongest close to the airfoil surfaces and dies away gradually far above and below.

7.3.3.8   The Airfoil Affects the Pressure over a Wide Area in What Is Called a Pressure Field. When Lift Is Produced, a Diffuse Cloud of Low Pressure Always Forms above the Airfoil, and a Diffuse Cloud of High Pressure Usually Forms below. Where These Clouds Touch the Airfoil They Constitute the Pressure Difference That Exerts the Lift on the Airfoil

Because the fluid moves as if it were a continuous material, the airfoil influences much more fluid than its surfaces touch, producing pressure changes over a wide area. When an airfoil produces lift, there is always a diffuse cloud of low pressure above the airfoil, and there is usually a diffuse cloud of high pressure below, as illustrated by the clouds of minus signs and plus signs in Figure 7.3.11. Note that the minus signs don't mean that the pressure is negative, just that it is lower than ambient. These pressure differences are generally largest at the airfoil surface, where they actually exert the lift force, and away from the airfoil they die away gradually in all directions: above, below, ahead, and behind. A spread-out pattern of pressure differences like this is often called a *pressure field*.

7.3.3.9   The Velocity Field and the Pressure Field Sustain Each Other in a Reciprocal Cause-and-Effect Relationship, an Interaction in Accordance with Newton's Second Law of Motion

The downward turning of the flow, the changes in flow speed, and the clouds of low and high pressure described above are all necessary for the production of lift. They support each other in a reciprocal cause-and-effect relationship, and none would exist without the others. The pressure differences exert the lift force on the airfoil, while the downward turning of the flow and the changes in flow speed sustain the pressure differences.

Although the clouds of low and high pressure depicted in Figure 7.3.11 don't have sharply defined boundaries, they are still effectively "confined" to a limited area, both vertically and horizontally. "Sustaining" the pressure differences essentially means maintaining this spatial "confinement" in both the vertical and horizontal directions. Neither direction can be said to be more important than the other. Both are essential.

To understand how this "confinement" of the pressure differences works, we start by imagining the flow to be divided into tiny "parcels" of fluid passing through the region around the airfoil. The interaction these parcels are involved in reflects *Newton's second law of motion,* a general relationship among mass, force, and motion. Every parcel is subject to Newton's second law because the fluid in the parcel has mass, and because the parcel's neighbors can exert a net force on it, through differences in pressure. When the pressure is higher on one side of a parcel than on the other, the forces pushing the parcel in opposite directions are not balanced, and there is a net force pushing the parcel in the direction from higher pressure to lower pressure. Examples of this are illustrated in Figure 7.3.12.

Net forces on fluid parcels can also come from internal friction due to the viscosity of the fluid. However, when the flow around an airfoil follows both surfaces all the way to the trailing edge, as it usually does, viscosity has only a small effect on the overall flow pattern, and for purposes of understanding lift, we need only consider the forces due to pressure differences.

Newton's second law tells us that when a pressure difference imposes a net force on a fluid parcel, it must cause a change in the speed or direction (or both) of the parcel's motion. But in fluid flows, this cause-and-effect relationship not a one-way street. The

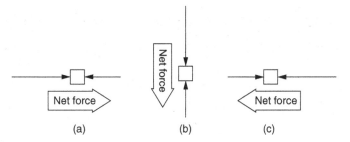

Figure 7.3.12 Examples of how pressure differences exert net forces on fluid parcels. (a) Pressure higher on left: Net force to the right. (b) Pressure higher above: Net force downward. (c) Pressure higher on right: Net force to the left

pressure difference causes a change in the parcel's motion, but the existence of the pressure difference depends on the parcel's motion. The relationship is thus mutual, or reciprocal: A fluid parcel changes speed or direction in response to a pressure difference, and its resistance to changing speed or direction (its inertia, due to its mass) sustains the pressure difference. The pressure difference can exist only if something is there to "push back," and what pushes back is the inertia of the fluid, which is why the mass of the fluid is so important.

To show how this applies in the airfoil flowfield, Figure 7.3.13 illustrates how typical fluid parcels are acted upon by the pressure in different parts of the field. In different parts of the high- and low-pressure clouds, the pressure is increasing or decreasing in different directions, exerting forces on the fluid parcels as shown by the arrows.

Fluid parcels directly above and below the airfoil see higher pressure above them than below and thus see a net force that is mostly downward. The downward force is resisted by the downward acceleration, or downward turning, of the flow. This interaction between pressure and downward turning is reciprocal: The pressure difference acting on a fluid parcel causes the parcel's path to be deflected downward, and the pressure difference is sustained because the fluid parcel has mass and therefore resists having its path deflected from a straight line.

On the upper left, where flow is entering the region of low pressure above the airfoil, the net force on a parcel is from left to right, which is resisted by an acceleration in the direction of the flow, or an increase in flow speed. On the upper right, where flow is leaving the region of low pressure, it sees a pressure difference in the opposite direction and is slowed down. Thus a parcel passing through the cloud of low pressure above the airfoil is speeded up and then slowed back down. On the other hand, a parcel passing through the cloud of high pressure below the airfoil experiences the opposite sequence: It is slowed down and then speeded back up.

These changes in flow speed follow the relationship that is often referred to as *Bernoulli's principle*, which can be expressed mathematically as *Bernoulli's equation*. Bernoulli's principle applies to regions of steady flow that have not been affected significantly by viscous friction. There is a small part of the airfoil flowfield in which Bernoulli's principle doesn't apply, and that is in the thin *viscous boundary layer* next to the airfoil surface and a thin *viscous wake* downstream, where the flow is strongly affected by viscosity and is usually turbulent. However, the pressure field is largely determined in the larger part of the flowfield in which viscous effects are small, outside of the boundary layer and wake. There, Bernoulli's

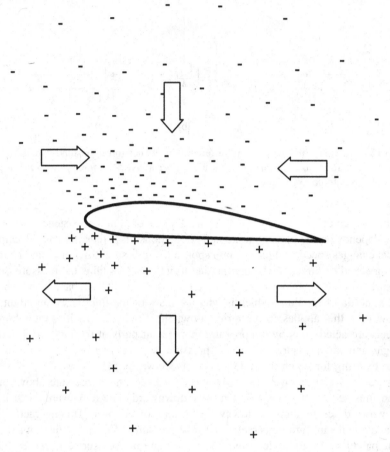

Figure 7.3.13 Pressure differences exert net forces on fluid parcels in different parts of the flowfield as indicated by the block arrows, and these forces are balanced by accelerations of the fluid parcels in the same directions

principle applies, according to which low pressure means high speed, and high pressure means low speed. This relationship is reciprocal: The differences in pressure in the horizontal direction cause the changes in flow speed, and the fluid's resistance to acceleration, because of its mass, sustains the pressure differences.

To summarize these interactions: *The downward turning of the flow provides the vertical "confinement" of the clouds of low and high pressure, and the changes in flow speed provide the horizontal "confinement."*

Thus downward deflection of the flow and different flow speeds above and below the airfoil are both essential parts of the reciprocal cause-and-effect relationship that sustains (or "confines") the pressure differences. The pressure differences cause the flow to change speed and direction, and the changes in flow speed and direction cause the pressure differences to be sustained. This circular cause-and-effect might seem a bit like "something for nothing" or "perpetual motion," but it's not. The pressure differences follow naturally from the

Newton's second law and from the fact that the flow close to the surface is forced to follow the predominantly downward-sloping contours of the airfoil. And of course the fact that the fluid has mass is crucial to the interaction.

In a way, it is surprising that mass is so important, given that air feels so light to our senses. However, air actually has more mass (and thus weight) than our intuition gives it credit for. In air at sea level, a volume of only about 14 cubic feet (a cube 2.4 ft on a side) weighs a pound. Thus even a modest volume of air contains significant mass, and an object moving through the air can influence a large mass of air in a short time. For example, the wing of a small airplane (a two-seat Cessna) flying at 100 mph causes significant changes in speed and direction in thousands of pounds of air every second.

7.3.3.10   The Roles of Camber, a Sharp Trailing Edge, and a Rounded Leading Edge

The downward-turning action that an airfoil produces is affected by both the airfoil shape and the angle of attack. The airfoil surfaces themselves need not be curved, and even the flat-plate shape of Figure 7.3.8a can produce downward turning when it has an angle of attack. However, adding camber to the airfoil shape, as in Figure 7.3.8b,d, aligns the airfoil surfaces better with the desired curved flow and enhances the downward-turning action. As a result, cambered shapes in which the upper surface is more convex than the lower surface produce more lift at a given angle of attack than shapes without camber.

What happens in the flow near the trailing edge plays an important role in controlling the downward-turning action. Figures 7.3.14 and 7.3.15 illustrate streamline patterns for low-to-medium angles of attack. Note that in both of these cases the flow from both the upper and lower surfaces leaves the airfoil smoothly from the trailing edge. If it didn't do this, the flow from one side would have to go around the trailing edge to the other side, as in Figure 7.1.3a. But viscosity generally prevents flow from going around a sharp edge in this way. So, with some help from viscosity, the trailing edge has the effect of directing

Figure 7.3.14   Streamline pattern at a low angle of attack: flow attaches near leading edge and follows the upper surface

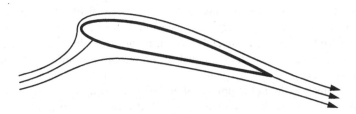

Figure 7.3.15   Streamline pattern at a medium angle of attack: flow attaches below the leading edge, flows around the leading edge, and follows the upper surface

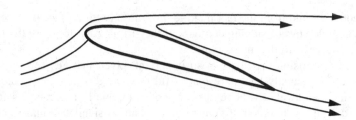

Figure 7.3.16 Flow around an airfoil at a high angle of attack breaks away from the upper surface, and the airfoil is said to be stalled

the flow as it leaves the airfoil to flow in the direction of the contours of the aft part of the airfoil. This tendency of the trailing edge to direct the flow is a major reason why angle of attack and camber are so effective in influencing lift.

The situation at the leading edge is very different. Referring again to Figures 7.3.14 and 7.3.15, note that the flow around the leading edge changes dramatically when the angle of attack is increased. At a low angle of attack (Figure 7.3.14) the flow approaches the leading edge almost head on, while at a higher (medium) angle of attack (Figure 7.3.15) it approaches the leading edge from below and flows around the leading edge from the lower surface to the upper surface. In fact, the reason for having a rounded leading edge is to accommodate the different flow patterns at different angles of attack with a minimum of disruption.

7.3.3.11 How Lift is Controlled

Lift depends on the speed of the flow. An increase in speed increases the lift by increasing both the amount of fluid that the airfoil influences in a given time and the downward acceleration imparted to each bit of fluid. These two effects multiply each other, so that lift is approximately proportional to the square of the speed, at a given angle of attack.

Lift can also be increased by an increase in either the camber or the angle of attack, as long as the flow follows both surfaces of the airfoil, as it usually does at low-to-moderate angles of attack, as shown in Figures 7.3.14 and 7.3.15. For a given airfoil shape, this increase in lift continues until an angle of attack is reached above which the effects of viscosity prevent the flow from following the upper surface. The flow then breaks away, as shown in Figure 7.3.16, which is called *stalling*. As an airfoil passes through the stalling angle of attack, the lift generally decreases dramatically, though it doesn't disappear altogether.

Controlling lift through angle of attack is an essential ingredient in achieving controlled flight over a range of speeds. For steady level flight, the total lift on an airplane must be kept equal to the weight. When an airplane speeds up, it reduces its angle of attack in order to keep the lift from exceeding its weight. When an airplane slows down, it increases its angle of attack in order to keep the lift equal to the weight, but this works only up to the stalling angle of attack. Thus an airplane can slow down only so much, to what is called the *stall speed*, which is the minimum speed at which an airplane can fly steadily.

Lift also depends on the density of the fluid. A decrease in density decreases lift at a given angle of attack and speed by decreasing the mass of the fluid the airfoil acts on. Thus

Figure 7.3.17   A control surface deflected upward to decrease lift

Figure 7.3.18   A control surface deflected downward to increase lift

Figure 7.3.19   A flap deflected downward to increase lift, with a slot to allow air to flow from the lower surface to the upper surface

when an airplane flies at high altitude where air density is lower, it must either fly faster or increase its angle of attack in order to maintain lift equal to the airplane's weight.

Lift can also be controlled by what is called a *control surface*, which is just an aft portion of an airfoil that is hinged so that it can be deflected up and down, as shown in Figures 7.3.17 and 7.3.18. Deflecting a control surface changes both the effective angle of attack and the camber of the airfoil and thereby changes the flow-turning action. Examples of control surfaces are the *elevators* on the horizontal tail, which can *pitch* an airplane nose up or nose down, the *ailerons* near the wing tips, which can *roll* an airplane to one side or the other, and the *rudder* on the vertical tail, which can *yaw* an airplane in one direction or the other.

A *flap* is similar to a control surface and is deflected downward at the trailing edge of a wing to increase both the lift at a given angle of attack and the maximum lift that can be generated without stalling. Flaps enable an airplane to take off and land at lower speeds than it could without flaps. A flap often has one or more *slots* that direct air from the lower surface to flow over the aft upper surface, as shown in Figure 7.3.19, which helps to delay stalling.

Leading-edge devices such as *leading-edge flaps* or *slats,* as shown in Figure 7.3.20, also help increase maximum lift and allow slower takeoff and landing speeds.

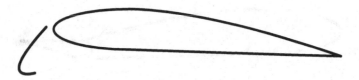

Figure 7.3.20   A leading-edge slat deflected downward to increase maximum lift

Figure 7.3.21   A spoiler deflected upward to disrupt the upper surface flow, decreasing lift and increasing drag

Spoilers are devices that hinge upward from the upper surface of a wing, as shown in Figure 7.3.21, disrupting the upper-surface flow by deflecting it upward, and causing a large decrease in lift and an increase in drag. When spoilers are used in flight, the purpose is not to decrease lift, but to increase drag and allow the airplane to descend steeply without speeding up. When spoilers are use in this way, the decrease in lift caused by the spoilers is compensated by an increase in angle of attack. When spoilers are used on the ground during the landing roll, the purpose is both to increase drag and to decrease the lift on the wing, putting more of the airplane's weight on the wheels to allow harder braking without skidding.

7.3.3.12   How the Main Popular Explanations Are Incomplete

The two most widely circulated explanations of lift are the *momentum based* (Newtonian) and the *Bernoulli based*. Both contain elements of the truth, but neither provides a complete explanation.

The momentum-based explanation generally goes as follows: When an airfoil produces lift, it deflects the flow downward, as indicated by the streamlines sloping downward to the right in Figure 7.3.10. Newton's second law tells us that to deflect the flow downward, the airfoil must push downward on the fluid. Then Newton's third law tells us that for every action there is an equal-and-opposite reaction, and the fluid must therefore push upward on the airfoil, and thus there is lift. This explanation is correct as far as it goes, but is incomplete in several ways. First, it doesn't point out that the force is actually transmitted to the airfoil by the pressure, and it doesn't explain how the airfoil can impart downward turning to a much deeper swath of the flow than it actually touches. Then although it hints at explaining how the pressure differences in the vertical direction are sustained, it doesn't explain at all how the pressure differences in the horizontal direction are sustained. That is, it leaves out the Bernoulli part of the interaction.

A Bernoulli-based explanation starts by arguing that the flow over the upper surface is speeded up, either because the path length over the upper surface is longer, or because of an "obstacle," "hump," or "Venturi" effect. Because of the higher speed, the pressure over the upper surface must be lower, according to Bernoulli's principle, and thus there is lift. Explanations of this type are incomplete in that they don't adequately or correctly explain what causes the flow to speed up. The longer-path-length explanation is simply wrong (see below, under "Popular misconceptions"). The "obstacle," "hump," or "Venturi" explanations are better, but only a little. They often mention "pinching" or "necking down" of the flow over the upper surface, but they don't provide a convincing physical explanation for the pinching. A common fault in all of these explanations is that they imply that a speed difference can arise from causes other than a pressure difference, and that the speed difference then causes, or at least implies, a pressure difference, according to Bernoulli's principle. This implied one-way causation is a misconception that we'll discuss below. Finally, a Bernoulli-only explanation doesn't really explain how the pressure differences in the vertical direction are sustained. That is, it leaves out the downward-turning part of the interaction.

There have been at least three schools of thought among proponents of the momentum-based and Bernoulli-based explanations. One is that only one or the other can be correct. Another is that both are correct, and that they apply to two different kinds of lift. A third is that both are correct, and either one suffices to explain lift in general. These are all misconceptions that we'll discuss next.

A third major category of explanation involves circulation around the airfoil. An explanation of this type starts with the observation that when an airfoil starts its motion through the fluid, a *starting vortex* is left behind. The formation of the starting vortex is accompanied by the establishment of a "circulatory flow," or *circulation,* around the airfoil, which is responsible for the lift according to an aerodynamic theory known as the Kutta-Joukowski theorem. This is not a proper physical explanation for two reasons. First, it depends on several advanced mathematical and aerodynamic theorems instead of direct physical arguments. Second, although the argument is mathematically and logically correct, its general progression does not reflect physical cause and effect, and is therefore misleading. It implies that lift is somehow caused by the formation of the starting vortex and the resulting circulation. The starting vortex and the circulation are actually more properly seen as byproducts of the lift than as causes.

7.3.3.13 Popular Misconceptions

We've seen that lift generation involves subtle cause-and-effect relationships, so it shouldn't be surprising that many of the attempts to explain it to a popular audience have made errors of one kind or another. To solidify our understanding and make it less likely we'll be taken in by incorrect ideas, let's identify some of the misconceptions and consider where they went wrong.

> *One-way causation*: This is a misconception we've already discussed, that a velocity difference can be deduced first, based on some argument that does not depend on the pressure, and that a pressure difference follows, according to Bernoulli's principle.

This implication that the causation runs in only one direction is not consistent with the physics of fluid flows. If you try to explain a speed difference without referring to the pressure difference, you'll inevitably get the reasons for the speed difference wrong. One example of this, an erroneous reason for high velocity over the upper surface of an airfoil, is the next item on our list.

> *Longer path length and equal transit time*: This is an argument that is widespread in explanations aimed at the layman. It is assumed that the upper surface of the airfoil is more convex than the lower surface, and that the path the fluid must follow around the upper surface is therefore longer than the path around the lower surface. It is further assumed that fluid parcels that are split apart at the leading edge to traverse the upper and lower surfaces must rejoin at the trailing edge. Thus fluid parcels negotiating both paths must do so in equal time, and we conclude that the velocity over the upper surface must be higher than that over the lower surface.

First, this isn't a proper kind of physical explanation. Just saying that something has to arrive somewhere at a particular time doesn't explain why the thing might speed up. To explain why something speeds up, we must identify and explain the force that makes it speed up. And this explanation is wrong on another level. There is no reason why fluid parcels that start together ahead of the airfoil must rejoin at the trailing edge, and in fact, they generally don't. A parcel that traverses the airfoil near the upper surface typically arrives at the trailing edge well ahead of one that traverses near the lower surface. So no difference in path length is required, and there are many situations in which lift is produced without a difference. And on airfoils where there is a difference, it is typically much too small to explain the speed difference that actually occurs when lift is produced.

> *Bernoulli is applicable, and Newton is not, or vice versa*: Some proponents of the Bernoulli-based explanations argue that lift is produced solely by a pressure difference, according to Bernoulli's principle, and that there is no downward momentum imparted to the fluid (Newton). Some proponents of momentum-based explanations argue the opposite: that imparting momentum is everything (Newton), and that the Bernoulli principle is not applicable.

We've seen that a pressure difference between the upper and lower surfaces and downward turning of the flow are both essential parts of the picture.

> *Bernoulli and Newton are both right, and they explain two different kinds of lift*: This line of argument maintains that "Bernoulli lift" and "reaction lift" represent two distinct physical mechanisms.

There is only one kind of lift, and explaining it requires both a pressure difference and downward turning.

> *Bernoulli and Newton are both right, and either one suffices*: According to this line of argument, the Bernoulli-based and the momentum-based explanations are just different but equivalent ways of looking at the same thing.

Again, we've seen that a complete explanation must include both the pressure difference and the downward turning of the flow.

> *Invoking the Coanda effect as the reason the flow is able to follow the curved surfaces of the airfoil*: Some explanations argue that viscosity plays a crucial role in enabling the flow to turn and follow the curved upper surface of the airfoil. They refer to this purported coupling between viscosity and flow turning as the *Coanda effect*.

This reflects a misunderstanding of the role of viscosity in fluid flows and of what the Coanda effect actually entails. As fluid flows over the surface of an airfoil, there is no direct coupling between viscosity and flow turning, and none is needed. Viscosity plays a significant role in lift generation only in the immediate vicinity of the airfoil trailing edge, by preventing the flow from going around the trailing edge from the lower surface to the upper surface. The real Coanda effect refers to the tendency of a turbulent jet flow with higher energy than the surrounding fluid to attach itself to an adjacent surface and to follow the surface even if it is curved. This is not so much a viscous effect as it is an indirect effect of the jet turbulence. It arises because of the tendency of jet flows to entrain surrounding fluid, and it plays no role in ordinary airfoil flows.

The low pressure on the upper surface pulls upward on the airfoil: Many popular explanations of lift describe the effect of low pressure in these terms. The idea of the airfoil being pulled upward has a strong intuitive appeal, but it is incorrect.

Pressure, especially in gasses such as air, is always a push, never a pull. The pressure on the upper surface of an airfoil pushes downward on the airfoil, but the higher pressure on the lower surface pushes upward harder, and the net effect is lift.

Not acknowledging the importance of angle of attack: Some explanations, such as the one based on the longer-path-length-and-equal-transit-time argument, never mention the angle of attack.

The angle of attack is a key factor that determines how much lift an airfoil produces at a given flow speed and is an essential ingredient in achieving controlled flight.

7.3.3.14 Why There Have Been So Many Misconceptions

Explaining lift in physical terms is more difficult than most people realize, and the difficulty is inherent in the basic nature of fluid mechanics. We are dealing with countless little parcels of fluid that move in coordination with their neighbors and exert forces on their neighbors, all while separately and simultaneously obeying Newton's second law. It simply isn't possible to look at an airfoil and deduce, by mental effort alone, what flow pattern satisfies the physical laws everywhere at once. There are too many simultaneous relationships to keep track of. And the difficulty is compounded by the circular nature of the cause-and-effect relationship between pressure and velocity.

This kind of complexity isn't easy to deal with mathematically, either. Mathematically expressing all the relationships a fluid flow must satisfy results in a set of partial-differential equations called the *Navier-Stokes equations*. By *solving* these equations we can *predict* in detail what the flow around an airfoil does and how much lift is produced. But solving the equations means mathematically determining how the pressure and the flow velocity vary throughout a large volume of space surrounding the airfoil. For any given flow situation it requires lengthy calculations that are practical only on a high-speed computer. Computer programs are available that can make such calculations routinely. However, all they do is provide a simulation of *what* would happen in a real flow; they don't provide a physical explanation of *how* it happens.

Simplified theories have also been developed, such as *potential-flow theory,* which ignores viscosity, and the *Kutta-Joukowski theorem,* which relates lift to a circulatory component of flow (circulation) around the airfoil, but these don't provide a direct physical explanation for lift either.

So we see that predicting the existence of lift using nothing but the properties of the fluid and the laws of physics would essentially require solving the NS equations or the potential-flow equation for the flow around the airfoil, which is not something we can do in our heads. Explaining what happens, with words and simple diagrams instead of laborious calculations, thus requires some prior knowledge of what the flow does. The explanation above started with knowledge of some basic features of the pressure and velocity fields around an airfoil, and then showed how the pressure field and the velocity field support each other in a manner consistent with the laws of physics, including the proper reciprocal cause-and-effect relationships.

Faulty explanations often assume too little prior knowledge and then try to do more than is logically possible by mental effort alone. As a result, they tend to leave important things unexplained (such as what really causes the high velocity over the upper surface) and to resort to logical fallacies such as one-way causation.

7.3.4 More Physical Details on Lift in 2D, for the Technically Inclined

The above explanation of the interaction between the pressure field and the velocity field explicitly distinguished between vertical and horizontal confinement of the pressure differences. Readers with technical backgrounds will recognize this as one way of explaining to a lay audience an interaction involving a vector equation. In this case, the equation is the Euler momentum equation in 2D. Of course this equation has two components, both of which must be satisfied, and thus the necessity to explain two components of the interaction.

Above, in trying to explain why there have been so many misconceptions, we claimed that predicting lift strictly from first principles can't be done just through mental effort, that it requires laborious computation. To avoid going into too much detail for a lay audience, we deliberately neglected to mention that simplified theories such as conformal mapping or linear theory can greatly reduce the required computational effort. This is a justified omission because the simplified theories are still heavily mathematical, and they provide no aid to physical understanding by a nontechnical audience. But we will make use of the linear theory to further our own understanding of airfoil pressure distributions in Section 7.4.1.

The main thrust of our basic explanation was that lift is transmitted to the airfoil by a pressure difference between the upper and lower surfaces and that it requires at the same time a downward turning of the air stream above and below the airfoil. We ignored the upward turning, both ahead of the airfoil and behind, that accompanies lift in general and can be seen in Figure 7.1.3b. Referring again to Figure 7.3.11, note that the flow approaching the front of the airfoil sees higher pressure below it than above, and thus must receive a net push upward. It therefore turns upward, as indicated by the upward-curving streamlines ahead of the airfoil in Figure 7.1.3b. Then comes the downward turning above and below the airfoil that we've already discussed and connected with the generation of lift. Finally, the air behind the airfoil feels lower pressure above than below, and is turned upward again, as shown by the curving streamlines behind the airfoil, but only enough to cancel the downward direction it acquired as it passed the airfoil.

The flow angle actually approaches zero far from the airfoil, behind the airfoil as well as ahead. Does this mean that the downward turning that we associate with the lift has been canceled by the upward turning? Well, yes, but only in terms of the local flow direction. The net momentum change imparted to the air by the airfoil is still downward and still accounts for the lift.

To explain how this can be, we start with the fact that lift requires circulation, according to the Kutta-Joukowski theorem. An airfoil flow with circulation effectively consists of a uniform flow with disturbances associated with a collection of bound vorticity superimposed on it. In the simplest way of deriving the Kutta-Joukowski theorem, we define a control volume as shown in Figure 7.3.22, with vertical outer boundaries at the front and back, extending to infinity vertically. Conservation of momentum requires that the lift on the airfoil be balanced by the forces and momentum fluxes at these outer boundaries. Because the boundaries are vertical, there is no net vertical force contribution by the pressure, and the lift must be accounted for by the net flux of vertical momentum into the control volume. It can be shown that as long as the two faces of the control volume enclose all of the bound vorticity, the flux of upward momentum into the front face and the flux of downward momentum out of the back face account for half the lift each, regardless of the distances from the airfoil to the faces. And from this, we also arrive at a simple derivation of the Kutta-Joukowski theorem, Equation 7.2.1.

So the flux of vertical momentum through either face of the control volume of Figure 7.3.22 remains the same, regardless of how far away from the airfoil the face is

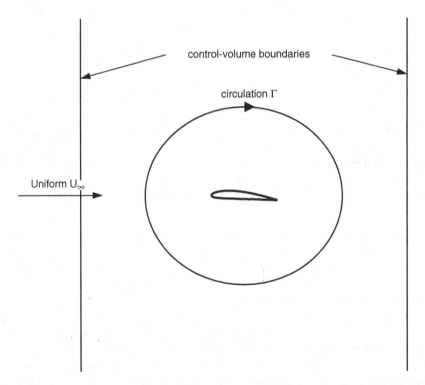

Figure 7.3.22 In the simplest derivation of the Kutta-Joukowski theorem, the farfield flow around an airfoil is modeled as a potential vortex superimposed on a uniform flow, and a control volume with vertical faces is used to assess the momentum flux due to lift

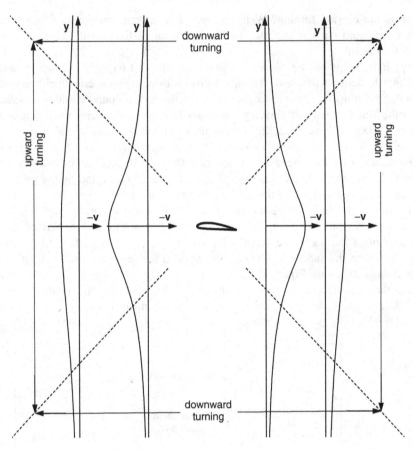

Figure 7.3.23 An illustration of the distributions of vertical velocity on vertical cuts at different distances ahead of an airfoil and behind. The vertical velocity is plotted as -v, so that in the upwash ahead of the airfoil the curves appear to the left of their respective axes, and in the downwash behind the airfoil, the curves appear to the right of their axes. The 45° dashed lines indicate the approximate dividing lines between upward turning and downward turning outside of the nearfield

placed. This is true in spite of the fact that the vertical velocity goes to zero at large distances. As either face is moved farther from the airfoil, the maximum vertical velocity decreases, but the distribution becomes more spread out vertically. This is illustrated in Figure 7.3.23 by plots of the downwash velocity $-v$ on vertical cuts at different distances upstream and downstream of the airfoil. Upstream of the airfoil, the curves are to the left of their respective axes, which, given that we are plotting $-v$, indicates upwash. Downstream of the airfoil, the curves are to the right of their axes, indicating downwash. With increasing distance from the airfoil either upstream or downstream, the vertical velocity in the middle of the distribution decreases, while the vertical velocity in the "tails" of the distribution, above and below the level of the airfoil, increases, and the integrated flux of vertical momentum remains the same.

Now it is clear how it can be that the upward turning of the flow (the reduction in the downwash velocity) aft of the airfoil trailing edge does not reduce the total downward momentum crossing successive vertical planes as we move downstream. It is because the upward turning is always offset by downward turning (the increase in downwash velocity) that is still taking place far above and below the airfoil. Outside of the near field, the boundary between the upward turning aft of the airfoil and the downward turning above and below is approximately along 45° lines, as indicated by the dashed lines in Figure 7.3.23. Thus along any vertical cut, at any distance behind the airfoil, we have both upward turning and downward turning taking place. We can look at the upward turning behind the airfoil as just part of the process by which the downward momentum becomes more spread out vertically.

Likewise, the upward turning ahead of the airfoil does not increase the total upward momentum crossing successive vertical planes because it is offset by downward turning that is already taking place far above and below. Across vertical planes ahead of the airfoil, the same total flux of upward momentum is always there, no matter how far ahead of the airfoil we look.

It seems counterintuitive that the airfoil can cause a fixed total flux of upward momentum at unbounded distances upstream. This finite influence on a flow integral at infinite distance upstream is of course an artificiality resulting from our implicit assumption that the airfoil has been in motion forever in an infinite atmosphere. If we assume instead a finite flight time since the start of the airfoil's motion, there would have to be a net starting vorticity in the field behind the airfoil, with opposite sign to that of the bound vorticity, and it would, in the integral sense, cancel the flux of upward momentum ahead of the airfoil. Or a finite altitude above a ground plane changes things even if the flight time is infinite. In that case, the reflection of the bound vorticity in the ground plane comes into play, canceling the integrated flux at very large distances. The effects of a ground plane and/or starting vortex are similar in the 3D case, as we'll see in Section 8.5.3. So in the real world, we don't have to worry about finite effects at infinite distances. But the dying off of the integrated momentum flux does require interference from a ground plane or a starting vortex, and in most situations these are far away. When the starting vorticity and the ground plane are both far away compared to the airfoil chord, there is a large domain around the airfoil in which the flow is practically indistinguishable from what it would be in an infinite domain and in which the total flux of upward momentum upstream of the airfoil is effectively constant.

So when an airfoil is effectively operating in free air, the increasing upwash in the flow approaching the leading edge does not represent a change in integrated momentum flux, just a redistribution of momentum from above and below. Anderson and Eberhardt (2001) misunderstand this redistribution process and argue that the increasing upwash ahead of the airfoil requires a downward force on the airfoil, which they call the "the additional load caused by the upwash" and which they claim offsets some of the lift. This leads them to an incorrect explanation for the reduction in induced drag of a wing flying close to the ground, which we'll discuss in Section 8.3.9. From the momentum analysis we've just discussed, we can see that there is no loading due to upwash on an airfoil.

Airfoil flow patterns in most practical flight situations are characterized by attached flow, in which the boundary layers on both surfaces are attached all the way to the trailing edge. The boundary layers then tend to have only minor effects on the global flow pattern, and the outer inviscid flow closely follows the predictions of potential-flow theory that we discussed in Section 7.1. The potential flow around a 2D airfoil is reversible, in that a flow that starts as a uniform onset flow from upstream becomes uniform again downstream, and the same flow pattern and pressures arise if the flow is run in the reverse direction. In a viscous flow in the attached-flow regime, the flow outside the boundary layer and wake still follows the reversible pattern quite closely. In addition to the near-reversibility of the general flow pattern, there is also very little permanent vertical displacement of streamlines between upstream and downstream. This is consistent with the view of a 2D airfoil flow as a uniform flow with a vortex superimposed, as in Figure 7.3.22.

There is one feature of 2D lifting flow that might appear at first glance not to be reversible, and that is that fluid parcels that start together ahead of the airfoil and split apart to flow above and below the airfoil experience a permanent longitudinal displacement between them after they have passed the airfoil, as we saw illustrated in Figure 7.3.2. This is, of course, a result of the nonzero circulation. But a little reflection should convince you that the longitudinal displacement needn't introduce anything irreversible. In the inviscid case, the displacement takes place entirely while the parcels are split apart by the airfoil and involves no slipping of adjacent layers of air relative to each other, either ahead of the airfoil or behind. The longitudinal displacement is therefore completely reversible in the 2D inviscid case.

Now let's consider further the issue of how lift varies with angle of attack. In Section 7.1, we noted that Newton's "bullet" theory predicts that lift varies as the square of the angle of attack, but that in continuum subsonic flow it tends to vary nearly linearly. This linear variation generally holds as long as the flow remains attached all the way to the trailing edge on both surfaces, so that the flow-turning action associated with angle of attack remains fully effective. The amount of flow turning varies directly with the angle of attack, but, as I stated in Section 7.1, angle of attack has very little effect on the amount of fluid per unit time that is subjected to the turning or is influenced significantly in any other way by the airfoil. Another way of looking at this is that the angle of attack affects the magnitude of the pressure difference between the upper and lower surfaces, but that the relative rate at which the pressure difference decays away from the surface is related primarily to the airfoil chord and depends very little on angle of attack. Therefore the vertical depth of the stream that passes through the region in which the pressure disturbance is significant in a relative sense (greater than 5% of the maximum, say) is essentially independent of angle of attack.

Let's look at some quantitative "data" (computational results) that illustrate this. Figure 7.3.24 shows isobar patterns in the inviscid flowfield for a thin symmetrical airfoil at angles of attack of 4° and 8°. Note that the lift at 8° is almost exactly twice that at 4°. Far from the airfoil, the pressure disturbance patterns are essentially the same, just with the disturbances for 8° being twice as strong as those at 4°. Thus the strength of the pressure disturbance is proportional to angle of attack, but the spatial "spreading" of the disturbance is not. What we see here is behavior that is nearly linear with angle of attack, something we'll consider further in the next section in connection with the linear theory.

So far we haven't said anything about the drag of an airfoil, nor have we considered in any detail how flow separation limits maximum lift. These effects depend not only on viscosity but on the detailed pressure distribution on the airfoil's surface, and we'll defer their discussion to the next section.

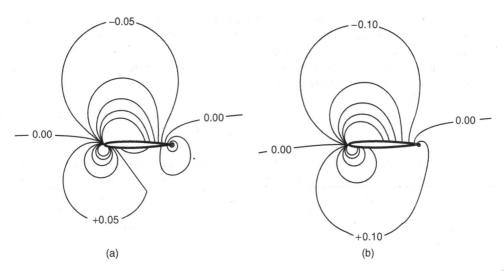

Figure 7.3.24 How the pressure distribution in the field around an airfoil is affected by a change in α, as illustrated by isobar patterns for a thin symmetrical airfoil (NACA 0010) at two angles of attack. Except for some differences due to thickness, doubling the angle of attack roughly doubles the strength of the pressure disturbances but does not much change the degree of "spreading" of the disturbances above and below the airfoil. Solutions calculated by the MSES code (Drela, 1993) in the inviscid mode. (a) α 4°, C_l 0.4761; Cp contour interval 0.05. (b) α 8°, C_l 0.9497; Cp contour interval 0.10

7.4 Airfoils

In the previous section, we defined an *airfoil* as a cross section of a *wing* of finite span. In the early years of the development of aerodynamics, this terminology was not yet standardized, and "airfoil" or "aerofoil" often referred to a complete 3D wing. Anderson (1997, pp. 288–289) provides an interesting look at the often-conflicting terminology that was used in the early days. In any case, in the previous section, we spent considerable effort just trying to explain how an airfoil produces lift, which is its primary purpose. Here we'll look into the many other important aspects of airfoils, including how pressure distributions and integrated forces depend on airfoil shape and angle of attack, factors influencing drag, factors influencing maximum lift, and considerations in airfoil design, including some specialized classes of airfoils.

7.4.1 Pressure Distributions and Integrated Forces at Low Mach Numbers

The pressure distribution on an airfoil surface, and the resulting integrated forces, over a wide range of flow conditions, can be predicted with good accuracy by computational fluid dynamics (CFD) calculations based on the NS equations or a coupled inviscid/viscous method. But for airfoils with nicely streamlined shapes at low Mach numbers and in the attached-flow regime, there are some general characteristics of the pressure distribution, the lift curve, and the pitching moment that we can predict without resorting to detailed flow solutions, but relying instead on a simplified linear theory. We'll begin this section by considering what the linear theory can tell us.

The linear inviscid theory was developed by Munk around 1920 (see the historical sketch by Giacomelli and Pistolesi, in Durand, (1967a) and formally applies in the limit of small thickness and angle of attack, but even for airfoils with practical amounts of thickness and camber, it is close enough to reality to provide useful insights. The beauty of the linear theory is that the effects of angle of attack, thickness, and camber can be treated separately and superimposed, which is not only computationally convenient, but provides a powerful way of thinking about how airfoil shape and angle of attack affect the pressure distribution.

In the linear inviscid theory, we assume incompressible potential flow and construct solutions by superposition of a uniform free stream and disturbances "produced" by elementary singularities. Superposition is allowed because the incompressible potential equation is linear to start with. To simplify the problem further we linearize the boundary conditions as well, which involves two things:

1. The flow-tangency boundary conditions that in higher-fidelity theories would be applied on the airfoil upper and lower surfaces as shown in Figure 7.4.1a, are applied instead along the x axis (y = 0), as shown in Figure 7.4.1b and
2. We ignore the longitudinal perturbation velocity u in calculating the velocity slope that is matched to the airfoil surface slope.

In keeping with applying the boundary conditions on the x axis instead of the airfoil surface, the sheet of distributed source strength and vorticity that "produces" the flow disturbance is also placed along the x axis. This sheet of singularity strength "causes" the perturbation velocities u and v to vary along the x axis and to have different values above and below the axis, which are taken to represent airfoil upper-surface and lower-surface values respectively. The equations that are solved for the distributions of source and vorticity strength are just the tangency condition illustrated in Figure 7.4.1b, imposed for both upper and lower surfaces, over the whole chord. The source and vortex singularities play complementary roles in influencing the perturbation velocities u and v. The averages and differences

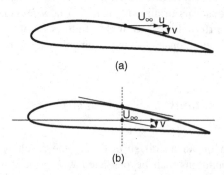

(a)

(b)

Figure 7.4.1 Flow tangency boundary conditions in airfoil potential flow, illustrated for the upper surface. (a) Full tangency condition applied at airfoil surface. (b) Linearized tangency condition applied on the x axis. Note that perturbation u is ignored

Table 7.4.1 Influence relationships in
linear airfoil theory

	Perturbation	
	Average	Difference
Source	u	v
Vorticity	v	u

in the perturbations between the upper and lower surfaces are influenced differently, and
the influence relationships are most easily illustrated in Table 7.4.1.

For the *analysis* problem (solving for the pressures for a given airfoil shape), we can
decompose the airfoil surface slope into contributions from the angle of attack, the shape
of the mean line, or camber line (halfway between the upper and lower surfaces), and the
shape of the thickness distribution (the difference between the upper and lower surfaces),
and we can then determine the first-order pressure coefficient $C_p = -2u/U_\infty$ for each part
separately, where u is the streamwise perturbation velocity evaluated on the chordline.
These components of the geometry and their corresponding contributions to the pressure
distribution are shown in Figure 7.4.2a–c for a NACA 4410 airfoil. Adding these pressure
contributions together, we get a prediction for the complete pressure distribution of the
airfoil, in Figure 7.4.2d. These plots follow the usual convention for plotting pressures in
aerodynamics, with C_p plotted on an inverted scale so that when an airfoil is lifting upward,
the pressure on the upper surface appears above the pressure on the lower surface.

The angle of attack and the mean-line shape affect only the average of the upper-and
lower-surface slopes and thus the average v perturbation, and they therefore affect only the
vorticity distribution, according to Table 7.4.1. As we saw in Section 3.3, the local strength
of a vortex sheet is equal to the difference (jump) in tangential velocity across the sheet.
In this case, the local strength of the vorticity distribution on the chordline is equal to the
perturbation u and is therefore also proportional to the pressure difference, or the local lift
loading.

The thickness shape affects only the difference in surface slopes and thus the difference
in the v perturbations, which is locally equal to the strength of the source distribution. The
source strength, and therefore the thickness distribution, affects only the average of the u
perturbations, not the difference, according to Table 7.4.1.

So we see that according to linear theory, the chordwise distribution of lift is affected
only by angle of attack and camber shape, not by thickness, and that the thickness shape
affects only the average velocity over the upper and lower surfaces. This is reflected in the
contributions to the pressure distribution shown in Figure 7.4.2. The pressure distributions
due to angle of attack in part (a) and camber shape in part (b) are pure lift-carrying pressure
differences (equal and opposite on upper and lower surfaces) with zero average perturbation
pressures. The pressure distribution due to thickness in part (c) has a nonzero average and
zero difference.

Now look in particular at the pressure distribution due to angle of attack, shown in
Figure 7.4.2a for an angle of attack of 2°. Note that the shape of this pressure distribution is
generic and that only the amplitude changes with angle of attack. The distribution is singular
at the leading edge (ΔC_p is infinite), but it is integrable, and the lift and moment at a given

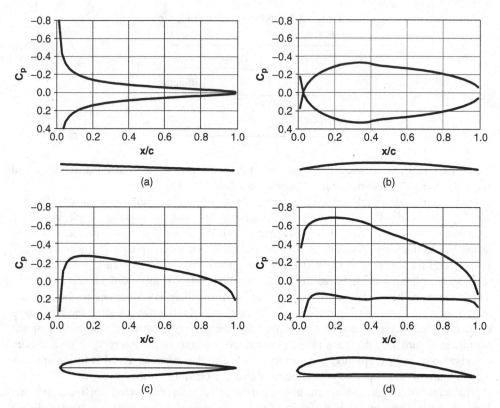

Figure 7.4.2 NACA 4410 airfoil at 2° angle of attack, shown decomposed into a chord line, camber (mean line), and thickness. The linear-theory contribution of each to the pressure distribution is also shown. (a) Angle of attack. (b) Camber line (NACA 44xx). (c) Thickness shape (NACA xx10). (d) All combined for NACA 4410 airfoil at 2° α

α can be calculated by integration. Because the ΔC_p varies linearly with angle of attack, the lift slope $dC_l/d\alpha$ is constant, and its value is 2π per radian. The lift is concentrated well forward, and the centroid, or the point about which the pitching moment is zero, is located at the quarter chord for any nonzero angle of attack. The quarter chord is thus a convenient reference point for pitching moments: Any nonzero pitching moment about the quarter chord can be due only to the camber shape and will not vary with α, according to the linear theory. This holds true with reasonable accuracy for airfoils with practical amounts of thickness and camber, in the absence of nonlinear compressibility effects, and up to angles of attack at which boundary-layer separation begins.

The pressure distribution due to a typical camber shape at zero α is shown in Figure 7.4.2b, in particular the NACA 44xx mean line. This mean line produces an incremental C_l of 0.455 with its centroid at 0.488 x/c, for a pitching-moment coefficient C_m of −0.108 about the quarter-chord.

Figure 7.4.2c shows the pressure distribution for the NACA xx10 thickness shape at zero α, which is, of course, the same upper and lower. Combining this with the α and camber contributions, (a) and (b), gives the complete linear-theory pressure distribution for the NACA 4410 airfoil at $2°$ α, shown in part (d).

Linear theory can also be applied to the "inverse" design problem (solving for the airfoil shape for a given pressure distribution). Here we would decompose the pressure distribution into a lift part (difference between upper and lower pressures, as shown in Figure 7.4.2 by the sum of (a) and (b), and a thickness part (average of upper and lower pressures, as in Figure 7.4.2c). The lift part determines the mean-line shape and the angle of attack, and the thickness part determines the thickness shape. Linearized design is almost never used in practice, because anyone who really wants to design an airfoil will invariably want to use a higher-fidelity theory. We'll discuss airfoil design further in Section 7.4.10.

So to summarize what the linear theory leads us to expect regarding the pressure distribution and forces on an airfoil:

1. The pressure difference between the upper and lower surfaces, and thus the lift, is due primarily to the angle of attack and the shape of the mean line and is much less affected by the thickness distribution.
2. The pressure disturbance averaged between the upper and lower surfaces is due mainly to the thickness distribution and much less affected by angle of attack and camber.
3. The integrated lift is linear with α, and the lift slope $dC_l/d\alpha$ is close to 2π per radian.
4. The pitching-moment coefficient C_m about the quarter-chord is nearly constant with α.

As we'll see, these expectations hold up reasonably well, at least qualitatively, for airfoils of practical thickness at low Mach numbers in the attached-flow regime.

Now let's look at some of the ways "real" airfoil flows differ from the predictions of the linear theory. First, even before we add viscous effects, just moving the inviscid-flow boundary conditions from the x axis to the actual airfoil surface and taking perturbation u into account in the boundary conditions have effects that increase with thickness, camber, and angle of attack. In Figure 7.4.3, pressure distributions from the linear theory are compared with inviscid-flow calculations from the MSES code (Drela, 1993) for two airfoils that differ by a factor of 2 in thickness, camber, and α: NACA 2405 at $1°$, and NACA 4410 at $2°$ (same case as in Figure 7.4.2). In both cases, the linear-theory underpredicts the pressure disturbances compared with the higher fidelity inviscid calculations, but much less so for the thinner airfoil than for the thicker one. The discrepancy seems to grow roughly quadratically with thickness, as one might expect for nonlinear effects. Still, the linear theory captures the general character of the pressure distributions.

Figure 7.4.4 compares the linear theory with inviscid MSES calculations for the NACA 4410 in terms of the change in the chordwise distribution of lift for a change in α from $0°$ to $2°$. Here we see that the linear theory captures the general character of the distribution, except that the real airfoil, with its blunt-nosed thickness distribution, has no singularity at the leading edge.

Viscosity can have significant effects on the pressure distribution, through the boundary-layer displacement effect. In Figure 7.4.5a, the inviscid MSES calculation of Figure 7.4.3b is compared with a corresponding viscous calculation at the same angle of attack for a chord Reynolds number of 10 million, assuming laminar-to-turbulent transition close to the leading edge. The boundary-layer displacement effect, which is at its strongest by far on the upper surface near the trailing edge, reduces the effective angle of attack and aft camber

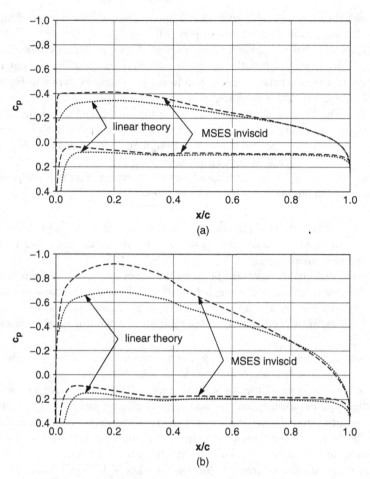

Figure 7.4.3 Comparison between the linear inviscid theory and incompressible inviscid flow with boundary conditions applied at the airfoil surface, calculated by the MSES code (Drela, 1993). Comparisons are made for two airfoil cases that differ by a factor of 2 in camber, thickness, and α, so that linear theory predicts a doubling of the pressure disturbances. (a) NACA 2405 at $1°$ α. (b) NACA 4410 at $2°$ α

of the airfoil, and changes the circulation, affecting pressures over the whole chord and noticeably reducing the lift. If the viscous solution is run so as to match the inviscid C_l instead of the inviscid α, the differences in the pressure distributions are reduced, as shown in Figure 7.4.5b. Note that due to the reductions in the effective camber and α, the viscous calculation requires a higher α to match the inviscid C_l. The effect of the higher α can be seen in the higher suction level on the forward upper surface.

The most dramatic local changes in pressure brought about by the boundary-layer displacement effect are seen in the neighborhood of the trailing edge. As we saw in Section 3.10, the inviscid flow off of a trailing edge with a nonzero wedge angle has a stagnation point, and the exact inviscid solution should thus have stagnation pressure ($C_p = 1.0$) at the trailing edge. The numerical solution plotted here doesn't reach stagnation pressure because the

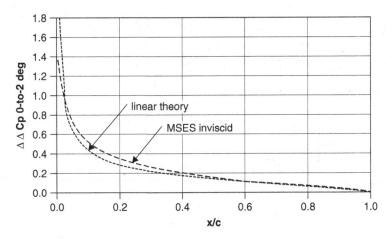

Figure 7.4.4   The effect on the chordwise distribution of lift of a change in α from 0 to 2°. The linear-theory prediction, which is independent of airfoil shape is compared with a calculation for a NACA 4410 airfoil, calculated by the MSES code (Drela, 1993) in the inviscid mode

pressure change approaching the stagnation is too rapid to be resolved by the numerical grid (note that when a potential flow turns a concave corner of less than 90°, stagnation pressure is approached with infinite slope, as in Figure 3.10.2c). But this issue is irrelevant in the real world with viscous effects. Real pressure distributions look very much like the viscous solutions plotted in Figure 7.4.5, in which the boundary-layer displacement effect has substantially reduced the pressure recovery near the trailing edge and eliminated the stagnation. This alteration of the pressure distribution is a significant contributor to the profile drag, as we'll see later. It is easy to be deceived by appearances, however. Although the pressure changes near the trailing edge are the most dramatic, the pressure contribution to the profile drag is subtle and depends on changes around the entire airfoil contour.

Now let's look at the effects of nonlinear boundary conditions and boundary-layer displacement on the lift-versus-α curve. Figure 7.4.6 shows the lift curve predicted by linear theory for the 44xx mean line, again showing the C_l shift of 0.455 at zero α and a slope $dC_l/d\alpha$ of 2π per radian. For comparison, MSES predictions are shown for the NACA 4410 and NACA 4420 airfoils, for both inviscid flow and all-turbulent flow at R 10 million. The inviscid MSES calculations show higher lift slopes than linear theory, increasing with increasing thickness ratio. The viscous calculations show significant losses of lift relative to inviscid. The calculation for 10% thickness ratio just happens to agree closely with the linear theory over the linear part of the lift curve, while the calculation for 20% thickness ratio is lower. At higher angles of attack, the displacement thickness of the boundary layer on the aft upper surface grows rapidly with angle of attack, causing the lift curve to become nonlinear and the lift slope to decrease. The maximum lift shown may be close to the maximum lift this airfoil could produce, but the MSES solution scheme based on simultaneous boundary-layer/inviscid-flow coupling did not in this case produce solutions showing maximum lift and beyond.

So far, we've seen two important effects of viscosity on the pressure distributions: The reduction in aft pressure recovery that contributes to profile drag, which we'll discuss in

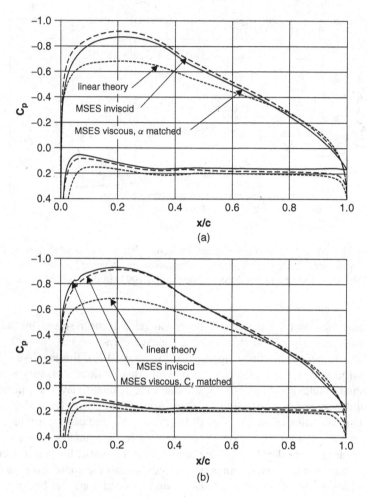

Figure 7.4.5   Effect of viscosity on the pressure distribution of NACA 4410 at 2° α, R $= 10^7$, tripped to turbulent at 5% chord, calculated by the MSES code (Drela, 1993). (a) Inviscid and viscous MSES calculations at 2° α. (b) Inviscid MSES calculation at 2° α; viscous calculation matches inviscid $C_l = 0.7385$ ($\alpha = 2.541°$)

Section 7.4.2, and a reduction in lift slope that increases with increasing thickness. The latter effect explains what Yates (1990) referred to as the "fatness paradox," in which the lift slope in inviscid flow increases with increasing thickness, while the lift slope in viscous flow decreases.

In Figure 7.4.4, we saw that a rounded leading edge removes the leading-edge singularity that would generally occur in the inviscid solution for the mean line alone. On an airfoil that has no sharp corner other than the trailing edge, flow attachment occurs at a stagnation point, as we saw in Section 5.2.1. A stagnation point is, of course, a feature that the linear theory cannot model accurately. At moderate angles of attack for most airfoils, the stagnation point occurs near the leading edge. Figure 7.4.7 shows the pressure distribution plotted versus arc

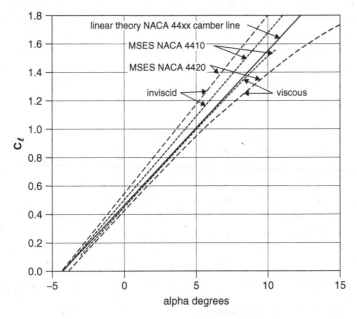

Figure 7.4.6 Lift curves predicted by linear theory and the MSES code (Drela, 1993) with and without viscosity for the NACA 4410 and 4420 airfoils

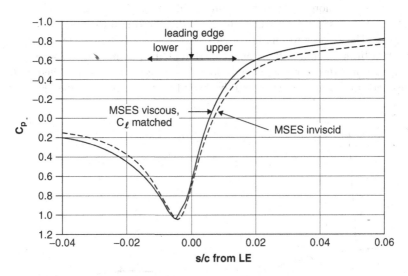

Figure 7.4.7 Pressure distributions in the neighborhood of the leading-edge stagnation point of a NACA 4410 airfoil. The inviscid MSES calculation is at 2° α, and the viscous calculation matches the inviscid $C_l = 0.7385$ ($\alpha = 2.541°$, as in Figure 7.4.5b)

length around the leading edge, calculated for the NACA 4410 at an angle of attack of 2°
in inviscid flow and at the matching C_l condition for viscous flow (same conditions as in
Figure 7.4.5b). There is a slight shift in the location of the stagnation point, but the shape
of the pressure distribution in the neighborhood of the stagnation point is affected very little
by viscosity. The value of C_p at stagnation is greater than 1 because of numerical issues
related to running a compressible code at a very low Mach number.

7.4.2   Profile Drag and the Drag Polar

In Chapter 6, we looked in considerable detail at drag in general, and we noted that although
it is not possible to rigorously decompose the drag force according to the flow mechanisms
that cause it, an approximate decomposition into lift-induced drag, shock drag, and viscous
drag is possible with the aid of simplified theoretical models. The problem we face here
is simpler: A 2D airfoil in 2D flow has no induced drag, and shock drag appears only at
transonic speeds when regions of supersonic flow appear in the field. We'll defer transonic
issues to Section 7.4.8 and look now just at the viscous drag in subcritical flow. As we
saw in Section 6.1.3, viscous drag has both a shear component and a pressure component.
In the context of airfoils and wings, the viscous drag is often called the *profile drag*.

Of course in an ideal 2D inviscid flow without shocks, the drag of an airfoil must be zero.
This is an example of D'Alembert's paradox, which we discussed in Sections 5.4 and 6.1.6.
Intuitively speaking, zero drag is even more surprising in the case of a lifting airfoil at an
angle of attack than it is in general. Consider the lifting airfoil sketched in Figure 7.4.8a.
Because of the general downward slope of the airfoil surfaces, due to the angle of attack,
the pressure difference between the upper and lower surfaces produces a net force that over
much of the airfoil chord is tilted backward as illustrated by the arrow above the airfoil.
If the integrated pressure force over so much of the chord is tilted backward, how can the
drag be zero? The answer to this lies with something that is loosely termed *leading-edge
suction*. On a lifting airfoil with a rounded leading edge, as in Figure 7.4.8a, the pressure
around the leading edge is predominantly lower than elsewhere, as indicated by the minus
signs, just enough to balance the general backward tilt of the lift over the rest of the chord.

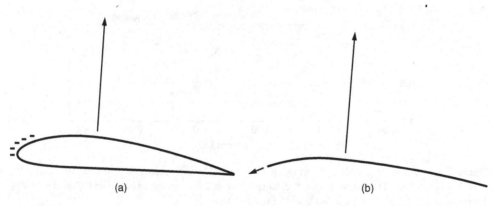

(a)   (b)

Figure 7.4.8   The concept of leading-edge suction. (a) Low pressure around a rounded leading edge.
(b) Singular suction force at the leading edge of an airfoil with zero thickness

On an airfoil with zero thickness, as in Figure 7.4.8b, there is only one angle of attack at which the flow attaches smoothly to the leading edge. At any other angle of attack, the potential flow around the leading edge has a singularity with infinite velocity, and produces a singular suction force parallel to the surface tangent at the leading edge, as indicated by the arrow. In 2D inviscid flow, this suction force must exactly cancel any net drag force resulting from the pressure differences over the rest of the chord.

In the real world, the viscous drag of a typical streamlined airfoil in attached flow tends to be small, on the order of 10^{-2} in terms of the drag coefficient. When we want to predict airplane performance or compare candidate designs, we would like to pin this small drag number down to within a small percentage, so we are looking at trying to predict the profile drag to within something on the order of a *drag count* (10^{-4}). Computational predictions at this level are at or beyond the limit of what is currently practical and are therefore problematic. When we are looking at increments on the order of a drag count, we should use only the best methods available, and we should apply them with great care. We'll address the CFD issues further in Chapter 10. Experimental measurements of drag at this level are likewise difficult. Only particular experimental setups and measurement methods are up to the task, and then only marginally so. Experimental uncertainties are often as large as the increments we would like to determine. These difficulties, however, don't keep us from drawing some general conclusions about the nature of viscous profile drag.

In Section 5.4 and in Section 6.1.3, we discussed the decomposition of drag into a surface shear stress, or *skin-friction* part and a *pressure* part, and noted that this decomposition involves none of the theoretical idealization that is required in the induced/shock/viscous drag decomposition. In attached flow, typical airfoil profile drag is mostly integrated skin friction, but pressure also makes a significant contribution, as a result of the boundary-layer displacement effect. A viscous-inviscid interaction code can do a reasonable job of predicting the boundary-layer displacement thickness, but an NS code is generally required if one is to calculate the effect on the pressure accurately enough to allow the viscous pressure drag to be calculated by direct pressure integration. In practice, accurate prediction of the pressure drag in an NS code requires a flow solution with an excellent degree of grid convergence. In viscous-inviscid interaction codes, the pressure drag cannot generally be calculated directly with sufficient accuracy, and the total drag must be inferred from less-direct flowfield manifestations. The wake development can be calculated into the farfield and used in the farfield wake-momentum formula, Equation 6.1.7, as is done in the MSES code (Drela, 1993), for example, or trailing-edge boundary-layer quantities can be used in the Squire-Young formula, Equation 6.1.8 for incompressible flow, or the compressible version by Cook (1971).

Now let's look at typical behavior of skin-friction and pressure drag, using MSES calculations for the NACA 4410, the airfoil whose pressure distributions we looked at in Figure 7.4.5. First, a typical distribution of C_f along the chord is shown in Figure 7.4.9, for the same flow condition as in Figure 7.4.5b. Flat-plate C_f from Equations 4.3.4 and 4.3.5 is shown for comparison. The artificially forced transition from laminar to turbulent at 5% chord is clearly seen. It is also clear that pressure gradients and edge velocities different from freestream make C_f on the upper and lower surfaces considerably different from that on a flat plate, and that C_f on the upper surface is higher than that on the lower surface, except near the trailing edge. In Figure 7.4.10 we look at the *drag polar* behavior, that is, how these things vary with C_l. The skin friction calculated by the code's boundary-layer

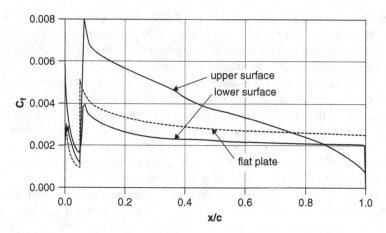

Figure 7.4.9 Skin-friction distribution for NACA 4410 calculated by the MSES code (Drela, 1993). $R = 107$, $C_l = 0.7385$ ($\alpha = 2.541°$). Here C_f is normalized by farfield q, not local edge q. C_f for a flat plate at the same Reynolds number is shown for comparison

equations was integrated directly, and the total drag was calculated from the farfield wake momentum. The difference was assumed to be the pressure drag and labeled as such. The total skin-friction drag is only slightly greater than that on a two-sided flat plate and varies little with C_l. Most of the variation of the drag with C_l is seen to come from the pressure drag. The variation is roughly parabolic at lower C_l, but it becomes strongly nonparabolic at higher C_l, as the boundary layer on the aft upper surface approaches separation.

In Section 6.2.1, we discussed the concept of a *form factor* that relates the actual profile drag to the skin-friction drag of a flat plate at the same chord Reynolds number and with laminar-to-turbulent transition at the same chord location. Handbook methods for drag estimation often use the form factor in estimating the profile drag of the airfoils that make up the wing. The flat-plate skin friction is easily calculated by Equation 4.3.2 for all-turbulent flow, and an adjustment can be made if a significant run of laminar flow is expected. The form factor itself poses more of a problem and is thus the weak point of the form-factor approach. To illustrate why, the form factor variation calculated from the MSES predictions of Figure 7.4.10a is plotted in Figure 7.4.10b. Like the pressure drag in Figure 7.4.10a, the form factor has a roughly parabolic variation at lower C_l and becomes strongly nonparabolic at higher C_l. Note that even at the drag minimum the form factor is greater than 1 by more than 20%, reflecting the fact that the viscous drag of an airfoil is typically significantly greater than the skin-friction drag of a flat plate. In handbook methods, only the parabolic part can be correlated reasonably well with the airfoil maximum thickness and maximum camber. The nonparabolic part depends on airfoil design details and is usually ignored. As a result, when it comes to making airplane performance predictions, relying just on the parabolic part of the airfoil polar that fits the low-C_l range is bad practice. Usually, the airfoil C_l that provides the highest airplane lift/drag ratio is determined by the rapid increase in profile drag on the nonparabolic part of the airfoil polar. Using the parabolic polar that fits the low-C_l range will generally lead to a choice of operating C_l that is too high and will also yield a seriously optimistic drag estimate.

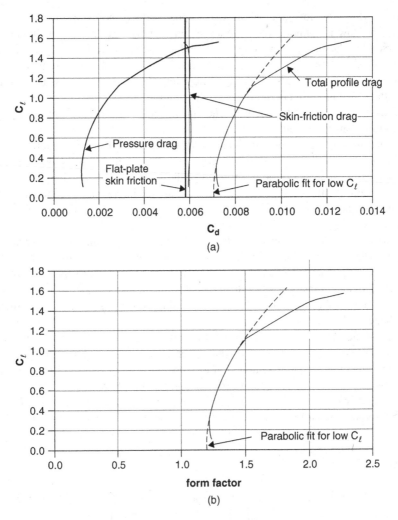

Figure 7.4.10   Drag predicted by the MSES code (Drela, 1993) for the NACA 4410, R = 10^7, trip at 5% chord. (a) Drag polar. (b) Form factor

7.4.3   Maximum Lift and Boundary-Layer Separation on Single-Element Airfoils

We now turn our attention to the problem of the maximum lift an airfoil can produce and how it is affected by the airfoil shape and pressure distribution. There are limits to the amount of lift that can be produced even in inviscid flow, as discussed by Smith (1975). Either the circulation becomes so large that the attachment and separation points merge, and it is assumed that larger circulation would be a mathematical artificiality, or the condition on the upper surface approaches vacuum. Smith points out that these limits are moot, however, because viscous effects impose much lower limits. As we noted in our basic explanation of lift in Section 7.3.3, the maximum lift of an airfoil is limited because "there is an angle of attack above which the flow cannot follow the upper surface and breaks away," as shown in Figure 7.3.16. Now it's time to look in more detail at what that means. The breaking away is of course boundary-layer separation, and it changes the basic topology of the flow, as we discussed in Sections 5.2 and 5.3. As we saw in Section 4.1.4, the immediate cause of any separation from a smooth surface in 2D flow must be an adverse pressure gradient. The angle of attack strongly affects the pressure distribution and is therefore a primary variable affecting separation. Separation also depends on the Reynolds number and the location of laminar-to-turbulent transition.

The progression of separation with angle of attack, and the resulting variation of lift with angle of attack, can follow a variety of scenarios, depending on the details of the airfoil shape and pressure distribution and on whether the boundary layer is laminar or turbulent at the separation point. We'll take a brief look at the interesting complexities that can accompany laminar separation before we take up the simpler case of turbulent separation that dominates at high Reynolds numbers.

If the Reynolds number is relatively low, or if the airfoil is thin enough, separation can often occur while the boundary layer is still laminar, and the progression with angle of attack can be complicated. The shear layer downstream of a laminar separation is generally highly unstable, and transition tends to happen within a short distance, but of course the distance depends on the Reynolds number. Depending on the length of the transition region and the shape of the airfoil, transition may or may not cause reattachment of the boundary layer to the surface. A laminar separation with subsequent turbulent reattachment is called a *laminar separation bubble* and generally has the structure illustrated in Figure 7.4.11. If the displacement-thickness Reynolds number at the separation point is above 500 (see Thwaites, 1958), transition is very quick and nearly always results in reattachment. The resulting bubble is typically only 1–2% of the chord in length and is called a *short bubble*. At lower Reynolds numbers, transition is less vigorous, and reattachment may or may not occur. If a bubble results, it is referred to as a *long bubble*.

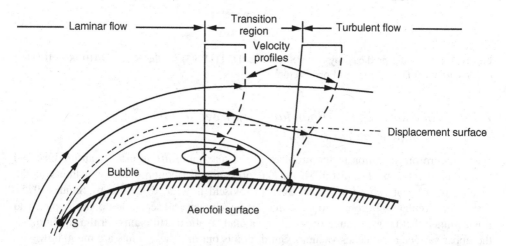

Figure 7.4.11 General structure of a laminar separation bubble with turbulent reattachment. From Thwaites, (1958). Used with permission of Dover Publications, Inc.

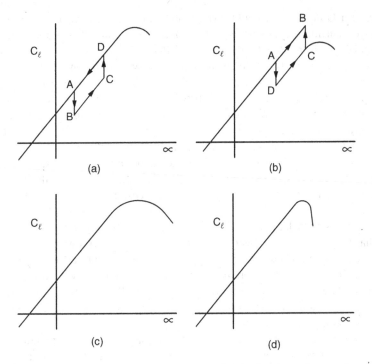

Figure 7.4.12 Lift curves showing different types of stall behavior depending on the progression of upper-surface boundary-layer separation. (a) Counter-clockwise hysteresis loop in the mid-C_l range associated with laminar separation near mid-chord. (b) Clockwise hysteresis loop at the top of the lift curve associated with the bursting of a laminar separation bubble near the leading edge. (c) Gentle turbulent stall: separation moves forward gradually, and lift increases somewhat after the first appearance of separation. (d) Sudden turbulent stall: separation jumps forward suddenly, and lift increases little or not at all after first separation

Most airfoil flows are unique, in that there is only one possible flow pattern for a given airfoil shape, farfield flow condition, and angle of attack. But in some situations when laminar separation bubbles are involved, there can be two possible flow patterns at the same angle of attack, and the lift curve can have a hysteresis loop. Figure 7.4.12a,b shows two possible types of hysteresis loop.

The counterclockwise loop in part (a) is associated with a laminar separation bubble near mid chord that first appears at an angle of attack below point A. As point A on the lift curve is approached from below, the boundary layer ceases to reattach, and the lift jumps down to point B on the lower branch. When the angle of attack is further increased to point C, transition moves ahead of the former separation point, the separation disappears, and the lift jumps up to point D. When the loop is traversed in the other direction, earlier transition continues to prevent the laminar separation, and the lift follows the upper branch back to point A.

A clockwise hysteresis loop like that shown in part (b) of Figure 7.4.12 is more common and is associated with laminar separation near the leading edge. Below point A on the lift curve, there is a laminar short bubble with reattachment. The laminar bubble thickens the boundary layer considerably, however, and reduces the turbulent boundary layer's ability to resist separation downstream. (We'll look at the effect of incoming boundary-layer thickness on turbulent separation in some detail in connection with turbulent stall below.) At point B, separation jumps forward suddenly from the trailing edge to the leading edge, leading to the precipitous drop in lift to the lower branch at point C. This is often described as the "bursting" of the laminar bubble. Now that the lift is reduced, and with it the circulation around the airfoil, the stagnation point is not as far aft on the lower surface as it was when the lift was on the upper branch. The distance around the leading edge from stagnation to separation is shorter, the displacement-thickness Reynolds number at separation is lower, and the ability to reattach is reduced. As angle of attack is reduced, the lift follows the lower branch to point D before the reattachment is reestablished, and the lift jumps back up to point A.

Obviously, flows with laminar separation bubbles can display complex behavior, and it turns out that they are not that easy to predict. Drela's ISES and MSES codes, which are based on viscous/inviscid interaction, and in which the integral-boundary-layer portion of the analysis includes a transition-prediction module, can predict such flows in some cases. At the current state of the art, NS codes based on the RANS equations with turbulence modeling don't routinely predict these effects because they don't typically incorporate any option for transition prediction.

At the high Reynolds numbers typical of large airplanes, laminar separation tends not to arise, and lift is limited by turbulent separation. In a "turbulent stall," lift can increase substantially or very little after separation first appears, depending on how the separation moves forward as angle of attack increases. As a result, the airfoil's lift curve can vary between the extremes illustrated in Figure 7.4.12c,d. Separation may move gradually forward from the trailing edge, producing a gradual reduction in lift slope and a broad maximum in the lift curve as in part (c), or it may jump forward suddenly, producing a precipitous drop in lift as in part (d). In the gradual scenario, the actual *stall,* or onset of the decrease in lift, is usually preceded by *buffeting* caused by flow unsteadiness related to the separation, and the loss of lift post-stall is gradual. In the sudden scenario, the stall comes without warning, and the loss of lift is abrupt, which can cause sudden changes in airplane handling characteristics. Obviously, flight safety favors a gradual stall, and we'll see later what pressure-distribution features can be designed into an airfoil to produce it. Even an airplane with a gradual stall, however, is deliberately flown into the buffeting regime only in special circumstances, typically flight testing and pilot training, and in this sense, the onset of flow separation provides one definition of the maximum *usable* lift coefficient. For this reason, and because it is easier to predict computationally than the actual maximum lift, we'll take the onset of flow separation as the threshold for purposes of this discussion.

By this criterion, the maximum lift depends on how low the average upper-surface C_p can be made before the upper-surface boundary layer can no longer make it to the trailing edge without separating. The "run" of the upper-surface boundary layer starts at the stagnation point of attachment, which under high-lift conditions is usually on the lower surface, back a short distance from the leading edge, so that the boundary layer must flow around the leading edge before it starts to traverse the upper surface. Then whether the boundary layer separates or not depends on the pressures it is subjected to around the leading edge and along the entire upper surface.

We saw what a typical upper-surface pressure distribution looks like in Figure 7.4.2: There is a decrease in pressure as we move away from the stagnation point of attachment near the leading edge, and a minimum pressure we'll call the *suction peak,* which can be very close to the leading edge or farther back, depending on the airfoil shape and the angle of attack. The suction peak is followed by an increase in pressure, or *pressure recovery* from the suction peak to the trailing edge. For an airfoil with nonzero thickness in attached viscous flow, the C_p at the trailing edge typically has a slightly positive value that depends primarily on the average thickness and is not very sensitive to the details of the thickness shape, for reasonable shapes.

The recovery region is where the pressure gradient is adverse, and that's where we'll concentrate our attention. In general, a streamlined body with thickness in attached flow must have a recovery region at the rear. In Figure 7.4.2c, we saw that a typical airfoil thickness distribution by itself contributes a modest amount to the recovery, and in Figure 7.4.2b, we saw that the typical camber distribution also contributes. For determining maximum lift, the contribution from angle of attack (Figure 7.4.2a) is crucial because it increases with angle of attack, making the pressure gradient over the entire recovery region increasingly adverse. Increasing α eventually causes the gradient to exceed what the boundary layer can tolerate, regardless of other details of pressure distribution.

To maximize the lift at the onset of separation, we want to make the average C_p on the upper surface as negative as possible, subject to the constraint imposed by the slightly positive C_p that we must reach at the trailing edge. We can decrease the average C_p either by lowering the minimum pressure or by keeping the pressure low over more of the chord, which delays the start of the recovery, or by a combination of both. Lowering the minimum pressure increases the pressure rise the boundary layer must withstand. Delaying the start of the recovery shortens the distance in which the recovery must take place and thickens the boundary layer at the start of the recovery, both of which tend to bring separation on sooner. The shape of the recovery pressure distribution is also an important factor.

There is no single, simple way to visualize how all of these factors come together to affect maximum lift. A conventional C_p plot normalized by the farfield dynamic pressure q_∞ has the advantages of directly showing the effects of the pressure distribution on C_l and of making it clear that all recovery pressure distributions are essentially "anchored" to a trailing edge C_p that is fixed within a narrow range. But conventional C_p at the suction peak varies widely, making it difficult to visualize how that part of the pressure distribution affects separation. The amount of pressure rise, or velocity drop, that the boundary layer can withstand is much more closely related to the conditions at the suction peak than to those in the farfield. In his classic Wright Brothers lecture on high-lift aerodynamics, Smith (1975) suggested that plotting recovery pressure distributions in what he called *canonical* form, in terms of a pressure coefficient relative to the suction-peak pressure and normalized by the peak dynamic pressure, reduces the variation in ΔC_p between the suction peak and separation and makes the separation trends easier to see. The canonical pressure coefficient $\overline{C}_p$ is thus defined and related to conventional C_p by

$$\overline{C_p} \equiv \frac{p - p_o}{\frac{1}{2}\rho u_o^2} = 1 - \left(\frac{u_e}{u_o}\right)^2 = \left(\frac{u_\infty}{u_o}\right)^2 (C_p - 1) + 1 = \frac{C_p - C_{po}}{1 - C_{po}}, \qquad (7.4.1)$$

where the subscript o denotes conditions at the suction peak, which we'll take as the start of the recovery. Note that $\overline{C}_p$ at the suction peak is zero by definition, and it takes on only

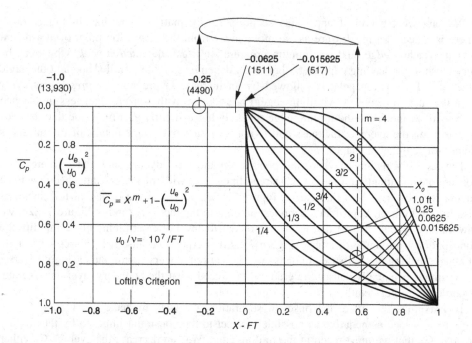

Figure 7.4.13 Separation loci for a family of power-law canonical pressure distributions preceded by different lengths x_0 of boundary-layer run ahead of the recovery. In parentheses are values of R_θ at x = 0. (From Smith, 1975; Figure 20.) The airfoil sketch at the top, and the circles and arrows pointing to it, were added to Smith's original figure to illustrate how these curves would apply to a hypothetical airfoil for the particular case of $x_0 = 0.25$ ft and $m = 1/2$

positive values in the recovery. Separation of a turbulent boundary layer generally occurs at $\overline{C_p}$ values between 0.4 and 0.9, a much smaller range than we would see in terms of conventional ΔC_p. Of course, to see how the separation trends affect C_l, we will still have to return to the conventional pressure distribution.

Let's look at some separation trends in terms of canonical pressure distributions. Figure 7.4.13, from Smith (1975), shows in a single plot how $\overline{C_p}$ at separation is affected by the shape of the recovery pressure distribution and the thickness of the boundary layer at the start of the recovery, which in these examples is determined by the length x_0 of an ideal constant-pressure boundary-layer run ahead of the start of the recovery. The longer the length x_0, the thicker the boundary layer at the start of the recovery. The results shown are for turbulent boundary layers and are based on numerical solutions to the boundary-layer equations with the algebraic eddy-viscosity model of Cebeci and Smith (1974) and with the given pressure distributions as boundary conditions. Because of the limitations of the simple turbulence model, these calculations are generally optimistic regarding to the amount of pressure recovery that is possible, especially in the cases where the recovery leading to separation is short when measured in terms of initial boundary-layer thicknesses, but the qualitative trends are realistic.

Three important trends are clear in Figure 7.4.13:

1. A *concave recovery* (a recovery pressure distribution characterized by m < 1, in which the adverse pressure gradient starts out strong and eases downstream) leads to separation in a shorter distance than does a *convex recovery* (m > 1),
2. In spite of the shorter distance to separation, a concave recovery produces greater pressure recovery than a convex recovery, and
3. A thin boundary layer at the start of the recovery (small x_0) results in greater pressure recovery, and a thick boundary layer results in less.

Smith (1975) showed that these results are not very sensitive to Reynolds number: Reducing the Reynolds number by a factor of 10 reduced the values of $\overline{C}_p$ at separation generally by less than 0.1. But for a given pressure distribution, separation is strongly affected by wing sweep, as we'll see in Section 8.6.2.

So viewing recovery pressure distributions in canonical form makes it easier to visualize how the initial boundary-layer thickness and the shape of the pressure distribution affect separation. But it isn't clear just from the canonical distributions what these boundary-layer separation trends imply regarding airfoil maximum lift. To visualize the effects on maximum lift, we must convert the canonical distributions back to conventional form and view them as candidate upper-surface pressure distributions of a hypothetical airfoil with a rapid flow acceleration near the leading edge and a region of constant pressure back to the recovery point, followed by the pressure recovery.

The conversion to a conventional pressure distribution goes as follows. We take x_0 in the canonical plot to correspond to the leading edge of our airfoil, and if we place the trailing edge at x_{sep} in the plot, we have separation beginning just at the trailing edge, which corresponds to our criterion for the maximum usable lift. The airfoil sketch at the top of Figure 7.4.13, and the circles and arrows pointing to it, were added to Smith's original figure to illustrate how this would work for a hypothetical airfoil in the particular case of $x_0 = 0.25$ ft and m = 1/2. In this case the leading edge lines up with x = −0.25 ft, and the trailing edge lines up with x_{sep} for m = 1/2, $x_0 = 0.25$. For general cases, each of the possible combinations of x_0 and x_{sep} then corresponds to particular locations of the leading and trailing edges on the plot, and a particular chordwise location x/c for the start of the recovery, often referred to as the *recovery point*. Then we assume that the conventional C_p at the trailing edge always has a slightly positive value, say 0.2, and that this corresponds to $\overline{C}_p$ at separation for the particular pressure distribution we're considering in the plot. Through Equation 7.4.1, this defines u_∞ for each case we consider, and thus defines the relationship between $\overline{C}_p$ and C_p. When we do this for any one of the canonical pressure distributions in Figure 7.4.13, and repeat it for various combinations of x_0 and x_{sep}, we get a family of conventional pressure distributions with different recovery points and different C_p levels at the suction peak. Strictly speaking, each of these pressure distributions corresponds to a different chord Reynolds number, because in Figure 7.4.13 it is the "unit" Reynolds number u_0/ν of the canonical x scale that is constant. But the Reynolds-number sensitivity is fairly weak, and this discrepancy shouldn't seriously distort the qualitative trends we're looking for.

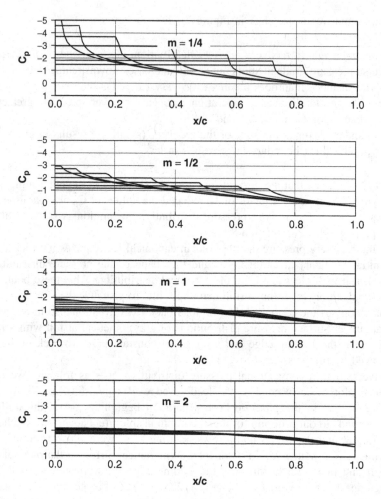

Figure 7.4.14 Families of conventional maximum-lift pressure distributions corresponding to some of the canonical pressure distributions of Figure 7.4.13; for each assumed location of the recovery point, the rooftop level was determined so that separation is reached at the trailing edge. All-turbulent flow was assumed ahead of the recovery

Results of this process are shown in Figure 7.4.14 for four of the canonical pressure distributions ranging from concave to convex. Several important trends are evident in these plots:

1. In the part of the pressure distribution ahead of the pressure recovery (often called the *rooftop*), higher suction levels (lower pressures) are possible with concave recoveries,
2. For a given shape of the recovery pressure distribution, the allowable suction level increases as the recovery point is moved forward, and
3. For a given suction level, a concave recovery allows a farther-aft recovery point and a longer rooftop.

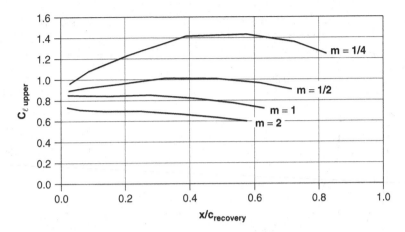

Figure 7.4.15 Upper surface C_l values for the pressure distributions of Figure 7.4.14, plotted versus the recovery-point location

A higher suction level and a longer rooftop both increase lift, so that in view of items (1) and (3) above, concave recoveries should be favorable. This is born out by the plots in Figure 7.4.14, which clearly show that concave recoveries produce higher lift, and by the corresponding upper-surface lift coefficients plotted in Figure 7.4.15.

It is also clear that for any given shape of recovery pressure distribution, high lift involves a compromise between having low pressure at the suction peak and having the low pressure extend over more of the chord. There is therefore an optimum location for the recovery point that yields the highest lift. Figure 7.4.15 shows that the optimum recovery point moves aft, and the maximum lift dramatically increases, as the recovery pressure distribution is made more concave. Remember that chord Reynolds number was not kept constant in these examples, and the trends are only qualitative approximations to what would happen for constant Reynolds number.

So in these examples, the highest maximum lift is achieved by the most concave recovery pressure distribution, or the most "rapid" recovery. Smith (1975) concluded that the highest possible lift that can be achieved without separation is achieved by the most rapid of all possible recoveries, the *Stratford recovery* (Stratford, 1959a, b), which is shown in canonical form in Figure 7.4.16 (Smith's Figure 21). In a Stratford recovery, an extremely strong pressure gradient at the start brings the boundary layer immediately to near-zero C_f, and the remainder of the pressure distribution is tailored to maintain the boundary layer in a constant state of incipient separation. The boundary-layer velocity profile in a Stratford recovery is nearly self-similar, and it is one of the family of *equilibrium* turbulent-boundary-layer flows that we discussed in Sections 4.3.2 and 4.4.2.

The potential for high maximum-lift coefficients led Liebeck (1973) to use the Stratford recovery pressure distribution as the basis for several low-speed airfoil designs. A drawback to this approach is that above the "design" angle of attack, separation jumps forward abruptly from the trailing edge, leading to the problem of abrupt stall that we've already discussed. As a result, Stratford-recovery airfoils have not found much application in aeronautical practice. Instead, designers developing airfoils for practical applications purposely back

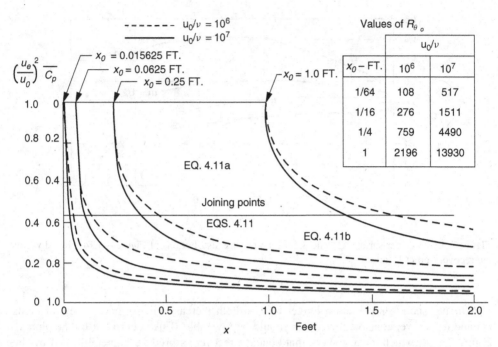

Figure 7.4.16 Canonical pressure distributions for Stratford's recovery at two different values of unit Reynolds number. (From Smith, 1975; Figure 21; including references to his equation 4.11)

away from the Stratford recovery by varying amounts, that is, they make the recovery less concave, depending on how gentle they want the stall to be and how much of a penalty in maximum lift they are willing to pay for it.

So far we've assumed that the boundary-layer development is all-turbulent from near the leading edge, a situation Smith referred to as a *turbulent rooftop*. Now let's look at the case of a *laminar rooftop*. A run of laminar flow ahead of the start of the recovery, that is, a laminar rooftop, will result in a thinner boundary layer at the start of the recovery than would be there if the boundary layer were all turbulent. In terms of the canonical pressure distributions in Figure 7.4.13, we saw that a thin boundary layer at the start of the recovery tends to delay separation. In terms of the conventional pressure distribution, as in Figure 7.4.14, the thinner boundary layer would allow either a higher suction level in the rooftop or a longer rooftop, either of which would increase maximum lift. Figure 7.4.17, from Smith (1975), illustrates this effect by comparing conventional maximum-lift pressure distributions with laminar and turbulent rooftops. These examples happen to show Stratford recoveries, but the effects of a laminar rooftop are reasonably generic. For a given recovery-point location, a laminar rooftop allows a higher suction level, and for a given suction level, a laminar rooftop allows a longer rooftop. In view of the potential for very high maximum lift coefficients, Liebeck (1973) assumed a laminar rooftop for several of his airfoil-design examples. For practical applications, a drawback to designing an airfoil for a laminar rooftop at the maximum-lift condition is that if the airfoil surface becomes contaminated, say by rain or insects, and the flow trips early to turbulent, the maximum-lift capability will be greatly reduced.

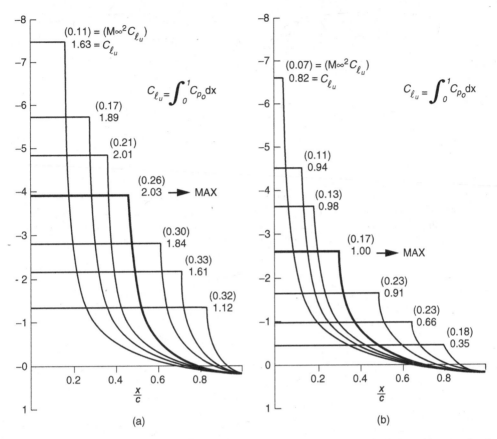

Figure 7.4.17 Conventional maximum-lift pressure distributions with Stratford recoveries at a chord Reynolds number of 5 million. (From Smith, 1975; Figures 24 and 25.). (a) Laminar rooftop. (b) Turbulent rooftop

To summarize, we've seen that high maximum lift in a single-element airfoil favors concave and even Stratford-recovery pressure distributions, and a laminar rooftop. In airplane applications, however, such features are to be avoided for handling characteristics and safety reasons, and besides, single-element airfoil design is usually dominated by considerations in other parts of the flight envelope. For many airplanes, maximum lift for takeoff and landing is enhanced by high-lift devices with slots, which brings us to our next topic.

7.4.4 Multielement Airfoils and the Slot Effect

Maximum-lift capability can be greatly enhanced if the airfoil has multiple elements with favorably configured slots between them. Figure 7.4.18 shows how slots are typically arranged on airfoils with slotted trailing-edge and leading-edge high-lift devices and between the mainsail/headsail combination often used on sailboats. Note that the slot widths (gaps) used in aeronautical applications are typically much smaller, relative to overall chord, than

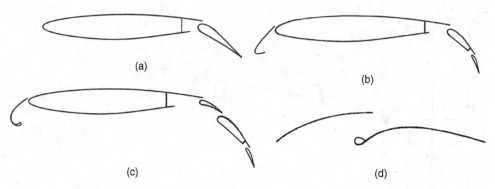

Figure 7.4.18 Sketches of multielement airfoils with slots, not to scale. (a) Single-slotted trailing-edge flap, as on many light airplanes. (b) Double-slotted trailing-edge flap and leading-edge slat, as on the Boeing 757 and Boeing 737NG. (c) Triple-slotted trailing-edge flap and a flexible (variable camber) leading-edge Krueger flap, as on the Boeing 747. (d) A mainsail/ headsail combination

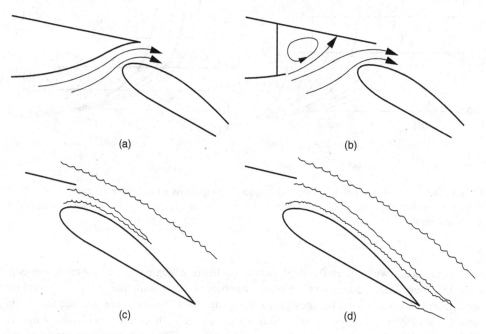

Figure 7.4.19 Typical features of the flow through an effective high-lift slot. (a) Smoothly faired slot. (b) Slot with a sharp-cornered cove with flow reattachment upstream. (c) Confluence of the wake of the forward element with the boundary layer of the aft element. (d) Confluence of the wake of the forward element with the wake of the aft element

those used in cloth sails. As we'll see, however, the physical mechanism for the effectiveness of the slot is largely the same in both cases.

For a slot to be effective in enhancing maximum lift, the general pattern of flow through it must be as shown in Figure 7.4.19, in which the flow passes smoothly through the slot, from the lower surface to the upper surface, and is directed along the upper surface of the aft element. A smoothly contoured slot as shown in part (a) naturally facilitates such flow, but even if the slot is preceded by a sharp-cornered *cove*, which is common in practical applications in which the slot opens when a high-lift device is deployed, the flow under high-lift conditions usually reattaches to the lower surface ahead of the slot, as shown in part (b). In both the smoothly contoured slot and the coved slot, much of the air passing through the slot is not in the boundary layers and has freestream total-pressure. This part of the flow through the slot and extending downstream in the region between the boundary layer on the aft element and the wake of the forward element is often referred to as a *potential-flow core*, or *potential core*, for short. The potential core can be thought of as ending where *confluence* occurs between the wake of the forward element and either the boundary layer or the wake of the aft element, as shown in Figure 7.4.19c,d. Identifying the beginning of confluence is a bit like identifying the "edge" of a turbulent boundary layer; both are difficult to define precisely in practice.

Historically speaking, explanations for how the *slot effect* enhances maximum lift have evolved considerably. Early on, the flow through the slot was typically seen as a *wall jet* that "energizes" the boundary layer downstream and prevents or delays separation on the downstream element, in the way we discussed in Section 4.5.3 (see also Prandtl and Tietjens, 1934, section 93). For the small gaps typical of aeronautical applications, this view has some merit. After all, the entire viscous layer downstream of the slot, which now includes the boundary layer on the aft element, what remains of the potential core of the slot flow, and the wake of the forward element, is still a boundary layer in the general sense, and the addition of the potential core does constitute a sort of "energizing" of the flow, relative to what would be there without the slot. The potential core does have higher energy than the rest of the boundary layer, albeit not higher than the freestream. But this view of the slot flow turns out to be disappointingly limited in what it can explain.

Much of what is known about the performance of wall jets in preventing separation applies to the case of an active jet in which the total-pressure in the jet core is well above that of the freestream (farfield). Active-wall-jet blowing tends to be implemented with a relatively narrow slot, high total-pressure in the jet, and a high jet velocity. In this situation, the jet mass flux required to control downstream flow separation tends to be relatively small, and momentum flux alone usually does a good job of correlating performance, as we noted in Section 4.5.3. On the other hand, a passive wall jet, in which the potential core is only at freestream total pressure, is not so simple and should not be expected to perform in a way that correlates neatly with simple jet-flow parameters. The sorts of relatively simple trends that can be seen in the behavior of ordinary boundary layers, as in Figures 7.4.13 and 7.4.15, don't apply to passive wall-jet flows either. If we are to really understand the effectiveness of passive slots, we must go beyond viewing the slot flow as a simple wall jet, even for cases with small gaps. And clearly, physical intuition indicates that the wall-jet model doesn't apply to the large gaps used with sails.

How the slot effect really works becomes much clearer if we view the boundary layers on the individual airfoil elements as ordinary individual boundary layers, a view that is valid as long as the potential core of the slot flow persists, and retains some validity even after some confluence has occurred. Looking at the problem in this way was first popularized in Smith's (1975) Wright Brothers Lecture. To the extent that this view holds true, it is logical to think of each airfoil element as an individual single-element airfoil, as far as its own boundary-layer development is concerned, with the only influence of the other airfoil elements coming through their effects on the pressure distribution. So our focus has shifted from seeing the slot effect primarily as a viscous-flow phenomenon (the wall-jet view) to seeing it in terms of a combination of viscous effects and the largely inviscid interaction by which the airfoil elements influence the pressure distributions imposed on their neighbors' boundary layers.

We can now look at the problem of how much lift an element of a multielement airfoil can carry in the same way that we looked at the maximum lift of a single-element airfoil in Section 7.4.3. There we saw that the main limitation on maximum lift is imposed by how much pressure rise (or velocity drop) the boundary layer on the suction side can withstand. We also saw that the boundary layer's ability to withstand the pressure rise is increased if the boundary layer is thin when it starts into the pressure rise. The slot effect works on both of these things: the amount of pressure rise imposed on each element, and the thickness of the boundary layer starting into the pressure rise.

So a slotted configuration enhances maximum lift by doing two things:

1. Starting a fresh upper-surface boundary layer at the leading edge of each element. For elements after the first, this means a thinner boundary layer at the start of the pressure rise than would be there if there were no slot.
2. Reducing the pressure rise imposed on the suction-side boundary layer of each element. Slots do this in either of two ways, or both simultaneously, depending on the situation of the given airfoil element. When there is an element ahead, it can provide *leading-edge suction-peak suppression,* essentially by providing some flow turning ahead of the leading edge of the current element, so that the flow doesn't rush around the leading edge as fast as it otherwise would. When there is an element behind, it can provide *trailing-edge dumping-velocity elevation,* effectively by placing the trailing edge of the current element in a high velocity region near the leading edge of the trailing element. Reducing the velocity at the leading edge of an element and elevating the velocity at the trailing edge both reduce the total velocity drop the boundary layer is subjected to.

These mutual-influence relationships by which the elements favorably affect each others' pressure distributions can be seen clearly in the pressure distributions of typical high-lift configurations. An example with a leading-edge slat and a double-slotted trailing-edge flap is shown in terms of the conventional pressure coefficient in Figure 7.4.20a. The slat does not have an element ahead of it to suppress its suction peak, but the downward deflection of the slat reduces the peak compared with what would occur on an airfoil without a deflected leading-edge device. The main element and the trailing-edge flaps have their suction peaks suppressed by the elements ahead of them. The trailing-edge dumping velocity on the slat is considerably elevated over what would occur on an isolated airfoil. Moving aft in the system, dumping velocities decrease until that on the aft-most trailing-edge flap reaches a level typical of an isolated single-element airfoil. Successive leading-edge peak velocities also decrease systematically. When we put these pressure distributions in canonical form, as in Figure 7.4.20b, we see that each element withstands a pressure rise comparable to what might occur on a single-element airfoil, which tends to support our view of the elements as separate airfoils.

Looking at the boundary layers on the individual elements in this way provides us with much of our explanation for the effectiveness of the slots, but of course the flowfield around

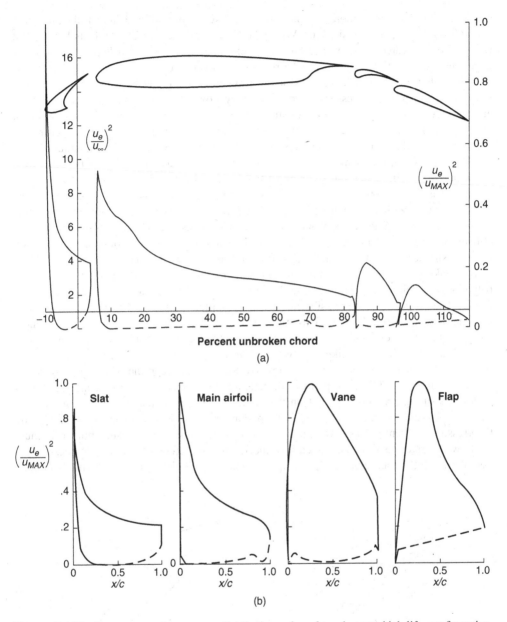

Figure 7.4.20 Geometry and pressure distribution of a four-element high-lift configuration. (From Smith, 1975; Figures 17 and 18.). (a) In terms of conventional pressure coefficient. (b) In terms of canonical pressure coefficient for the individual elements

a multielement system is more complex than the flow around individual airfoils, and there are additional issues and limitations. While the boundary layers on all elements but the last benefit from elevated trailing-edge dumping velocities, the wakes that leave these trailing edges must complete the remainder the pressure recovery, reaching freestream pressure or slightly higher before they leave the system. In other words, the wakes of forward elements must withstand adverse pressure gradients as they flow over the upper surfaces of the aft elements. A wake is typically more resistant to flow reversal in an adverse pressure gradient than a boundary layer is, so that flow reversal in the wake is not a common occurrence. But even if a wake doesn't have reverse flow, it can adversely affect the maximum lift of the system. Its displacement thickness in an adverse pressure gradient can become quite large, and it can influence the pressure distribution and separation behavior of the airfoil element it is flowing over. And unless the slot gap upstream is large, the wake will merge to some extent with the boundary layer of the element underneath. If too much of this *boundary-layer confluence* takes place, the separation resistance of the entire viscous layer will be reduced. Thus the choice of slot gap is a trade: Greater suction-peak suppression favors a smaller gap, but boundary-layer confluence penalizes a smaller gap.

The wakes of forward elements, and the interaction of the wakes with the separation behavior of aft elements, can affect the lift curve of a multielement airfoil in various ways, and it is safe to say that the possibilities are not completely understood. For example, there is some evidence from 2D RANS solutions and from experiments that some three-element airfoils (slat, main element, and single-slotted aft flap) can have a counterclockwise hysteresis loop in the lift curve like what we saw for a single-element airfoil in Figure 7.4.12a. In the multielement case, this seems to involve toggling between the two basic flow states illustrated in Figure 7.4.21. In part (a), there is mostly attached flow on the aft flap, which imposes a strong pressure gradient on the combined wakes of the slat and main element, causing a large displacement thickness and perhaps reverse flow in some cases. In part (b), the pressure gradient is less severe, and the forward-element wake remains thinner, because the flow separates near the leading edge of the aft flap. A different scenario that doesn't necessarily involve a double-valued lift curve can lead to a so-called "inverse-Reynolds-number

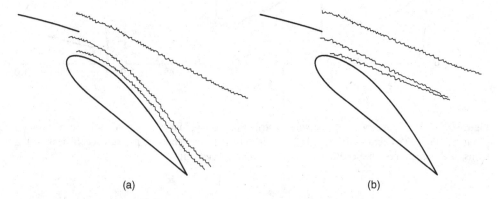

(a)   (b)

Figure 7.4.21   Illustration of flow states in the interaction of forward-element wakes with the separation on aft elements. (a) Mostly attached flow on aft flap and thick forward-element wake. (b) Separated flow on aft flap and thin forward-element wake

effect," in which an increase in Reynolds number results in a thinner forward-element wake, which increases the adverse pressure gradient imposed on the aft flap, causing the flow on the flap to separate earlier, thus reducing maximum lift.

Now let's look again at sail combinations like that shown in Figure 7.4.18d. These appear at first glance to be very different from aeronautical high-lift systems, in that the gap between elements is typically so much larger. But the mechanisms by which the slot does its job are much the same. This has not always been correctly understood by the sailing community. According to older folklore among sailors, the slot directs a "high-velocity jet" of air along the lee (suction) side of the main (aft) sail to "energize" the boundary layer. This is reminiscent of the aeronautical community's older wall-jet interpretation of the slot effect, which we've seen is not a very good explanation in the case of small gaps, and which I've already argued is even less appropriate in the case of large gaps. The correct view of the slot effect in the case of typical sail configurations recognizes that the velocity through the slot is only modestly elevated above freestream, and that the main benefit the slot provides for the mainsail is suppression of the leading-edge suction peak. Gentry (1971) provided the sailing community with one of the earliest correct interpretations of the slot effect.

7.4.5 Cascades

A cascade is a stacked array of nominally identical airfoils. The common examples are turning-vane arrays, as used in wind-tunnel corners, and rows of compressor blades, turbine blades, and stators, as used in turbomachinery. These arrays obviously serve a variety of purposes, but what they have in common is that they all involve turning the flow. For stationary arrays (turning vanes and stators), flow turning is the primary function, while for moving arrays (compressor and turbine blades), flow turning is the means to an end: imparting work or extracting work. In the case of moving blades, imparting or extracting work requires a force component in the direction of the blades' motion. The cascade must therefore change the component of the stream's momentum in that direction, which requires turning the flow.

Because the blade elements of a cascade are usually not spaced far apart relative to their chords, the turning that can be accomplished is typically much greater than would be possible with isolated airfoils. The typical wind-tunnel turning-vane array, for example, turns the flow 90°, as shown in Figure 7.4.22. The reason such large turning angles are possible can be looked at in two ways. First, a global view is that the force generated by each blade is directly "responsible" only for turning a streamtube of limited depth, so that the required loading on each blade is not too large. A more local view is that in spite of the large camber that such blades typically have, loadings are kept low, and stalling is avoided, by the inviscid interference provided by neighboring blades.

The geometry of a cascade array repeats in a spatially periodic pattern, and for theoretical purposes, we can usually assume that the resulting flow pattern is also spatially periodic. To solve for the detailed flowfield, we need only solve for the flow around a single blade, with periodic boundary conditions representing the effects of the other blades, as shown in Figure 7.4.23a. The ISES airfoil code (Drela and Giles, 1987) has options for analyzing and designing cascade blades, based on this kind of treatment of a single element in the array. For viscous calculations, the inlet flow angle must be specified, which is analogous to specifying the angle of attack of a single airfoil. The outlet flow then depends on the

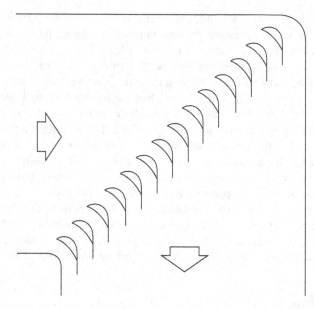

Figure 7.4.22   Wind-tunnel turning vanes for a 90° turn. Profiles sketched from Gelder *et al.*, (1986); published by NASA

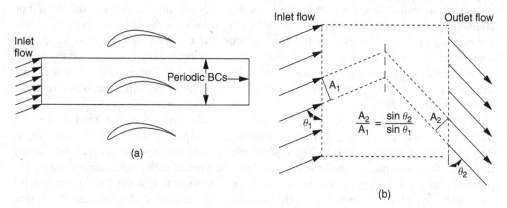

Figure 7.4.23   Simplified views of a cascade. (a) Dealing with a single element and imposing periodic boundary conditions. (b) Illustration of a view of integrated performance in terms of uniform farfield flow, and the relationship between inlet and outlet flow angles and streamtube areas

solution for the flow around the blades. In inviscid calculations, either the inlet or outlet flow angle may be specified, but not both.

For purposes of analyzing the integrated performance of an array, we can simplify things further by taking a control-volume view in which the disturbances from individual blades are assumed to have died out, and the entry and exit flows are taken to be uniform. Practically speaking, this requires moving only about one spacing length upstream and downstream of the leading and trailing edges of the elements. The details of the flow around the blades can then be ignored, and the cascade can be viewed as a control volume with a steady, uniform farfield flow entering at a particular angle and speed, and a uniform exit flow leaving at a different angle and speed, as shown in Figure 7.4.23b. A streamtube passing through the cascade has its cross-sectional area changed by an amount that is dictated by the entry and exit angles, as indicated in the diagram. If the blades operate in the attached-flow regime, so that viscous losses are small, the flow can be approximated as inviscid. If we are in the reference frame in which the blades are stationary, the entry and exit flow conditions must then be related by the 1D isentropic flow relations we discussed in Section 3.11.1, and can be determined based on the area change only, and thus on the entry and exit angles only. Then given the flow conditions and the flow angles, we can apply conservation of momentum in control-volume form to calculate the forces per unit span on the blades in both the vertical and horizontal directions.

In inviscid flow without shocks, this force is the cascade counterpart to the lift of a conventional airfoil. If we are in a reference frame in which the blades are moving, the isentropic relations do not apply because we must take into account the work done by the blades, and the exit flow conditions and the blade forces are then implicitly related. Otherwise the analysis is the same as in the stationary-blade case. For details of such analyses, see Hill and Peterson (1992).

At the control-volume level, when only the entry and exit flow conditions are known, the blade shape and details of the flow are not determined. Determining a blade chord and spacing and a detailed shape that would accomplish the specified flow turning is a design problem that usually has multiple solutions. As the chord is reduced relative to the spacing, however, the loading on each blade increases, and in the case of viscous flow, there is a limit beyond which no blade shape exists that can carry the required load.

There is thus a minimum chord that the blades must have (relative to the periodic spacing) that depends on the required entry and exit flow conditions. The problem of determining the minimum chord is very similar to that of determining the maximum lift of a conventional airfoil, and the physics issues are essentially the same. In subsonic flow, the lift capability of a blade is limited by boundary-layer separation on the suction side, so that much of the discussion in Section 7.4.3 applies. In this regard, the pressure level at the trailing edge is very important, just as it was in the case of a conventional airfoil. The trailing-edge pressure of a cascade blade is closely tied to the exit pressure of the cascade, just as the trailing-edge pressure of a single-element conventional airfoil is closely tied to freestream pressure.

Because exit pressure plays such an important role, there is a substantial difference between the load-carrying capabilities of compressor cascades and turbine cascades. When the exit pressure is higher than the inlet pressure, as in a compressor stage, the suction-side boundary layer is not able to recover from as low a pressure as it would otherwise be able to, and the load-carrying capacity of the blades is reduced. Another way to look at this is that the passages between blades are required not just to turn the flow, but to slow it down as well (i.e., to act as diffusers). An exit pressure that is lower than the inlet pressure, as in a turbine stage, has the opposite effect and enhances the load-carrying capacity of the blades. This is the main reason why turbine blades can sustain higher loadings than do compressor blades. In the design of gas-turbine engines, this difference is usually used to allow the turbine section to have far fewer stages than the compressor section.

7.4.6 Low-Drag Airfoils with Laminar Flow

We've already noted that most of the drag of an airfoil in attached flow consists of the integrated surface shear stress, or skin friction. Early in the history of our understanding of boundary layers, it became known that laminar skin friction is lower than turbulent skin friction under the same conditions, and that keeping the boundary layer laminar over part of the airfoil chord should lead to lower drag. And laminar-flow stability theory predicted that a favorable pressure gradient would delay the growth of small disturbances in the boundary layer and thus should lead to a longer run of laminar flow.

All of this was known by the late 1920s, but it was not applied immediately. The stability theory had not yet been tested against experiment, and it was not yet known what it would take to turn it into a practical transition-prediction method. Recall from Section 4.4.1 that stability-based e^n transition prediction wasn't developed until the early 1950s (Smith, 1981). And in the 1920s airfoil design was still a cut-and-try art.

But long before the e^n method was available, it had been observed that under the right conditions (a smooth airfoil surface and a low freestream turbulence level) transition often did not take place until the favorable pressure gradient at the front of the airfoil ended, and it was reasonable to expect that extending the favorable pressure gradient would extend the run of laminar flow. In the 1930s, a group at the NACA, under the direction of Eastman Jacobs (see Anderson, 1997), carried out a long program of airfoil development aimed at putting this idea into practice. This was one of the earliest practical applications of *inverse design* (see Section 7.4.10), in which an airfoil is purposely designed to produce a particular pressure distribution. The objective of this effort was of course to extend the regions of favorable pressure gradient and produce long runs of laminar flow. Jacobs' group also developed a low-turbulence wind tunnel in which to test the new airfoils.

In the earlier attempts that resulted in the NACA 2-through-5 series airfoils, the designs were carried out by approximate methods. The resulting airfoils demonstrated long runs of laminar flow at some conditions, but the approximate design methods did not control the pressure distribution near the leading edge well enough to produce good performance over adequate ranges of lift coefficient (see Abbott and von Doenhoff, 1958).

Then in the late 1930s, Jacobs' group undertook the design of the NACA 6-series airfoils, using a more accurate design method. The inverse design was actually applied only to a family of thickness distributions. These were then combined with various mean lines designed by linear theory to produce uniform load and therefore not grossly change the pressure gradients that had been designed into the thickness distributions. Each thickness distribution was designed to produce a favorable pressure gradient back to some intended point on the chord and to maintain the favorable gradient on the suction side up to a design angle of attack. Thus each thickness distribution had a range of angle of attack over which the pressure gradient would be favorable, and over which a *low-drag bucket* could be expected in the airfoil's drag polar.

In wind-tunnel testing of the 6-series airfoils, the expected runs of laminar flow and resulting low drag levels were verified. Figure 7.4.24 shows the shape and calculated inviscid pressure distributions of a 6-series symmetrical airfoil, 63_2-015. Compared with a conventional airfoil of that time, the nose radius is smaller, and the maximum thickness is farther aft. The purposely designed region of favorable pressure gradient is clearly seen in the pressure distribution. The low-drag bucket in the measured drag polar can be seen in Figure 7.4.25. The width (range of C_l) of the low-drag bucket depends on design details

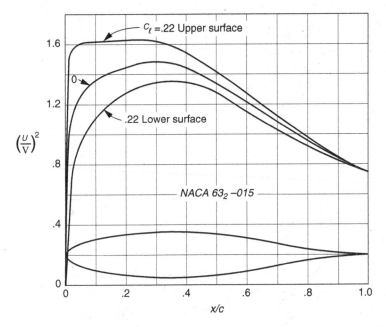

Figure 7.4.24 Shape and calculated inviscid pressure distribution of the symmetrical NACA low-drag laminar-flow airfoil, 63_2-015. From Abbott and von Doenhoff, (1958). Used with permission of Dover Publications, Inc.

but is constrained by airfoil thickness. Making the bucket wider generally requires a thicker airfoil, which exacts a penalty in drag level. There is thus a trade between the width and the depth of the bucket. Adding camber (a cambered mean line) to the airfoil shape moves the low-drag bucket to higher lift coefficients, but because the width of the bucket depends primarily on the thickness distribution, the width tends to stay about the same. Drag polars are shown for some cambered 6-series airfoils in Abbott and von Doenhoff (1958).

The success of the 6-series airfoils in the wind tunnel led to attempts to exploit their low drag levels on several military airplanes during the World War II years, the North American P-51 Mustang being the prime example. Achieving the long runs of laminar flow for which the airfoils were designed, however, requires low levels of surface waviness and roughness, something that was not routinely achieved with the conventional metal construction used on wings at the time. In these applications, the 6-series airfoils probably did not reach the low drag levels seen in the wind-tunnel data. True laminar-flow drag levels were probably not routinely achieved in practice until laminar-flow airfoils were used on high-performance gliders, beginning in the 1950s.

As we discussed in Section 4.4.1, the verification of the stability theory wasn't widely known until after World War II, the e^n transition-prediction method wasn't developed until the early 1950s, and it wasn't widely used until the 1970s. This was not much of a hindrance to the early development of laminar-flow airfoils because successful airfoils could be developed without detailed stability calculations. On a wing without much sweep, over a wide range of practical Reynolds numbers from about a million to more than 10 million, a moderate favorable pressure gradient generally suffices to enable a long run of laminar

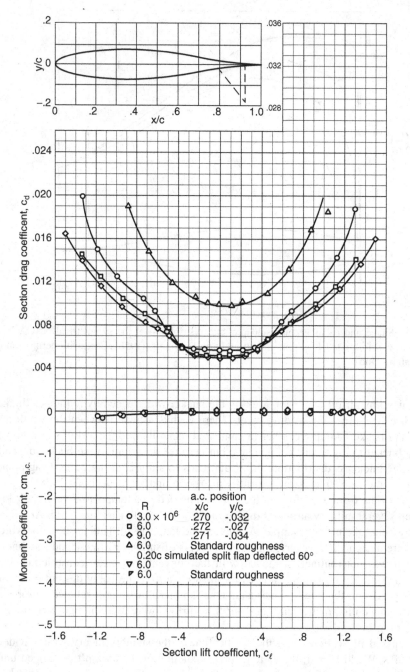

Figure 7.4.25 Measured drag polars for the symmetrical NACA low-drag laminar-flow airfoil, 63_2-015, showing the *low-drag bucket*. From Abbott and von Doenhoff, (1958). Used with permission of Dover Publications, Inc.

flow, with transition at or near the end of the region of favorable gradient. When a wing has substantial sweep, however, the situation is more complicated and tends to require more analysis, including detailed stability calculations, as we'll discuss in Section 8.6.4.

On airfoils like the 6-series, designed for low-drag laminar flow at moderate lift coefficients, the favorable gradient usually disappears at higher angles of attack, and transition moves forward, well before maximum lift is reached. Thus premature loss of laminar flow due to poor surface quality or contamination does not usually entail a reduction in maximum lift, unlike the Liebeck airfoils designed for maximum lift with laminar flow that we discussed in the Section 7.4.3. The small leading-edge radius typical of low-drag laminar airfoils tends to produce a pronounced suction peak at the leading edge at high angles of attack, however, and thus can reduce the maximum lift dictated by turbulent stall.

In the years since the 6-series airfoils were designed, the technology of low-drag laminar-flow airfoils has become much more sophisticated, especially for high-performance gliders. Part of the focus of these developments has been to ensure that transition happens where it is wanted and is not delayed so far that laminar separation occurs before transition. Features like a "transition ramp," which is a region of mild adverse pressure gradient carefully designed to trigger transition before the start of the hard aft recovery, have been used. Other airfoils have been designed to make use of roughness strips to trip the flow (Althaus, 1981), or a "pneumatic turbulator" (Horstmann and Quast, 1982), a row of tiny air jets that are supplied by a passive ram-air inlet and can be turned on over the range of C_l where artificial tripping is desired and can be sealed closed when a longer run of laminar flow is desirable. Laminar flow on swept wings poses additional problems that we'll discuss in Section 8.6.4.

7.4.7 Low-Reynolds-Number Airfoils

Though low-Reynolds-number airfoils cover a wide spectrum, the single issue of laminar separation is pervasive across the entire range. Laminar separation is dealt with by a variety of strategies, from simply living with it to taking purposeful measures to prevent it or to control its effects. At the low end of the spectrum are rubber-powered indoor model airplanes that use cambered-membrane airfoils at chord Reynolds numbers in the range of 3000–6000. In this regime, the laminar boundary layer on the upper surface typically separates at around 70% of the chord and does not reattach, and maximum sectional L/D is around 12 at a C_l of 0.8–0.9. At the high end are airfoils designed for human-powered airplanes, as, for example, the MIT Daedalus (Drela, 1988), on which the chord Reynolds number on the inboard wing was around 500 000, with a maximum sectional L/D of around 138 at a C_l of 1.5. This Reynolds number is still well below those for which conventional low-drag airfoils are usually designed, and transition tends not to happen ahead of laminar separation unless artificial tripping is used. To avoid having to install tripping devices on the delicate plastic-film covering of the Daedalus wing, Drela chose to accept the presence of a laminar separation bubble on the upper surface and to use the bubble, in effect, as the tripping mechanism. Drela found that through proper design of the pressure distribution, the bubble could be kept short, and the drag penalty associated with the presence of the bubble could be kept small. The "signature" of the separation bubble on the upper surface (the kink in the pressure distribution, with a short plateau preceding it, and a precipitous pressure rise following) can be seen in the pressure distributions in Figure 7.4.26.

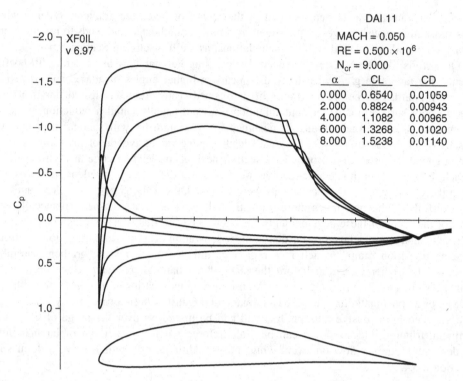

Figure 7.4.26 Shape and pressure distribution of DAE 11 designed by Drela (1988) for the wing center section of the Daedalus human-powered airplane, showing the indication of a laminar-separation bubble on the upper surface. From Drela, (1988). Used with permission of AIAA. XFOIL calculation by Steven R. Allmaras

7.4.8 Airfoils in Transonic Flow

As Mach number increases from zero, the pressure disturbances produced by an airfoil, when expressed in terms of the conventional C_p, increase gradually, but the general character of the pressure distribution does not change. Then when the freestream Mach number is high enough that the peak local velocity outside the boundary layer approaches sonic, the physics begins to change character. As the local Mach number passes through sonic, the local relationship between pressure and streamtube area reverses, as we saw in connection with 1D isentropic flow in Section 3.11.1, and the airfoil pressure distribution changes dramatically.

Figure 7.4.27a shows this progression for the NACA 4412, a conventional airfoil not designed for performance at transonic speeds, illustrated in terms of pressure distributions at several Mach numbers at the same moderate C_l of 0.72, as calculated by the MSES code (Drela, 1993). At low freestream Mach numbers, the upper-surface suction peak is broad and centered at about 20% chord. As the freestream Mach number increases and the local velocity at the suction peak approaches sonic, the suction peak moves aft. Sonic C_p values for the different Mach numbers are indicated by horizontal dashed lines. The freestream Mach number at which the peak local velocity first hits sonic is often

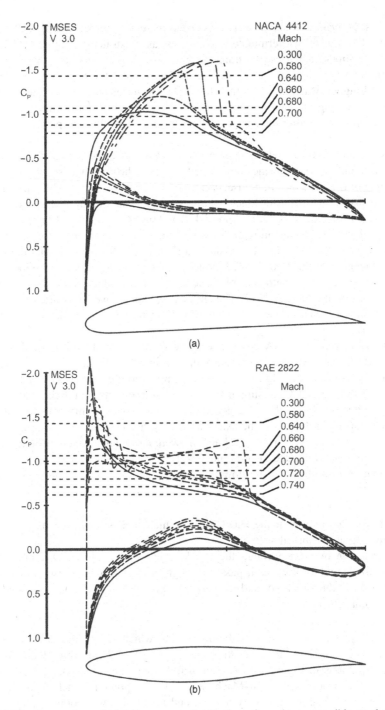

Figure 7.4.27 Progression of airfoil pressure distributions from incompressible to the transonic regime at constant lift coefficient of 0.72, $R = 10^7$, calculated by the MSES code (Drela, 1993). Sonic C_p values for the different Mach numbers are indicated by horizontal dashed lines. (a) NACA 4412, representative of older conventional airfoils. (b) RAE 2822, representative of an early generation of modern transonic airfoils

referred to as the *critical Mach number*. As speed increases further, a region of supersonic ("supercritical") flow forms, terminated by a shock, as illustrated in Figure 3.11.1a. As the strength of the shock increases, the transonic *drag rise* begins. For low-speed airfoils like this one, the margin between the critical Mach number and the beginning of drag rise is small. At still higher Mach numbers than those shown in this series of solutions, the shock becomes strong enough to cause separation, dramatically reducing maximum lift. Note that the high velocities preceding the shock are concentrated over the mid portion of the airfoil chord.

Modern transonic airfoils delay the drag rise to well above the critical Mach number by delaying the rapid increase in the local Mach number ahead of the shock. This requires shifting some of the lift from the midsection to the front and the rear, which means making the pressure distribution at low Mach numbers "peaky" at the front and introducing an increased degree of "aft loading." Figure 7.4.27b shows the calculated progression of the pressure distribution for an airfoil representative of an early generation of this general design approach, the RAE 2822 (Cook, McDonald, and Firmin, 1979), an airfoil with the same 12% maximum thickness as the NACA 4412. Note that the peak upper-surface velocity remains near the leading edge well into the transonic regime rather than being located near mid-chord, as it was on the NACA 4412. The shock first appears well forward and then moves aft rather than first appearing near mid chord and not moving much, as was the case with the NACA 4412.

Richard Whitcomb of NASA developed some of the earliest airfoils of this type. His first such airfoil was designed to operate with the shock quite far aft and incorporated a slot in a manner similar to a high-lift slotted flap, but smoothly faired as in Figure 7.4.19a. Subsequent airfoils aimed for a similar overall pressure-distribution architecture, but were unslotted (Whitcomb, 1974). For this general design approach, Whitcomb coined the term "supercritical airfoil," referring to the fact that the drag rise was delayed well past the critical Mach number. Whitcomb's supercritical airfoils were more aggressively aft-loaded than the RAE 2822 example we looked at in Figure 7.4.27b.

The formation of a shock in the flow field causes drag to increase through two mechanisms:

1. There is the dissipation in the shock itself, which makes itself felt as an increase in pressure drag at the airfoil surface.
2. The increase in pressure through the shock, even when it does not cause separation, thickens the boundary layer as it passes through the shock, which also increases its rate of growth from the shock aft, and increases the pressure drag due to the boundary-layer displacement effect.

The results of this can be seen in Figure 7.4.28, which shows the drag-rise curves of the NACA 4412 and the RAE 2822. At a constant C_l of 0.72 for which we compared the pressure distributions in Figure 7.4.27, both airfoils maintain a reasonably low and nearly constant drag level across the low-Mach range, until a point is reached for each airfoil where the drag rise begins, gradually at first, and then quickly becoming quite steep. For the NACA 4412, the drag rise begins at a Mach number of about 0.60, while for the RAE 2822 it is delayed to about 0.68. Over the first few tens of counts of drag rise, shock drag typically accounts for roughly two thirds of the increase over the low-Mach drag level.

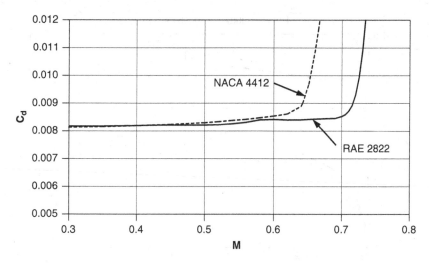

Figure 7.4.28   Drag-rise curves for NACA 4412 and RAE 2822 airfoils at C_l of 0.72, calculated by the MSES code (Drela, 1993)

The other roughly one third is accounted for by an increase in viscous drag associated with the increasing punishment imposed on the boundary layer as the shock gets stronger.

We've seen how transonic flow brings with it a huge change in the pressure distribution on an airfoil. It also brings a dramatic increase in the sensitivity of the pressure distribution to small changes in the effective shape of the airfoil, and thus much more sensitivity to the boundary-layer displacement effect that we discussed in Section 4.1.3 and that we saw illustrated for subcritical airfoil flow in Figure 7.4.5. We make a similar comparison for transonic flow in Figure 7.4.29. The boundary layer displacement thickness is largest on the upper surface near the trailing edge, where it affects the circulation and the overall lift, and it does so much more strongly than it did in the subcritical case in Figure 7.4.5.

The interaction between the shock and the boundary layer plays an important role in the physics of transonic airfoil flows. We can see some of the effects of this interaction in the pressure distributions through the shocks in Figure 7.4.29:

1. The pressure rise through the shock is "smeared" over some distance along the airfoil surface. The pressure rise thickens the boundary layer, as we've already mentioned, and the thickening of the boundary layer acts like adding a curved "ramp" or "wedge" to the effective shape of the airfoil, which has the effect of making the shock "fan out" at its base into compression waves (Figure 7.4.30) and smearing the pressure rise over a distance roughly proportional to the incoming boundary-layer thickness. This smearing of the shock pressure rise is clearly visible in the surface pressure distribution in the viscous case in Figure 7.4.29.

2. The pressure downstream of the shock is not as high as that which a normal shock would produce at the same upstream Mach number. This is another result of the effective viscous "ramp" or "wedge" associated with the thickening of the boundary layer. In addition to causing the shock to fan out; the "wedge" causes it to become oblique, so that the normal-shock relations don't apply, and the post-shock pressure is moved closer to the sonic level, C_p^*, as can be seen in the viscous case in Figure 7.4.29.

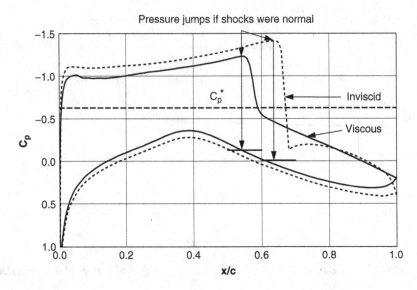

Figure 7.4.29 Effect of boundary-layer displacement on the pressure distribution of the RAE 2822 airfoil in transonic flow, $R = 10^7$, $M = 0.74$, $\alpha = 2.221°$, calculated by the MSES code

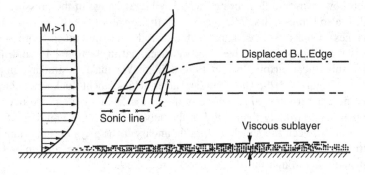

Figure 7.4.30 Sketch of transonic shock/boundary-layer interaction. From Sirieix, Delery, and Stanewsky, (1980). Used with permission of Springer Science+Business Media

Other interesting aspects of transonic airfoil shocks can be seen in Figure 7.4.29. The shock pressure jumps that would have occurred, had the shocks been normal shocks with the upstream Mach numbers M_1 consistent with the pressure distributions, are indicated by arrows. In the viscous case, the pressure rise falls far short of the normal-shock value because of the "wedge" effect we just discussed. In the inviscid case, the pressure rise falls a little bit short, but for a different reason. A shock impinging on a smooth surface in inviscid flow must impinge normal to the surface, and the pressure rise should be the normal-shock pressure rise. But when the surface is curved, there is a singularity at the point where the shock impinges (the Zierep singularity, see Oswatitsch and Zierep, 1960), and the flow downstream of the shock very quickly reexpands to a higher Mach number. The numerical solution for the inviscid case in Figure 7.4.29 didn't have enough resolution

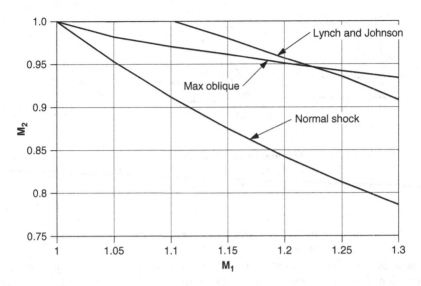

Figure 7.4.31 Correlations of downstream Mach number in transonic shock/boundary-layer interactions. (From Jou and Murman, (1980 ("Max oblique") and Lynch and Johnson, 1987)

to show the singular dip to the normal-shock value. There is also some numerical smearing of the shock in the inviscid case.

In the viscous shock/boundary-layer interaction, the downstream pressure varies somewhat as a function of the strength of the shock. By examining experimental data, Jou and Murman (1980) concluded that the downstream pressure corresponds closely to that of an oblique shock on a wedge with the maximum deflection angle for an attached shock (see Section 3.11.2 for a discussion of oblique shocks). Lynch and Johnson (1987) proposed a similar kind of correlation for the downstream pressure, with a slightly different result. The two are compared in Figure 7.4.31 in terms of the inviscid Mach number M_2 downstream of the shock inferred from measured surface pressures. It is clear that transonic airfoils don't "shock down" to anywhere near the normal-shock level, and for the weak-shock range that is of interest in cruise ($M_1 < 1.20$) the downstream Mach number M_2 will be between 0.95 and 1.00.

As either Mach number or angle of attack increases, the shock generally becomes stronger, and when it is strong enough it causes immediate separation of the boundary layer. Because the pressure rise through the shock is smeared over a distance that scales roughly with the thickness of the incoming boundary layer, the severity of the pressure gradient through the shock varies so as to compensate for the effects of Reynolds number, with the result that shock-induced separation has practically no Reynolds-number dependence. Separation first appears at an upstream Mach number M_1 very close to 1.30, practically independent of Reynolds number, and with only a slight dependence on the shape factor of the incoming boundary layer. For M_1 between 1.30 and 1.40, the boundary layer reattaches to the airfoil surface, and the separation is thus limited to a local bubble beginning at the base of the shock, as illustrated in Figure 7.4.32. For M_1 above 1.40, there is usually no reattachment, and the final flow pattern in Figure 7.4.32 is generally associated with severe buffeting. A more

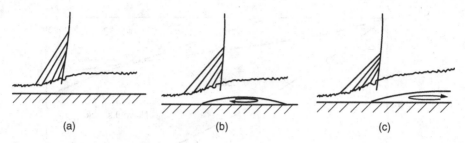

(a)   (b)   (c)

Figure 7.4.32 Illustration of the progression of separation at the foot of the shock on an airfoil. (a) $M_1 <= 1.30$: Attached flow. (b) $1.30 < M_1 <= 1.40$: Separation bubble at foot of shock. (c) $M_1 > 1.40$: Separation without reattachment

detailed discussion of transonic shock/boundary-layer interaction can be found in Sirieix, Delery, and Stanewsky (1980).

The progression of separation at the foot of the shock described above can be complicated by interaction with separation starting at the trailing edge and moving forward. Because these two types of separation can progress either separately or simultaneously, the number of possible scenarios is large. Which one actually happens depends on the details of the particular airfoil's pressure distribution. Pearcey, Osborne, and Haines (1968) exhaustively described the possibilities and divided them into those with separation at the foot of the shock only (Model A) and those with aft separations or combinations (Model B). Their diagram of the possibilities is shown in Figure 7.4.33.

The problem of designing an airfoil for good performance under transonic conditions, for example, achieving a high sectional L/D at a high Mach number, is effectively that of designing for the highest C_l at which a reasonably low drag level can be had. Thus it is similar to the problem of designing a single-element airfoil for maximum lift in subcritical flow that we discussed in Section 7.4.3, where we saw that an optimum design represents a tradeoff between increasing the magnitude of the negative C_p in the "rooftop" on the forward part of the chord and delaying the start of the pressure recovery. The limiting constraint in the general high-lift case was delaying boundary-layer separation to the trailing edge. In the transonic case, we have the same boundary-layer separation constraint, but we also have the requirement for "reasonably low" drag, which is effectively an active constraint as well. In particular, that means keeping shock drag under control over the required range of operating conditions while trying to maximize usable lift.

It is possible to design an airfoil with the required region of supersonic flow and a shock-free recovery at one operating condition. But shock-free design has not proven to be practical because it provides low drag over too narrow a range of Mach and C_l. At conditions other than the design condition, a shock-free airfoil tends to have a stronger shock and significantly higher drag than a conventional transonic airfoil would. The conventional compromise is to design for an upper-surface pressure distribution with the general features illustrated in Figure 7.4.34, including a weak shock, over the desired transonic part of the operating envelope. Shock drag increases rapidly for upstream Mach numbers above about 1.15–1.20, which effectively limits the value of negative C_p that can be tolerated just ahead of the shock. Note that a shock of this strength is not even close to causing separation (recall Figure 7.4.32a).

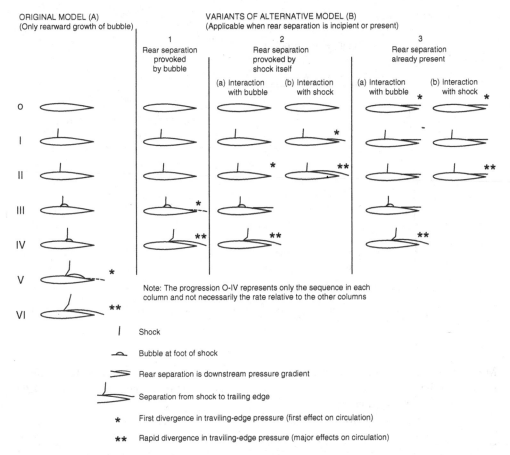

ORIGINAL MODEL (A)
(Only rearward growth of bubble)

VARIANTS OF ALTERNATIVE MODEL (B)
(Applicable when rear separation is incipient or present)

Note: The progression O-IV represents only the sequence in each
column and not necessarily the rate relative to the other columns

| Shock
Bubble at foot of shock
Rear separation is downstream pressure gradient
Separation from shock to trailing edge
* First divergence in traviling-edge pressure (first effect on circulation)
** Rapid divergence in traviling-edge pressure (major effects on circulation)

Figure 7.4.33 Possible sequences in the development of separation at the foot of the shock and separation moving forward from the trailing edge, shown as successive stages as either Mach number or angle of attack increases. Turbulent boundary layers are assumed. From Pearcey, Osborne, and Haines, (1968). Published by AGARD

Downstream of the shock, it seems to be advantageous to provide a slight re-expansion before the final aft pressure recovery begins. This increases lift and is probably also favorable to the boundary layer. Although the shock does not cause separation, it is sufficiently traumatic to the boundary layer that it, seems to be helpful to provide the re-expansion to allow the boundary layer to "relax" before the start of the final aft pressure recovery. Downstream of the re-expansion, a slightly convex recovery is favored because it also increases lift. This differs from what we saw in Section 7.4.3, where maximum lift in subcritical flow favored a concave recovery. In that case, a concave recovery allowed more negative C_p at the start of the recovery, while in the transonic case, negative C_p is limited by shock drag.

It is also favorable to design the supersonic rooftop with a slight built-in deceleration or adverse pressure gradient. This increases lift by allowing more-negative C_p near the leading edge, and it facilitates a benign progression of shock position and strength with

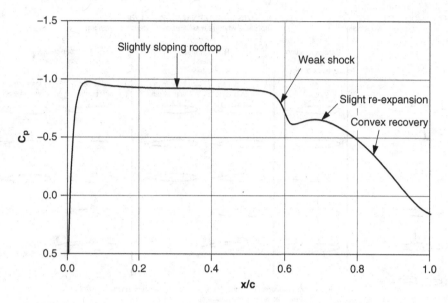

Figure 7.4.34 Illustration of desired features of an upper-surface pressure distribution for efficient transonic cruise

Mach and C_l. When all these considerations are taken into account, it seems to be optimum to position the shock just aft of mid chord at the C_l at which the maximum sectional L/D is desired at the targeted cruise Mach number.

Practical considerations for design are discussed in Section 7.4.10.

7.4.9 Airfoils in Ground Effect

The presence of a ground plane below an airfoil alters the flow relative to free-air conditions at the same angle of attack. The effects generally become stronger with decreasing height above the ground plane, and the height must typically be less than about one chord for the effects to be large.

In the simplest model, which should be valid for large h/c, the lift is represented by a vortex at the airfoil quarter-chord, and the effect of the ground is represented by an image vortex below the ground. The presence of the image reduces the effective freestream velocity seen by the airfoil from U_∞ to a lower value U_{eff} and thus reduces the lift at a given angle of attack. If we ignore any change to the effective camber of the airfoil and assume that the airfoil shape and angle of attack determine Γ/U_{eff}, we obtain the following expression for the lift relative to its free-air value at the same angle of attack:

$$C_l/C_{l\infty} = 1/[1 + (C_{l\infty}c)/(8\pi h)]^2. \qquad (7.4.2)$$

Note that the C_l ratio depends on both $C_{l\infty}$ and h/c, and that the effect disappears (the ratio goes to one) as either h/c $\to \infty$ or as $C_{l\infty} \to 0$. The predicted reduction in lift is plotted as the solid curve in Figure 7.4.35 for $C_{l\infty} = 1.30$.

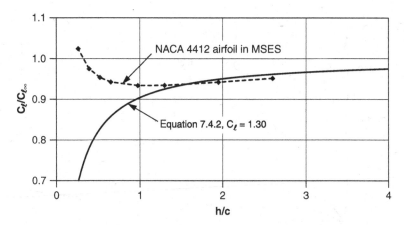

Figure 7.4.35 Effect of a ground plane on the lift of an airfoil at a fixed angle of attack. Comparison of the simple model of Equation 7.4.2 with MSES calculations for the NACA 4412 airfoil at $\alpha = 7.622°$ ($C_l = 1.30$ out of ground effect), R = 10 million, M = 0.10, with viscous effects included on the airfoil surface but not on the ground plane

Of course, a real airfoil has bound vorticity distributed along the chord and has thickness, and thus the ground alters the effective camber and angle of attack as well as U_{eff}. These effects will be more complicated and will not generally disappear as $C_{l\infty} \to 0$. As an illustration of what these more-realistic effects can look like, Figure 7.4.35 also shows results calculated by the MSES code for the NACA 4412 airfoil, with viscous effects included on the airfoil surface but not on the ground plane. At the larger heights, the lift loss is roughly the same as that predicted by the simplest model, Equation 7.4.2. At the lower heights, the lift reduction reverses, and lift increases, presumably because the ground increases the effective camber or angle of attack. Such deviations from the simple model will of course depend on the chordwise lift distribution. And for small heights, the boundary layer on the ground plane could alter these results significantly. (Note that a local unsteady boundary layer will develop even on a stationary ground plane in zero wind, because of the local unsteady velocities produced as the airfoil passes by.)

There is a widespread notion that ground effect produces an "air cushion" of high pressure under the airfoil, which increases lift. In the NACA 4412 example in Figure 7.4.35, lift was indeed increased for h/c less than about 0.3. Pressure distributions for three of these cases are plotted in Figure 7.4.36. At h/c = 1.95, there is a decrease in lift, coming from a pressure increase that is greater on the upper surface than on the lower surface, consistent with the simplified model that explains the effect in terms of a reduction in effective freestream velocity. At the lowest h/c of 0.26, the pressure increases have grown larger on both surfaces, but the increase on the lower surface is now larger than that on the upper surface, so that there is a small lift increase. The change in loading is not concentrated toward the leading edge, and thus is consistent with a change more in the effective camber rather than in the angle of attack. From what we see in this case, the "air cushion" image isn't a bad one, as long as you keep in mind that much of the effect on lift from the lower surface is offset by a pressure increase on the upper surface that is almost as large as that on the lower surface.

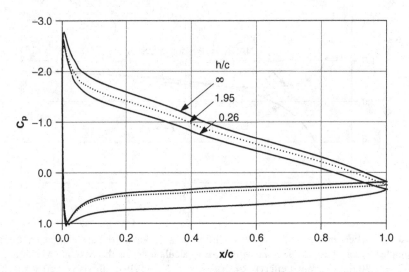

Figure 7.4.36 Effect of a ground plane on the pressure distribution on an NACA 4412 airfoil for three of the cases in Figure 7.4.35

In the case of a 3D wing, the 3D effects of the ground plane often overwhelm the 2D effects we've just discussed. We'll look at the 3D effects on lift in Section 8.2.4 and the 3D effects on induced drag in Section 8.3.9.

7.4.10   Airfoil Design

Different mission performance requirements place different amounts of importance on aerodynamic performance measures such as high maximum lift, low drag, and high-Mach-number capability. In our discussions so far, we've considered these aerodynamic objectives separately, looking at some of the purely aerodynamic considerations that apply to designing a low-speed airfoil for maximum lift, a low-drag airfoil with laminar flow, or an airfoil for high L/D at high subsonic Mach numbers. Of course, practical airfoil design should take into account much more than just these single objectives. An airfoil on a practical flight vehicle should perform well not just at one flight condition, but should provide favorable performance over the full range of conditions encountered during the vehicle's mission. The aerodynamic design should therefore take into account multiple operating conditions, including those in which high-lift devices may be deployed. Then there are nonaerodynamic requirements. In addition to providing aerodynamic performance, an airfoil must enclose wing structure, fuel, and mechanical systems. Sometimes an airfoil optimized for purely aerodynamic considerations will already meet these other requirements, but usually not. In the usual case, the best shape is a compromise between aerodynamic and nonaerodynamic objectives and constraints.

In the early era from the 1800s to the 1930s, airfoil design was a cut-and-try affair, with the airfoil shape as the input, and the performance, as determined by wind-tunnel testing and flight experience, as the output. Progress was slow, and the performance levels that were achieved were not high by today's standards.

We've seen that the lift and drag performance of an airfoil is largely determined by the behavior of the boundary layer, which means that the pressure distribution is hugely important. In the second era of airfoil design, from the 1930s to the 1990s, controlling the pressure distribution through *inverse design* (solving for the shape that yields a specified pressure distribution) led to major advances in the ability to target particular kinds of performance. An early example we already considered is the NACA 6-series airfoils designed for low-drag laminar flow. In the early years, the state of computational methods was such that inverse design was limited to incompressible inviscid flow, but this sufficed for low-Mach-number applications. It was not until the 1970s that methods became available for routinely computing inviscid transonic flows. Turning this capability into a practical design tool required adding viscous effects, through boundary-layer coupling, because the pressure distribution in transonic flow is very sensitive to small changes in the effective shape of the airfoil, as we saw in Figure 7.4.29. The application of viscous transonic inverse-design methods led to dramatic improvements in transonic airfoil performance.

The inverse-design method in general entails some interesting difficulties. One has to do with the nature of the inverse problem itself, and that is that only particular target pressure distributions lead to realizable airfoil shapes that are not either open or crossed over at the trailing edge. One way of dealing with this problem is to allow the solution algorithm to alter the target pressures in some arbitrary way, for example, by adding a linear ramp or some other distribution to the pressures, and to solve for how much of the alteration is needed to make the resulting geometry realizable. How close the resulting design comes to achieving the desired pressure distribution then depends on how close to realizable the target was. Another approach is called *modal inverse design*, which starts with a "seed" airfoil shape that is then modified by the addition of particular shape functions, or "modes," whose amplitudes are solved for by optimization, so as to minimize some measure of the deviation between the actual pressure distribution and the target pressure distribution. The shape-changing modes are purposely defined so as not to be able to "open" the trailing edge, and the resulting airfoil automatically has a realizable shape. Both of these methods of dealing with the realizability problem are available as options for doing inverse design in the ISES and MSES codes (Drela, 1993).

Since the 1990s, CFD-based optimization has replaced inverse design as the method of choice for designing an airfoil, or at least for the final refinement of a design. The objective function that is minimized can include almost any weighted combination of aerodynamic performance measures and airfoil shape characteristics. The aerodynamic performance objective almost always includes multiple operating conditions, since airfoils optimized for only one condition usually perform poorly at "off-design" conditions. Constraints can also be imposed on the aerodynamic characteristics (pitching moment, for example) or on the shape (maximum thickness or some other measure of the structural fitness of the shape, for example). Often the things we might ordinarily think of as constraints can be handled equally well as part of the objective instead.

With the advent of optimization, it might seem that airfoil design has become a "pushbutton" activity: Define the objective and the constraints, and turn the code loose to produce the best possible airfoil. But airfoil optimization is not usually that simple in practice. Numerical optimizers have an uncanny ability to find and exploit weaknesses in the formulation of the problem, often with the result that the optimized solution is practically useless. The weakness can be in the physical modeling, as in some flaw in the physical fidelity of the CFD method, or it can be flaw in the choice of objectives or constraints. It usually takes considerable skill to use airfoil optimization in a way that produces an excellent design for the intended purpose.

Regardless of whether inverse design or optimization is used to generate a design, there are lessons that crop up repeatedly and seem to be universal. One is that the pressure distribution on the upper surface has much greater leverage over the performance (drag or maximum lift) than the pressure distribution on the lower surface does. In the case of drag, this is presumably because of the higher average velocity over the upper surface under lifting conditions. In the case of maximum lift, it is because boundary-layer separation on the upper surface is determined by the upper-surface pressures. Thus for either low drag or high lift, a deviation from the aerodynamic optimum pressure distribution on the upper surface carries a heavier penalty than a similar deviation on the lower surface. The upper surface pressure distribution is therefore more or less "inviolable," while the lower-surface pressure distribution is more "negotiable" and can be compromised, for structural or other nonaerodynamic reasons, with only minimal aerodynamic performance penalty.

7.4.11 Issues that Arise in Defining Airfoil Shapes

Defining the shape of an airfoil would seem to be a simple proposition: Either use analytic functions, as in the NACA 4- and 5-digit series, or use a string of defining points combined with some method of interpolating smoothly between them. But interesting issues can still arise that are worth discussing here.

Decomposing an airfoil shape into a *camber line*, or mean line, and a *thickness shape* was common in much of the early work on airfoils, including the various NACA series. As we saw in Section 7.4.1, the aerodynamic effects of camber and thickness (and angle of attack) are exactly separable only in the linear limit of small surface slope. Still, it was common practice to design camber lines and thickness shapes separately as late as the NACA 6 series of low-drag laminar airfoils.

The traditional way of lofting an airfoil by combining a camber line and a thickness shape is described in Abbott and von Doenhoff (1958). The leading and trailing edges are defined as the ends of the camber line, and the chord line is defined as the straight line joining the leading and trailing edges. The thickness shape is added normal to the camber line, as shown in Figure 7.4.37. The circle defined by the leading-edge radius has its center not on

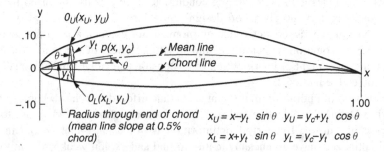

Figure 7.4.37   Traditional way of combining a camber line and a thickness shape. From Abbott and von Doenhoff, (1958). Used with permission of Dover Publications, Inc.

the chord line, but on a line drawn tangent to the camber line at the leading edge. Thus if the slope of the camber line at the leading edge is nonzero, the center of the circle will be displaced vertically from the chord line, and the nose of the airfoil will protrude slightly ahead of the "leading edge."

There are two formal problems with this procedure:

1. Some camber lines have infinite slope at the leading edge, in which case the procedure has to be "fudged" as explained in Abbott and von Doenhoff (1958), and
2. An airfoil surface defined in this way can "double back" on itself if the half-thickness exceeds the local radius of curvature of the camber line, but this is something that practically never occurs.

A more basic objection to the traditional recipe is that there's no real reason to use such a complicated process. It certainly isn't justified by linear theory, since for the small surface slopes assumed in the theory there's no significant difference between adding the thickness normal to the camber and simply adding the y coordinates. And for airfoils with practical thickness and camber, the difference is typically significant only in the neighborhood of the leading edge, where linear theory doesn't apply anyway. So it makes just as much sense theoretically, and much more sense practically, just to add the y coordinates.

Now that most new airfoils are designed by CFD calculations, either by inverse design or optimization, as we discussed in Section 7.4.10, we tend not to deal with separate camber lines and thickness shapes anyway. If we forego the decomposition into camber and thickness, defining an airfoil shape is just defining a function y(x) that is double valued so as to have upper and lower surfaces. We generally like an airfoil to have a well behaved pressure distribution in subsonic flow, which dictates some desirable properties for y(x). In potential flow, a discontinuity in either slope or curvature produces a singularity (infinite slope) in the pressure distribution (This is discussed in Section 9.2.1, and two cases are sketched in Figure 3.10.2b,c). In real, viscous flow, a discontinuity in slope or curvature doesn't produce singular behavior, but it can still produce a large local pressure gradient, which is best avoided. It is thus prudent to require that the surface be continuous to the second derivative (C2). This is not a difficult requirement to meet.

What can cause some difficulty, however, is handling the leading-edge region. The usual blunt leading edge has a combination of concentrated high curvature and a leading edge point where the slope is infinite, which is not that easy to represent with simple conventional curve-fitting procedures.

A general method of representing body shapes has been proposed by Kulfan (2008) that provides one way of dealing with the leading-edge issue. The function y(x) defining the cross section of a 2D body like an airfoil is defined as the product of a *class function* that gives the body its basic character, and a *shape function* that defines the specific shape. The idea is that with the class function providing the basic blunt leading-edge shape, the shape function should be a function that is easier to curve-fit. A simple class function for an airfoil with a blunt leading edge and a wedge trailing edge is $\sqrt{x}(1 - x)$, where x goes from 0 to 1 from the leading edge to the trailing edge. The $\sqrt{x}$ provides a blunt leading edge with a finite leading-edge radius, and the $(1 - x)$ provides a wedge trailing edge, as shown in Figure 7.4.38. The airfoil shape is then given by

$$y = f\sqrt{x}(1 - x) + y_{TE}x,$$ (7.4.3)

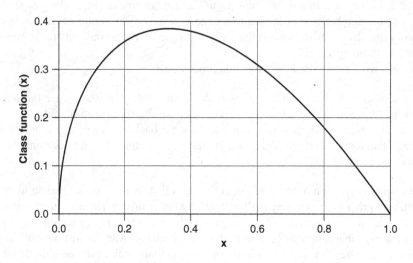

Figure 7.4.38 The class function $\sqrt{x}(1-x)$ for fitting airfoil shapes according to Equation 7.4.3

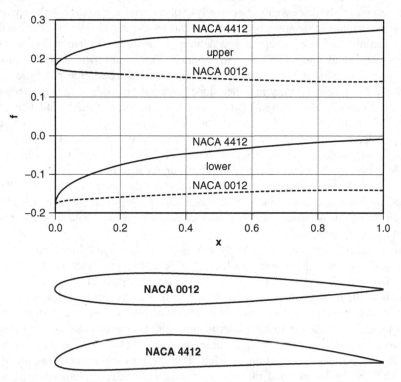

Figure 7.4.39 Airfoil shapes for NACA 0012 and NACA 4412 and corresponding shape functions $f(x)$ defined by Equation 7.4.3

with separate shape functions f_u and f_l for the upper and lower surfaces. Separate values of y_{TE} for the upper and lower surfaces define the thickness of the trailing edge. The value $f(1)$ determines the slope at the trailing edge, and $f(0)$ determines the leading-edge radius. Thus to have continuity of curvature at the leading edge requires $f_u(0) = -f_l(0)$. Figure 7.4.39 shows the airfoil shapes and the functions $f(x)$ defined by them by Equation 7.4.3 for the symmetrical NACA 0012 and cambered NACA 4412 (lofted by adding y coordinates, not by the more elaborate procedure). For the symmetrical case, $f(x)$ is well behaved and would be easy to curve-fit to high accuracy. In the cambered case, however, things are not so well behaved. Note that $f(0)$ is the same regardless of camber, since the leading-edge radius is determined only by the thickness shape. But the NACA 44xx camber line has nonzero slope at the leading edge, which causes $f(x)$ to have infinite slope there, though $f(x)$ is reasonably well behaved everywhere else. So when the camber line at the leading edge has nonzero slope, Equation 7.4.3 doesn't completely eliminate the difficulty of fitting the leading edge, and special measures must be taken to handle the infinite slope there.

8

Lift and Wings in 3D at Subsonic Speeds

In this chapter, we extend the discussion of lift from 2D to 3D take up the topics of the flow around a 3D wing, the lift distribution on a 3D wing, the induced drag, wingtip devices, and the manifestations of lift in the extended flowfield. Finally, we'll delve into some of the interesting issues that arise when wings are swept.

8.1 The Flowfield around a 3D Wing

The flow around a 3D wing must differ in some basic ways from the flow around a 2D airfoil, simply because of the finite span and the resulting flow gradients in the spanwise direction. In this section, we'll first describe the general features of 3D wing flowfields, and then we'll look at how they can be explained. The classical approach (starting with the early work of Prandtl and others, see historical sketch by Giacomelli and Pistolisi, in Durand, 1967a) looks at the distribution of bound vorticity and the vorticity in the wake and deduces the velocity field everywhere else using the Biot-Savart law. Though this yields correct results if the correct vorticity distribution is used, it is not a real physical explanation in the cause-and-effect sense, as we argued in Sections 3.3.9 and 7.2. So we'll also seek an explanation based on the local balance of force and acceleration, that is, the interaction of the pressure and velocity fields. We constructed explanations of this type for the generic flow around an obstacle in Section 5.1 and the flow around a 2D lifting airfoil in Section 7.3.3. Even in those relatively simple flow situations, however, we found that qualitative arguments alone did not enable us to predict the flow a priori. Instead, we had to settle for explaining things "after the fact," based on prior knowledge of what the pressure and velocity fields look like. In the case of a 3D wing, we will also have to rely heavily on prior knowledge of the flow structure.

8.1.1 General Characteristics of the Velocity Field

The flow around a 3D wing is similar in some ways to the flow around a 2D airfoil, so to start the discussion, let's review the relevant features of the flow in the 2D case. In the flow

Understanding Aerodynamics: Arguing from the Real Physics, First Edition. Doug McLean.
Images and Text: Copyright © 2013 Boeing. All Rights Reserved. Published 2013 by John Wiley & Sons, Ltd.

over a lifting 2D airfoil, the velocity disturbances produced by the airfoil die out rapidly in all directions, including downstream. Downstream of the airfoil, the only significant velocity "signature" of the lift production is the downwash field, which carries a flux of downward momentum across any vertical plane, corresponding to half of the lift. With increasing distance downstream, this downwash spreads out rapidly in the vertical direction and becomes very diffuse, but the flux of downward momentum remains constant. As we discussed near the end of Section 7.3.4, the flow around a 2D airfoil in the inviscid case is reversible, in that an onset flow that is uniform in the limit far upstream becomes uniform again in the limit far downstream, and the same flow pattern and pressures would arise if the flow were run in the reverse direction. In a viscous flow in the attached-flow regime, the flow outside the boundary layer and wake still follows the reversible pattern quite closely. In addition to the near-reversibility of the general flow pattern, there is also very little permanent vertical displacement of streamlines between upstream and downstream (none in the inviscid, shock-free case). And, of course, all these flow features are by definition uniform in the spanwise direction.

Now consider a 3D wing of finite span, with moderate-to-high aspect ratio, operating in the attached-flow regime. At any station along the span of the wing other than very close to the tips, the chordwise distributions of pressure and boundary-layer development are not much different from those of a 2D flow, and the streamlines projected in a longitudinal plane look qualitatively like the flow around a 2D airfoil, as illustrated in Figure 8.1.1. But this projected view misses an important aspect of the 3D flow; that is, that the streamlines of the 3D flow don't generally lie in planes as those of a 2D flow do. A key part of what distinguishes the 3D flow from the flow around a 2D airfoil is a significant out-of-plane component to the motion. This out-of-plane component would be difficult to discern if it was shown in a perspective view like Figure 8.1.1, but it is important nonetheless.

We can visualize the 3D flowfield more clearly in terms of velocities projected in planes perpendicular to the freestream, as illustrated in Figure 8.1.2. This cross-stream velocity field develops in conjunction with a pressure field that is nonuniform in the spanwise direction.

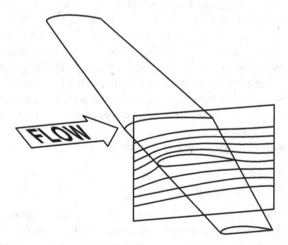

Figure 8.1.1 Flow around a 3D wing viewed in terms of streamlines projected in a plane perpendicular to the span is similar to flow around a 2D airfoil

Figure 8.1.2   Flowfield around a lifting wing illustrated by velocity vectors in a cross-flow plane. This general flow pattern is well established around the wing itself and persists for long distances downstream

The general pattern is characterized by downward flow in the area between the wingtips, upward flow outboard of the tips, outboard flow below the wing, and inboard flow above the wing. Note that these lift-induced velocities are not concentrated closely just around the wing itself or the wingtips, but are spread fairly diffusely over a wide area of the flowfield.

The streamwise development of the cross-stream velocity field in the 3D flow is quite different from anything in the development of a 2D airfoil flow. In the flow more than about one wingspan ahead of the wing, the velocity disturbances are small and are distributed diffusely, as they are in the 2D case. As we approach closer to the wing, a pronounced upwash appears ahead of the leading edge, as in the 2D case. As we pass behind the leading edge and over the wing itself, the general flow pattern shown in Figure 8.1.2 becomes well established. Behind the wing, the flow pattern continues to evolve, with velocities increasing in parts of the field and decreasing in others, but continuing to look qualitatively like Figure 8.1.2. At a distance on the order of a wingspan behind the wing, the flow will have settled into an asymptotic pattern, and then it changes only very slowly over long distances downstream. This is a key distinction between 3D and 2D: While the velocity disturbances in the 2D case begin to decrease immediately behind the airfoil and become very small and very diffuse far downstream, the cross-stream velocity field that develops in 3D persists for very long distances downstream. Another way to look at this distinction is in terms of reversibility: Both the 2D and 3D cases start with a uniform onset flow upstream, but while the 2D flow becomes uniform again downstream, the 3D flow becomes nonuniform, with persistent transverse velocities even at very large distances downstream. It is therefore not even close to being reversible like the flow around a 2D airfoil.

At the location of the wing itself, we have a well-established flow pattern in which the wing is flying through air that is already moving generally downward between the wingtips. Thus the wing can be thought of as flying in a downdraft, or *downwash,* of its own making. At the location of the wing itself, the downwash can in general vary considerably, both spanwise and chordwise. But on a high-aspect-ratio wing, we can simplify the picture: We can think of the downwash at the location of the wing as consisting of two parts, a 2D part that would be there if the local airfoil section were in a 2D flow at the same sectional lift (not the same geometric angle of attack), and a 3D part that is a result of finite span. In the limit of high aspect ratio, the 3D part of the downwash is constant along the chord at a given span station. The 3D downwash can thus be seen as a *downward shift in the apparent angle of attack* of each airfoil section along the wing, often called the *induced angle of attack*. For positive total lift, the effect of the induced angle of attack, integrated over the span, always corresponds to a reduction in the apparent angle of attack of the wing.

One consequence of the apparent downdraft in which the wing is flying is that a 3D wing generally requires a higher geometric angle of attack to achieve the same lift coefficient as a corresponding 2D airfoil, a fact we'll make use of when we attempt to explain features of the 3D flowfield. And because the downwash increases with angle of attack and thus subtracts progressively more from the lift, the lift slope of a 3D wing is generally less than that of a 2D airfoil section.

The other important result of the downwash is that the total apparent lift vector is tilted backward slightly. This backward component of the apparent lift is called *induced drag*, and the work done against it is reflected in the kinetic energy of the large-scale flow pattern. We'll discuss the physics and the theory of induced drag further in Section 8.3.

In Figure 8.1.2, we saw that the spanwise velocity components behind the wing are in the outboard direction below the wing and in the inboard direction above the wing. There is thus a mismatch, or jump, in the spanwise velocity, and this jump constitutes a vortex sheet that is shed from the trailing edge and convected downstream. The development of this *vortex wake* is our next topic. The induced drag and the presence of vorticity that is convected downstream are both earmarks of the general irreversibility of the 3D flow pattern.

8.1.2 The Vortex Wake

The trailing vortex wake is a distinctive feature of the lift-induced flowfield, and it plays a prominent role in discussions of induced drag and in the quantitative theory. The nature of the vortex wake and its role in induced drag have been a source of some serious misunderstandings, so we'll take extra care in the following discussion to point out the common misconceptions, to help ensure that we develop a correct understanding.

As we noted above, the vortex wake starts as a vortex sheet shed from the trailing edge of the wing as a byproduct of the establishment of the flow pattern shown in Figure 8.1.2. It is a necessary part of the flowfield because the wing cannot produce the general flow pattern of Figure 8.1.2 without also producing the jump in spanwise velocity between the streams that pass above and below the wing. Even if we model the flow around the wing as inviscid, a vortex sheet must be shed if the lift is nonzero. Milne-Thomson (1966, Section 3.31) describes the shedding of a vortex sheet from a body in 3D inviscid flow as the "bringing together of layers of air which were previously separated, and which are moving with different velocities." Of course, if a shed vortex sheet is wetted on both sides by air that has come from the freestream without any change in stagnation pressure or stagnation temperature, the velocity magnitude on both sides must be the same. So by "different velocities" Milne-Thomson means different flow directions. Farther along in the discussion, we'll attempt to explain how those different flow directions arise in the case of a lifting 3D wing.

Milne-Thomson's quote above provides interesting food for thought and merits a little digression. He's given us an intuitively appealing way to think of what's happening when the vortex sheet leaves the trailing edge, but "previously separated" in this context is ambiguous. If all it means is that the layers of air were *not together* prior to being joined, then there's no problem. But it could also be taken to imply that the layers of air coming together at the trailing edge were separated *from each other* at some location upstream, presumably where they attached to the wing at the attachment line. This more specific meaning wouldn't be precisely correct. On a wing of finite span, a layer of air can come only very close to attaching to the surface, but can't actually attach in a rigorous sense. Recall from our discussion in Section 5.2.2 that a finite attachment line is at best a band of approximate attachment, and that only discrete filaments of flow can actually attach to the surface at singular points. Often there is only one point of attachment, like the nodal point marked "N" on the nose of the fuselage of the simple wing-body combination sketched in Figure 5.2.4. In this case, strictly speaking, the layers of air that join at the trailing edge originated from the filament that attached at the nose of the fuselage, not from layers that were split apart by the wing. At any short distance above and below the trailing edge, however, we find layers of air that were similarly close together when they passed near the leading edge. When such layers come close to "joining" at the trailing edge, they generally will have experienced considerable spanwise displacement relative to each other, in addition to the longitudinal displacement that we discussed in connection with 2D flow in Section 7.3.1.

And now to return to our main topic. We've seen that the vortex wake starts its life as a free vortex sheet that seems to originate from the trailing edge of the wing. But the trailing edge cannot be the actual origin of the vorticity in the wake. Because vortex lines cannot end at a solid surface with a no-slip condition (except at singular points of attachment or separation, as we saw in Section 3.3.7), the vorticity in the wake must originate in the viscous or turbulent boundary layers on the upper and lower surfaces of the wing. Where does this lead? If we look at all of the vorticity present in the 3D flow in the boundary layers and the wake, we see a very complicated picture, but it can be simplified greatly if we boil it down to its essentials. With regard to the global flowfield, what really matters is the net vorticity at any station on the wing or wake, as seen in a local plan view. At a station on the wing planform, the net vorticity would thus be defined by integration through both the upper and lower surface boundary layers; and at a station on the wake, it would be defined by integration through the entire viscous layer. The complicated distribution of vorticity through the viscous layer at a given station is thus replaced by a single vector value that is much easier to visualize.

Viewed thusly in terms of net vorticity, the shed vortex sheet is actually a continuation of the bound vorticity associated with the lift of the wing, which we discussed in Section 7.2 in connection with lift in 2D. This view in terms of net vorticity was option 3 of the ways of looking at bound vorticity that we identified in that discussion, illustrated in Figure 7.2.1c. In the 3D case, as the lift per unit span decreases in the outboard direction along the span, the circulation and total bound vorticity flux must also decrease. The vorticity representing this loss in total strength cannot just disappear and is shed from the trailing edge into the flowfield, supplying the vorticity that forms the vortex wake.

Now imagine the net vorticity on the wing and in the wake as an array of vortex filaments. These filaments take on a general horseshoe shape, as shown in Figure 8.1.3. Since each filament of this system forms a horseshoe, if we take a cut through the wake anywhere downstream of the wing, at the plane marked A, for example, the filament will be cut in two places, and the flux of vorticity passing through the cut will be equal and opposite

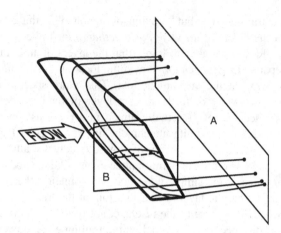

Figure 8.1.3 Bound and trailing vorticity of a lifting wing viewed as vortex filaments. The plane marked "A" illustrates how these filaments would be cut by a transverse plane behind the wing, and the plane marked "B" illustrates how they would be cut by a longitudinal plane through the wing

in the two places. (Recall from Section 3.3.7 that a vortex filament is a construct that carries the same flux of vorticity along its entire length.) Thus it is clear from the horseshoe configuration of the vortex system that the total vorticity flux passing through a cut through the whole wake is zero. We could arrive at the same conclusion by placing a closed contour in the cut such that it encloses the entire wake, and invoking Stokes's theorem on a capping surface that bulges out ahead of the wing and thus cuts none of the vorticity. The circulation around the closed contour must then be zero, and there can then be no net vorticity flux through any cut across the entire wake.

Another conclusion that follows from the horseshoe configuration of the system is that if we take a cut through the wing anywhere along the span, at the plane marked B in Figure 8.1.3, for example, the total fluxes of vorticity shed from the trailing edge on opposite sides of the cut will be equal and opposite, and their magnitudes will equal the flux of bound vorticity at that span station. Then, as a special case, we can say that when the lift distribution is laterally symmetrical, the total vorticity flux of the sheet shed from each side must equal the flux of bound vorticity at the center.

Like the boundary layers in which it originated, the wake shear layer is a real physical shear layer filled with small-scale turbulent motions. The idealized inviscid theories model the shed vortex wake as a thin vortex sheet of the kind we discussed in Sections 3.3.7 and 3.3.8, and illustrations often show it that way for simplicity. In all of the discussion that follows, "vortex sheet" can be thought of as referring to either a real physical shear layer or to an idealized thin sheet.

The development of the vortex sheet after it leaves the trailing edge is illustrated in Figure 8.1.4. The vortex lines in the sheet leave the trailing edge and follow the general direction of the flow downstream. In the case of the ideal thin vortex sheet, the vortex lines are aligned with the mean of the velocity vectors above and below the sheet, as in Figure 3.3.8c. Within the first couple of wingspans downstream, the sheet generally rolls up toward its outer edges to form two distinct vortex cores. (This is the general pattern

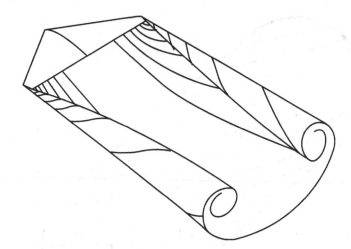

Figure 8.1.4 Development of the vortex wake downstream of a lifting wing. The lines drawn on the sheet are the vortex lines of a continuous distribution of vorticity in the sheet

for a wing in the "clean" condition, flaps up. The flaps-down pattern is more complicated, with cores forming behind flap edges as well behind the wingtips.) Although the vortex cores are distinct, they are not as concentrated as they are sometimes portrayed, because a considerable amount of air that was initially nonvortical is entrained between the "coils" of the spiral formed by the sheet during rollup.

In simplified theoretical models, such as the Trefftz-plane theory we'll discuss in the next section, the vortex lines are assumed to stream straight back from the trailing edge in the direction of the freestream, and the deformation and rollup of the wake are not represented. The classic argument is that the assumption of a nondistorting wake is valid in the limit of zero lift on the wing. For finite lift, however, it is kinematically impossible for the wake sheet to remain undistorted. Obviously a nonuniform downwash field behind the wing will distort the wake, and downwash fields behind lifting wings are generally nonuniform. Even in the case of an elliptic spanwise load distribution, which we'll see in the next section ideally produces a uniform downwash "contribution" from the trailing vortex wake, the "contribution" from the bound vortex is nonuniform, and the sheet must still distort and ultimately roll up. Even if we could find a situation in which the downwash was uniform, however, it would still be impossible for the wake to remain undistorted. This is sometimes attributed to an "instability" (Milne-Thomson, 1966, for example, in Section 10.4 refers to the wake sheet as "unstable" but does not provide a detailed explanation, and in Section 3.31 hints that viscosity might play a role). Spalart (1998), however, has shown that it is not an instability, but a result of the basic kinematics of the sheet, associated with the singularity in its strength at the edge, due to the usual infinite slope of the loading at the tip. In the real world, of course, there is no singularity, but there still tends to be a high concentration of vortex strength, and real wake sheets still roll up at their edges.

Note in Figure 8.1.4 that in the early phase of wake rollup, the vortex lines are swept outboard toward the rolling-up edges of the vortex sheet. As the sheet rolls up into coils, the vortex lines become helical, and when rollup is complete, the vortex lines in the outer part

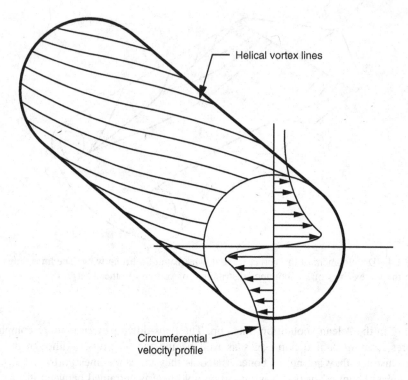

Helical vortex lines

Circumferential
velocity profile

Figure 8.1.5 Helical vortex lines and the associated velocities in the rolled-up vortex cores. The helical vortex lines line up closely with the helical streamlines

of the core appear as illustrated in Figure 8.1.5. In the idealized inviscid world, however, the vortex cores would appear as tightly wound spirals, continually stretching and tightening. In the real world, the coils of the spiral merge by turbulent diffusion, and diffuse vortical cores are formed in which the vortex lines are still helical. The helical vortex lines are very closely aligned with the helical streamlines. This is consistent with Crocco's theorem (Equation 3.8.8) and with the fact that the total-pressure loss associated with the viscous drag has diffused throughout the wake, so that the local total-pressure deficit is relatively small.

To get an idea of how the vorticity should be distributed within the cores, consider the initial distribution of vortex strength in the sheet that leaves the trailing edge and is eventually "wound up" into the cores. Figure 8.1.6 illustrates the distributions of bound and shed vorticity for a typical wing, showing that the shed vorticity is most intense at the tips and is much weaker inboard. Based on this, we should expect intense vorticity in the center of the rolled-up core and much weaker vorticity in the outer part.

Now let's look at the details of the velocity field in the mature wake far from the wing. The first feature we must note is that the pair of vortex cores descends slowly relative to the flight path of the airplane. This is often attributed to "mutual induction," but it is better to think of it simply as convection by the downwash behind the wing, which persists far downstream because there is practically nothing acting to stop it. We'll consider the cause-and-effect issues further in Section 8.1.4. As the cores descend, they carry with them a "descending oval" of fluid, as illustrated in Figure 8.1.7.

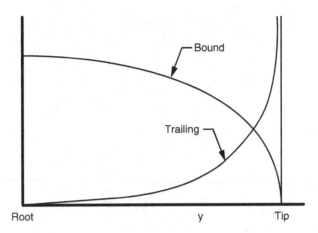

Figure 8.1.6 Sketch of typical distributions of the fluxes of bound and shed (trailing) vorticity, showing that shed vorticity is heavily concentrated near the tip

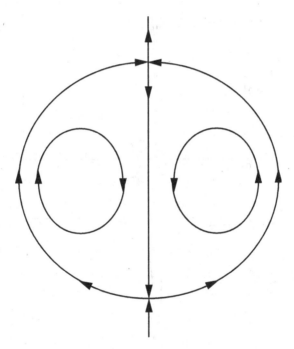

Figure 8.1.7 Sketch of the descending "oval" associated with the mature, rolled-up vortex wake, in terms of streamlines in the reference frame descending with the oval

The flow within much of the vortex core in each half of the descending oval is nearly axisymmetric. Figure 8.1.8 shows how the vertical velocity, the vorticity, and the pressure are distributed spanwise along a horizontal line through the centers of the cores (based on rollup calculations by Spalart and consistent with measurements by Widnall; see Spalart, 1998). Only the right half of the field is shown, with a symmetry plane assumed at $y/b_o = 0$.

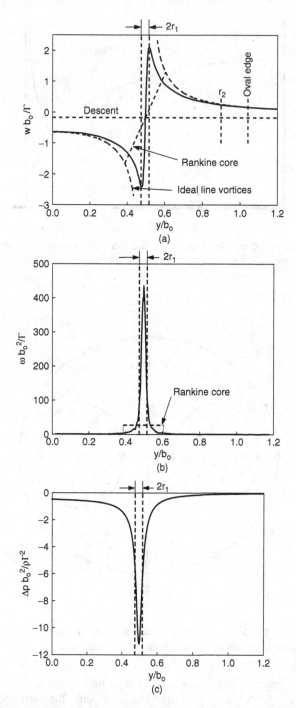

Figure 8.1.8 Details of the flow field of the mature vortex wake, shown as distributions along a horizontal cut through the middle of the descending oval. (a) Vertical velocity w. (After Spalart (1998).) (b) Vorticity, estimated from w, assuming axisymmetric flow in core. (c) Pressure, estimated from w, assuming axisymmetric flow in core. (From Spalart, 1998)

For comparison, dashed curves show what these distributions would be if all of the vorticity were concentrated in a pair of line vortices, and the flow everywhere else were irrotational (potential flow). Another dashed line shows what the distributions would look like for a Rankine core, with the core in solid-body rotation with constant vorticity, a model we'll look at again in Section 8.3.3 in connection with induced drag.

In the real flow (solid curves), the peak circumferential velocity (Figure 8.1.8a) is quite high and occurs at a small radius r_1 from the center of the core. Significant vorticity persists out to a much larger radius r_2, which extends almost to the center plane and to the boundary of the oval. This persistence of the vorticity is clearly seen in the fact that the velocity profile does not fair in to the potential-flow curve until it reaches r_2, but it is hard to see in the plot of the vorticity distribution (Figure 8.1.8b) because the scale was chosen to show the very high vorticity at the center of the core. The intense vorticity concentrated in the central peak and the lower levels in the rest of the core are consistent with the vorticity distribution in the initial sheet that feeds the core, as we expected from our earlier discussion. The plot of the pressure distribution (Figure 8.1.8c) shows that very low pressures are concentrated only in the intense central core.

The low pressure in a vortex core is accompanied by low temperature, which often causes condensation of water vapor, making the core visible. How much of the core is visible under such conditions depends on the situation. In a newly forming core just downstream of a wingtip or flap end, usually only a central portion of the core is marked by condensation, making the core appear more compact than it really is. In the farfield, the situation is more complicated. Often, engine exhaust has been rolled up into the cores, carrying with it water vapor and condensation nuclei (soot) from the engines into the outer parts of the cores. Under such conditions, nearly the entire turbulent wake of the airplane may be visible. But the picture can change over time, as the condensation evaporates, as it appears to be doing in the photos in Figure 8.1.10.

, The vortex cores are often referred to as "wingtip vortices," though we can see from the foregoing that this is a bit of a misnomer. While it is true that the cores line up not very far inboard of the wingtips, the term "wingtip vortices" implies that the wingtips are the sources of all of the vorticity. Actually, as we saw in Figure 8.1.4, the vorticity that feeds into the cores generally comes from the entire span of the trailing edge, not just from the wingtips. Though it is difficult to tell from the curve in Figure 8.1.8b, the concentrated peak of high vorticity inside of r_1 in Figure 8.1.8b accounts for only about 30% of the total vorticity in the core.

Figure 8.1.9 illustrates another feature of the velocity field associated with the rolled-up wake. In the direction parallel to the core axes, there is usually an axial "jet" in the downstream direction, away from the wing.

The decay of the trailing vortex cores, if it were by viscosity alone, would be extremely slow, and even though the cores are turbulent at all but the lowest Reynolds numbers, turbulent transport is suppressed by flow curvature, and the decay of the vortices is still very slow. At the scale of a large airplane, the vortices would persist for hundreds of miles behind the wing if viscosity and small-scale turbulent diffusion were the only dissipation mechanisms. In actuality, the vortices typically persist for something more on the order of 10 miles, and the eventual breakup of the wake is not by small-scale turbulence but by large-scale motions and distortions of the vortices, resulting from slow-growing instabilities such as the Crow instability (Crow, 1970). Figure 8.1.10 shows a progression of breakup

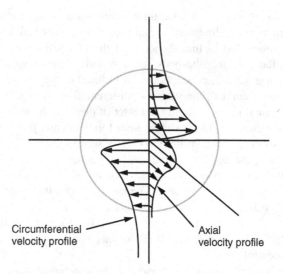

Circumferential
velocity profile

Axial
velocity profile

Figure 8.1.9 Sketch of a "jet" of axial velocity in a vortex core

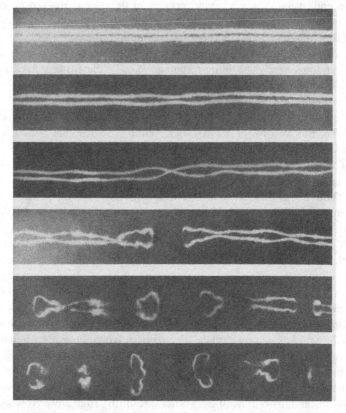

Figure 8.1.10 Progression of vortex breakup initiated by the Crow instability, made visible by condensation. Wake of a B-47 photographed at 15-second intervals. (From Crow, 1970)

initiated by the Crow instability, made visible by water condensation. Spalart (1998) refers to this process as *collapse*, to contrast it with the much slower process of *decay*. We'll consider the breakup process further and speculate on what the whole flowfield looks like in a time-averaged sense on a very large scale, in Section 8.5.5.

8.1.3 The Pressure Field around a 3D Wing

If we take longitudinal vertical cuts through the flowfield as we did to look at streamlines in Figure 8.1.1, but look at the pressure field instead, what we see looks qualitatively like what we saw for a 2D airfoil in Figures 7.3.11 and 7.3.24. This is a view in which 3D effects are difficult to discern. For purposes of understanding the 3D flow, the view in cross-stream planes is more informative. Consider pressure distributions in a succession of cross-stream planes: one a short distance upstream of the wing, one through the middle of the wing, and one immediately downstream of the wing. These are illustrated in Figure 8.1.11 for both the 2D and 3D cases. The cuts shown for the 2D case are just cross-sections of the generic 2D lifting pressure field we considered in Figure 7.3.11. In making these sketches, I've assumed that the maximum chord and load per unit span at the center plane of the 3D wing are the same as for the 2D airfoil, so that the center section of the 3D wing matches the lift coefficient of the 2D airfoil, not the angle of attack. Because of the downwash in the 3D case (which is one of the things we'll be seeking to explain in Section 8.1.4), the 3D wing will need a higher angle of attack than the 2D airfoil, and eventually we'll come back around to seeing this reflected in our explanation of the velocity field.

Now note in Figure 8.1.11 that the pressure distributions in the 3D case show distinct effects of finite span. As the lift decreases outboard of the center section, a combination of the intensity and vertical extent of the pressure distribution must decrease, depending on the planform and lift distribution of the wing. In this case, we show the vertical extent decreasing, as would be the case if the reduction in local lift load were due mostly to a reduction in chord. Note that there is also a kind of "3D relief" effect, in which the pressure disturbances off the surface inboard are "dragged down" closer to the smaller disturbances outboard. As a result, the vertical extent of the pressure distributions in 3D is smaller everywhere along the span, even at the center section, than it is for the 2D airfoil. This more rapid "dying off" of the pressure disturbances away from the wing in 3D is seen at all stations: ahead of the wing, at the wing, and behind the wing.

8.1.4 Explanations for the Flowfield

Now that we have a qualitative description of the flowfield around a 3D lifting wing, we'd like to explain physically how the flow does what it does. One of the main things we'll want to explain is why the velocity disturbance downstream, which dies off rapidly in the case of a 2D airfoil, persists over long distances in the case of a 3D wing. As we've already noted, the classical approach to this is to describe the distribution of the vorticity, both the bound vorticity and the vorticity in the wake, and to use the Biot-Savart law to infer what the velocity field does. Of course the Biot-Savart law is applicable, and all of the features of the cross-flow velocity field near the wing that we saw in Figure 8.1.2 are "explainable" as being "induced" by the bound vorticity and the shed vortex sheet, mostly the part in the near field downstream of the trailing edge. Likewise, the velocity field that persists far downstream,

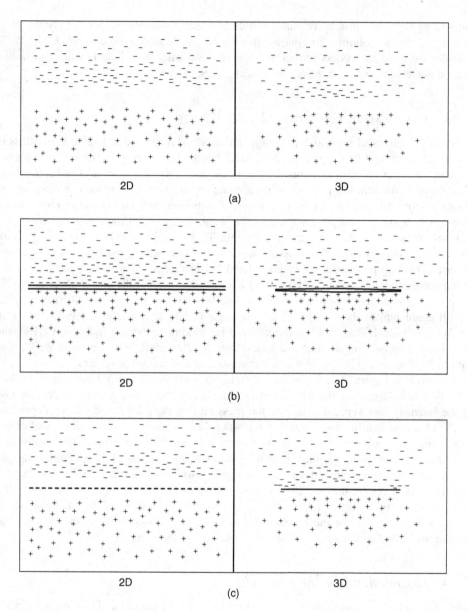

Figure 8.1.11 Gross features of pressure distributions in cross-planes in lifting flows. (a) Upstream. (b) Cut through the wing. (c) Downstream

and differs from Figure 8.1.2 only in some details, is consistent with "induction" by the rolled-up vortex wake shown in Figure 8.1.8. An apparent strength of the vorticity-based approach is that the convection of a somewhat compact vortex wake downstream provides an "explanation," of sorts, for the persistence of the velocity disturbance. A weakness is that the "explanation" that it provides is incomplete in that we had to know or assume a priori how the vorticity is distributed. And we must also remember that even if we know the vorticity distribution, appealing to Biot-Savart only gives a correct *description* of the flowfield, and that it does not *explain* it directly in terms of physical cause and effect. The description using Biot-Savart is convenient as a mental crutch and for quantitative purposes, but it is prone to misinterpretation in terms of the induction fallacy, as discussed in Section 3.3.9. The correct view is that the vorticity is not the cause of the flowfield but is more of a passive result of other things that are happening in the flowfield.

As we noted in the introduction to Section 8.1, a real physical explanation must involve the interaction of the pressure and velocity fields. In the case of a 2D airfoil in Section 7.3.3, the vertical component of the velocity and its interaction with the pressure field played a prominent role in our explanation of the flow. Now let's see how far this kind of thinking can take us in the 3D case, in explaining the cross-flow velocity components illustrated in Figure 8.1.2. Of course, now we'll have to explain the evolution of the spanwise velocity component, in addition to that of the vertical component.

We'll approach the problem by thinking in terms of individual fluid parcels passing through different parts of the pressure field sketched in Figure 8.1.11, and of the pressure gradients the parcels are subjected to during their passage. The most obvious conclusions we can draw have to do with those major portions of the field where one component of the pressure gradient maintains the same sign throughout a parcel's passage through the region. In these situations, the corresponding velocity component is set in motion and not stopped, and we should expect that part of the motion to persist downstream. By inspection of the pressure distributions in Figure 8.1.11, we can see that this mechanism is consistent with the spanwise velocities in the outward direction below the wing and in the inward direction above the wing, and with the upwash outboard of the tips. So these features of the flowfield of Figure 8.1.2, which persist far downstream and have no counterparts in the 2D case, seem to be explainable in terms of simple gross features of the 3D pressure field associated with the lift.

The downwash between the tips is the only major feature of the pre-rollup velocity field not yet explained, and it is more complicated. The vertical component of the pressure gradient, which drives this part of the motion, reverses sign twice for fluid parcels passing above or below the wing, just as it does in the 2D case. In our explanation of the 2D case in Section 7.3.4, we saw that the pressure field participates in a delicate balancing act that results in downwash that decays to zero far downstream of the airfoil. In the 3D case, on the other hand, we know the downwash persists over long distances downstream. With pressure fields that are qualitatively so similar, that is, with two reversals of the gradient in both cases, how do we account for the dramatic difference in the resulting downwash fields? To answer this question, we have to look at the interactions in both cases in more detail.

In the 2D case, there is both upward and downward turning taking place in the flowfield ahead of the airfoil. In connection with Figure 7.3.23, we noted that vertical cuts through the field ahead of the airfoil see the same net flux of vertical momentum across them, corresponding to half the lift, which doesn't change from one cut to another. However, if we limit our attention to a streamtube that passes close to the airfoil above and below, we see that the pressure gradient ahead of the airfoil turns the flow upward, then the gradients above and below the airfoil turn the flow downward, and finally the gradient behind the airfoil turns the flow upward again, canceling the local downwash velocities in an asymptotic sense far away from the airfoil. The upward turnings ahead of the airfoil and behind are just enough to cancel the downward turning that takes place as the flow passes close to the airfoil surfaces.

In the 3D case, the downward turning immediately above and below the wing is stronger than it is in the 2D case, for the same lift. This is because the more rapid dying off of the pressures above and below the airfoil means the vertical pressure gradient near the wing surface is stronger than in 2D. More rapid downward turning, resulting in larger downwash by the time the trailing edge is reached, is also consistent with the fact that the 3D wing requires a higher angle of attack to achieve the same lift. The airfoil pressure field also dies out more rapidly ahead of the airfoil and behind, which results in less upward turning of the flow in those regions. So in the 3D case, we have more downward turning above and below the wing, and less upward turning ahead and behind, with the result that some downwash persists in a central portion of the field behind the wing, that is, between the wake-vortex cores.

To complete this explanation, we must point out that all of these effects of the pressure gradients on the cross-flow velocities constitute only one side of the interaction. Remember that cause-and-effect is a two-way street and that the velocity changes, or accelerations of the flow, are both caused by the pressure gradients and also serve to sustain the pressure gradients. This is the same point that we made a major issue of in our explanation of 2D airfoil flow in Section 7.3.3. There we talked about "confinement" of the "clouds" of high and low pressure and how vertical and longitudinal accelerations of the flow provided that confinement. That description of vertical and longitudinal confinement also applies in the 3D case, but the spanwise component of acceleration also comes into play: The outboard acceleration of the flow beneath the wing and the inboard acceleration above the wing provide spanwise confinement of the pressure differences around a 3D wing.

In the 3D wing flow, the vertical pressure gradients above and below the airfoil are sustained by the downward turning of the flow, just as we noted that they are in 2D. However, a feature of the 3D pressure field that is not so easy to explain in simple qualitative terms is the "3D relief effect" that we described in Section 8.1.3, in which the vertical extent of the pressure distribution in 3D is lower than in 2D for the same chord and lift per unit span. Reducing the vertical extent of the pressure distribution means an increase in the vertical pressure gradient close to the wing surface and a reduction farther from the surface. It is a result of the 3D flow's freedom to accelerate spanwise, but not a simple result to explain.

It is also interesting to note how the pattern of horizontal and vertical velocities that we've just explained fits together in terms of conservation of mass. Referring to Figure 8.1.2, note that the horizontal velocities converge toward the center plane above the wing and diverge from the center plane below the wing. The downwash between the tips thus "exhausts" the converging flow above and "feeds" the diverging flow below. Around the tips, the opposite occurs: divergence above and convergence below, which are "relieved" by the upwash outboard of the tips. So we see that for the horizontal velocities that were set in motion by the wing to persist far downstream, they must be accompanied by downwash between the tips and upwash outboard, and must therefore be part of a general circulatory pattern behind each half of the wing. And, of course, each of these circulatory regions must have vorticity (half of the vortex wake) somewhere inside it.

The final features needing an explanation are the axial jets in the rolled-up vortex cores that we saw in Figure 8.1.9. It has been shown that far downstream of the wing the component of the velocity disturbance parallel to the core axes is nonzero only within the vortical cores (Spalart, 2008). Within the cores, it is clear from Figure 8.1.9 that the axial velocity disturbance is "explainable" as being "induced" by the circumferential component of vorticity associated with the helical configuration of the vortex lines, which, as we've already noted, lines up closely with the helical configuration of the streamlines. A direct physical explanation starts with the observation that balancing the centrifugal forces associated with the circumferential velocities requires a radial pressure gradient, and therefore substantially lower than ambient pressure within the cores, as we saw in Figure 8.1.8c. The air in the cores started at ambient pressure upstream of the wing and, having entered the low-pressure region within the cores, has experienced a net acceleration in the axial direction. And again, our caveat regarding one-way cause and effect applies, and we must remember that the accelerations and the pressure gradients share a mutual interaction.

8.1.5   Vortex Shedding from Edges Other Than the Trailing Edge

So far we've considered the flow over wings of moderate-to-high aspect ratio, assuming that all of the significant vorticity shedding is from the trailing edge. But vorticity shedding is not always confined to trailing edges. On many wings, especially those with squared-off or nearly squared-off tips, the shedding at the tip starts well forward of the trailing edge, on the nearly streamwise "edge" at the tip. In this case, the vorticity can roll up over the wing upper surface, as shown in Figure 8.1.12a. A similar pattern is common on the squared-off

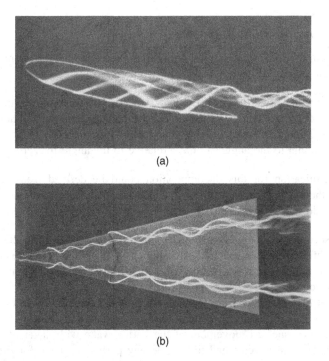

(a)

(b)

Figure 8.1.12   Vorticity shedding from edges other than the trailing edge. (a) The nearly streamwise tip "edge" of a rectangular wing (dye visualization in a water tunnel by Werle, 1974). Photo by Werle, (1974). Courtesy of ONERA. (b) The highly-swept leading edge of a low aspect-ratio delta wing (dye visualization in a water tunnel by Werle, 1963) Photo by Werle, (1963). Courtesy of ONERA

outboard edges of deployed trailing-edge flaps. On high-aspect-ratio surfaces, such details near the tips have only small effects on the overall development of the vortex wake and the global flowfield. However, on a low-aspect-ratio wing, shedding from edges other than the trailing edge can dominate the development of the flow. For example, on low-aspect-ratio delta wings at moderate-to-high angles of attack, most of the vorticity is shed from the highly swept leading edge, and large, partly-rolled-up vortex coils occupy much of the area over the wing upper surface, as shown in Figure 8.1.12b.

8.2 Distribution of Lift on a 3D Wing

In the steady attached-flow regime, the lift distribution on a 3D wing can generally be predicted reasonably accurately by high-fidelity computational fluid dynamics (CFD) with turbulence modeling. The importance of viscous effects in these predictions varies greatly, depending on the conditions. Under transonic conditions, the displacement effect of the boundary layer is very important, and the accuracy of predictions is often limited by our inability to model the turbulent boundary layer sufficiently accurately. At low Mach number and high Reynolds number, the displacement effect of the boundary layer has a smaller effect on the pressure distribution, and an inviscid solution can provide a reasonable prediction of the lift distribution. Here, we'll consider what we can learn with help from simplified inviscid theories.

8.2.1 Basic and Additional Spanloads

If the aspect ratio of a 3D wing is reasonably high, and the lift coefficient isn't too high, the spanload can usually be decomposed into two parts:

1. The *basic spanload* at zero total lift, which depends on the planform, the shapes of the local airfoil sections, and the *twist distribution* of the wing, which we'll define below, and
2. The *additional spanload* due to angle of attack, which depends only on the planform and is proportional to the angle of attack relative to the angle for zero total lift.

A formal justification for this decomposition could be derived from either linearized lifting-surface theory or lifting-line theory, which are both discussed briefly below. Less formally, it should hold provided that

1. The airfoil sections all along the span, except near the tips, behave like 2D airfoil sections that feel the effects of finite span only through changes in their effective angles of attack, due to the local 3D downwash, which is where our assumption of high aspect ratio and low loading comes in;
2. The sectional lift curves are linear, which we found in Section 7.4 to be approximately true for 2D airfoils in the attached-flow regime in the absence of transonic effects; and
3. Nonlinear effects, such as movement of the vortex wake with angle of attack are negligible. Note that we haven't had to assume any particular shape for the vortex wake, only that any effects of movement of the wake are negligible.

Note also that it should be permissible for the wing to be nonplanar, that is for it to have dihedral or nonplanar tip devices. We define the "twist distribution" as the distribution along the span of the orientations of the zero-lift lines of the sections, though sometimes in other contexts the term is used to describe the incidences of the sectional chord lines. Thus if the wing is shaped so that the sectional zero-lift lines of all of the airfoil sections are parallel, the wing is considered to be *untwisted,* and the basic spanload at zero total lift will be zero all along the span. If the orientations of the sectional zero-lift lines vary along the span, the wing is said to be *twisted,* and there will be positive and negative loads on different parts of the span when the total lift is zero, which constitutes a nonzero basic spanload, and there will be nonzero vorticity shed into the wake. Now as the angle of attack is changed from the zero-lift value, assumptions (2) and (3) above guarantee that both the additional wake vortex strengths and the additional sectional loadings all along the span will vary linearly. Because the local 3D downwash changes with angle of attack, the sectional lift slope at each station along the span will be different from what it would be for that airfoil section in 2D. And as we saw in Section 8.1, the overall lift slope of the 3D wing will be less than that of a 2D airfoil.

The basic and additional spanloads, and their sum, are illustrated for a typical twisted, unswept wing in Figure 8.2.1. For an untwisted wing, the basic spanload would be zero everywhere, and only the additional spanload would be nonzero. In this case, the wing was assumed to have a typically small amount of *washout* (i.e., it is twisted leading-edge down outboard) so that the basic spanload is negative outboard. Figure 8.2.2 shows the same spanload decomposition for a comparable swept wing, illustrating how aft sweep tends to shift the additional spanload outboard.

This effect is usually explained in terms of vortex "induction," as illustrated in Figure 8.2.3. At any station on the wing outboard of span station A, the wing "feels" more upwash from the trailing vorticity inboard and less downwash from the trailing vorticity outboard than it would if the wing were unswept As a result, the wing outboard of A feels less downwash than it would in the unswept case. This effect is just as easily (and better)

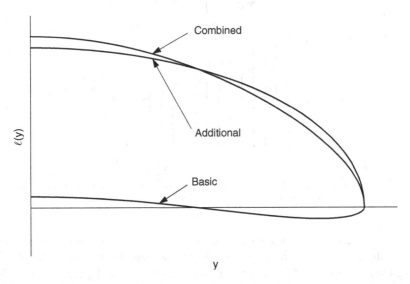

Figure 8.2.1 Illustration of the spanload decomposition for a typical unswept wing, assumed to have a small amount of *washout* (twist leading-edge down outboard)

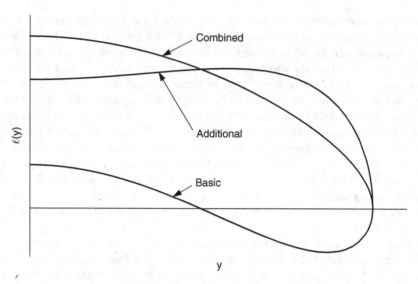

Figure 8.2.2   Illustration of the spanload decomposition for a typical aft-swept wing, assumed to have the substantial *washout* (twist leading-edge down outboard) that is typically required to achieve a favorable total spanload on a swept wing

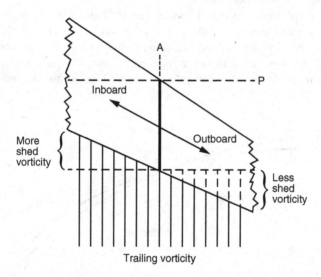

Figure 8.2.3   Illustration of how a section of an aft-swept wing is "influenced" by more shed vorticity inboard and less shed vorticity outboard compared with an unswept wing

explained in terms of the pressure field. Consider the cross-stream plane P cutting the wing near where the leading edge crosses station A. The pressure field in this plane would look like a part-span version of that shown for the full 3D wing in Figure 8.1.11b. The flow approaching the wing outboard of station A thus experiences a vertical pressure gradient like that outboard of a wingtip, that is, low pressure above and high pressure below, and is accelerated upward (more so than it would be in the unswept case, because of the influence of the wing inboard). Again we conclude that the wing outboard of A feels less downwash than it would in the unswept case. This 3D effect of the planform can be quite strong. In fact, for sweep angles typical of swept-wing transport airplanes, the outboard half of the wing actually feels a 3D upwash instead of a downwash.

An untwisted aft-swept wing would have a spanload like the additional spanload illustrated in Figure 8.2.2, which is undesirable for several reasons: It produces unnecessarily high induced drag, it leads to excessive bending moments on the structure at high-load conditions, and it produces a tendency for the tips to stall first, which is bad for airplane handling characteristics. To avoid these effects, an aft-swept wing must generally be designed with considerable washout, and thus have a basic spanload with a strong download outboard, to have an advantageous total spanload like that shown in Figure 8.2.2. The favorable spanload that is achieved in this way persists only over a limited range of angle of attack.

8.2.2   Linearized Lifting-Surface Theory

A linearized version of the incompressible inviscid theory is sometimes useful for illustrating trends and providing insight into the behavior of 3D wings, though it suffers a significant loss in physical fidelity. We assume the airfoil sections are thin, and the angle of attack is small, just as we did in 2D (see Section 7.4.1). The flow disturbance produced by the wing is represented by singularities distributed over the chord plane, and the no-through-flow boundary conditions on the wing's upper and lower surfaces are approximated by velocity-slope conditions applied at the chord plane, ignoring the perturbation u, as in 2D. A complication not present in 2D is that the vortex wake must also be modeled. This is done on the assumption that the wake is confined to a sheet that does not distort, and the vortex lines in the sheet stream straight back from where they are shed from the trailing edge, an assumption we'll see again and discuss further in connection with the Trefftz-plane theory of induced drag in Section 8.3.4. A derivation of the integral equations of the theory is given by Ashley and Landahl (1965). We'll not go into the details here; we'll limit our discussion to the general conclusions to be drawn from the theory.

Of course, linearity allows solutions to be constructed by superposition, just as in 2D, and we can look at the effects of various geometry features separately. In 2D we identified separate effects of camber, thickness, and angle of attack. In addition to these three effects of airfoil section shape and orientation, in 3D we have the effects of the wing planform, that is, the distribution of chord along the span, and the sweep, if any. As was the case with regard to spanload decomposition in Section 8.2.1, it should be permissible for the planform to be nonplanar, that is, for the wing to have dihedral and for the dihedral angle to change along the span, which would require the wake sheet to be correspondingly "bent" in rear view. However, references on the theory usually assume that the wing is confined to a single plane, as in Ashley and Landahl (1965).

The three basic sectional effects have different relationships to the effects of planform. Sectional camber and angle of attack both affect lift, and therefore they affect the distribution of vorticity in the wake, which by "induction" affects the velocity perpendicular to the chord plane at other locations on the span. Because of this, sectional camber and angle of attack have effects that are not just local, but spread over the entire planform in a way that depends on the details of the planform. The effects of section thickness are less strongly coupled to the planform. If the wing has a high aspect ratio in addition to being thin, the effects of thickness become effectively local in the limit, depending only on the local streamwise distribution of thickness and the local sweep of the planform, in a manner consistent with the "simple sweep theory" that we'll discuss in Section 8.6.1. In 3D linearized solutions, just as in 2D, airfoil thickness does not affect the distribution of lift.

So in the linear limit, the distribution of lift on a 3D wing depends only on the planform and the distributions of sectional camber and angle of attack. The total lift varies linearly with α, just as it does in 2D, but due to 3D effects the lift curves at different stations along the span can have different slopes and intercepts. The lifting-surface theory predicts both the spanwise and chordwise distributions of load. The downwash is not assumed to be constant in the chordwise direction, so that downwash can affect not just the local effective angle of attack, but also the local effective camber. Still, because of the general linearity that is assumed, the spanload can be decomposed into a basic part at zero lift and a part proportional to angle of attack, as in Section 8.2.1. There we assumed that the aspect ratio is high, and the local downwash affects only the local angle of attack. Here we assume that disturbances are small, and we needn't assume high aspect ratio.

8.2.3 Lifting-Line Theory

The simplest way to predict just the spanload of a 3D wing is the so-called *lifting-line theory*, in which the chordwise distribution of the load is ignored. The lift is assumed to be concentrated in a single bound vortex, called the *lifting line*, generally located along the quarter-chord line of the planform, and the vortex-wake sheet is assumed to stream straight back from that, as illustrated in Figure 8.2.4. The bound vortex strength is related to the local lift per unit span using the Kutta-Joukowski theorem, Equation 7.2.1, and as the local lift changes along the span, the change in bound vortex strength is shed into the wake, in keeping with Helmholtz's second theorem (Section 3.3.7). Thus the distribution of vortex strength in the wake sheet is equal to the spanwise rate of change of the bound vorticity. In the early theory developed by Prandtl and his colleagues, the lifting line is assumed to be straight, so that the bound vortex at one part of the span has no influence on the downwash on other parts. The 3D downwash is thus assumed to be only that which is "induced" by the

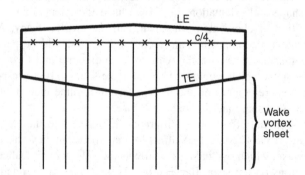

Figure 8.2.4 Arrangement of the bound vortex at c/4 and the trailing vortex lines in the early development of *lifting-line theory*. Control points (x) are placed on the bound vortex

trailing vortex wake, and it is evaluated at the upstream end of the vortex wake, which is by definition on the lifting line itself. Local sections of the wing are assumed to function as 2D airfoils with known sectional (2D) lift curves, with each section operating at an effective angle of attack modified by the local 3D downwash angle.

The original theory was justified by the informal physical arguments I just outlined. Later, Van Dyke (1964) used the method of matched asymptotic expansions to show that lifting-line theory represents a formally valid approximation in the limit of high aspect ratio and small loading. Early lifting-line theory was used not only to predict spanload, but also induced drag, which we'll consider in Section 8.3.

In the more general case in which the lifting line is not straight, the "contribution" of the lifting line itself to the downwash must be taken into account. The original formulation of lifting-line theory, in which the downwash is evaluated on the lifting line itself, then breaks down because a curved lifting line has infinite self-induced velocity. One way to get around this problem is to introduce a different kind of boundary condition, in which the downwash angle "induced" by the bound and trailing vorticity is evaluated at a *downwash line* located off of the lifting line and is made to account for both the 3D part of the downwash and the effective sectional angle of attack. This calls for setting the downwash angle equal to the angle of the sectional zero-lift line, and placing the downwash line at the 3/4-chord location, as illustrated in Figure 8.2.5. The 3/4-chord location is chosen because the downwash there, in the 2D case, is equal to the angle of attack of the zero-lift line, provided the 2D lift-curve slope has the linear-theory value of 2π, a result known as *Pistolesi's theorem*. (The reader can easily verify this using the Kutta-Joukowski theorem, Equation 7.2.1 and the definition of circulation.)

It is clear from Figure 8.2.5 that when the downwash line is located off of the lifting line, the calculation of the downwash "induced" by the trailing vortex lines requires accounting for the additional chordwise segment between the downwash line and the lifting line. In the Weissinger "L" method (Weissinger, 1947), a simplified approximate accounting for the additional segment is used (also see Ashley and Landahl, 1965).

Of course when the lifting line is not straight, the bound vortex influences not just the local 3D downwash angles, but the effective local freestream velocity magnitudes as well. This leads to what is often called a *nonlinear lift* effect because 3D "induction" now affects

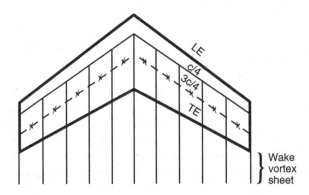

Figure 8.2.5 Arrangement of the bound vortex at c/4, the trailing vortex lines, and the *downwash line* at 3c/4 in later developments of lifting-line theory. Control points (x) are placed on the downwash line

the local lift through both the local downwash (and thus the local effective angle of attack and the local Γ) and the local effective U_∞. The local-U_∞ effect can be thought of as either affecting the lift for a given Γ or the Γ required to produce a given lift. This nonlinear effect is generally ignored in lifting-line calculations for several reasons. First, it keeps the equation system linear. Second, it wouldn't be consistent to include this nonlinear effect while assuming the crude lifting-line model for the vortex wake. Finally, the effect has been found to be small for wings of reasonably high aspect ratio (see Eppler, 1997, for example).

In numerical implementations of lifting-line theories, the vortex wake is usually discretized as an array of line vortices of finite strength, and the bound vortex is assumed to be straight and to have constant strength between the intersections with the trailing vortices. The boundary condition is enforced at discrete *control points* between the trailing vortices. In most methods intended for application to nonstraight lifting lines, the control points are placed at the 3/4-chord location as indicated in Figure 8.2.5. Discrete methods have been proposed, however, in which the control point is placed on the lifting line, as in Figure 8.2.4, even though the lifting line is not globally straight (Phillips and Snyder, 2000, for example). The problem that this incurs is hidden from view because the discrete straight lifting-line segments artificially mask the problem of infinite velocity that we discussed above. But the problem is still there in the limit as the segment length goes to zero. Thus locating the control points away from the bound vortex is still the only way to have a general formulation that doesn't behave badly as the discretization is refined.

Even when a downwash line separate from the lifting line is used, lifting-line theory in effect assumes that the downwash due to finite span doesn't vary much in the chordwise direction, over the whole chord of the section at any given station along the span. This is not a bad assumption for high-aspect-ratio wings with reasonably straight quarter-chord lines, and in such cases, the theory can provide fairly accurate results. However, for swept wings, which generally have a pronounced kink in the quarter-chord line at the center station, the assumption is poor for the inboard part of the wing, and Thwaites (1958) goes so far as to state that lifting-line theory is "completely unjustified" for swept wings. Still, it is often used for swept wings anyway, and semi-empirical adjustments to improve its accuracy in the neighborhood of the kink in the lifting line have been proposed, as in Barnes (1997).

8.2.4   3D Lift in Ground Effect

In Section 7.4.9, we saw that as an airfoil in 2D flow gets closer to a ground plane, the lift is first reduced and then increased. For a 3D wing, a ground plane has a 3D effect on lift, which often overwhelms the 2D effects. A ground plane in 3D also affects the induced drag, as we'll see in Section 8.3.9.

When a wing flies close to the ground, the no-through-flow condition at the ground forces the flowfield around the wing to change in a way that generally increases the lift at a given angle of attack or reduces the angle of attack required for a given lift. One way to look at this is that the ground has the effect of inhibiting vertical velocity throughout the field and therefore reduces the 3D downwash in which the wing is flying.

A second way to look at it that also provides a basis for simplified quantitative calculations is to invoke the idea of images. A simple way to ensure that the no-through-flow condition at the ground is satisfied is to place an image of the vortex system below the ground, as shown

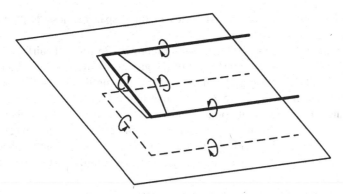

Figure 8.2.6   Bound and trailing vortex system of a wing flying in ground effect, and its image under the ground plane

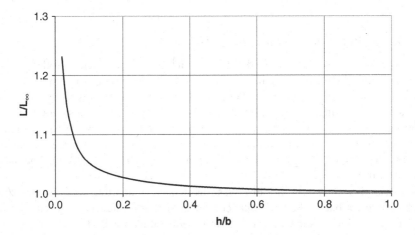

Figure 8.2.7   The increase in lift of a wing in ground effect. Results of a lifting-line calculation for a rectangular planform AR = 10, no twist. $C_l = 0.81$ out of ground effect

in Figure 8.2.6. Then part of the downwash "induced" by the real vortex system can be seen to be canceled by the upwash "induced" by the image system. A single horseshoe vortex and its image could be used for this purpose, but the physical fidelity would be poor. A model with a more realistic distribution of the shed vorticity, but ignoring rollup, like that used in lifting-line theory (Section 8.2.3) or an inviscid panel method (Chapter 10), would provide somewhat better fidelity, and the effect on lift could still be readily calculated. The change (increase) in lift at a fixed angle of attack for a planar wing with a rectangular planform of aspect ratio 10 and no twist is shown in Figure 8.2.7, as calculated by a panel method with a nondistorting wake sheet. In this example, the 2D lift decrease is more than offset by the 3D lift increase due to the reduction in downwash. We can look at finite span as having two effects that work in the same direction. First, finite span introduces 3D downwash, which is reduced by the presence of the ground. Second, finite span also attenuates the 2D effects of the ground because an image bound vortex of finite span has less "influence" than one of infinite span.

Comparing Figure 8.2.7 for a 3D wing with Figure 7.4.35 for a 2D airfoil, it appears that for wings of ordinary aspect ratio the effect of the ground on lift is dominated by finite-span effects. Note that 2D lift in Figure 7.4.35 is decreased until h/c goes below 0.3, which corresponds to h/b = 0.03 in Figure 8.2.7. Above this value, the 3D effect in Figure 8.2.7 clearly dominates. Presumably at some point below h/c = 0.3, the 2D effect would begin to contribute more to the lift increase than does the 3D effect, but this range is seldom of practical interest. For a wing with significant dihedral, or sweep combined with angle of attack, such low values of h/c would not be reachable over much of the span, even without a landing gear.

Depending on the geometry of the wing, ground effect can change the shape of the spanload, and if the wing is swept, it can cause substantial changes in the pitching moment. A swept-wing airplane with an aft tail can experience complicated changes in its lift curve and pitching moments as functions of height when in close proximity to the ground plane.

8.2.5   Maximum Lift, as Limited by 3D Effects

In Sections 7.4.3 and 7.4.4, we looked at how the maximum lift of 2D airfoils, both single-element and multiple-element, is limited by boundary-layer separation. It turns out that sectional maximum lift, as limited by boundary-layer separation, is also generally the limiting factor for 3D wings. However, in the late 1950s, there was considerable interest in flap systems that used active jet blowing to control separation ("blowing BLC") and, when the blowing was very strong, to directly enhance the circulation around the airfoil (the "jet flap"). In such cases, the 3D downwash field can become the factor that limits the maximum lift of a 3D wing.

As we saw in Section 8.1, the downwash associated with finite span has the effect of tilting the lift vector back. The horizontal component is felt as induced drag, and the vertical component is reduced to something less than the magnitude of the force. Furthermore, the magnitude of the force for a given circulation (bound vortex strength) is reduced because the vortex wake is generally tilted downward, so that the velocity "induced" by it has a forward component that subtracts from the effective freestream velocity. Thus when we try to increase the circulation on a 3D wing, by whatever means, the tilting back of the force vector and the reduction of the effective freestream velocity both increase, and presumably at some point the vertical component (the lift) should stop increasing, thus defining a maximum achievable lift limited by 3D downwash.

Davenport (1960) looked at three highly idealized models for this effect that had been proposed by others and found that their predictions varied widely depending on their assumptions about the wake. By their nature, such theories predict maximum lift proportional to span, independent of wing area. Thus when normalized by wing area, they all predicted C_{Lmax} proportional to aspect ratio. However, the constants of proportionality ranged from about 0.8 to 2.0. Davenport proposed a model of his own that gave a result at the high end of this range, but also concluded that the effect depends strongly on the details of the flow, especially as reflected in the tilt of the wake near the airfoil. In any case, the range of C_{Lmax} predicted by these models is so high as not to be achievable without some form of active flow control.

8.3   Induced Drag

In Section 8.1.1, we looked at the flowfield around a lifting wing of finite span, and we saw how the lift vector is tilted back, making a contribution to drag that we call induced drag. In this section, we delve into the related quantitative theory, which we should note at the outset requires some degree of idealization. Recall that in Section 6.1.3 we discussed how it is essentially impossible to decompose the total drag force on a body rigorously into separate contributions based on the different flow mechanisms responsible. In the theories of induced drag in this section, we'll sidestep that issue by assuming that the flow is inviscid and that there are no total-pressure losses through shocks, so that the induced drag is the only drag "component" present. So we must keep in mind that quantifying induced drag as a separate "component" of the drag force is an idealization.

But assuming inviscid flow in the theory doesn't cost us as much in terms of accuracy as one might think initially. We can use induced-drag theory without necessarily assuming that the entire flowfield must be consistent with inviscid flow. For example, in theories in which the lift distribution on the wing is an input, we can use a lift distribution consistent with the real flow, including viscous and transonic effects. Because we reintroduce realism in this way, the conclusions we draw from induced-drag theory can be reasonably accurate in most of the more general situations we'll encounter in practice. Just keep in mind that the theory of induced drag generally ignores some physical complications and incurs at least some small error as a result.

8.3.1   Basic Scaling of Induced Drag

By appealing to the idealized lifting-line model for the flow around a simple wing illustrated in Figure 8.2.4, we can deduce how induced drag should scale with the lift, the flow conditions, and the dimensions of the wing. If Γ is the centerline circulation, the total lift will go as $\rho U_\infty \Gamma b$. As we argued in Section 8.1.1, the wing is flying in a downwash field of its own making, as illustrated in Figure 8.1.2. We'll therefore assume that the induced drag is given by the lift tilted back through an average downwash angle ε, or $D_i \sim L\varepsilon$. For the simple straight lifting line in Figure 8.2.4, the lifting line "induces" no downwash on itself, and we need only consider that "induced" by the trailing vortex wake, for which the downwash velocity goes as Γ/b, and ε goes as $\Gamma/U_\infty b$. Combining these so as to eliminate Γ, we get

$$D_i \propto \frac{L^2}{\rho U_\infty^2 b^2} \propto \frac{L^2}{q b^2}, \tag{8.3.1}$$

which illustrates some important trends. Induced drag increases rapidly with increasing lift and decreases rapidly with increasing span. Flying at high altitude (small ρ) or low speed increases the induced drag. Induced drag is the one major part of the drag of an airplane that decreases with increasing speed, in contrast with the viscous drag that we considered in Chapter 6, which tends to increase roughly as U_∞^2. Given these two opposing trends, the drag tends to be dominated by induced drag at low speeds and by viscous drag at high speeds, with a drag minimum in between, as illustrated in Figure 8.3.1. In this illustration, the ideal $1/U_\infty^2$ and U_∞^2 dependences were assumed, so that the minimum total drag occurs where each component contributes half the total. For real wings, the nonideal behavior of the profile drag tends to drive the minimum drag to a higher speed (lower C_L), as we

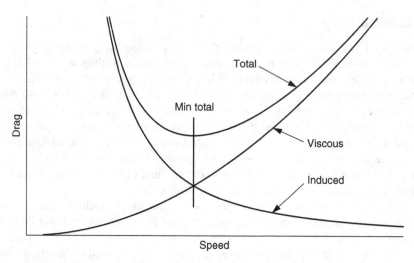

Figure 8.3.1 Schematic drag-versus-speed curve for an airplane, illustrating the induced and viscous contributions

discussed in Section 7.4.2. Nonideal behavior of the induced drag can shift the minimum in either direction.

Note that as a result of our lifting-line assumptions the induced drag depends on the span of the wing and not the area. This also holds in the Trefftz-plane theory, which we'll discuss in Section 8.3.4 and which provides sufficient accuracy for nearly all practical predictions of induced drag. So for practical purposes, induced drag does not depend at all on wing area. This sets induced drag apart from the lift, viscous drag, and pitching moment, which tend to be proportional to area, and introduces a practical problem that has been a source of some confusion. Most of our dealings with these forces are in terms of the dimensionless coefficients C_L, C_D, and C_M. Because all of the raw dimensional forces other than the induced drag are roughly proportional to qS, nondimensionalizing by qS is the only choice that makes sense. But then to be consistent, we must nondimensionalize induced drag the same way, getting

$$C_{Di} \sim C_L^2 \frac{S}{b^2} = \frac{C_L^2}{AR}. \tag{8.3.2}$$

So unfortunately, nondimensionalizing induced drag by wing area makes it look (misleadingly) as if the induced drag depends on wing area, or aspect ratio. The appearance of aspect ratio in such formulas is a red herring, an artifact of a nondimensionalization that is more appropriate for other quantities.

8.3.2 Induced Drag from a Farfield Momentum Balance

We can derive a very general formula for the induced drag based only on the farfield flow, with only minimal assumptions about how the flow behaves:

1. The flow is steady and inviscid, and density variations in the farfield can be ignored, so that we can use the steady, incompressible form of Bernoulli's equation.

2. Velocity disturbances in the farfield tend to zero except in the neighborhood of a vortex wake that is convected indefinitely downstream but does not spread out without bound in the other directions. This is consistent with the general character of the vortex wake that we saw in Section 8.1.2.
3. The wake flowfield is established during the wake rollup process, which is effectively completed within the relative nearfield of the airplane, so that the wake becomes unchanging from there downstream, except for a downward drift.

We orient the x-axis in the flight direction and tie our reference frame to the airplane in steady flight, so that the freestream velocity is U, and the velocity everywhere else is (U, V, W) = (U + u, v, w). We expect the wake far downstream to carry a significant u disturbance, including a nonzero integrated u, and thus a net flight-direction mass flux (remember the axial jets in the rolled-up vortex wake in Figure 8.1.9). This flight-direction mass flux will enter into the momentum balance in any control volume with a far-downstream boundary, and it must come from somewhere. A wake with a jet, as in Figure 8.1.9, requires flow to converge toward the region where the wake is forming, while a wake with a velocity deficit (negative u) would require flow to diverge. From far enough away, this will look like a single sink or source located in the neighborhood of the airplane, because we have assumed that wake development is completed not far from the airplane. In our derivation of the simplified formula for viscous drag (Equation 6.1.3) we also had to account for a balancing mass flux, in that case a source.

In the momentum balance for a general control volume surrounding the airplane, the pressure and momentum-flux disturbances in the neighborhood of where the wake leaves the downstream boundary are obviously significant. The significance of disturbances elsewhere is not quite as obvious. The velocity and pressure disturbances associated with the source or sink are spread diffusely in all directions and die off with increasing distance, but it turns out that they don't die off fast enough that their integrated effects can be neglected, no matter how far away we put the boundaries. Having to deal with the source or sink terms complicates the analysis a bit, and the source or sink ends up dividing up its contributions through the pressure and the momentum flux differently depending on the shape of the control volume. The total contribution of the source or sink to the inferred drag must of course be the same regardless of how the control volume is shaped. We can get the right answer for the total contribution of the source or sink and simplify the algebra considerably, if we assume a particular kind of shape for the control volume.

We give the control volume a general cylindrical shape as in Figure 8.3.2, with an upstream boundary, a downstream boundary, and lateral/top/bottom boundaries that simply need to form a general cylinder (not necessarily circular) parallel to the x-axis.

The cross-section shape in the y-z plane can be anything, as long as the cross section is large enough that the downstream boundary captures all of the significant pressure and velocity disturbance associated with the vortex wake. Now we take the upstream and downstream boundaries to large distances compared to the other dimensions of the control volume, but not so large that downward drift of the vortex wake carries it out through the cylindrical boundary instead of the downstream boundary. To proportion the control volume in this way requires a small inclination angle for the vortex wake. This requires that the spanloading on the wing producing the wake not be too large, which is more restrictive than our original assumptions (1–3). This does not mean we are assuming that u, v, and w are small in

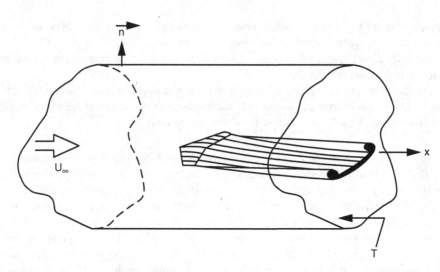

Figure 8.3.2 Cylindrical control volume for deriving Equation 8.3.5 for the induced drag from a far-field momentum balance

the vortex wake. The assumptions regarding the shape of the control volume lead to the following simplifications:

1. The cylindrical part of the boundary makes no contribution to the momentum balance through the pressure, because its normal is everywhere perpendicular to the x-axis.
2. At the upstream and downstream boundaries we can neglect the pressure and velocity disturbances due to the source or sink.
3. The strength of the source or sink can be calculated from the integrated u deficit or excess at the downstream boundary, designated T, due solely to the vortex wake:

$$\dot{m} = \rho \iint_T (U - u_{wake})\, ds. \qquad (8.3.3)$$

The total effect of the source or sink on the momentum balance is due to the momentum flux through the cylindrical boundary and is given by $\dot{m}U$.

With these simplifications, the momentum balance becomes

$$D_i = \iint_T (p - p_\infty)\, ds - \rho \iint_T (2Uu + u^2)\, ds - \dot{m}U. \qquad (8.3.4)$$

After we use the steady, incompressible Bernoulli equation to express the pressure in the wake terms of $U, u, v,$ and w, and simplify, we obtain

$$D_i = \frac{\rho}{2} \iint_T (-u^2 + v^2 + w^2)\, ds. \qquad (8.3.5)$$

The negative sign of the u^2 term is a little disturbing at first glance, because it means that the integrand is not positive-definite and raises the concern that the drag might not be always be positive. A rigorous argument by Spalart (2008) indicates that this is not a problem. An informal argument that reaches essentially the same conclusion goes as follows:

The components v and w are proportional to the vorticity in the wake, while u is proportional to both the vorticity and to the inclination of the helical vortex lines, which is proportional to v and w. So u is higher order in the vortex strength than v and w. Because we haven't assumed small disturbances in the wake, this does not guarantee that u^2 is small compared with the other terms, but it does indicate that u^2 will never outweigh the other terms and that the drag will always be positive.

In the next two theoretical models that we'll consider, we'll ignore the downward drift of the wake and the associated general downward tilt of the vortex lines. We'll also ignore any other deviation of the vortex lines from the freestream direction, as, for example, in the helical alignment of the vortex lines in the cores shown in Figure 8.1.5. The farfield wake then has no u disturbance associated with it, and the u^2 term in Equation 8.3.5 is zero. We can then interpret the induced drag as being accounted for by the kinetic energy left behind in successive slices of the flow in the wake.

8.3.3 Induced Drag in Terms of Kinetic Energy and an Idealized Rolled-Up Vortex Wake

Here we make the same basic assumptions as we did leading to Equation 8.3.5, and we calculate the velocities in Equation 8.3.5 using the Biot-Savart law, based on simple assumptions regarding the distribution of vorticity in the wake. We ignore the tilt of the wake and the circumferential component of the vorticity in the vortex cores. The simplest model for a rolled-up vortex wake is a pair of line vortices, but the kinetic energy integral in the neighborhood of a line vortex is infinite, which rules this model out for evaluating induced drag in terms of kinetic energy. For the kinetic energy to be finite, the vorticity must have finite strength and must therefore be spread out over a finite area. The simplest vortex model that does this is the so-called Rankine vortex we described in Section 3.3.8 and Figure 3.3.8f, in which the vorticity is assumed constant inside of a circular core of radius r_c, outside of which the flow is assumed irrotational. So a simple model for the rolled-up wake behind a lifting wing consists of two Rankine vortices of strength Γ_o and radius r_c with their centers separated by a spanwise distance b_o, as shown in Figure 8.3.3. Now three parameters, Γ_o, b_o, and r_c, are sufficient to determine the drag. If Γ_o, b_o, and the drag are known, r_c can be adjusted to match the drag. This idea of adjusting the core radius to match the drag began with Prandtl (see Spalart, 2008).

One problem with a wake model based on Rankine vortices is that the combined flowfield associated with two such vortices is not consistent with maintaining the circular boundaries of the cores over time. (The flow outside the Rankine cores should be the same as the flow around two point vortices, and this flow does not have circular streamlines of radius r_c centered on the vortices.) This has not discouraged use of the model, however, and the problem has been sidestepped in at least two different ways, neither of which is entirely satisfactory.

Spreiter and Sacks (1951) sought to work around the problem by taking advantage of the fact that the streamlines of the flow around two point vortices are actually circular, just not all centered on the vortices. They noted that only the streamline circles of zero radius are centered on the vortices and that as we look at larger radii, the streamline circles have centers shifted increasingly outboard. They therefore placed the Rankine cores so that their boundaries matched the streamline circles of radius r_c, so as to align the core boundaries

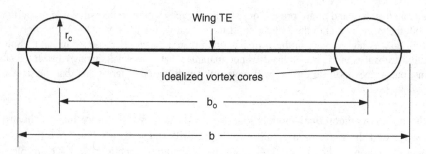

Figure 8.3.3 Idealized model for the rolled-up vortex wake used in the theories of Spreiter and Sacks (1951) and Milne-Thomson (1966) (drawn to scale for $r_c/b = 0.0855$, $r_c/b_o = 0.1089$)

with streamlines of the irrotational flow surrounding the cores. The trouble with this is that when the cores are positioned in this way, their centers don't coincide with the locations of the original point vortices, so that the flow they "induce" isn't consistent with the assumed flow outside the cores.

Milne-Thomson (1966) also assumed the flow outside the cores is consistent with two point vortices but did not shift the centers of the Rankine cores outboard from the locations of the point vortices, arguing that this was valid as long as the core radius is small compared to the separation. However, the core radius he finally deduces turns out not to be that small. So this is also an inconsistency, just a different one from that of Spreiter and Sacks.

In spite of the inconsistencies of both of these models, the calculations for the kinetic energy can be carried out. For the Milne-Thomson version, the result is

$$D_i = \rho \frac{\Gamma_o^2}{2\pi} \left(\frac{1}{4} + \log \frac{b_o}{r_c} \right), \tag{8.3.6}$$

where Γ_o is the circulation of the cores. To see what this implies about the size of the cores behind a typical wing, assume an elliptic load distribution on a wing of span b, for which the classical theory of Section 8.3.4 gives

$$D_i = \frac{\pi \rho \Gamma_o^2}{8} \tag{8.3.7}$$

and

$$b_o = \frac{\pi}{4} b. \tag{8.3.8}$$

Then the core radius is given by

$$\frac{r_c}{b} = 0.0855. \tag{8.3.9}$$

The model is thus kinematically inconsistent. But the more serious problem with Rankine vortex cores is that they don't model the flow in the rolled-up wake behind a wing well at all, as we saw in Figure 8.1.8.

8.3.4   Induced Drag from the Loading on the Wing Itself: Trefftz-Plane Theory

Here we seek to infer induced drag from the loading on the wing itself, without laboriously computing the entire flowfield. This requires making simplifying assumptions not just about the farfield wake, but regarding the development of the wake all the way from where it leaves the wing trailing edge. *Trefftz-plane theory* does this by ignoring the rollup of the trailing vortex sheet and assuming instead that the vortex lines stream straight back in the freestream direction from where they are shed at the trailing edge, as illustrated in Figure 8.3.4. This theoretical framework was established in the early 1900s (Prandtl and Tietjens, 1934) and is still in use today. The formal justification is that neglecting the deformation of the wake should be valid in the limit of small loading or high aspect ratio. The practical justification is that the resulting theory gives reasonably accurate results for practical loadings and aspect ratios.

A real vortex wake, outside of the rolled-up portion at the outboard edge, is a finite-thickness shear layer that aligns itself with the flow on both sides, has no flow through it, and supports very little pressure difference across it. At the next level of idealization, this shear layer would be modeled as a thin vortex sheet that is also a stream surface of the flow, having, by definition, no flow passing through it and no force on it. No force means there can be no pressure jump across the sheet. In a shock-free inviscid flow, no pressure jump means there can be no jump in velocity magnitude across the sheet, only a jump in velocity direction. No jump in velocity magnitude in turn requires that the vortex lines in the sheet be aligned parallel to the average of the velocities on the two sides of the sheet (If the vortex lines were not aligned in this way, there would have to be a jump in velocity magnitude, as we noted in Section 3.3.8, and thus also a jump in the pressure, violating our no-force condition). So a vortex wake modeled as a force-free vortex sheet must satisfy requirements on both the shape of the sheet and on the alignment of the vortex lines in the sheet.

Unlike this ideal force-free vortex sheet, the assumed wake sheet in Trefftz-plane theory, made up of vortex lines aligned with the free-stream direction, will generally have a nonzero component of velocity perpendicular to the vortex lines, both from flow passing through the sheet and from vortex lines not being appropriately aligned within the sheet. Whenever

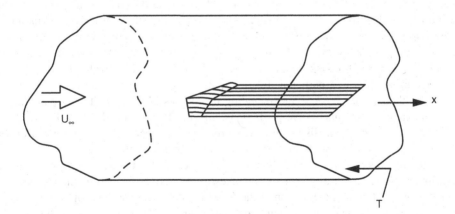

Figure 8.3.4   Assumed wake model in the Trefftz-plane theory of induced drag, with the vortex lines running straight back in the freestream direction

there is a velocity component perpendicular to vortex lines, and the vortex lines are not being convected with the flow, the Kutta-Joukowski theorem requires that there be a "lift" force perpendicular to both the velocity and vorticity vectors. Thus the nonzero component of velocity perpendicular to the vortex lines results in local forces exerted on the wake that are small for a typical high-aspect-ratio wing, but nonzero. Because these forces must be perpendicular to the vortex lines, they make no contribution in the drag direction. So the wake in Trefftz-plane theory is not force free, but it is drag free.

Kroo (2001) argues that because the wake assumed in Trefftz-plane theory is drag free, Trefftz-plane theory incurs no error in the drag calculation, and that it should therefore give the same result for the induced drag as a calculation that takes into account realistic distortions of the wake. I would argue that a drag-free wake does not guarantee this level of *correctness,* but only *consistency* in the sense that the total drag in the field does not contain a contribution from drag on the wake. Consistency in this sense then means that the farfield drag determination would agree with the drag determined by surface-pressure integration, if the flow is inviscid and shock-free, but it does not preclude an error in the drag, due to the incorrect positioning of the wake. The error due to the simplified wake model will generally be nonzero, though in most practical cases it is likely to be small. This is an issue we'll take up again in Section 10.4.3 when we discuss 3D CFD codes based on potential-flow theory, in which the wakes of lifting surfaces are often modeled in a way that is equivalent to the assumptions of Trefftz-plane theory.

Trefftz-plane theory is applicable to a lifting system that can be made up of one or more lifting surfaces, each of which can be either planar or nonplanar. The theory takes the spanwise distribution of lift to be known and determines the induced drag either locally in terms of the backward tilt of the lift vector distributed along each lifting surface, or globally in terms of the total kinetic energy in a cross-flow plane (the *Trefftz plane*) far downstream. The local determination defines both the spanwise distribution of induced drag along the surfaces and the total, while the global determination defines only the total. The total induced drag determined either way is the same.

The flowfield velocities that are used in calculating the drag do not come from solving the equations of motion in the flowfield, but are inferred from the idealized model of the vortex wake, through the Biot-Savart law. The theory thus depends on inferring velocity from vorticity, which is justified mathematically, but obscures the physical cause-and-effect relationships. As a result, the physical understanding provided by the theory is not all that we might hope for. While the theory makes very clear the relationship between the downwash distribution and the distribution of drag, it does not provide much intuitive physical understanding as to why a particular lift distribution produces a particular downwash distribution. Although the physical understanding it provides is minimal, Trefftz-plane theory is very valuable for its quantitative predictions, and we depend heavily on it, for nearly all predictions of induced drag. Historically, its predictions have been found to agree reasonably well with drag measurements both in the wind tunnel and in flight.

Note that in our simplified model of the wake in Figure 8.3.4, the wake sheet forms a general cylinder ("general" in the sense of not necessarily circular) defined by the trailing edge and having generators in the direction of the freestream. The distributions of vortex strength in the wake sheet and the bound vorticity on the wing are defined consistently with the Kutta-Joukowski theorem (Equation 7.2.1), and Helmholtz's second theorem (Section 3.3.7), in the manner we described in connection with the lifting-line theory in Section 8.2.3. Given the spanwise distribution of lift and the geometry of the trailing edge, the distribution of vorticity in the wake is defined, and the downwash at each station along the span of the wing itself can be calculated using Biot-Savart. This determines the local backward tilt of the lift vector and the local induced drag. In the literature, this local determination of the drag seems to be discussed only in the context of straight, unswept wings, for which the contribution to the downwash from the bound vorticity on other parts of the span can be ignored. For example, Milne-Thomson (1966), in his Section 11.22, evaluates the downwash at the trailing edge, which is assumed to be straight, using only the contribution of the trailing vorticity. To extend this treatment to swept wings or wings that are otherwise not straight would entail the same inaccuracies we discussed in connection with lifting-line theory in Section 8.2.3. Because of these inaccuracies, local determinations of induced drag are not usually pursued for wings that are not straight, and for most purposes, it is only the total induced drag that matters anyway.

The total induced drag can also be determined from the flowfield far downstream using Biot-Savart and Equation 8.3.5. One way of expressing the result is through the following integral expression:

$$D_i = \frac{\rho}{2} \int v_n \Gamma(\ell) \, d\ell, \qquad (8.3.10)$$

where the integration is over the line or curve where the undeformed trailing vortex sheet intersects the Trefftz plane, and v_n is the velocity "induced" by the wake perpendicular to the line or curve. The factor of 1/2 is required when v_n is evaluated far downstream, where the wake vortices appear infinite, instead of at the wing itself, where the wake vortices appear only semi-infinite. Only the trailing vortex wakes enter into this integration for the total induced drag, and given our idealized model for the wakes, the distribution of vortex strength depends only on the spanwise distribution of lift and on the shapes of the lifting surfaces as seen in rear view, the so-called "Trefftz-plane view." Thus for a given spanwise distribution of lift, the total induced drag is independent of the fore-and-aft arrangement, including the sweep, of the parts of the lifting system. This result was originally derived with reference to a biplane and is referred to as *Munk's stagger theorem* (see Kroo, 2001). Fore and aft disposition affects both the spanwise distribution of induced drag on individual lifting surfaces and the distribution of drag among multiple lifting surfaces, but not the total, provided the spanwise distribution of lift is held constant. Note that this doesn't generally apply when lifting surfaces of fixed shape are moved fore and aft relative to each other because in that case the lift distributions generally change. To keep the lift distributions constant as surfaces are moved fore and aft, as required by the stagger theorem, generally requires changes in twist and/or camber.

Numerical methods for calculating the rollup of the wake sheet in potential flow that in principle predict induced drag with higher fidelity than Trefftz-plane theory are available, but they are not that widely used when the objective is to study induced drag. CFD methods based on Euler or Navier-Stokes equations predict the entire flowfield in detail, including the rollup of the vortex wake. However, flow solutions provided by these methods present us with the same problem we encountered with the real flow; that is, how do we define what part of the total drag is induced drag? It is telling that when users of high-fidelity CFD codes want a separate number for the induced drag predicted by their solutions, they usually plug their calculated lift distributions into Trefftz-plane theory.

8.3.5   Ideal (Minimum) Induced-Drag Theory

What is the minimum induced drag that a wing can have? This question makes sense only if we constrain the total lift to a nonzero value, because if the load is zero everywhere, the induced drag is zero. So for a given configuration of lifting surfaces as viewed in the Trefftz plane, Equation 8.3.10 defines an optimization problem that can be solved for the minimum induced drag, and for the spanload that goes with it, provided we constrain the total lift. When total lift is the only constraint, and minimum induced drag is the only objective, the results of this optimization are called the *ideal induced drag* and the *ideal spanload*. There are various ways this optimization problem can be solved, sometimes analytically, but usually numerically. We'll forego the details here and discuss only the general results and conclusions of the theory.

One general conclusion has to do with the *normalwash* in the Trefftz plane. If the spanload is ideal, the component of the wake-induced velocity in the direction perpendicular to the wake cut in the Trefftz plane is related to the local dihedral angle θ of the cut:

$$\mathbf{v_n} = \text{constant} \times \cos(\theta). \qquad (8.3.11)$$

This relation was derived by Munk (1921) and is often referred to as *Munk's minimum-induced-drag criterion*. A normalwash distribution obeying this rule is illustrated in Figure 8.3.5.

For a planar wing, θ is zero, the normalwash is constant and is the same as the down-wash, which, according to the theory, requires an elliptic lift distribution, as illustrated in Figure 8.3.6.

In this case, we have the classic result for the induced drag:

$$D_i = \frac{L^2}{\pi q b^2}, \qquad (8.3.12)$$

or in dimensionless terms:

$$C_{Di} = \frac{C_L^2}{\pi \text{AR}}. \qquad (8.3.13)$$

This, of course, reflects the same scaling we deduced in Equations 8.3.1 and 8.3.2 and fills in the constant of proportionality for one particular case. The dimensionless form,

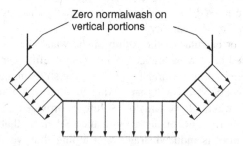

Figure 8.3.5   Illustration of the *normalwash* rule, Equation 8.3.11, associated with an ideal spanload. The vectors drawn here show only the normal component of the velocity

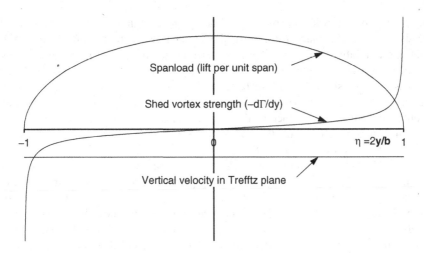

Figure 8.3.6 Elliptical spanload and constant downwash of an ideally loaded planar wing

Equation 8.3.13, is the better known of the two, which is unfortunate. It gives the misleading impression that aspect ratio plays an important role in induced drag. As we saw earlier, the aspect ratio in Equation 8.3.13 is really a red herring, an artifact of the nondimensionalization. It is clear from Equations 8.3.1 and 8.3.12 that the induced drag force depends on span, not on aspect ratio.

So the well-known elliptic spanload is "ideal" for a planar (flat) wing. For nonplanar configurations, the ideal spanload is not generally elliptic, but it is easily calculated for a given geometry. With a vertical winglet added, for example, the ideal spanload shows less lift inboard and more lift outboard, relative to elliptic, with a certain optimum distribution on the winglet itself, as shown in Figure 8.3.7. Note that "lift" in this context refers to the aerodynamic force perpendicular to the wing locally, which in the case of the vertical winglet

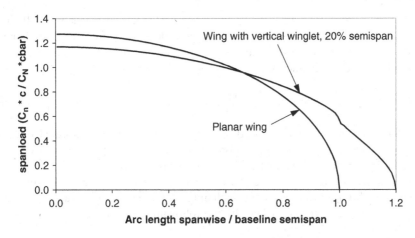

Figure 8.3.7 Ideal spanload for a wing and vertical winglets, compared with the elliptic ideal loading of a planar wing

is a horizontal force inward. Also note that according to Equation 8.3.11 the normalwash produced in the Trefftz plane by a vertical winglet is zero. If the wing and winglet are unswept, the normalwash at the surfaces themselves is half what it is in the Trefftz plane, because the wing and winglet "see" only a semi-infinite wake sheet, whereas the Trefftz plane "sees" a wake sheet that is effectively infinite both fore and aft. This means that the winglet itself experiences no sidewash, or that the load on the winglet cancels the sidewash that would be there in the absence of the winglet. This is a point we'll consider further in Section 8.4.2 in connection with the common misunderstanding that wingtip devices like winglets generally produce induced thrust.

Relative to the "ideal" spanloads we've discussed here, the spanloads used on real wings are usually modified somewhat to reduce bending loads at highly loaded structural critical conditions and allow a lighter wing structure, at the expense of a slight increase in drag in cruise. The presence of a fuselage and wing-mounted engines also tends to alter the spanloads on real wings, an effect for which we'll discuss one idealized model next.

The fact that downwash is constant for an elliptically loaded planar wing, both at the wing itself if it is unswept, and in the Trefftz plane, might lead us to expect that the vortex wake would remain undistorted, as Trefftz-plane theory assumes. This expectation is unrealistic for more than one reason. Remember from our discussion of the vortex wake in Section 8.1 that Spalart (1998) showed that it is not possible for the wake to remain undistorted at the outer edge. And the "induction" by the trailing vortex sheet is not the only contribution to the downwash. The bound vorticity "induces" a nonuniform downwash that distorts the wake in the nearfield of the wing. So uniform downwash in the Trefftz plane doesn't mean that the wake remains undistorted. And there is another important way in which the undistorted wake assumed in the Trefftz-plane theory is unrealistic. The farfield descent rate of an undistorted wake sheet of an elliptically loaded wing, implied by its own self-induced velocity, is much higher (a factor of $\pi^2/2 \approx 4.9$) than the descent rate of the real rolled-up wake.

8.3.6   Span-Efficiency Factors

Equations 8.3.12 and 8.3.13 apply to an ideal elliptically loaded planar wing. If the wing is nonplanar or the spanload is nonideal, the induced drag will differ from the planar ideal. It is often convenient to relate the actual drag to the planar ideal through the *induced-drag span-efficiency factor*, e, defined by

$$C_{Di} = \frac{C_L^2}{\pi \mathrm{ARe}}. \tag{8.3.14}$$

For a planar wing, the actual induced drag is always greater than or equal to the ideal, so that e is always less than or equal to one. For a nonplanar wing, the induced drag can be less than the planar ideal, so that e can be greater than one.

Another "efficiency factor," called the *Oswald efficiency factor*, e_o, takes into account the variation with C_L of the total drag, including the viscous profile drag. It is defined in practice by fitting the total drag polar with the following expression:

$$C_D = C_{D_o} + \frac{C_L^2}{\pi \mathrm{ARe}_o}. \tag{8.3.15}$$

Recall from Section 7.4.2 that the sectional profile drag polar of a 2D airfoil typically has a roughly parabolic variation in the low-to-moderate range of C_L. The integrated profile drag of a 3D wing tends to behave similarly, so that the inclusion of the profile drag makes the coefficient of the C_L^2 term in Equation 8.3.15 larger than it would be for induced drag alone. Thus for wings with ordinary airfoil sections, e_o is usually smaller than e.

An important point to note about the two efficiency factors is that e_o is defined such that it can in general be determined by a curve-fit of a known total-drag polar, though the value generally depends on what part of the polar is fitted, while e cannot generally be determined just from the total drag. Sometimes this point is overlooked, and e_o is confused with e. An example of this is seen in the claims of low induced drag that were made for the Winggrid wingtip device (La Roach and La Roach, 2004), which consists of a cascade of constant-chord airfoils appended to the tip of a conventional wing. The cascade almost certainly has high profile drag at low C_l s, and as a result, a range of C_l over which the profile drag decreases with increasing C_l. This would make that portion of the total polar shallower than it would be with induced drag alone, rather than steeper, as would usually be the case with an ordinary single-element airfoil. The author apparently fitted this portion of the polar and obtained a high value of e_o, from which he inferred a low value of induced drag that is probably not realistic.

8.3.7 The Induced-Drag Polar

If we assume that e is constant, independent of C_L, Equation 8.3.14 implies that the induced-drag polar is a simple parabola, with zero induced drag at zero lift. But for this to be true the wing would have to maintain the same spanload shape as C_L varies, which a real wing doesn't generally do. The spanload of a real wing can have a complicated variation with angle of attack, due to transonic and viscous effects, and the induced-drag polar will be correspondingly complicated. But even in the absence of nonlinear effects, the spanload of a twisted wing changes shape with angle of attack, and it is instructive to look at the trends predicted by simplified theories. Combining Trefftz-plane theory for the total induced drag with the decomposition of the spanload into basic and additional parts (Section 8.2.1) leads to a prediction of the induced-drag polar (Rubbert, 1984). In terms of the bound circulation, the spanload decomposition can be expressed as

$$\Gamma = \Gamma_o + \frac{d\Gamma}{d\alpha}(\alpha - \alpha_o), \qquad (8.3.16)$$

where Γ_o and α_o are the circulation distribution and angle of attack of the wing when the total lift is zero. Substituting this into the total-induced-drag integral, Equation 8.3.10, and doing some rearranging yields the result that the induced-drag polar is parabolic:

$$C_{D_i} = C_{D_{io}} + A_1 C_L + A_2 C_L^2, \qquad (8.3.17)$$

for a general twisted wing, where $C_{D_{io}}$ and A_1 are zero if Γ_o is zero everywhere. Thus for an untwisted wing, we would have only

$$C_{D_i} = A_2 C_L^2. \qquad (8.3.18)$$

Equation 8.3.17 can be rearranged in terms of $C_L(C_{Dimin})$ and C_{Dimin}, the lift and drag coefficients at the minimum of the polar:

$$C_{Di} = C_{Dimin} + \frac{[C_L - C_L(C_{Dimin})]^2}{\pi AR e_{NT}}, \qquad (8.3.19)$$

where e_{NT} is the span-efficiency factor of the corresponding untwisted wing.

Figure 8.3.8a illustrates the parabolic induced-drag polar of a swept, twisted wing. Note that in this polar, the C_{Dio} term is significant, but the A_1 term is relatively small, which seems to be typical of most real wings. The spanload-efficiency factor e in Figure 8.3.8b varies dramatically with C_L, showing what a serious mistake it is to assume e is constant, even though e_{NT} is constant. And this is another example in which confusing e_o with e and determining e by a curve-fit to the C_L^2 term in the total drag would be a mistake in that it would miss the C_{Dio} part of the induced drag.

8.3.8 The Sin-Series Spanloads

There is a simple trigonometric series for representing the spanload of a planar wing that has some very useful properties (see Durand, 1967b). The lift per unit span is represented in terms of a series with coefficients A_i:

$$l = 2\rho U_\infty^2 b A_1 \Sigma (A_i/A_1) \sin(i\psi), \qquad (8.3.20)$$

or in dimensionless form:

$$Cl\, c/\bar{c} = 4(b/\bar{c})\, A_1 \Sigma (A_i/A_1) \sin(i\psi), \qquad (8.3.21)$$

where ψ is the transformed spanwise coordinate:

$$\psi = a\cos(-2y/b). \qquad (8.3.22)$$

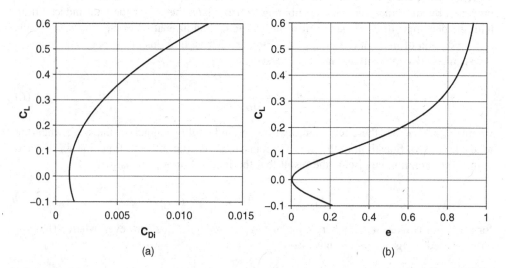

Figure 8.3.8 Induced-drag behavior of a swept, twisted wing according to Equation 8.3.17. (a) The induced-drag polar. (b) The corresponding span-efficiency factor e

The integrated lift involves only the first term of the series:

$$L = (\pi/2)\rho U_\infty^2 b^2 A_1, \tag{8.3.23}$$

or in dimensionless form:

$$CL = \pi \, AR \, A_1. \tag{8.3.24}$$

Only the odd-numbered terms are usually used, as they are the ones that are symmetrical about the center plane. The first three odd-numbered terms are plotted in Figure 8.3.9. The first term represents an elliptic loading, while all the higher terms carry no net lift, which is why they don't appear in Equation 8.3.24.

When the spanload expression 8.3.20 is introduced into Equation 8.3.10 for the induced drag, the result is

$$D_i = (\pi/2)\rho U_\infty^2 b^2 A_1^2 \Sigma i (A_i/A_1)^2, \tag{8.3.25}$$

or in dimensionless form:

$$C_{Di} = (C_L^2/(\pi \, AR)) \Sigma i (A_i/A_1)^2, \tag{8.3.26}$$

so that the span efficiency factor is given by

$$e = 1/\Sigma (i(A_i/A_1)^2. \tag{8.3.27}$$

The wing-root bending moment M_{BR} due to the lift distribution also has a simple expression:

$$2M_{BR}/Lb = (4/(3\pi)) \, [1 + (3/5)(A_3/A_1) + \ldots] \tag{8.3.28}$$

The first two odd-numbered terms of the series provide a convenient way to generate span-load shapes that are compromised from elliptical to reduce bending moments and thus reduce wing structural weight. Using just the first two terms gives the least increase in induced

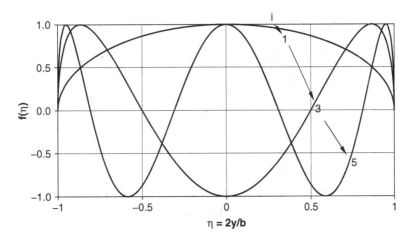

Figure 8.3.9 The first three odd-numbered terms in the sin-series expansion for representing spanloads

drag for a given reduction in root moment. A family of such spanloads is illustrated in Figure 8.3.10, and the e-versus-bending moment curve is plotted in Figure 8.3.11. Note that induced drag increases (e decreases) whether root bending moment decreases or increases, consistent with fact that the baseline spanload is the elliptic optimum. Note also that e decreases only quadratically with the deviation from the optimum, something we'll see again in connection with the effect of a tail or canard on induced drag in Section 8.3.11. The first two odd-numbered terms also provide a convenient way of sketching spanloads of different shapes for illustration purposes, which was used in generating the sketches in Figures 8.2.1 and 8.2.2.

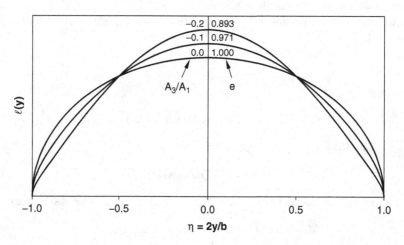

Figure 8.3.10 A family of spanloads compromised from elliptic to reduce wing-root bending moment using the first two odd-numbered terms in the sin-series expansion

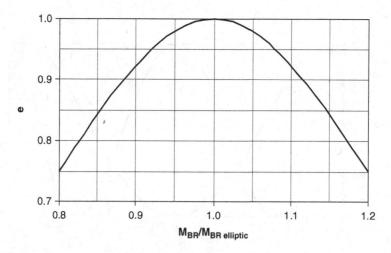

Figure 8.3.11 The variation of e with M_{BR} for spanloads using the first two odd-numbered terms in the sin-series expansion

8.3.9   The Reduction of Induced Drag in Ground Effect

When a wing flies close to the ground the no-through-flow condition at the ground forces the flow field around the wing to change in a way that reduces the downwash in which the wing is flying. This has the general effect of increasing the lift at a given angle of attack, as we saw in Section 8.2.4, and reducing the induced drag for a given lift.

For calculating this reduction in induced drag, it is helpful to invoke the idea of images, just as it was for calculating the increase in lift in 3D ground effect in Section 8.2.4. A typical result of a panel-method calculation is shown in Figure 8.3.12. Note that both C_{Di} and the change ΔC_{Di} due to ground effect are of order C_L^2, so that the drag reduction expressed as a ratio $C_{Di}/C_{Di\infty}$ doesn't depend on C_L.

It's clear from Figure 8.3.12 that the ground produces substantial percentage reductions in induced drag when h/b < 1. This can be important in many types of flight on minimal power, as, for example, in human-powered flight and the long-distance flight of some birds over water. The idea of specifically designing a transport airplane to take advantage of ground effect has been explored, especially for flight over water. The *Ekranoplans* (Russianized French for "screen planes") developed in the former Soviet Union were the largest and most highly developed of such craft (see Scott, 2003). The operational disadvantages of flying very close to the surface (relatively low cruise speed, the possibility of conflicts with surface vessels, and sensitivity to weather) have discouraged widespread applications.

In Section 8.2.4, we saw that the change in lift in ground effect on a 3D wing tends to be dominated by the 3D change in downwash, and of course the reduction in induced drag we've discussed in this section is also a 3D effect. But there is occasionally confusion in this regard. One example is the explanation of the induced-drag reduction proposed by Anderson and Eberhardt (2001). This explanation never refers to the finite span of the wing, but relies instead on a 2D reduction in a purported "loading due to upwash," which we showed in Section 7.3.4 doesn't exist. And Anderson and Eberhardt compounded the error by trying to use a 2D argument to explain a 3D effect.

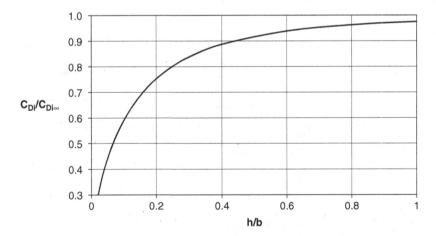

Figure 8.3.12   The reduction in induced drag in ground effect at fixed lift. Results of a panel-method calculation for a planar wing with elliptic planform, AR = 10, no twist

8.3.10   The Effect of a Fuselage on Induced Drag

So far, we've considered 3D lifting flow and induced drag only for lifting surfaces by themselves. The presence of a fuselage that typically encloses or replaces part of the wing complicates the situation in several ways. The pressure field produced by the wing is imposed on the fuselage, so that the fuselage also produces lift, but generally significantly less than a continuation of the wing would produce in the absence of the fuselage. Even if the fuselage sheds no vorticity, so that the bound vorticity and circulation of the wing carry across the fuselage undiminished, the fuselage needn't produce the corresponding lift loading because the Kutta-Joukowski theorem doesn't apply to such a low-aspect-ratio body. The presence of a fuselage also affects the downwash field and the kinetic energy left behind in the farfield, and thus affects the induced drag.

In early simplified theories, the fuselage was modeled as an infinite cylinder, extending forever fore and aft (Lennertz, 1927, and Pepper, 1941). Such models by definition fail to deal with one of the most important issues from a practical standpoint, which is the effect of the aft closure of the body and the distortion of the vortex wake that it causes.

A highly simplified model for the fuselage-closure effect was developed by Nikolski (1959). In this model, the trailing-vortex lines shed from the trailing edge of the wing are assumed to follow streamlines of the body-alone flowfield, an assumption that should be valid in the limit of small lift loading, just as in conventional Trefftz-plane theory. For the most common implementation of the theory, we further assume:

1. The body is axisymmetric and at zero angle of attack,
2. The wing is "planar" with its trailing edge in the same horizontal plane as the body axis, and
3. The velocity disturbance due to the body at the location of the wing is negligible.

Given assumptions 1 and 2, the trailing-vortex sheet is in the horizontal plane of symmetry of the body flowfield and therefore remains planar. The vortex lines simply "neck in" with the flow closing in around the body, as shown in Figure 8.3.13. A vortex line that leaves

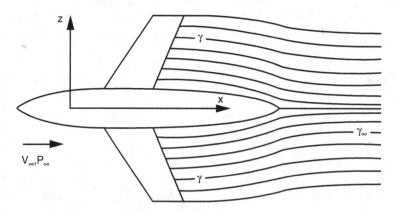

Figure 8.3.13   Trailing-vortex lines in Nikolski's model for the effect of fuselage closure on induced drag. From Nikolski, (1959). Published by NASA

the trailing edge at a distance y (z in Nikolski's drawing, Figure 8.3.13) from the axis ends up far downstream at a distance y' such that the area of the circular streamtube of radius y' is the same as that of the original annular streamtube between the body radius r_b and y. This leads to $y'^2 = y^2 - r_b^2$ as the rule that defines the distribution of shed vorticity in the Trefftz plane as a function of the distribution shed from the trailing edge. Thus the wake in the Trefftz plane looks as if it were shed by a hypothetical wing alone with a reduced span $b' = \sqrt{b^2 - r_b^2}$. To complete the model, we assume that the lift and induced drag of the wing-body combination are the same as for this hypothetical wing alone, because both configurations produce the same wake in the Trefftz plane.

Because we've assumed that both configurations produce the same total lift, the difference between the lift on the hypothetical wing alone and the lift on the exposed wing (the wing outside the body) can be interpreted as an estimate of the "carry-through" lift induced on the body by the wing. This relationship between the loadings is illustrated in Figure 8.3.14, where the carry-through lift on the body is shown as a constant, because the theory defines only the total body lift, not its "spanwise" distribution on the body.

Implementing the model for an arbitrary spanload on the wing of a wing-body combination, especially if the model is extended to a wing that is nonplanar, generally requires numerical integration to obtain the carry-through lift and the induced drag, as is done in the WINGOP code (Craig and McLean, 1988).

For the ideal planar case, analytic expressions have been derived. The ideal spanload for the planar case is elliptic in the Trefftz-plane (the hypothetical wing alone), and when this is mapped back to the physical wing, the loading also elliptic, centered on the airplane axis, with only a truncated part of the elliptic load showing up on the exposed wing. The spanloads plotted in Figure 8.3.14 show this ideal case in a quantitatively correct way for $2r_b/b = 0.2$. The body carry-through lift predicted by the ideal relationship turns out not to differ much from the result of Lennertz for the infinite-cylinder body, as shown in Figure 8.3.15.

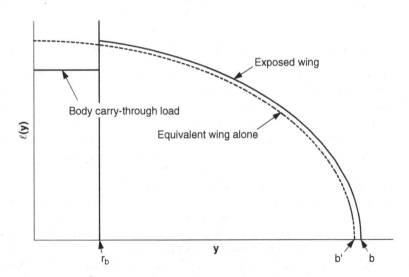

Figure 8.3.14 Illustration of the spanload on a wing-body combination and the hypothetical equivalent wing alone, according to Nikolski's (1959) model

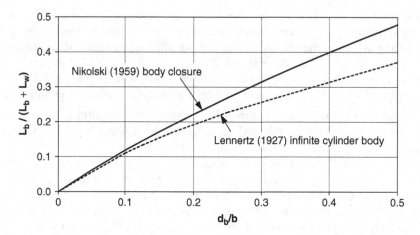

Figure 8.3.15 Comparison of fuselage carry-through lift predicted by Nikolski's (1959) theory for the effect of fuselage closure and Lennertz's (1927) theory that represents the fuselage as an infinite cylinder

For the ideal case, the induced drag is given by

$$D_i = \frac{L^2}{\pi q (b^2 - d_b^2)}, \tag{8.3.29}$$

where d_b is the body diameter. Comparing this with Equations 8.3.13 and 8.3.14, we see that it corresponds to an induced-drag efficiency factor given by

$$e = 1 - \frac{d_b^2}{b^2}. \tag{8.3.30}$$

8.3.11 Effects of a Canard or Aft Tail on Induced Drag

True flying-wing airplanes are relatively rare. At least two lifting surfaces, with some longitudinal distance between them, make it much easier to satisfy requirements for *longitudinal trim* (zero pitching moment) and *longitudinal static stability* (the tendency to return to the trimmed condition after a disturbance). Thus most airplanes are configured with a main wing and a smaller auxiliary lifting surface, either a forward *canard*, or, more commonly, an aft horizontal tail (*stabilizer*). And thus the question arises how the lift loads on these additional surfaces affect the induced drag.

First, let's look at how the trim and stability requirements affect the lift loads the auxiliary surfaces must carry. Longitudinal static stability depends on how the aerodynamic pitching-moment of a configuration changes with angle of attack, which is determined primarily by the planforms and positions of the lifting surfaces and to a lesser degree by the fuselage and engine nacelles. Of course, it is the moment about the center of mass, or center of gravity (CG), that matters, so the other important factor in longitudinal stability is the location of the CG relative to the aerodynamic configuration. The requirement for positive longitudinal stability generally places an aft limit on the CG location, while the requirement to be able to trim the airplane over a range of angles of attack without exceeding the maximum-lift capability of either surface generally defines a forward limit. So stability and trim together define a usable range of CG locations. Within the usable CG range, trimming the airplane for a particular total lift coefficient will require a particular division of the lift between the two lifting surfaces.

For a canard configuration to be stable, the CG must generally be so far forward that the canard must lift upward to trim the airplane at a positive total lift coefficient, and thus *trimmer* is an alternative name for a canard surface. The lift on an aft tail can be in either direction, depending on how far aft the CG is located. The relatively large horizontal tails of some free-flight model airplanes provide ample stability with the CG far aft, so that an upward load on the tail is required for trim. Most full-sized airplanes have relatively small tails and farther forward CG locations, so that the tail must carry a downward load. Of course, the down load must be offset by increased lift on the main wing.

Proponents often argue that the canard configuration is superior because it is better to have both surfaces lifting up than to have one lifting up and one lifting down, but if we look just at induced drag, we find no support for this argument. For example, consider the simplest idealized comparison we can make:

1. The total lift on the airplane is fixed,
2. The lifting surfaces are coplanar, so that the total induced drag of each configuration is the same as if the total spanload acted on a single lifting surface, and
3. The canard and aft tail to be compared have the same span and carry the same spanload, just of opposite signs, and the loadings on the main wings have the same offset from the elliptic ideal, also of opposite signs, as illustrated in Figure 8.3.16.

The ideal wing alone with an elliptic loading has the minimum induced drag. The total spanloads for the aft-tail and canard configurations represent the same perturbation from the ideal, just in opposite directions. If we represent the total spanloads by the sin series of Section 8.3.8, Equation 8.3.20, both must have the same coefficient A_1 because the total lift is the same. For lateral symmetry, only the odd-numbered coefficients are nonzero, and the deviations of the spanloads from elliptic are thus represented by odd A_i's for $i \geq 3$. Because the deviations are of opposite sign, corresponding A_i's are of opposite sign, and according to Equation 8.3.25 the total induced drag is the same. Here I've glossed over the difficulty the sin series would have in representing the kink in the total spanload at the tail or canard tip, but the conclusion is probably still valid.

The above argument indicates no fundamental advantage of a canard over an aft tail in terms of induced drag. But it overlooks some significant practical issues. First, the uploads on canards are typically much larger than the downloads on aft tails. And an elliptic-looking loading on the main wing would be difficult to achieve in the presence of the downwash from a coplanar canard. Besides, it would be better to allow the downwash from the canard to depress the loading on the inboard part of the main wing, and thus to produce a total load closer to elliptic. Another issue is that many real-world configurations are not coplanar, that is there is often a vertical gap between the lifting surfaces.

A nonzero vertical gap between the lifting surfaces has a significant effect on their combined induced drag. How the total induced drag varies as a function of gap depends on

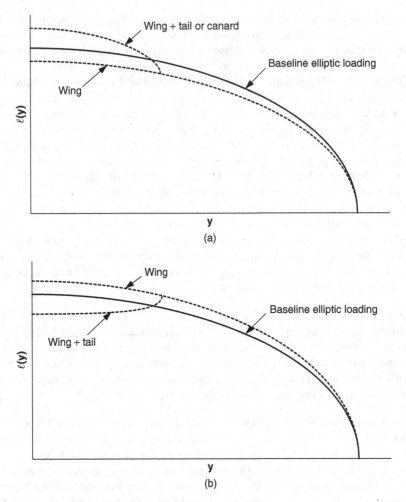

Figure 8.3.16 Illustration of the simple argument for the induced-drag equivalence of a canard and an aft tail. (a) Uploaded tail or canard. (b) Downloaded tail

how the spanloads on the two surfaces vary. For a look at the gap effect, there are several options regarding the spanloads:

1. Assume some arbitrary, fixed spanloads, say elliptic on both surfaces,
2. Assume fixed planforms and twist and camber distributions for the surfaces and solve a lifting-surface analysis problem for the spanloads that go with them, or
3. Optimize the loadings on both surfaces for minimum drag, with a constraint on total lift and on either the percentages of the lift carried by the two surfaces or on the total pitching moment, for whatever gap is chosen.

Fixed geometry (option 2) is no longer very relevant to modern design practice. Optimized loadings (option 3) are the most interesting, and we'll compare them with imposed elliptic loadings (option 1) for theoretical interest.

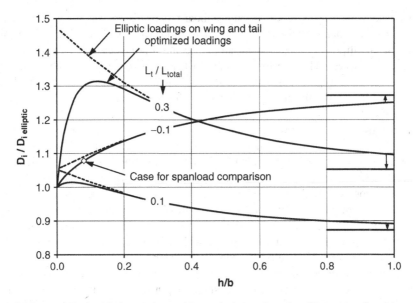

Figure 8.3.17   Variation of induced drag with vertical gap for aft tails or canards with optimized spanloads with a constraint on the percentages of the total lift carried by the two surfaces. Results for elliptic loadings imposed on both surfaces are shown for comparison. Tail or canard span is 0.40b in all cases

Figure 8.3.17 shows Trefftz-plane induced drag as a function of vertical gap for cases in which the second lifting surface has 40% of the span of the wing. In one, the small surface carries a down load equal to 10% of the total lift, as might be the case with an aft tail. The other two cases are up loads of 10% and 30% of the total, spanning a range that might apply to a canard. The basis of comparison $D_{i\,elliptic}$ is the ideal induced drag of a planar wing carrying the same total lift, with no load on another surface. The calculations are based on Trefftz-plane theory and were carried out in the Boeing WINGOP code (Craig and McLean, 1988). We can make the following observations:

1. For gap ratios greater than 0.2–0.3, there is practically no difference in drag between elliptic loadings and optimized loadings.
2. At the large-gap end, the drag approaches the limiting the case of infinite gap, or two surfaces flying in isolation, for which the optimum is to have each surface carry an elliptic loading; and the total induced drag is just the sum of the two ideal induced drags. These limits are indicated by horizontal lines to the right in Figure 8.3.17.
3. At zero gap, the optimized drag goes to an idealized limiting case in which the total loading is elliptic, and the total induced drag is just the ideal induced drag of a planar wing. The spanload on the canard or tail is not uniquely defined in this case; it has only to satisfy the assumed total canard or tail load required for trim. Then whatever the spanload on the canard or tail is, the wing must carry a load distribution that compensates for it, so that the total load is elliptic. This is an ideal that is not practically realizable, especially if the canard or tail has a spanload with infinite slope at its tip, as is usually true for all practical purposes. Imagine an elliptically loaded canard that sheds the usual

vortex wake with concentrated vorticity outboard that rolls up. The wing would somehow have to capture that vorticity and cancel it in order to leave behind a wake equivalent to that of an elliptically loaded wing alone. So the limiting case of zero gap gives us a well-defined theoretical value for the drag, just not a practically realizable one.

4. The 0.1 and −0.1 loading cases with elliptic spanloads go to the same drag level at zero gap, somewhat above the ideal, consistent with our earlier argument that they should differ from ideal by the same amount.

5. For gap ratios less than about 0.2, there is an advantage to optimizing the spanloads, relative to elliptic, but it is small for the smaller tail or canard loadings. Shortly we'll compare the optimum and elliptic spanloads for the small-download case indicated by the diamond symbol, typical of many aft-tail configurations, where the predicted advantage of optimization is only a couple of percent. For large canard loadings, a very small gap is predicted to be good, if the loadings are optimized, and the predicted advantage for optimization is large. But remember that the zero-gap optimum requires the aft surface to "capture" and cancel the vorticity from the forward surface. In this case, only a fraction of the advantage of optimization is probably realizable.

6. Having a second surface of small span carry part of the load can reduce the induced drag only if the load on the second surface is small and upward, and the vertical gap is larger than about 0.2. Most practical, stable canards are probably closer to the 30% load case shown, for which the second surface exacts a sizeable penalty in induced drag unless the vertical gap is very large.

Note that according to Munk's stagger theorem these results are independent of the longitudinal positions of the lifting surfaces. Also note that they should be taken only as indicating qualitative trends, given that they ignore the effects of rollup of the forward surface's vortex wake.

It is interesting to note that for an aft-tail configuration with a vertical gap typical of a large low-wing airplane, the spanload on the wing that gives minimum total induced drag is significantly altered from elliptic. This effect is illustrated in Figure 8.3.18 for a vertical gap of 8% of wingspan (the case indicated by the diamond symbol in Figure 8.3.17). For comparison with the optimum spanload, an elliptic spanload and the optimum spanload for the idealized case of zero gap, assuming elliptic spanload on the tail, are shown. The calculated Trefftz-plane optimum for the spanload on the wing is a kind of washed-out compromise between these limiting cases. Note that on the tail there is very little difference between the optimum spanload and elliptical. Also recall from Figure 8.3.17 that the difference in drag between the optimum and elliptic spanloads is small.

In some treatments of this topic, the induced drag of a two-surface configuration is decomposed into four pieces. Because the downwash experienced by either surface is a first-order quantity, it can be decomposed into a "self-induced" contribution from the same surface and a contribution "induced" by the other surface. The total induced drag can therefore be expressed as the sum of two "self-induced" parts and two parts "induced" on one surface by the other. This approach was convenient in the old days when simplified assumptions were often used in estimating the "shared" parts. Now that numerical Trefftz-plane calculations are so easy to do on a computer, it is just as easy to skip the decomposition and calculate the total drag using Equation 8.3.10. However, there is some intuitive appeal to thinking of the drag in terms of its separate parts. For example, for a downloaded aft tail

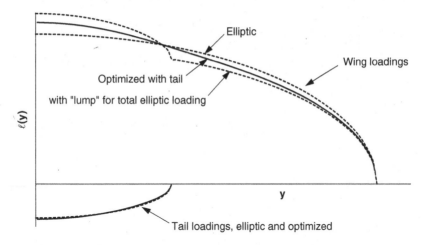

Figure 8.3.18 Optimum spanloads for a wing and a tail or canard with 40% span ratio and 8% vertical gap and downward load equal to 10% of the total lift. An elliptic loading on the wing and an optimum loading for zero gap (an elliptic loading with a "lump" in the middle, so as to have an elliptic total loading) are shown for comparison. The wing spanloads all include an excess over the total lift, to make up for the download on the tail

flying in the downwash from the wing, the contribution "induced" by the wing on the tail is a small thrust that offsets some of the "self-induced" parts of the drag.

8.3.12 Biplane Drag

For a given span and total lift, a biplane with some vertical gap between the wings has lower ideal induced drag than a monoplane, according to Trefftz-plane calculations by Munk and Prandtl (see Thwaites, 1958). Their calculations covered a range of span ratios between the two elements of the biplane, but here we'll look just at the case where the two elements are of equal span. In that case, the ideal spanload is the same for both elements, with each carrying half the total lift, and the ideal induced drag depends on the vertical gap as shown in Figure 8.3.19. Note the substantial drag reductions for relatively small gaps, 17% drag reduction for a 10% gap, for example. Also note that the curve is headed for an asymptote at 50% drag reduction as the gap goes to infinity. In this large-gap limit, we have two elliptically loaded wings that don't significantly interfere with each other, each carrying half the lift and a quarter of the induced drag of the monoplane, for a total of half the induced drag of the monoplane.

Given this induced-drag advantage of a biplane over a monoplane of the same span, why aren't more airplanes configured as biplanes? The reasons are several. The biplane arrangement generally incurs higher profile drag and manufacturing cost. Structural weight can either favor a biplane or penalize it, depending on the structural arrangement. If a biplane is configured with the same span and wing area as a monoplane, the biplane wings would have only half the chord of the monoplane, and for the same airfoil thickness ratio would have only half the physical thickness. The smaller thickness would entail a large structural weight penalty unless external bracing (struts and wires) were used. Of course

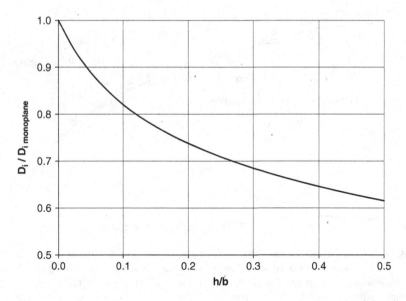

Figure 8.3.19 Ideal induced drag of a biplane of equal spans as a function of vertical gap, relative to the induced drag of an elliptically loaded monoplane of the same span. (Calculated by Munk and Prandtl, from Thwaites, 1958)

most actual biplanes use external bracing, saving considerable structural weight but incurring a substantial viscous-drag penalty.

In the early days of aviation, the preference for biplanes was largely driven by structural weight, as influenced by the airfoil technology of the time. Most of the airfoil data that were available early on were taken in wind tunnels like those built by the Wright brothers, in which models were very small and chord Reynolds numbers were very low. Data at very low Reynolds numbers generally indicate a heavy drag penalty for airfoil thickness, and early designers therefore assumed that airfoils should be quite thin, which favored the externally braced biplane arrangement.

Apart from the drag of struts and wires, the airfoil profile drag of a biplane is an interesting issue and is the subject of some popular misconceptions. When two airfoils are placed in vertical proximity, there is a 2D inviscid interference effect that reduces the lift at a given angle of attack, compared with that of the isolated airfoils. When there is no longitudinal stagger between the wings, this is partly a local-q effect, similar to the one we assumed in our simple model for 2D ground effect in Equation 7.4.2, and partly an induced-camber effect. Some commentators (Garrison, 2008, for example) infer from this lift loss a serious loss in sectional "efficiency" for biplanes. However, inferring an "efficiency" loss from the inviscid lift loss seriously overstates the case. It's true that the lift loss implies a substantial reduction in sectional L/D at fixed angle of attack, but the full reduction needn't be accepted because the lift loss is easily compensated by an increase in angle of attack. And presumably we could even compensate for the details of the induced-camber effect by redesigning the airfoils to produce something close to the same pressure distribution as an isolated airfoil, at least at one operating condition. What cannot be compensated without thinning the airfoils is the induced thickness effect, that is, the effective thickness of the airfoils is increased by their proximity. Indeed, when the viscous drag polar of a biplane pair of airfoils is calculated, the drag is shifted upward by a modest amount relative to the isolated case at the same chord Reynolds number, consistent with an increased effective thickness. The actual loss in maximum sectional L/D is not nearly as gross an effect as that implied by the lift loss at constant angle of attack. In practical design applications, chord Reynolds number is also an issue. A biplane will usually have smaller chords than a comparable monoplane, and the lower Reynolds number will incur some increase in profile drag.

8.4 Wingtip Devices

The idea of a beneficial wingtip appendage or "device" has been around since the early twentieth century, when theoretical calculations first indicated that a vertical endplate added to a wingtip would reduce the induced drag. Early on, however, reality did not live up to the theoretical promise. The simple flat endplate turned out to be a disappointment in practice because the added viscous profile drag more than offsets the saving in induced drag, and the device fails to produce a net benefit. Whitcomb (1976) seems to have been the first to recognize that it is possible to reap the induced-drag benefit of an endplate, and at the same time to realize a net benefit, by keeping the additional profile drag to a minimum through good aerodynamic design practice. The direct result of Whitcomb's work is the classic near-vertical winglet. Less directly, Whitcomb's paradigm of applying good design practice has also contributed to the development of concepts other than the winglet. Both winglets and tapered horizontal span extensions (raked tips) have been put into commercial service, and several other device concepts have also been proposed and brought to varying levels of development (see Figure 8.4.1).

From an aerodynamicist's point of view, the motivation behind all wingtip devices is to reduce induced drag. Beyond that, as Whitcomb showed, the designer's job is to configure the device so as to minimize the offsetting penalties, so that a net performance improvement is realized. For any particular airplane and tip device, the performance-improvement can be measured relative to the same airplane with no tip device.

In Section 6.1.3, we noted that it is not possible to decompose the drag exactly into component parts, but that with the help of idealized theoretical models, it is possible to estimate an induced-drag component. In Sections 8.1 and 8.3, we discussed the physics and the theory of induced drag, including a correct understanding of the role of the vortex wake. Because the vortex wake has been the source of so much confusion regarding how wingtip devices work, we'll take the time to set the record straight on that score before proceeding.

8.4.1 Myths Regarding the Vortex Wake, and Some Questionable Ideas for Wingtip Devices

With our background so far, we are ready to discuss two common misunderstandings regarding the nature and role of the vortex wake. I'll refer to these as the "compactness myth" and the "induction myth," and after defining them and explaining where they go wrong, I'll discuss some of the erroneous tip-device ideas that arise from them.

The compactness myth is simply the idea that the vortex wake consists of vortex cores that spring from the wingtips and are quite compact from the start. Illustrations that present

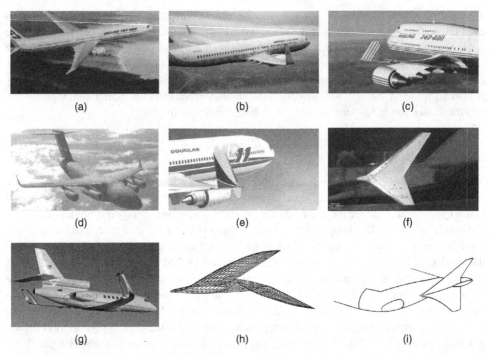

Figure 8.4.1 An assortment of wingtip-device concepts. (a) Raked tip. (b) Blended winglet. (c) 747-400 canted winglet. (d) C-17 canted winglet. (e) MD-11 style up/down winglet. (f) MD-12 style up/down winglet. (g) API spiroid. Used with permission of Aviation Partners, Inc. (h) Tip feathers. (i) Tip fence

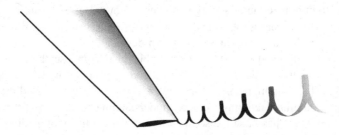

Figure 8.4.2 A misleading view of a compact "wingtip vortex." From Larson, (2001), drawn by John MacNeill. Used with permission

misleading views of the vortex wake, such as the one in Figure 8.4.2, are common, and they have helped to perpetuate the myth. The water-vapor condensation trails that can sometimes be seen streaming from flap edges or wingtips under humid conditions can also be misleading. These trails tend to mark only an inner portion of the core and give the impression that the core is more compact than it really is. The correct view, as we saw in Figure 8.1.4, is that the vortex wake starts as a sheet and that the wake rollup process generally produces cores that are relatively diffuse. A very compact vortex would require shedding all of the vorticity from the tip itself, which would in turn require a uniform spanload. A uniform spanload simply can't be generated by a wing of any reasonable shape, given the strong downwash that such a loading would entail near the tips. The compactness myth is a simple misunderstanding that by itself wouldn't cause much harm, but when it is combined with the induction myth, the potential for serious mischief arises.

The induction myth is more complicated and involves a serious misunderstanding of cause and effect. The trailing vortex sheet and the rolled-up vortex cores are often seen as the direct cause of the velocities everywhere else in the flowfield and thus also the cause of induced drag, but this view is mistaken. It is true that when a 3D wing produces its characteristic large-scale flow pattern, as illustrated in Figure 8.1.2, there must be a vortex sheet shed from the trailing edge, but the vortex sheet is not a direct physical cause of the large-scale flow; it is more of a manifestation. The induction myth reflects a common misunderstanding of what the Biot-Savart law represents, a misunderstanding we discussed in Section 3.3.9.

So what kind of mischief can result from the combined compactness and induction myths? The induction myth leads us to think of induced drag as being "caused" by the vortex wake, and thus to think that by doing something very local to change the flow in the core of the "tip vortex" we can have a large effect on the induced drag. To compound the error, the compactness myth leads us to think we can influence the induced drag by acting just on a very small part of the flow. This kind of thinking has spawned many questionable ideas for novel wingtip devices. A common theme is to provide an inlet that swallows the tip vortex itself, or some of the flow that would otherwise become part of the vortex, and to exhaust it straight back, presumably with its swirl and thus its vorticity removed. A schematic illustration of a device of this kind is shown in Figure 8.4.3. Specific proposed devices based on similar ideas are subjects of U.S. patents by Loerke (1937), Frakes (1984), and Hugues (2005). A naive view might lead one to expect such a device to produce dramatic reductions in the strength of the tip vortex and in the induced drag. However, there is no reason to expect that any such device can provide a reduction in induced drag beyond what can be explained as the result of an increase in physical span when the device is added.

The device concept of Figure 8.4.3 has a fatal flaw. It is based on thinking that we can alter the global flow pattern of Figure 8.1.2 by tinkering with the tip vortex, without having to change the overall distribution of lift on the wing. This is wrong, of course. First, we shouldn't expect to be able to alter the general velocity field remotely through the vorticity in a limited region, because as I argued in Section 3.3.9, the vorticity at one point does not directly "induce" velocity elsewhere. Second, unless the overall distribution of lift on the wing is changed, the global flow pattern cannot be significantly altered.

But how can this be? If Biot-Savart is correct, and a device like that shown in Figure 8.4.3 succeeds in swallowing part of the vortex and "straightening" it out, why isn't the global flow pattern changed? The answer is that a relatively small device can only rearrange the vorticity locally; it cannot significantly change the total vorticity flux, as measured, say, by the circulation around each half of the vortex wake. Generally speaking, in this kind of device it is kinematically impossible to eliminate the vorticity flux from a streamtube without producing a compensating vorticity flux adjacent to it.

Let's look at this in a little more detail. Consider a device that swallows an entire vortex into an inlet. Ahead of the inlet, the vortex retains its full vorticity flux and circulation. Inside the duct however, the circulation of the entire captured streamtube must be zero, even before the flow has gone through any straightener. This follows from the no-slip condition on the

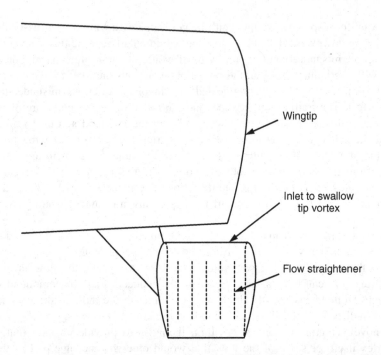

Figure 8.4.3 Schematic plan view illustrating a class of tip device that cannot work as intended

duct wall. The vortex's original vorticity can still be there in the middle of the duct, but its net vorticity flux is exactly canceled by the net vorticity flux that has been automatically generated in the duct-wall boundary layer. This process involves dynamics, of course, but the net vorticity flux cancelation is a kinematic necessity arising from the no-slip condition. So the internal flow ahead of any straightener contains a vortex in the middle and an equal net vorticity flux of opposite sign in the wall boundary layer. An internal flow straightener could now, in principle, zero out all of this internal vorticity locally, but the net internal vorticity flux was already zero ahead of the straightener. Now consider the flow outside the device. On any closed contour outside the device's external boundary layer, we should see a circulation approximately the same as that of the original vortex. The net vorticity flux associated with this circulation has been automatically generated in the external boundary layer, also by kinematic necessity. Finally, when all of this rearranged vorticity, both internal and external, is shed into the wake, the circulation of the wake and the large-scale flow field outside it will not have been greatly changed, and there will have been no dramatic reduction in induced drag. This is similar to what we noted in Section 6.1.10, that it is impossible for a propeller to produce a net axial vorticity in its slipstream.

8.4.2 The Facts of Life Regarding Induced Drag and Induced-Drag Reduction

We have seen that induced drag is a result of large-scale air motion produced by the lifting system. This motion is not physically "induced" by the vortex wake, but is a response to the lift force and depends on the overall lift distribution. When we try to go beyond this on an intuitive level, we are limited to very general observations, such as that increasing the span of a wing generally reduces induced drag. For anything more specific, especially regarding any device other than a simple span extension, we must rely on quantitative predictions, usually from Trefftz-plane theory. Unfortunately, as we noted in Section 8.3.4, the theory does not generally provide for a simple intuitive understanding of how the details of a particular configuration or lift distribution will affect the drag.

Based on our general appreciation of the physics, we can anticipate that drag-reduction devices need to be fairly large as viewed in the Trefftz plane, because any significant reduction in induced drag requires changing the global flowfield associated with the lift, so as to reduce its total kinetic energy. We know that we can't do this just by tinkering with the "tip vortex" and thus that having a significant effect on the drag requires a significant change in the way the lift is distributed spatially. If our starting point is a wing on which the lift is already advantageously distributed, the only way to improve will be to provide a significant increase in the horizontal span or to introduce a nonplanar element that has a similar effect. The quantitative theory tells us that the effect on drag will be roughly proportional to the horizontal and/or vertical span of the device and that a small device can therefore produce at most a small drag reduction.

There is a common misunderstanding that a wingtip device reduces drag by producing thrust on the surfaces of the device itself. For example, there is the popular explanation that likens a winglet to a sailboat beating into the wind, usually accompanied by a diagram showing the lift vector on the winglet tilted forward by the strong sidewash directed inboard above the wingtip, as in Figure 8.4.4. A corollary is that drag-reduction effectiveness is enhanced if the winglet is mounted well aft on the wingtip, where the sidewash is stronger. This general picture of winglet effectiveness has been put forward by Hackett (1980) and McCormick (1995), among others. It is intuitively appealing, but it is flawed.

First, the idea that there is a favorable sidewash at the winglet location is based on the flowfield (sidewash) that would be there in the absence of the winglet. This view is valid

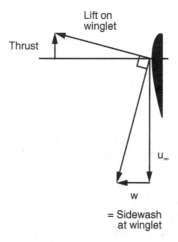

Figure 8.4.4 Misleading view indicating that a winglet produces thrust in the presence of an inboard-directed sidewash "induced" by the wing. Actually, for an unswept wing with a vertical winglet, the optimum loading produces zero sidewash at the winglet

only in the limit of small loading on the winglet and is not even close to being correct for practical levels of winglet loading. When a winglet is advantageously loaded, the flowfield is altered considerably, and the resulting force on the winglet can be very different from what the undisturbed flowfield would lead you to expect. For example, when a wing and vertical winglet are unswept and are carrying their ideal spanload, the sidewash all along the span of the winglet is canceled, and the winglet itself feels no induced thrust or drag. This is required by Munk's minimum-induced-drag criterion from Section 8.3.5, which tells us that any vertical portion of an optimally loaded system must see zero sidewash in the Trefftz plane. With no sweep, the sidewash at the lifting line itself is half what it is in the Trefftz plane, and is therefore still zero on a vertical winglet. So in this case, all of the drag reduction due to the winglet is felt on the horizontal wing, and the sailboat analogy misses the mark badly.

So it is a mistake to try to understand a tip device in terms of the flowfield that would be there in the absence of the device, and it is also a mistake to expect the benefit of a tip device to come just from thrust on the device itself. As we saw above, in an optimally loaded unswept wing/winglet combination, the winglet produces no thrust. And the idea of a winglet as a thrust producer fares even worse in the case of a forward-swept wing: An optimally loaded winglet on a forward-swept wing can produce a large drag on its own surfaces and still produce a net drag reduction. In general, the sweep of the wing has a strong effect on the thrust or drag felt by a tip device. Let's look at this sweep effect in more detail.

According to Trefftz-plane theory, the total induced drag of a wing/winglet combination depends only on the arrangement of the trailing edges as viewed in the Trefftz plane and on the spanload (Section 8.3.4). Trefftz-plane theory is not exact, but for high-aspect-ratio surfaces it is a reasonable approximation. Total induced drag should thus be largely independent of sweep, provided that the spanload is fixed. On the other hand, the induced drag or thrust felt by the winglet itself depends strongly on the general sweep of the whole lifting system. In fact, the *distribution* of induced drag on the whole system, whether it has a tip device or not, is strongly affected by sweep. Induced-drag distributions that illustrate this are shown in Figure 8.4.5. Ideal spanloadings were assumed for a planar wing and for a wing with 20%-semispan vertical winglets, as shown in Figure 8.3.7. The Boeing WINGOP code (Craig and McLean, 1988) was used to carry out numerical lifting-line calculations of the induced-drag distributions for cases in which the wing and winglet are both unswept, and for cases in which both are swept back by 30°, and swept forward by 30°. Small spikes in the distributions at tips are due to the lifting-line numerics, and larger spikes at the winglet junction reflect shortcomings of lifting-line theory (Getting rid of the spikes at the winglet junction would require spreading the bound vorticity out chordwise, as in a lifting-surface theory). In spite of these anomalies, the general trends can be discerned.

In the unswept baseline case (Figure 8.4.5a), the induced-drag distribution is elliptic, as expected. In the unswept winglet case, the drag distribution on the wing is as we might expect based on the spanload in Figure 8.3.7, and the induced drag on the winglet is effectively zero, as we expect based on the theoretical discussion of Section 8.3.5. The strongest reduction in induced drag comes from the inboard part of the wing, and there is a small portion of the wing near the junction where the induced drag is increased.

In the swept cases (Figure 8.4.5b,c), the distributions are dramatically different. For aft sweep without a winglet, there is a large thrust on the outboard half of the wing, more than

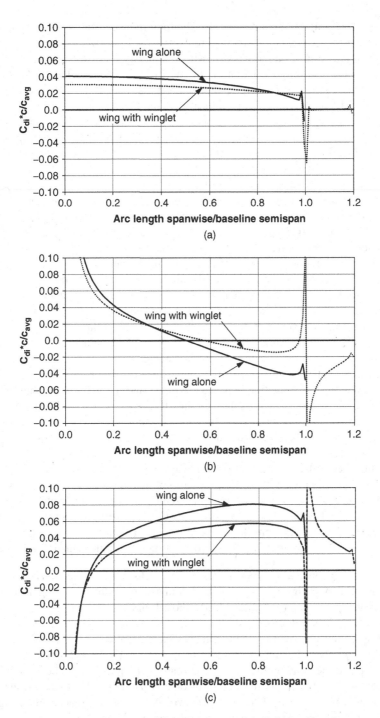

Figure 8.4.5 Induced-drag distributions calculated by numerical lifting-line theory, assuming the ideal spanloadings for a planar wing and a wing with 20%-semispan winglets, as shown in Figure 8.3.7. Lift equivalent to AR = 10.0, C_l = 1.0. (a) Wing and winglet both unswept. (b) Wing and winglet both swept 30° aft. (c) Wing and winglet both swept 30° forward

offset by the large drag on the inboard half. The total induced drag, which is the same as in the baseline unswept case, is a small difference between the two. In the aft-swept winglet case, there is a large thrust on the winglet, which is mostly offset by a reduction in thrust on the outboard wing. The drag reduction on the inboard wing isn't as large as in the unswept case. So in the aft-swept case, there is significant thrust on the winglet, but looking just at that thrust would lead you to grossly overestimate the benefit.

With forward sweep, we see generally the opposite of what we saw with aft sweep, except that the crossover from thrust to drag is farther inboard. Without a winglet, there is a very large thrust inboard of 10% semispan, more than offset by high drag on the rest of the wing. When a winglet is added, it carries a high drag on its own surfaces, which is more than offset by a reduction in drag over most of the span of the wing. In this case, the sailboat analogy, or any idea of the winglet as a thrust producer, is highly inappropriate.

Trefftz-plane theory tells us that we can reduce the ideal induced drag by increasing the vertical height of the lifting system, as well as by increasing the horizontal span. A vertical fin or winglet that adds vertical height to the system will reduce the ideal induced drag if it is placed anywhere along the span of the wing off of the airplane center plane, but it is most effective by far when it is placed at the station of maximum span; that is, at the tip. This is one example of the more general problem of minimizing the ideal induced drag of a lifting system with given maximum horizontal span and vertical height; that is, a system that must fit within a given rectangular box in the Trefftz plane. Figure 8.4.6 illustrates a series of such configurations, in order from lowest ideal induced drag to highest. The configuration with the lowest drag is the box wing, which has lifting surfaces along all four edges of the box. Note that any retreat from either the corners or the edges of the box (e.g., a "blending" region in the junction between a winglet and the wing in the third example, or the retreat from the outer edge of the box by the "feathers" in the fourth example) increases ideal induced drag, but that there may be compensating advantages such as avoiding the viscous-drag penalty associated with sharp-cornered intersections or reducing the wetted area of the surfaces.

Ideal-induced-drag theory is useful for guidance as to how to achieve a large reduction in induced drag, but the benefits that it implies are not generally achievable in practice. First, the induced-drag reduction that can actually be achieved in most applications typically falls significantly short of ideal. In addition, the actual induced-drag reduction is always offset by other factors that detract from the net benefit to the airplane.

Several factors can contribute to the shortfall in induced-drag reduction relative to ideal:

- **Spanloads compromised to save weight**: As we discussed in Sections 8.3.5 and 8.3.8, spanloads of real wings are usually compromised to reduce bending loads and save structural weight, by carrying more load inboard and less load outboard than the ideal spanload. (This is in addition to the fact that the spanload with minimum induced drag for a wing in the presence of a down-loaded horizontal tail typically already has less-than-elliptic outboard loading, as we saw in Figure 8.3.18.) If the baseline wing without a tip device and the wing with a device added are both optimized to the same weight/drag objective, they will both carry reduced loads outboard compared with their respective ideal loadings, and a tip device of a given size will have less leverage in reducing induced drag than it would if both configurations had ideal loadings.
- **Twist distribution of the existing wing in retrofit applications**: If the tip device is to be retrofitted to an existing wing, the jig-twist distribution of the existing wing is nearly

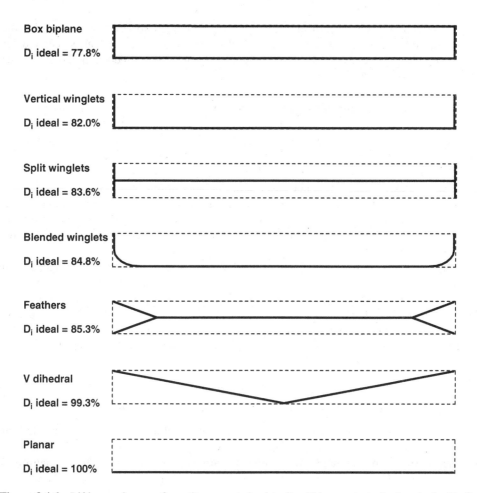

Figure 8.4.6 Lifting-surface configurations constrained to fit within a rectangular box in the Trefftz plane (height equal to 10% of full span), shown in order from the lowest ideal induced drag to the highest

always preserved for cost reasons, and as a result the best spanload that can be achieved with the device installed will be farther from ideal than the load on the original wing was. This penalty for keeping the existing twist distribution is larger when the baseline wing is more lightly loaded outboard than ideal, which is usually the case.

- **Aeroelastic effects in retrofit applications**: In the case of an aft-swept wing, aeroelastic effects may add to the shortfall in drag reduction. Increasing the outboard loading (using a tip device) tends to increase wing bending at cruise conditions, which on an aft-swept wing washes the wing out, reducing the drag benefit of the tip device compared with what it would have been on a rigid wing. The amount of additional wing washout and the resultant reduction of the induced drag benefit will depend on the flexibility of the wing, on the weight distribution of the airplane (payload and fuel), and on whether the structure is beefed up for the addition of the tip device.

- **Trim effects**: In the case of an aft-swept wing, adding a tip device that increases the loading outboard will increase the download on the horizontal tail that is required for trim, which in most cases will offset some of the induced-drag reduction and add to the offsetting profile-drag increase.

The addition of a wingtip device generally adds wetted surface area and thereby increases viscous drag, and there may also be junction flows or areas with unfavorable pressure distributions that further increase the viscous drag. The redistribution of the spanload that the device produces can change the shock drag on the rest of the wing, but this effect can go in either direction and is usually not large. In any case, the induced-drag reduction is nearly always partly offset by a net increase in the other drag components.

Any practical device that reduces induced drag generally increases bending moments on the entire wing at the cruise condition and at the critical flight conditions that determine the design of the wing structure. The addition of a wingtip device therefore often requires beefing up the wing structure, which adds weight, over and above the weight of the device itself, and subtracts from the net benefit of the device. This trade between drag reduction and weight increase is discussed further in Section 8.4.5. When a tip device is included in the design of an all-new wing, this structural-weight penalty must generally be paid in full. On an existing airplane, flight testing will sometimes have established that the wing has excess structural margin that can be "used up" by the addition of a tip device. The presence of an existing excess structural margin can thus reduce or even eliminate the required beefing up of the existing structure.

8.4.3 Milestones in the Development of Theory and Practice

Lanchester, the British aeronautical pioneer, had developed a qualitative understanding of the 3D flow around a lifting wing, including the vortex wake, by 1895 (Lanchester, 1907). A quantitative understanding of induced drag was first provided by the Trefftz-plane/lifting-line theory, developed by Prandtl in 1910 (Prandtl and Tietjens, 1934) and elaborated by several others in the following years. Even now, well into the era of CFD, our conceptual understanding of induced drag depends almost entirely on this early theoretical work. The conceptual touchstones include

- Ideal-induced-drag theory, including the elliptic ideal spanload (for minimum induced drag) and the simple formula for predicting the ideal induced drag of a planar wing.
- The prediction that multiple and/or nonplanar lifting surfaces, including endplates, could have lower induced drag than a simple planar wing of the same maximum horizontal span.
- Munk's stagger theorem (see Kroo, 2001), and the general prediction that, for a given Trefftz-plane geometry and spanload, the induced drag is independent of the fore-and-aft disposition of the lifting surfaces.

We've already discussed ideal-induced drag theory and pointed out that it provides a method by which the potential induced-drag-reduction effectiveness of various lifting-surface geometries can be compared fairly. One of the best-known examples of such studies is by Cone (1962), in which he used a physical analog apparatus based on a rheoelectric analogy to "solve" the ideal-induced-drag optimization problem and compare many different

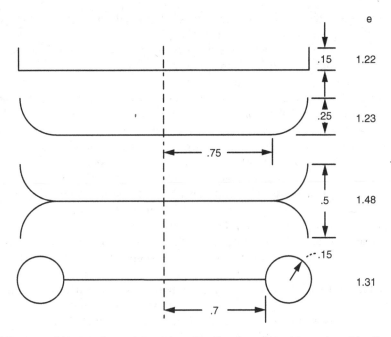

Figure 8.4.7 Some lifting-surface configurations (Trefftz-plane shapes) investigated by Cone (1962). (Published by NASA)

Trefftz-plane shapes, a sampling of which is shown in Figure 8.4.7. He found that practically any nonplanar shape that adds vertical height near the tip would reduce ideal induced drag. Results like these provided the basis for the development of many of the tip-device concepts discussed in the next section.

The realization that non-planar lifting systems could generally reduce drag did not immediately lead to successful applications, however. For example, simple flat endplates were tried numerous times in the years after it was first predicted that endplates would reduce induced drag, but in practice they never produced a net drag benefit. Their induced-drag reduction tended to fall short of ideal, and it was always more than offset by the increase in viscous drag due to added wetted area and corner flows. I think it is likely that the failure for so many years to find a better configuration than the flat endplate can be blamed on an "endplate paradigm" based on a particular way of looking at the Trefftz-plane theory. The reasoning leading to the endplate paradigm is as follows. In the limit as the vertical span of an endplate becomes large, the ideal spanload on the horizontal wing becomes uniform. One way to achieve this situation is to have 2D flow over the wing, enforced by endplates that are flat and large in chord as well as span. The resulting paradigm is that an endplate should always be flat and have a large chord.

What the endplate paradigm fails to recognize is that to realize the ideal induced drag of an endplated configuration, only the spanwise distribution of load on the endplate needs to be ideal, and it doesn't matter how the load is distributed longitudinally. A vertical tip device of small chord can achieve the same induced-drag reduction as a large endplate of the same span just by carrying the right spanload. Whitcomb (1976) seems to have been the first to recognize this and to realize that an effective endplate is just another part of the lifting system; that is, a lifting surface that should be carrying a spanload close to ideal, just like the rest of the wing. Of course, to keep the viscous drag of a lifting surface low, the surface should have an efficient aerodynamic cross section; that is, an airfoil, and the chord of the surface should be sized consistent with the efficient load carrying capacity of the section. This is just good aerodynamic design practice of the kind that has always been applied to wings, and Whitcomb's contribution was to apply it to what had formerly been seen as just an endplate. While the direct result of Whitcomb's work was the classic near-vertical winglet, his general idea of applying good design practice to keep the profile drag low has also contributed to the development of concepts other than the winglet.

The trade between drag reduction and structural weight was not addressed explicitly in most early work, but it began to attract attention with the development of the winglet. Whitcomb's work on winglets suggested that for a given increase in bending moment on the inboard wing, a near-vertical winglet offers nearly twice as much drag reduction as a horizontal span extension. This suggestion was not based on theory, but on the results of wind-tunnel tests in which the winglet and horizontal span extension configurations were supposed to be sized so as to be equivalent in terms of root bending moment. However, judging by the results of many later studies, the horizontal span extension Whitcomb used for comparison was not as large as it should have been. Furthermore, the horizontal span extension was not as well optimized aerodynamically as the winglet. Both of these factors contributed to an overly optimistic assessment of the winglet.

Later, a systematic theoretical investigation of the question was published by Jones and Lasinski (1980). They used Trefftz-plane theory to calculate induced drag, and bending moment integrated over the span as an indicator of likely structural weight. (This is only a rough indicator because for real wings at the high-g conditions that are a factor in determining the required structure, the spanload shape is usually different from that at cruise, and the effective structural depth of real wings is not usually constant along the span.) Starting with a baseline elliptically loaded wing of a given span, they added horizontal and vertical tip extensions of varying length. For each device length, they optimized the spanload to minimize induced drag, subject to the constraint that the integrated bending moment, or "structural weight," was the same as that of the baseline. As the size of the extensions increased, the spanloads had to become increasingly non-ideal (farther from the pure induced-drag optimum) to meet the constraint, and yet drag was still reduced. Repeating the calculations with root bending moment instead of integrated bending moment as the constraint produced essentially the same results. The calculations indicate that horizontal span extensions and vertical winglets offer essentially the same maximum induced-drag reduction when the spanloads are constrained so that there is no increase in "structural weight." They also indicate that to achieve a given level of drag reduction, a vertical winglet must be nearly twice as large as a horizontal span extension. In Section 8.4.5, we'll see similar results for some variations on the basic vertical winglet.

8.4.4 Wingtip Device Concepts

Whitcomb's breaking of the "endplate paradigm" has led to the development of a variety of wingtip devices that can be effective in reducing total drag, some of which were shown in Figure 8.4.1. It is assumed from here on that Whitcomb's basic idea of applying good

Table 8.4.1   Basic strategies for practical tip devices

Increasing horizontal span	Straight tip extensions
	Tapered tip extensions
Going non-planar	Bending (winglets)
	Bending with blending (blended winglets: reduced wetted area and junction drag)
	Splitting (split winglets and feathers: vertical height with less wetted area penalty)
	Splitting and rejoining (spiroids)
Part-chord devices (less chord than the baseline wingtip)	Trapezoidal baseline tip often has more chord than needed to carry spanload
Pronounced tapering	Reduces wetted area and critical structural loads
	Applicable to all tip-device concepts
Additional sweep (e.g., raked tips)	Increases aeroelastic washout and reduces wing weight
	Aeroelastic relief not very effective for a vertical winglet, but sweeping a winglet aft can make transonic tailoring easier

aerodynamic design practice will be adhered to in executing any of the concepts. We'll start our discussion of this assortment of competing concepts by listing the basic strategies (Table 8.4.1) that are used in various combinations by the different devices.

8.4.5   Effectiveness of Various Device Configurations

For comparisons of the potential performance advantages of various tip-device configurations to be fair, the devices in question must be comparably optimized. We've seen that much of the early theoretical work was highly idealized and concentrated on induced drag only. For that case, the appropriate optimization is the minimization of induced drag alone, which defines the "ideal" induced drag and the "ideal" spanload. Ideal-induced-drag theory is useful for initial screening of concepts and for understanding basic trends, and we'll look at such comparisons below, but it is not a realistic optimization target for real-world tip devices. The spanload of a real transport-airplane wing, with or without a tip device, is not generally optimized for minimum induced drag, but instead is optimized for a favorable trade between total drag and structural weight. This is still spanload optimization, just to a bottom-line performance objective such as fuel-burn or maximum range, rather than to an esoteric aerodynamic target.

The usual procedure in design studies is to define the general configuration of a candidate tip device in terms of its planform and dihedral angle(s) and then to estimate its performance through analysis. A step that should always be included in this process is the optimization of the spanload. When the planforms and airfoil cross sections of the wing and the tip device are given, the spanload is controlled by the twist distribution. (Recall that in Section 8.2.1 we defined "twist distribution" to refer to the spanwise distribution of the orientation of the zero-lift lines of the sections. Thus what we refer to here as the "twist distribution" would encompass the overall incidence setting of a tip device, or, for example, the "toed-in" or "toed-out" setting of a winglet, as well as the variation in incidence along the span of the device.) If both the wing and the tip device are all new, the twist distribution of the entire system is open to optimization. In derivative or retrofit applications, the jig twist distribution of the existing wing is generally fixed, and the twist distribution of only the tip device itself can be optimized. If the twist distribution of the existing wing was optimized for operation without a tip device, the benefit available from the addition of the tip device will usually be substantially less than it would have been if the wing could have been reoptimized. In the discussion that follows, it is assumed that the twist distribution of each candidate tip device has been optimized to an appropriate bottom-line performance objective, taking into account structural weight and any other factor that affects the objective, subject to whatever constraints are applicable. For real-world design studies, computational tools are available that can carry out this optimization to different levels of physical fidelity, depending on the purpose of the study.

Device size is always an important design decision, and it affects all devices in essentially the same way. Figure 8.4.8 shows figurative trends in drag, structural weight, and fuel burn as functions of device size for a generic tip device, assuming that the baseline airplane with no device has no excess structural margin, and that for each size the device has been optimized for some bottom-line measure of airplane performance, such as fuel burn. For simplicity, the drag and weight effects are assessed separately first (the drag reduction is plotted for constant weight) and then combined in their overall effect on fuel burn. Note that induced drag cannot be reduced below zero, no matter how large the device is made and that wetted area and profile drag will continue to increase with increasing size. Thus the drag reduction

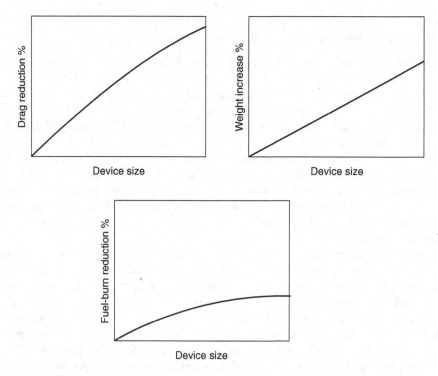

Figure 8.4.8 Effects of device size on performance for a generic wingtip device, assuming no excess structural margin in baseline wing

at constant weight must eventually show a diminishing rate of return with increasing size, as indicated by the decreasing slope of the drag reduction curve in the first frame. Because of the ever-increasing profile drag, eventually the drag reduction would reach a maximum and then decline. But for this pure aerodynamic trade at constant weight, the maximum drag reduction would typically occur well outside the practical range of device size. In the real world, the effects of the increasing structural weight would intervene long before that.

For reasonable device sizes, the weight increase tends to be roughly linear with size. When the weight increase is taken into account, the fuel-burn reduction generally reaches a maximum well before the maximum in pure aerodynamic drag reduction at constant weight. The optimum device size defined by the maximum fuel-burn reduction depends on the relative importance of drag and weight, which depends on the length of the mission. If maximum range is the objective, the situation is even more complicated because the critical condition that limits maximum range can be different depending on the details of the baseline airplane design. Compared to a device optimized for fuel-burn, a device optimized for range can be anywhere from considerably smaller to considerably larger, depending on the details. Thus the optimum device size depends on many factors, including what performance objective is sought.

Now let's look at the comparative advantages of horizontal span and vertical height. Earlier we mentioned that Whitcomb's work on winglets (Whitcomb, 1976) suggested a rule of thumb to the effect that for a given increase in bending moment on the inboard wing, a near-vertical winglet offers nearly twice as much drag reduction as a horizontal span extension. This rule of thumb has not been borne out by studies since then. Jones and Lasinski (1980) indicated that horizontal span extensions and vertical winglets offer essentially the same maximum induced-drag reduction when the impact on "structural weight" due to bending loads is constrained to be the same.

In Figure 8.4.9, we look at the problem a different way but reach a similar conclusion. While Jones and Lasinski constrained the root bending moment and looked at non-ideal induced drag, here we look at how ideal induced drag is reduced as a function of the increase in root bending moment. In these ideal-induced-drag calculations carried out by the author, induced drag was minimized for a fixed total lift, and root bending moment was free to increase by different amounts for the different devices. The three winglets indicated by the diamond symbols have the same 20% semispan height and thus the same ideal-induced-drag reductions as the corresponding ones shown in Figure 8.4.6. The difference here is that the increase in root bending moment is also shown, on the horizontal axis. Two main conclusions are:

1. For a horizontal span extension to produce the same ideal-induced-drag reduction as a winglet, it needs to be only about half as large (a little more than half compared with a sharp-cornered vertical winglet, and a little less than half compared with a blended or split winglet). A horizontal span extension thus needs less additional wetted area than a winglet for the same induced-drag reduction, and taking profile drag into account would move the total-drag curves for horizontal and vertical devices closer together than the induced-drag curves.
2. For a given increase in root bending moment, the three types of winglets produce roughly the same ideal-induced-drag reduction, about 17% more reduction than a horizontal extension. This is a much smaller difference than the factor of nearly two claimed by Whitcomb (1976). And again, taking profile drag into account makes the total-drag difference even smaller.

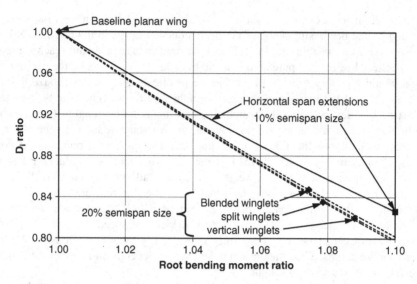

Figure 8.4.9   Reduction in ideal induced drag versus increase in root bending moment for the three types of winglets shown in Figure 8.4.6, compared with horizontal span extensions. The lines indicate the trends as the sizes of the devices are increased from zero, and the solid symbols are for two particular sizes that yield roughly comparable drag reductions: 10% semispan horizontal span extensions, and 20% semispan winglets (height measurement)

This comparison is highly idealized and should be taken only as a rough indication of the trends. As we noted above, real tip devices are not generally optimized for ideal induced drag. Also, wing root bending moment associated with the cruise spanload is only a very crude indicator of structural-weight impact. Still, a general conclusion is that horizontal span extensions and vertical winglets have very similar performance potential in terms of the trade between drag reduction and weight increase. In general, in the choice between winglets, horizontal span extensions, and other tip-device configurations, there is no clear-cut favorite for all applications. In terms of the basic physics, the benefits they offer tend to be comparable. Which choice is favored for a particular application depends on the details of the baseline airplane design, both aerodynamic and structural, and on the mission objective. And the differences between the choices are usually not large.

Theoretically, there is a small general advantage for devices in which the lifting surface splits into two branches. In general, splitting allows a given level of induced-drag reduction to be achieved with less additional wetted area than a non-split configuration would require. For a feather configuration with a small included angle (see Figure 8.4.6), the wetted-area increase is only a little more than that of a horizontal span extension of the same projected span, but the induced-drag reduction is considerably greater. Of course, the weight increase is also considerably greater, so that the advantage in terms of drag versus weight is not large. If we increase the included angle between the feathers to 180°, we have split vertical winglets of equal span. Comparing this with a simple vertical winglet of the same total height, we find that the optimum spanloads for both are about the same just inboard of the junction. In the case of the split winglets, this load splits evenly between the two branches, so that if we size the chord of the device according to the load carried, the split winglets need only half as much chord as the single winglet. The split winglets produce about 90% of the induced-drag reduction of the single winglet (compare the "split winglets" with the "vertical winglets" in Figure 8.4.6) with only about half as much additional wetted area. A drawback to split winglets in practice is that the span of the lower winglet is often limited by ground clearance.

8.5 Manifestations of Lift in the Atmosphere at Large

In previous sections, we looked at the flow around a 3D wing, the lift distribution on a 3D wing, and the theory of induced drag. In this section, we look at the manifestations that the lift on a wing has in the atmosphere on a large scale and at some of the theoretical issues that arise in representing those manifestations.

We start with a look at how the flowfield associated with the lift satisfies Newton's laws in an integrated sense. Highly simplified models suffice for this purpose, but the mathematics involved is tricky, and some erroneous results have been widely propagated. A secure understanding of these issues thus requires addressing how the erroneous analyses went wrong. When the smoke has cleared, we'll find that there is no net downward momentum imparted to the atmosphere as a whole and that the lift is reacted by pressure differences on horizontal planes above and below the wing, or on the ground plane, if there is one. We'll also consider how conservation of momentum applies to control volumes that don't encompass the entire atmosphere. To keep things simple, we'll look only at rectangular volumes with horizontal and vertical surfaces. We'll find that the lift can show up at the boundaries either as pressure differences on the horizontal surfaces or as fluxes of vertical momentum mainly through the vertical surfaces, or as combinations of the two, depending on the proportions of the control volume.

Finally, the actual process by which the vortex wake disappears far downstream has received little coverage in the usual aerodynamics sources, and we'll attempt to at least fill in a rough picture of what happens there.

8.5.1 The Net Vertical Momentum Imparted to the Atmosphere

There is a widespread notion that an airplane in steady level flight continuously imparts net downward momentum to the atmosphere. Contributing to this notion is the fact that the airplane is continuously adding to the *impulse* of its vortex-wake system and the fact that there is net flux of downward momentum across vertical planes behind the airplane, issues we'll deal with in the following sections. We also saw that a 2D airfoil imparts a downward momentum change to the air, but confusing the issue in that case is the fact that the downward momentum behind the airfoil accounts for only half the lift. However, appearances are deceiving, and the notion that the airplane leaves behind a continuously increasing net downward momentum is mistaken.

For our purpose of assessing the net vertical momentum, a very simple model of the flowfield produced by an airplane in steady level flight suffices. That is the classical, highly idealized model in which the vortex wake is represented by a pair of parallel line vortices, and the sinking and any other deformation of the wake are ignored. Assuming the airplane has been flying for a finite time since an impulsive start, we then have a vortex system consisting of a planar, rectangular loop, as shown in Figure 8.5.1. This model is unrealistic

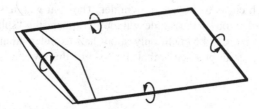

Figure 8.5.1 Idealized rectangular-loop model for the vortex system of a 3D wing flying for a finite time

in several ways, especially in the vicinity of the starting vortex. Spalart (2008) has pointed out that a simple flow structure for this part of the vortex loop is not sustainable over time, where in the real world the vortex system would quickly distort, become unsteady, and disperse. However, we'll find that the starting vortex isn't a player in the overall vertical-momentum balance, which indicates that this oversimplification does no harm for the level of analysis we're attempting here. And in Section 8.5.5, we'll discuss how the vortex wake really terminates downstream.

Given our representation of the vortex-wake system as a closed vortex loop, a relevant concept is that of the *impulse of a vortex loop*. It is known that any incompressible potential flow can be established by the application of an impulsive pressure field at an initial instant, starting from an initial condition in which the fluid is at rest (Milne-Thomson, 1966, Section 3.31). In the case of a closed vortex loop like the one in Figure 8.5.1, in a field that would otherwise be at rest, an impulsive pressure field that could initiate the flow would have to impose a net *impulse* on the fluid, proportional to the circulation and the area of the loop, and in the direction perpendicular to the plane of the loop (Milne-Thomson, 1966, Section 10.21). This is what is called the impulse of the vortex loop. In our idealized model, the rectangular loop of Figure 8.5.1 is attributed to an airplane flying with lift L for time t. The impulse of this vortex loop, and the actual mechanical impulse imposed by the airplane on the atmosphere (lift times time), are both equal to Lt. They both increase with time as the airplane continues to fly and the area of the vortex loop grows. So we have a one-to-one correspondence between the mechanical impulse imposed by the airplane and the impulse of the vortex loop.

Thinking in intuitive physical terms, we might also expect the impulse imposed by the airplane on the air (the product Lt) to produce a net vertical momentum in the atmosphere that grows with time. This expectation is not satisfied by the mathematics, however. Wigton (1987, private communication) showed that the volume integral of the vertical velocity inferred from the vortex system, over the whole infinite atmosphere, is nonconvergent. The integrand (the vertical velocity) doesn't decrease fast enough with increasing distance (only as r^{-3}), and the flowfield associated with the vortex loop thus contains infinite amounts of both upward and downward momentum. An attempt to define a value for the integral can lead to any answer between minus infinity and plus infinity, depending on how the integration is sequenced. Thus the vertical-momentum content of the infinite atmosphere at any given time is at best undetermined, even though the mechanical impulse imposed by the airplane and the impulse of the vortex loop are well defined.

This unsatisfying result is an artifact of our idealized assumptions: incompressible flow and an infinite atmosphere. In the real world, no disturbance generated by the airplane would propagate faster than the speed of sound, and for finite flight time all flowfield integrals would be well defined, so presumably we could resolve this difficulty by doing an unsteady compressible analysis. But there is an easier way. The problem is resolved if we consider flight in a semi-infinite atmosphere bounded by a ground plane. The simplest way to impose the no-through-flow condition at the ground plane is to place an image of the "real" vortex system below the ground, as shown in Figure 8.5.2. This is the same thing we did in Figure 8.2.6 to model flight close to the ground, but this time our purpose is to model flight many wingspans from the ground. At large distances, the velocity disturbances associated with the vortex loop and its image die off as r^{-4}, which is fast enough for the vertical-momentum integral to converge. If we expected to see a net downward momentum equal to Lt, the result comes as a surprise: The value of the integral over the semi-infinite space above the ground is zero, which means that the airplane imparts no net downward momentum to the atmosphere in steady level flight over a ground plane, regardless of height above the ground. If momentum conservation is to be satisfied, there is only one other way the lift can be balanced, and that is by the pressure, which is our next topic.

8.5.2   The Pressure Far above and below the Airplane

Our finding of zero net vertical momentum shouldn't have been a surprise. Prandtl and Tietjens (1934) showed how in steady level flight the lift is balanced by an overpressure on the ground under the airplane, so that of course there is no need for net momentum transfer. They assumed a rectangular vortex system and its image, like that shown in Figure 8.5.2, and showed that when the height is large compared with the wingspan, there is no significant pressure disturbance at the ground associated with the trailing vortices, but that there is a disturbance at the ground associated with the bound vortex and its image, given by

$$\Delta p = \frac{Lh}{2\pi R^3},\qquad(8.5.1)$$

where h is the airplane's height above the ground and R is the radial distance from the airplane to the point in question on the ground. The disturbance is thus distributed in a

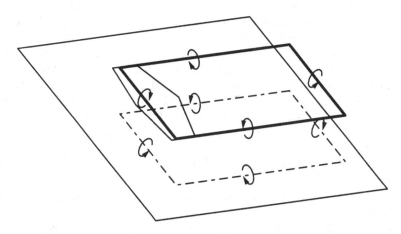

Figure 8.5.2   The rectangular-loop vortex system of Figure 8.5.1 reflected in a ground plane

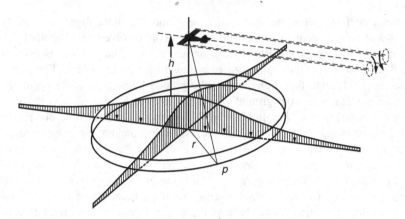

Figure 8.5.3 The airplane's pressure footprint on the ground in steady, level flight. From Prandtl and Tietjens, (1934). Used with permission of Dover Publications, Inc.

pattern of circular isobars centered directly under the airplane. The overpressure dies off radially in a bell-shaped distribution, as shown in Figure 8.5.3, and Prandtl and Tietjens showed that the integrated pressure disturbance accounts for all of the lift. The gross features of the pressure footprint predicted by the simple incompressible model are not altered by compressibility at subsonic speeds. For steady flight, the footprint remains centered under the airplane regardless of the Mach number, as long as it is subsonic. This steady-flow pressure distribution is not like an acoustic radiation pattern, which would lag behind the airplane at high subsonic Mach numbers.

The presence of an integrated pressure disturbance imposed by the airplane on the ground plane has an easy intuitive interpretation: The ground plane is supporting the weight of the airplane in addition to the weight of the atmosphere. The typical local magnitude of the pressure disturbance at the ground is extremely small. For example, for a 747 weighing 800 000 pounds flying at 35 000 ft AGL, the maximum pressure disturbance directly under the airplane is 1.0×10^{-4} pounds per square foot or 5×10^{-8} atmospheres. And, of course, the area covered by the disturbance is typically quite large. The disturbance is greater than half its maximum value inside a circle on the ground of radius 0.766 h, which for our 747 at 35 000 ft covers an area of 81 square miles.

Above the airplane, the integrated pressure disturbances "due to" the bound vortex and its image cancel each other, which is consistent with having all of the lift accounted for by the pressure disturbance on the ground below. The other place we should look for manifestations of lift is under the starting vortex, which wasn't mentioned by Prandtl and Tietjens and which in the real world would be out of the picture in any reasonable length of time. We'll look at it anyway, just to understand the full implications of our vortex-loop model. One might expect that there would be a "suction" footprint under the starting vortex, a mirror image of the pressure footprint under the airplane, because the starting vortex has a rotation opposite to that of the bound vortex. But it turns out that the integrated pressure disturbance under the starting vortex is zero. This is an instance in which being in the right reference frame is crucial.

The Prandtl and Tietjens analysis of the pressure footprint is carried out in the reference frame moving with the airplane, in which the flow under airplane is steady, and the steady form of Bernoulli's equation holds. In this reference frame, there is a substantial freestream velocity, and far below the airplane, where the disturbances are small, the pressure disturbance goes as $\rho u_\infty u$, that is, the pressure disturbance is first order in the velocity disturbance u. To use the steady Bernoulli equation in the analysis of the flow under the starting vortex, we must move to the reference frame of the air mass, in which the starting vortex is not moving. In this frame, there is no freestream velocity, and the pressure disturbance goes as ρu^2, second order in the disturbance velocity, so that the integrated pressure disturbance vanishes in the limit of large height.

So we see that for steady level flight in a semi-infinite atmosphere the lift is transmitted to the ground by the spread-out pressure disturbance of Figure 8.5.3. And none of the conclusions we've drawn here would change much if we introduced a more realistic model for the flow and the development of the wake near the airplane.

Now let's return briefly to the case of the infinite atmosphere. The Prandtl and Tietjens analysis of the pressure on the ground plane has interesting consequences when it is applied to the situation without a ground plane. If we remove the ground plane by removing the image vortex system, the pressure disturbance at the original location of the ground plane is reduced to half what it was when the ground plane was there. And in the absence of an image vortex system, the integrated pressure disturbance above the airplane is not canceled. So looking at horizontal planes above and below, we now have a positive integrated pressure disturbance below the airplane accounting for half the lift, and a negative disturbance above the airplane accounting for the other half, as illustrated in Figure 8.5.4a. This holds regardless of how far above and below the airplane we place the planes. The transport of vertical momentum through these horizontal planes above and below the airplane is second order in the velocity disturbances, so that its integrated effect vanishes as we move the planes far away. All of this indicates that in an infinite atmosphere, pressure differences on horizontal planes far above and below the airplane account for the lift, and that there is no net transfer of momentum to the atmosphere. This of course implies that the "physically correct" value for the volume integral of the vertical momentum in the atmosphere should be zero, though we found it before to be mathematically indeterminate.

8.5.3 Downwash in the Trefftz Plane and Other Momentum-Conservation Issues

We've seen that no net vertical momentum is imparted to the atmosphere as a whole by an airplane in steady level flight. However, there are regions of both upward and downward momentum in the field, and it is instructive to look at the momentum balance at the boundaries of control volumes that don't encompass the entire atmosphere. We'll limit our attention to rectangular volumes with vertical and horizontal faces and start by considering fluxes of vertical momentum across vertical planes at various stations in the flow. One example is the Trefftz plane downstream of the airplane, which we used as the downstream boundary of a control volume in our derivation of the general farfield integral for the induced drag, Equation 8.3.5. Then, of course, it was the flux of flight-direction momentum we were interested in, while now it is the flux of vertical momentum.

There is a classical argument for the downward momentum at the Trefftz plane in an infinite atmosphere that goes as follows. A rectangular vortex loop is assumed, as in Figure 8.5.1, and the Trefftz plane is placed between the bound and starting vortices, far

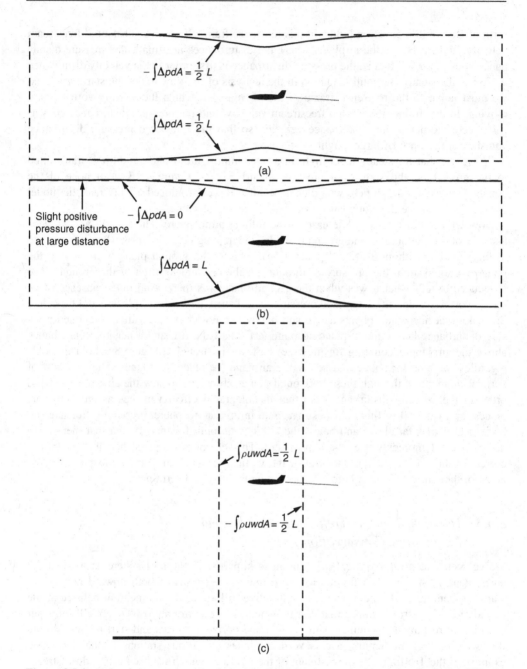

Figure 8.5.4 Balancing momentum, pressure, and lift in rectangular control volumes of different proportions. (a) $\Delta x \gg \Delta z$ with no ground plane, or with a ground plane far away relative to any dimension of the control volume. (b) $\Delta x \gg \Delta z$ with a ground plane as the bottom boundary of the control volume. (c) $\Delta z \gg \Delta x$ with or without a ground plane

from both, and is pierced by the trailing vortices. At sufficiently large distances, the bound and starting vortices contribute nothing, and only the trailing vortices have to be considered. The expression for the w disturbance "induced" by a trailing-vortex pair of infinite length is then integrated over the Trefftz plane, and the result is found to be (Thwaites, 1958, p. 303, for example)

$$\iint_T w\,dz\,dy = \frac{-L}{\rho u_\infty}. \tag{8.5.2}$$

A widely held interpretation of this result is that there is a flux of downward momentum through the Trefftz plane, consistent with the lift on the wing, and that it is "induced" by the trailing vortices.

However, it turns out that this classical result is incorrect. Wigton (1987, private communication) found that the w integral 8.5.2 is nonconvergent (w dies off only as r^{-2}), and he observed the classical symptom of nonconvergence, that is, that the value obtained depends on the order of integration. Reversing the order from that shown in Equation 8.5.2 yields a value of zero. Wigton also found that if he used the w expression for trailing vortices of finite length, consistent with a finite vortex loop as in Figure 8.5.1, the integral converges, and the correct answer is zero, regardless of the length of the vortices or the order of integration:

$$\iint_T w\,dy\,dz = \iint_T w\,dz\,dy = 0. \tag{8.5.3}$$

So Wigton concluded that the classical argument is wrong and that the part of the velocity field "induced" by the trailing vortices results in no net flux of vertical momentum through the Trefftz plane. Lissaman (1996) independently came to the same conclusion.

The story of this error represents an interesting cautionary tale. Clearly, incorrectly evaluating an improper integral is an easy trap to fall into. You can apply the standard procedures for evaluating integrals and, without making any procedural error, obtain a wrong answer. Equation 8.5.2 fortuitously gave the "expected" result, which apparently fooled more than one prominent aerodynamicist. And the incorrect result has been used in other analyses, rendering them incorrect as well. Sears's analysis of the pressure in the Trefftz plane is an example we'll discuss in Section 8.5.4.

The trailing-vortex result of Equation 8.5.2 is not the only error Wigton found. He also found that the classical argument that the bound and starting vortices contribute nothing is wrong. For any Trefftz plane between the bound and starting vortices, each "induces" an integrated downward momentum corresponding to half the lift, regardless of distance. At large distances, the w disturbances "induced" by the bound and starting vortices are very small, but are spread over a very large area. Thus the Trefftz plane in an infinite atmosphere does see a flux of downward momentum corresponding to the lift after all, but it is distributed very differently from the way it was in the classical picture, in which it was assumed to be associated with the trailing vortices. Lissaman (1996) also independently found this result. It applies only to a vertical plane between the bound and starting vortices. For a vertical plane either ahead of the airplane or behind the starting vortex, the flux of vertical momentum is zero, because the integrated contributions from the two vortices cancel.

This finding of an integrated downward momentum in some vertical planes and none in others appears to contradict our finding of zero for the volume integral of the vertical momentum, even for an infinite atmosphere (in Sections 8.5.1 and 8.5.2). If we view the infinite-atmosphere case as a series of "Trefftz-plane" slices stacked in the flight direction, we see no net downward momentum in the slices ahead of the airplane or behind the starting vortex, and nonzero net downward momentum in the slices between the airplane and the starting vortex. If we were to naively sum the stack, we would conclude that the atmosphere contains a net downward momentum equal to Lt. But this naive summing is just another example of an incorrect evaluation of an improper integral, in this case the 3D volume integral of w in an infinite domain. Remember that we found that this integral is nonconvergent and therefore indeterminate mathematically. Physically speaking, however, our argument based on the pressure indicated that the airplane imparts no net vertical momentum to the atmosphere. So the physically "correct" value for the volume integral of the vertical momentum in an infinite atmosphere is still zero and is not contradicted by the Trefftz-plane findings.

So how do we reconcile a net flux of downward momentum in the Trefftz plane with the findings that there is no net vertical momentum imparted to the atmosphere by the lift and that the lift is reacted by pressure differences on horizontal planes? If we look at the momentum balance in control volumes of different proportions, as illustrated in Figure 8.5.4, it appears that all of this can be reconciled, though we'll find that the lift is accounted for either by pressure or by momentum flux, depending on the proportions of the control volume.

Consider rectangular control volumes with dimensions Δx, Δy, Δz. Assume that the spanwise dimension Δy is very large compared with the other two, so that fluxes of vertical momentum through the side boundaries effectively vanish. Now, keeping Δx and Δz much smaller than Δy, we can define two limiting shapes for the control volume in side view, tall and slender, or wide and flattened. Looking at these limiting cases with and without a ground plane gives us the three combinations shown in Figure 8.5.4. Unlike Wigton's (1987, private communication) analysis above, we put the starting vortex out of the picture by assuming that it is far downstream relative to any dimension of our control volume, so that its "contribution" to any integrated momentum fluxes effectively vanishes. There are only three distinct combinations in Figure 8.5.4 because, with the starting vortex out of the picture, it doesn't matter whether there is a ground plane in the case of the tall, slender control volume in Figure 8.5.4c. The momentum balances then work out as follows:

1. $\Delta x \gg \Delta z$ with no ground plane, or with a ground plane far away relative to any dimension of our control volume, so that the image vortex system has negligible "effect": As Δx grows large relative to Δz, the integrated fluxes of vertical momentum through the upstream and downstream boundaries effectively vanish, and the lift is accounted for by the pressure disturbances on the top and bottom boundaries, as in Figure 8.5.4a. The pressures on the top and bottom boundaries account for half the lift each, regardless of the relative distances above and below the airplane. If one of the boundaries is moved farther away, the pressure disturbances become weaker but are more spread out, so that the integral is unchanged.

2. $\Delta x \gg \Delta z$ with a ground plane as the bottom boundary of the control volume: As in (1) above, the integrated fluxes through the upstream and downstream boundaries effectively vanish, and the lift is accounted for by the pressure disturbance on the ground, as in Figure 8.5.4b. This is similar to what we found in Figure 8.5.4a, except that the ground plane has the effect of doubling the pressure disturbance there, so that it accounts for all of the lift, and the pressure footprint is the same one illustrated in Figure 8.5.3.

On the upper boundary, the disturbances "due to" the bound vortex and its image are of opposite signs and cancel in an integrated sense. But they do not cancel locally. There is a central region directly above the airplane where the "contribution" of the bound vortex dominates, and the pressure disturbances are negative. At larger horizontal distances, the "contribution" of the image vortex dominates, and the pressure disturbances are positive. The positive disturbances are much weaker than the negative disturbances, but they can cancel the positive disturbances in the integral sense because they are spread over a much larger area (note that the distribution is axisymmetric, and the integration is therefore radius-weighted).

3. $\Delta z \gg \Delta x$ with or without a ground plane: As Δz grows large relative to Δx, the integrated pressure forces on the top and bottom boundaries effectively vanish, and the lift is accounted for by the fluxes of vertical momentum through the upstream and downstream boundaries, as in Figure 8.5.4c. Because we have put the starting vortex out of the picture, the total flux of vertical momentum is split equally between upward momentum upstream and downward momentum downstream instead of showing up only as downward momentum downstream, as it did in Wigton's (1987, private communication) analysis. This is the same distribution of integrated momentum flux that we found in 2D in Section 7.3.4.

I've presented these arguments in an arm-waving way, but I'm confident they represent correct limiting cases. They demonstrate that there is no inherent contradiction in seeing the lift manifested both as pressure disturbances on horizontal planes and as fluxes of vertical momentum through vertical planes. And there is no inconsistency between these manifestations and our earlier finding of no net vertical momentum in the atmosphere.

8.5.4   Sears's Incorrect Analysis of the Integrated Pressure Far Downstream

Sears (1974) used conservation of momentum in control-volume form, combined with the classical Trefftz-plane integral for the induced drag, Equation 8.3.10, to deduce that the integrated pressure in the Trefftz plane behind a lifting wing in inviscid flow is higher than ambient.

On an intuitive level, this result is surprising, because we tend to think of induced drag as being associated with the kinetic energy left behind in the downstream flowfield, and we tend to associate excess kinetic energy with a velocity magnitude higher than freestream, and pressure lower than ambient. We also tend to associate the swirling motion about the vortex cores with lower-than-ambient pressure. But our naive intuition in this case is mistaken. In inviscid incompressible flow, the pressure perturbation goes as $-2Uu - u^2 - v^2 - w^2$, so that even though v and w tend to be larger in magnitude than u near the vortex wake, the $-2Uu$ term is of higher order and can dominate. When downwash occurs near a tilted wake, u can be negative, and the perturbation pressure can be positive, contrary to our naive intuition. So there can be regions of positive perturbation pressure behind a lifting wing, and we can't rule out Sears's result just on the basis of arm waving and intuition.

But it turns out that Sears's conclusion is actually wrong. Given his flowfield model, there is a region of higher-than-ambient pressure in the Trefftz plane, but the integrated pressure is lower than ambient, not higher. So where did the analysis go wrong?

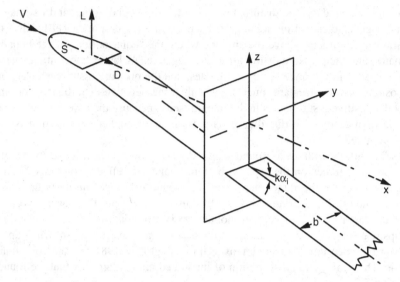

Figure 8.5.5   Sears's (1974) model using a planar, tilted vortex wake, and Trefftz plane perpendicular to flight direction

Sears assumed that the wing is elliptically loaded and that the vortex wake remains planar but is tilted downward as shown in Figure 8.5.5, at an angle consistent with what this wake model gives for the downwash far downstream of the wing. He assumed that far downstream the velocity disturbances are those "induced" by the vortex wake perpendicular to its own plane, and he substituted these velocity disturbances into the integral expressions governing the integrated momentum balance. The tilt angle of the wake thus enters directly into his analysis, which is unfortunate, as the tilt angle of an undistorted tilted wake is much too large, $\pi^2/2 \approx 4.9$ times that of a realistic rolled-up wake, as we saw in our discussion of ideal induced drag in Section 8.3.5.

But even being this far off in the tilt angle of the wake isn't enough to account for the wrong sign in the final result. That, it turns out, is the result of a mathematical error. One of the integrals appearing in Sears's analysis is the integral of w, that is, the integrated down-wash in the Trefftz plane, associated with the trailing vortex wake, for which he assumed the incorrect value from Equation 8.5.2, $-L/(\rho u_\infty)$. We saw earlier that this integral is nonconvergent for a wake of infinite length, but that its value is zero for a wake of finite length, regardless of the length. Had Sears used the correct value of zero for the w integral, he would have obtained the negative value $-D_i$ for the integrated pressure in the Trefftz plane, instead of the erroneous $+D_i$, where D_i is the induced drag.

8.5.5   The Real Flowfield Far Downstream of the Airplane

Most of our discussion so far has been based on a vortex-wake system consisting of a rectangular loop terminated at the downstream end by a starting vortex, though we noted that the assumption of a persistent starting vortex is unrealistic. Of course, in the real world, the starting vortex would be left on the runway where the airplane took off, and it would dissipate rapidly through viscous interaction with the ground. And there would be diffuse spanwise vorticity left behind as the bound vorticity changed during climb and the transition to cruise. But even if the flight started impulsively at cruise altitude, the starting vortex could not persist in the simple form we assumed. So what really happens far downstream, as the vortex-wake system collapses and the flow disturbances associated with the lift die out?

For purposes of this discussion, we'll skip the complexities of takeoff and climb. We'll assume the airplane has been in steady level flight at high altitude for a long time, and we'll seek a "steady state" understanding of the flowfield far downstream, in which a starting vortex is no longer in the picture. In Section 8.1, we described some features of the flow, up to the point where the wake "collapsed" into an unsteady pattern of loops and eddies. Here, we'll attempt to deduce what the time-averaged structure of the flow must be throughout this process of collapse and beyond, to where the velocity disturbances die out. The final stages of this progression are not of any practical importance, because the mean velocities become very small. It's just an interesting and informative exercise to figure out what the laws of motion require.

For a long distance downstream of the airplane, the rolled-up vortex wake is relatively compact and retains the descending-oval form shown in Figure 8.1.7, and the velocity and vorticity distributions shown in Figure 8.1.8 change very little. The collapse of the wake starts with the slow growth of instabilities such as the Crow instability, which leads eventually to breakup into loops and eddies, as shown in Figure 8.1.10. This process is highly variable in response to atmospheric conditions, and it changes over short and long time scales. Downstream of breakup, we have a turbulent flow with motions over a wide range of length scales, spread over a much wider area than the original descending oval occupied.

Of course, this breakup into a diffuse turbulent flow doesn't mean the vorticity in the upstream wake has disappeared. Remember that Helmholtz's second theorem tells us that vortex lines can't end in the interior of the field. The vortex lines from the wake upstream haven't ended in the field, but in the instantaneous turbulent flow downstream of breakup they have become an unsteady, chaotic tangle. Does this mean the flow is just an unorganized turbulent jumble that is otherwise featureless? No. Averaged over time, the flow must have an overall structure consistent with its origins upstream. The time-averaged flow must obey all the same kinematic rules as any other flow, which means that the abundant time-averaged vorticity that was present in the wake upstream cannot end in the interior of the field any more than the instantaneous vorticity could. And the time-averaged flow must obey the RANS equations, with the instability motions and turbulence supplying the Reynolds stresses (see Section 3.7) that transport time-averaged momentum.

So to see the underlying structure of what happens during and after breakup, imagine time-averaging the flow in the reference frame moving with the airplane, so as to take out the unsteady motions on the time scales of the instability and the eventual turbulence. This is formally the same process we used in Reynolds averaging for turbulence modeling (Section 3.7), but in this case the flow we're applying it to is different from the more usual applications of Reynolds averaging. For one thing, we'll be averaging out part of the motion (the highly coherent motion associated with instability growth) that doesn't look like ordinary turbulence. And over much of the flowfield far downstream, the mean-velocity disturbances are smaller than the turbulent fluctuations and are effectively invisible before we do the averaging. Furthermore, we'll do this strictly as a thought experiment, since no such calculation has actually been done, to my knowledge. We'll do the averaging in our imaginations and use physical reasoning to try to deduce what the qualitative structure of the time-averaged flow must be.

First, we'll assume that the airplane and the mean flow are symmetrical about the $y = 0$ plane. We'll also assume that at the start of the instability-growth and breakup process the wake is in the form of a descending oval as in Figure 8.1.7 and that the vorticity is still concentrated in two distinct cores, as in Figure 8.1.8b. The net vorticity and circulation of the core on either side of the symmetry plane must then be the same as what was shed into the nearfield from the corresponding half of the wing. One of the main things we'll be looking to explain is where all this mean-flow vorticity "goes" in the farfield, consistent with the requirement that vortex lines can't end in the interior of the field.

Something we know from observations is that things change slowly in the flight direction. Even though the breakup process produces much faster change than what ordinary turbulent diffusion could produce, change in the mean flow is still slow in the sense that the flight-direction distance over which it takes place is long compared with any other dimension of the wake. Thus we have an effectively "slender" flow in which mean-flow gradients in the direction of the wake axis play a much smaller role in the development of the flow than do gradients in other directions. In this sense, the flow in cross-flow planes is similar to the limiting case of 2D planar unsteady flow, and as an aid to understanding, it is useful to look at what can happen to the mean vorticity in simple 2D cases.

In 2D planar flow, the mean vorticity vector can have only one nonzero component, perpendicular to the plane of the flow, so we can think of the vorticity as a scalar. We'll start with the example of a single axisymmetric vortex core in an infinite 2D domain. At an initial instant, we have some axisymmetric distribution of vorticity, as shown in Figure 8.5.6a, and as time goes on it remains axisymmetric but becomes more diffuse, as in Figure 8.5.6b. The integrated vorticity and circulation remain constant, even as the vorticity spreads over a very large area and becomes locally very weak.

Our second example is a pair of counter-rotating vortices of equal strength separated by a plane of symmetry, like our vortex wake. In this case, each core has its own total vorticity and circulation in its own half of the domain, but because the two cores have vorticity of opposite signs, the total vorticity and circulation of the entire flow are zero. Starting with two cores that are initially distinct, as in Figure 8.5.7a, viscous or turbulent diffusion eventually brings the cores into "contact" in the sense that the vorticity gradient becomes significant across the plane of symmetry, as in Figure 8.5.7b. Because there isn't any no-slip surface anywhere in the field to interact with, vorticity can be destroyed only by interaction with vorticity of opposite sign, and that is what happens here. Vorticity is diffused toward the symmetry plane and annihilated there by interaction with the opposite-sign vorticity on the other side, and the circulation on each side of the symmetry plane is reduced. In this example, not only does the vorticity become more diffuse and locally weaker, but the total vorticity on each side of the symmetry plane tends to zero with time. In this 2D domain, vorticity disappears in time, without having vortex lines end in space, so Helmholtz's second theorem is not violated.

What does this imply for our real 3D flow? As we've already observed, the 3D mean flow throughout the region of breakup and decay is "slender" and therefore behaves in a nearly 2D fashion, but it is steady by definition in the reference frame of the airplane. Imagine cutting this flow with a constant-x plane fixed to the atmosphere, so that in the airplane reference frame, the cutting plane moves away from the airplane at the speed of

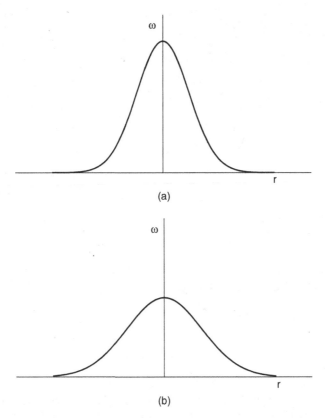

Figure 8.5.6 Diffusion of vorticity in a single axisymmetric vortex core in 2D unsteady flow. (a) Concentrated at an initial instant. (b) More diffuse at a subsequent time

the freestream. For the flow that we observe in the cutting plane, the movement in x is analogous to the advance of time in the 2D flow we just discussed, and much of what happens will be similar in the two cases. In the 3D case, the vigorous "turbulent" motions of the instability growth will diffuse the mean vorticity in all directions, including toward the plane of symmetry, just like what we saw in 2D in Figure 8.5.7b, only faster. As our cutting plane continues to move downstream, vorticity diffusing toward the symmetry plane will effectively disappear there, and the integrated vorticity passing through the cutting the plane on each side of the symmetry plane will decrease, as it did in the 2D case.

The analogy between our 3D steady mean flow and the 2D unsteady case isn't perfect, however, and there must be some differences in the details. In the 2D unsteady case, the vorticity disappears in time, while in the 3D steady case, it can't do that. In the steady case, for the vorticity piercing a cutting plane to decrease as the plane moves downstream, Stokes's theorem (see Section 3.3.5) requires that some of the vortex lines must cross the symmetry plane and connect with the vortex lines on the other side. Because only the perpendicular component of the vorticity can be nonzero at the symmetry plane, the vortex lines must turn perpendicular to the plane as they approach and cross it. Also, since we've assumed that the mean flow is symmetrical, having vortex lines cross the symmetry plane

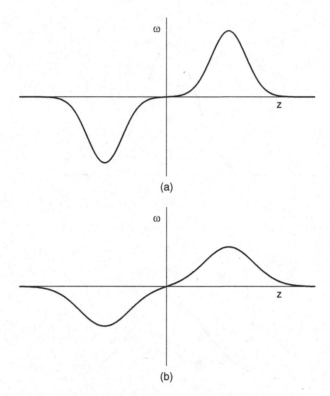

Figure 8.5.7 Diffusion of vorticity in a pair of counter-rotating vortices in 2D unsteady flow. (a) Vorticity in distinct cores at an initial instant. (b) Vorticity diffused into contact with the symmetry plane at a subsequent time

constitutes a *reconnection* of vortex lines that are already connected upstream through the bound vorticity on the wing, and such reconnections thus form closed vortex loops.

The overall structure this leads to is illustrated schematically in Figure 8.5.8, which shows a plan view of the wake with the streamwise scale greatly compressed so as to fit the entire development on the page. To illustrate the general topology of the vortex lines clearly, we've removed the circumferential component that made the vortex lines appear helical in Figure 8.1.5. In a process that starts with the growth of instability motions and continues through the breakup and postbreakup turbulence, the vortex core on each side of the symmetry plane spreads both inboard and outboard. The vorticity that diffuses toward the symmetry plane turns and reconnects with vorticity on the other side. The wake spreads and becomes more diffuse, and at the same time, the net vorticity (circulation) on each side decreases through the process of reconnection.

So far our argument for reconnection has been indirect: We expect vorticity from both sides to disappear at the symmetry plane, and the only way it can do that is through reconnection. Now let's try to identify a more direct physical mechanism. Reconnection requires the production of a nonzero spanwise component $\omega_y = -\partial w/\partial x + \partial u/\partial z$ at the symmetry plane. The rotation associated with ω_y must be opposite to that of the bound vorticity on the wing, as illustrated in Figure 8.5.9, and from the sense of the rotation

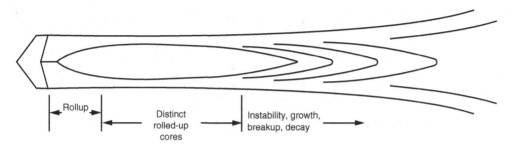

Figure 8.5.8 Plan view of the farfield development of the vortex wake of a lifting wing, showing spreading out of vortex cores, vortex-line reconnection, and the formation of closed loops. The streamwise scale is highly compressed to allow showing the entire development in one view

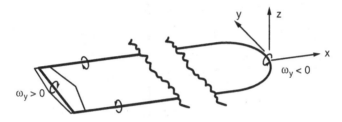

Figure 8.5.9 Illustrating the rotation of the bound vorticity and the reconnecting vorticity at the plane of symmetry far downstream in the vortex wake

arrows, we can see that this dictates negative ω_y and thus either positive $\partial w/\partial x$ or negative $\partial u/\partial z$ or both. A positive $\partial w/\partial x$ is fairly easy to explain. Looking at the vertical-velocity distribution in the wake oval in Figure 8.1.8a, we can see that when vorticity diffuses out from the center of a vortex core the associated shear-stress gradient has the effect of reducing the circumferential velocities around the core. Thus when vorticity reaches the symmetry plane, it should have the effect of reducing the downwash there. A gradual reduction in the magnitude of the downwash (the negative w) as we move downstream constitutes a positive $\partial w/\partial x$, which is consistent with reconnection. We know that things change slowly in the x direction, so that dw/dx is small, and the corresponding ω_y is weak, which is consistent with a reconnection process that is spread out over a long distance.

As all of this happens, the cross-sectional shape of the wake probably changes in other ways. In addition to the mean vorticity diffusing toward the symmetry plane, some vorticity likely diffuses outward beyond the boundary of the oval and is convected upward relative to the oval, forming a vertical "curtain," as illustrated in Figure 8.5.10.

This distortion of the vorticity distribution relative to the oval has been observed in 2D unsteady calculations for laminar vortices in a stratified atmosphere (Spalart, 1996), and Figure 8.5.10 is based on the assumption that the same qualitative pattern will occur in the "turbulent" mean flow during instability growth and breakup. The mean velocity distribution in the "curtain" is similar to that in the viscous wake of a 2D body, as in Figure 5.3.1. In the reference frame of the atmosphere, the fluid in a viscous wake flows toward the body that generated the wake, just not as fast as the body is moving away. Similarly, the fluid

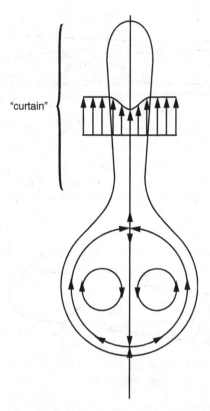

Figure 8.5.10 Change in cross-sectional shape of the vortex wake as vorticity diffuses beyond the boundary of the descending oval. In the reference frame of the descending oval, the distribution of vertical velocity in the "curtain" above the oval is similar to that in a viscous wake

in the "curtain" is descending relative to the atmosphere, just not as fast as the wake oval from which it came. The velocities in the oval are easier to visualize in the reference frame descending with the oval, and in this frame the vertical velocity in the "curtain" has a conventional-looking wake velocity profile, as illustrated in Figure 8.5.10.

As we move downstream, and the wake becomes more diffuse, the net vorticity on each side decreases, and the descent rate of the oval decreases as well. Thus in side view, the wake should take the general form shown in Figure 8.5.11, in which the oval spreads out, and the "tail" grows taller.

To summarize what we have deduced so far: a time-averaged vortex-wake system as sketched in Figures 8.5.8 and 8.5.11, consisting of closed loops that include the bound vorticity and the trailing vortices, and are closed downstream in a long and diffuse region of time-averaged vortex reconnection. Presumably all of the mean vorticity from the wake upstream eventually disappears through reconnection, though how much of this is complete before the wake encounters the ground is unknown. The ground, being a surface with a no-slip condition and thus capable of generating surface shear stresses, could serve as a sink for some of the vorticity. Of course, much of this is impractical to verify experimentally because the mean-velocity disturbances become much too small to be detected against the

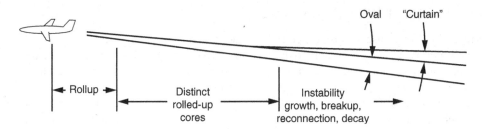

Figure 8.5.11 Side view of the farfield development of the vortex wake of a lifting wing, including the "curtain." The streamwise scale is highly compressed to allow showing the entire development in one view

background turbulence in the atmosphere. The downstream reconnection of the vorticity is topologically similar to a starting vortex, but it is spread over a very large area, especially in the flight direction. The mean vorticity in the reconnection region is convected freely with the flow, so it is moving downstream at roughly freestream velocity. But at the same time, diffusion of the vorticity by the turbulent motions is such that the entire time-averaged structure moves with the airplane and is steady in the frame of the airplane, unlike the field around an idealized starting vortex.

The fact that the mean vorticity in the wake disappears in a process that is steady in the reference frame of the airplane raises an interesting question: Does this process produce an integrated pressure disturbance at the ground? In Section 8.5.2, we argued that a starting vortex at high altitude would not do so because the idealized starting vortex was assumed to be steady in the reference frame of the atmosphere, not that of the airplane. Now, in our "real" situation, we have reconnecting mean vorticity (nonzero ω_y) imbedded in a mean flow that is steady relative to the airplane, and we might therefore expect it to produce an integrated pressure disturbance. But this expectation would be incorrect. The spanwise vorticity ω_y in the region of vorticity reconnection is not bound vorticity like that on the wing; it is free vorticity that is convected with the flow. Free vorticity is by definition force-free. Only bound vorticity that experiences a lift force could result in an integrated pressure disturbance. The flow in this region appears steady in the frame of the airplane only because of convection and diffusion, not because the vorticity is bound to a surface carrying a force. As each filament of ω_y is convected downstream, a new one replaces it by convection, and turbulent diffusion maintains the ω_y gradients that allow the whole field to appear steady even though ω_y varies with x.

So much for what happens to the vortex wake. After this process is effectively complete, and the vortex wake has decayed into insignificance, is anything left other than undif-ferentiated turbulence? Well, if the flowfield hasn't yet interacted significantly with the ground so as to exchange a significant integrated shear force with it, the remaining viscous wake must have a momentum flux deficit corresponding to the total flight-direction aerody-namic force on the airplane, including the induced drag. After all, the assumptions behind Equations 6.1.2–6.1.6 are still satisfied. For an airplane in steady, level, powered flight, the total flight-direction force, including thrust, is zero, and the wake would have no momentum-flux deficit. But for a glider, there would be a deficit (in the flight direction, not the horizontal direction), and after the disappearance of the vortex wake, the wake would have become a viscous wake with a deficit that represents both induced drag and viscous drag.

8.6   Effects of Wing Sweep

Wing sweep can be used to provide better longitudinal stability and trim for a tailless airplane (flying wing), but it is most commonly used to increase Mach-number capability, whether the airplane has a tail or not. Sweep is so effective at increasing critical Mach number that it has become ubiquitous on wings designed to operate at high subsonic speeds. It works because compressibility effects in the wing flowfield are determined more by the component of Mach number perpendicular to the isobars of the flow than by the magnitude of the Mach number. On a swept wing, the isobars tend to sweep more or less with the general sweep of the wing, which can make the perpendicular component of Mach number significantly smaller than the magnitude. This mechanism was first explained by Busemann in 1935 and was classified as a German military secret shortly thereafter. In 1945, it was independently discovered by R.T. Jones at the NACA. See Anderson (1997) for an account of the history of the idea and its incorporation into airplane design. We'll discuss the idealized theoretical argument in Section 8.6.1.

Sweep can also benefit an airplane by providing *static aeroelastic relief.* Under the high g-loadings that often determine the required strength of the wing structure, a wing generally tends to bend upward. Even if there is no torsional deflection of the structure, a simple bending deflection, when combined with aft sweep, produces an effective geometric twist, leading-edge-down on the outboard wing. The result is reduced loading outboard and reduced bending moments compared with what would happen without the bending. An aft-swept wing can therefore usually have a lighter structure than a comparable unswept wing.

In Section 8.2.1, we discussed the effects of planform on the spanload of a 3D wing, and in Figure 8.2.2, we saw how aft sweep moves the additional load due to angle of attack outboard. Because of this effect, aft-swept wings must generally be twisted leading-edge down, or washed out, to have an advantageous total spanload like that shown in Figure 8.2.2. A favorable spanload that is achieved in this way persists only over a limited range of angle of attack.

8.6.1   Simple Sweep Theory

The original idealized argument for the effectiveness of sweep is amazingly simple, thus the name. It can also be called *infinite-span yawed-wing theory.* The theory is exact for inviscid flows that satisfy the steady Euler equations and the geometric assumptions described below, and for real flows over real wing geometries, it can still be close enough to the truth to provide useful insight. The theory not only shows why sweep is effective in increasing critical Mach number; it provides a simple way of understanding the kinematics of swept-wing flows, and it predicts the "2D" effects of sweep on sectional lift and pitching moment. It does not predict the effects of sweep on profile drag or maximum lift, because these are viscous effects.

We assume the wing is a cylinder of airfoil cross-section and infinite span, with its generators parallel to the y'-axis, which is yawed at an angle Λ relative to the y-axis as shown in Figure 8.6.1. Thus in the x',z' system, the wing is a 2D airfoil, and in the x,y,z system, it is just a yawed version of the airfoil. The freestream velocity U_∞ is taken to be in the x direction and is assumed uniform far from the airfoil. Note that the sweep angle Λ is thus measured in the x-y plane, which contains U_∞. Tying our definition of Λ to

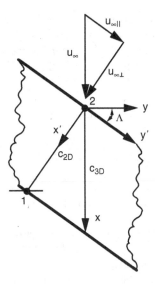

Figure 8.6.1 Assumed wing geometry and flow situation in *simple-sweep theory*

U_∞ in this way makes the formulas we'll arrive at below simpler than they would be if we used a sweep angle measured in the more conventional way, in the chord plane of the wing, which moves with angle of attack (α). Thus if we want to keep the conventional sweep angle constant, we must allow Λ to vary with α. But this doesn't matter for small α. Note that the physical principle behind simple sweep theory doesn't require a small-α assumption. In what follows we will assume small α only for relationships that involve the chord. Otherwise, the only limiting assumption is our unconventional definition of Λ.

Because the span is infinite, and the airfoil contours and freestream conditions are unchanging in the y' direction, we can assume that all flowfield quantities must be unchanging in the y' direction. Thus there can be no pressure gradient in the y' direction, and the steady inviscid momentum equation dictates that the velocity component in the y' direction must be constant throughout the field, with the freestream value $U_{||}$.

Now comes the insight that makes this theory so beautifully simple: The solution for the yawed-wing flow in the x, y, z coordinate system is just the solution for the 2D flow in the x', z' coordinate system, with the constant velocity component $U_{||}$ added to it. The swept flow and the unswept flow are the same flow, just viewed in different reference frames, related by a Galilean transformation in which the constant velocity is added. This, of course, means that the pressures throughout the field and the velocity components in the x', z' plane are the same, and this is reflected in the conversion formulas we'll discuss next.

The dimensionality of the infinite-span yawed-wing flow is a mixture of 3D and 2D. The flow is 3D in the sense that it has three non-zero velocity components, but it is quasi-2D in the sense that all flow quantities are independent of y' and can therefore be completely defined as functions of just two independent spatial coordinates, x', z'. Though we are now dealing with the inviscid case, this dimensionality situation is the same as what we discussed in connection with infinite-span viscous (boundary-layer) flow in Section 4.3.5. There we also had three non-zero velocity components, and everything could also be described in terms of only two spatial coordinates. But the inviscid case is simpler in one respect: The inviscid velocity in the direction parallel to the yawed wing is constant everywhere, defined by the freestream velocity and the sweep angle, while in the boundary-layer case the parallel velocity component has a viscous velocity profile that goes to zero at the wall.

Note that the chord lengths c_{2D} and c_{3D} in Figure 8.6.1 are depicted in the same plane as the freestream velocity vector, as they would appear at zero α. Thus based on Figure 8.6.1, we have formulas relating geometry parameters in the 2D and 3D systems, applicable to the view at zero alpha:

$$c_{3D} = \frac{c_{2D}}{\cos \Lambda}. \tag{8.6.1}$$

And because the thickness of the airfoil is fixed, we have

$$\text{Sectional thickness ratio}: \frac{t}{c_{3D}} = \frac{t}{c_{2D}} \cos \Lambda. \tag{8.6.2}$$

Formulas for converting dimensionless force and moment coefficients between the 2D unswept flow and the yawed-wing flow are simply derived, given that the pressures are the same, and only the reference chord and reference free-stream velocity change. The theory guarantees that the pressures will be the same only if the flow is inviscid, so that the formulas for forces and moments are valid, strictly speaking, only for inviscid flow. This is especially limiting in the case of drag. The theory says nothing about how sectional viscous drag should vary with sweep, so the drag formula is limited to the shock drag, the only drag component that can be present in inviscid infinite-span flow. The shock drag differs from the lift, following a $\cos^3$ rule rather than $\cos^2$, because the shock can produce a pressure force only in the x' direction, and resolving it in the drag (x) direction in the swept case introduces an additional factor of cosine. These formulas can be used for rough estimates of the effects of sweep in viscous flows, but accuracy is not guaranteed because differences in viscous effects between the unswept and swept cases will usually change the pressure distribution. Here we have not assumed small α, only the definition of Λ that we discussed previously. Denoting the yawed-wing case by the subscript 3D, we obtain:

$$\text{Freestream Mach number}: \quad M_{3D} = \frac{M_{2D}}{\cos \Lambda} \tag{8.6.3}$$

$$\text{Pressure coefficient}: \quad C_{P_{3D}} = C_{P_{2D}} \cos^2 \Lambda \tag{8.6.4}$$

$$\text{Sectional lift coefficient}: \quad C_{l_{3D}} = C_{l_{2D}} \cos^2 \Lambda \tag{8.6.5}$$

$$\text{Sectional shock} - \text{drag coefficient}: \quad C_{d\,shock_{3D}} = C_{d\,shock_{2D}} \cos^3 \Lambda \tag{8.6.6}$$

$$\text{Sectional moment coefficient}: \quad C_{m_{3D}} = C_{m_{2D}} \cos^2 \Lambda \tag{8.6.7}$$

$$\text{Moment} - \text{coefficient slope}: \quad C_{mCl_{3D}} = C_{mCl_{2D}} \tag{8.6.8}$$

Dealing with the angle of attack in this framework requires some care. There are two ways of handling angle of attack, and if executed correctly, they both yield the same result. Defining angle of attack as a rotation about the y'-axis is the simpler choice because it preserves the y' axis as the direction in which all flow quantities are constant, and there are thus no additional issues to account for. For small angles of attack, simple geometry then gives:

$$\text{Sectional angle of attack}: \quad \alpha_{3D} = \alpha_{2D} \cos \Lambda \tag{8.6.9}$$

This result is correct, but perhaps not entirely satisfying intuitively because a rotation about the y'-axis isn't the usual way we change angle of attack. For an airplane, a change in angle of attack is a rotation about a pitch axis parallel to the y axis. We can impose an airplane-type angle-of-attack change on the yawed wing, but then there is an additional complication we have to account for: the spanwise direction of the wing has been rotated relative to the original direction of the spanwise component of freestream velocity, $U_{||}$. Referring to Figure 8.6.1, if we rotate the wing through an angle $\Delta\alpha_{3D}$ about an axis through point 1 and parallel to the y axis, the 2D alpha change is, for small angles:

$$\Delta\alpha_{2D} = \frac{\Delta z_2}{c_{2D}} + \frac{\Delta w_{||}}{U_\perp} \tag{8.6.10}$$

where Δz_2 is the vertical movement of the leading edge at point 2, and $\Delta w_{||}$ is the change in the velocity perpendicular to the wing chord plane due to the rotation of the wing's spanwise direction relative to $U_{||}$. Given that

$$\Delta z_2 = \Delta\alpha_{3D}c_{3D}\cos^2\Lambda, \tag{8.6.11}$$

and

$$\Delta w_{||} = \Delta\alpha_{3D}U_{||}\sin\Lambda, \tag{8.6.12}$$

we then have

$$
\begin{aligned}
\Delta\alpha_{2D} &= \Delta\alpha_{3D}\left[\frac{c_{3D}\cos^2\Lambda}{c_{3D}\cos\Lambda} + \frac{U_{||}\sin\Lambda}{U_\perp}\right] \\
&= \Delta\alpha_{3D}\left[\frac{\cos^2\Lambda + \sin^2\Lambda}{\cos\Lambda}\right] \\
&= \frac{\Delta\alpha_{3D}}{\cos\Lambda},
\end{aligned}
\tag{8.6.13}
$$

which is the same result as Equation 8.6.9. So we have the same relationship between the effective 2D and 3D alpha changes regardless of whether the rotation is about the yawed-wing axis or an airplane pitch axis.

Given Equations 8.6.5 and 8.6.9, we obtain our final conversion formula, valid for small α:

$$\text{Sectional lift} - \text{curve slope}: \quad C_{\ell\alpha 3D} = C_{\ell\alpha 2D}\cos\Lambda \tag{8.6.14}$$

8.6.1.1 Some General Sweep Effects Deduced from Simple Sweep Theory

In a yawed-wing flow, the critical Mach number is determined by the critical Mach number of the 2D airfoil in the perpendicular flow, M_{2Dcrit}. The 3D critical Mach number is thus increased by the sweep, going as the inverse of the cosine of the sweep angle according to Equation 8.6.3. While sweep increases critical Mach number, it decreases the lift-curve slope and the lift coefficient at a given angle of attack according to Equations 8.6.14 and 8.6.5. Again, simple sweep theory says nothing about profile drag or maximum lift, because these are determined by viscous effects.

Although the yawed-wing flow is the same as a 2D airfoil flow viewed in a different reference frame, it looks more complicated, belying its simple underlying flow structure. First, the stagnation point of attachment near the leading edge of the 2D airfoil is not a stagnation point in the yawed flow because of the nonzero spanwise velocity $U_{||}$. Thus instead of a leading-edge stagnation point, we have a *leading-edge attachment line,* with flow along it and diverging away from it, as shown in Figure 8.6.2. We considered the general properties of attachment-line flows in some detail in Section 5.2.2. On a yawed wing, the maximum surface pressure is generally reached on the attachment line, but because of the spanwise velocity, it is lower than freestream stagnation pressure. The boundary layer that develops along the attachment line is different from a stagnation-point boundary layer in ways we'll discuss in Section 8.6.2.

Another major facet of flowfield behavior related to sweep is the variation in flow direction that takes place in response to pressure changes. Figure 8.6.3 illustrates how this works ideally, according to simple sweep theory. In this diagram, y' is the direction of the generators of the yawed wing, x' is the perpendicular direction on the wing surface, and $U_{||}$ and $U_{\perp}$ are the corresponding velocity components at some point along the wing surface. In sweep theory, the spanwise velocity $U_{||}$ is constant, while the perpendicular velocity $U_{\perp}$ changes in response to the pressure. Figure 8.6.3 shows several representative velocity vectors consistent with different values of $U_{\perp}$ and with the requirement that $U_{||}$ is constant. It is clear that there is a one-to-one correspondence between $U_{\perp}$ and the velocity magnitude and a one-to-one correspondence between the velocity magnitude and the flow direction. And, of course, for the steady inviscid flow assumed in simple sweep theory, there is then a one-to-one correspondence between the flow direction and the pressure. When the pressure is high and the velocity magnitude is low, the flow is in the outboard direction, with the attachment-line flow in the spanwise direction as the limiting case (i.e., the minimum velocity magnitude and the maximum pressure). When the pressure is at the freestream value, the flow is in the freestream direction (assuming that we can ignore the slope of the airfoil surface). When the pressure is low and the velocity magnitude is high, the flow is in the inboard direction. In the limit, flow in the x' direction perpendicular to the generators would require infinite perpendicular velocity.

When such changes in direction in the inviscid flow outside the boundary layer take place, they constitute flow curvature. In local physical terms, the curvature is a result of the fact that the sweep skews the pressure gradient so that it is not lined up with the local flow direction, and an acceleration perpendicular to the streamlines is the result. This of course has consequences for the flow in the boundary layer, as we saw in Section 4.1.2 and will discuss further in Section 8.6.2.

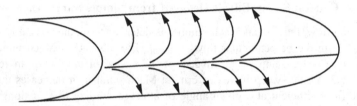

Figure 8.6.2 Flow pattern around the *leading-edge attachment line* of a swept wing

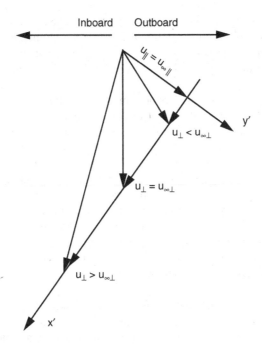

Figure 8.6.3 Illustration of the relationship between the velocity magnitude and the flow direction at some point along the surface of a swept wing, according to simple sweep theory. Several possible velocity vectors are shown, along with their parallel and perpendicular components

8.6.2 Boundary Layers on Swept Wings

The development of the boundary layer on a swept wing can be thought of as beginning at the leading edge, but the flow situation is more complicated than that at the leading edge of a 2D airfoil. The inviscid attachment-line flow near the leading edge of a swept wing is essentially a 2D stagnation-point potential flow with a non-zero spanwise velocity added to it, as we saw in Figure 8.6.2. So unlike the stagnation-point boundary layer on a 2D airfoil, the *attachment-line boundary layer* on a swept wing has a non-zero edge velocity. And because of this, the attachment-line boundary layer is similar to boundary layers in many other situations, in that it can be either laminar or turbulent. However, a notable difference is that, unlike most other boundary layers, the attachment-line boundary layer tends not to grow thicker in the flow direction. On a wing where the leading-edge sweep is constant along the span, the spanwise velocity and the pressure along the attachment line are nearly constant. (According to simple sweep theory, they would be exactly constant.) A 2D boundary layer developing under these conditions would be a flat-plate boundary layer, the first case of "simplified" boundary-layer development we discussed in Section 4.3, and the boundary layer would grow thicker in the flow direction. In the attachment-line boundary layer on a swept wing, this tendency of the boundary layer to grow is offset by the strong divergence of the flow, and the boundary layer remains thin.

To understand how flow divergence has this effect, consider first the flat-plate boundary layer. The shear force exerted on the flow by the surface (the skin friction) is a retarding force that produces an increasing momentum-flux deficit in the boundary layer, which shows up as a positive rate of growth, in the flow direction, of the momentum thickness. The skin friction in the attachment-line case is of similar magnitude, but the resulting momentum-deficient air is carried away by the flow divergence instead of building up as increasing momentum thickness.

The divergence effect can be understood mathematically through plane-of-symmetry boundary-layer theory, which is illustrated in Figure 4.3.10. The assumptions of the theory are satisfied exactly in the infinite-span yawed-wing case, because the attachment line is straight in that case. In the more general case of a typical 3D swept wing, the attachment line generally has some spanwise curvature due to the detailed shaping and aeroelastic deflection of the wing. However, the radius of curvature is usually so large compared with the thickness of the attachment-line boundary layer that the effects of curvature are negligible, and the plane-of-symmetry theory incurs very little error in practical leading-edge attachment-line flows. In Section 4.3.4, we discussed the applicable simplifications to the equations of motion. In the momentum-integral equation for such flows, Equation 4.3.13, the divergence effect shows up as the integral of the divergence derivative $\partial w/\partial z$ weighted by the velocity defect $u_e - u$, which is consistent with our intuitive idea of momentum-deficient air being carried away by the flow divergence.

On the attachment line in the infinite-span case, the skin friction and the divergence integral term are in equilibrium, which is consistent with zero $d\theta/dx$, and with spanwise invariance of everything in the flow. On the attachment line of a typical 3D swept wing, conditions vary along the span, but because the distances over which changes take place are typically so much greater than the boundary-layer thickness, the boundary layer locally achieves very nearly the same equilibrium as it would in the infinite-span case. An example of this is illustrated in Figure 8.6.4, which shows two sets of numerical solutions for the turbulent attachment-line boundary layer on the swept wing of a 727-200 (McLean, 1977). One set is based on the spanwise-varying attachment-line boundary-layer equations, marched from root to tip, and the other is based on infinite-span equations and local conditions at selected stations on the span. The calculated momentum thickness and skin friction vary considerably along the span, mainly as a result of the varying leading-edge radius. In spite of this, there is practically no difference between the two sets of calculations, showing that this turbulent boundary layer is very nearly in local infinite-span equilibrium, or in other words, that the quantitative state of the boundary layer is dominated by the local flow divergence and skin friction, and carries practically no memory of its history upstream (inboard). The same holds for a laminar boundary layer. We'll discuss the factors determining whether the flow is laminar or turbulent next.

So the attachment-line boundary layer at any spanwise location along a typical swept wing is practically identical to what the boundary layer would be on an infinite-span attachment line under the same local conditions. And the infinite-span attachment-line boundary layer is especially simple to characterize. In either the laminar or turbulent case, the state of the boundary layer depends on a single parameter, a Reynolds number based on a combination of the velocity along the attachment line and the divergence derivative (Cumpsty and Head, 1967):

$$C^* \equiv \frac{u_e^2}{v_e \dfrac{\partial w_e}{\partial z}}, \tag{8.6.15}$$

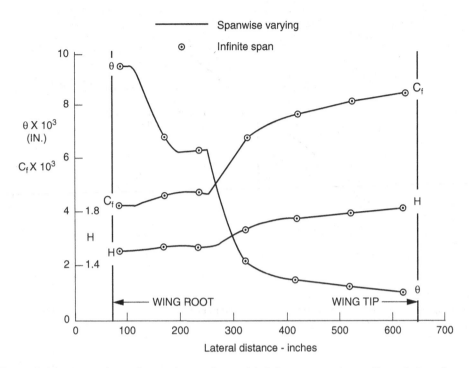

Figure 8.6.4 Comparison of spanwise-varying and infinite-span attachment-line solutions for turbulent flow on 727–200 leading edge. Outer-flow boundary conditions are from a panel-method solution for $M_\infty = 0.70$, $C_L = 0.342$; the boundary-layer calculations are for a unit Reynolds number of 278 000 in^{-1}. (From McLean, 1977)

where the nomenclature corresponds to what we used for the plane-of-symmetry boundary layer in Figure 4.3.10. More-recent references on the subject (Poll, 1989) use $\overline{R}$ instead, which is equivalent:

$$\overline{R} \equiv C^{*\,1/2}. \tag{8.6.16}$$

The momentum thickness depends on $\overline{R}$ as shown in Figure 8.6.5, where the laminar case is defined by a similarity solution, and the turbulent case by numerical solutions using the Mellor-Herring algebraic eddy viscosity (McLean, 1977). Figure 8.6.6 shows a comparison between these calculations and the experimental data of Cumpsty and Head (1969) for the spanwise velocity profile in the turbulent case. Note in Figure 8.6.5 that the boundary layer is always laminar for $\overline{R} < 260$ and always turbulent for $\overline{R} > 600$, and that in between, it can be either laminar or turbulent. It is sometimes useful to look at these thresholds in terms of R_θ of the laminar attachment-line boundary layer: The boundary layer is always laminar for $R_\theta < 100$, and it cannot be laminar for $R_\theta > 240$. We'll discuss laminar-to-turbulent transition issues further in Section 8.6.4.

Away from the attachment line, the boundary layer on the rest of the wing's upper and lower surfaces is more complicated and 3D. In the outer inviscid flow, the streamline curvature we mentioned earlier is consistent with a pressure gradient that is skewed relative to the local flow direction and therefore has a component perpendicular to the local flow.

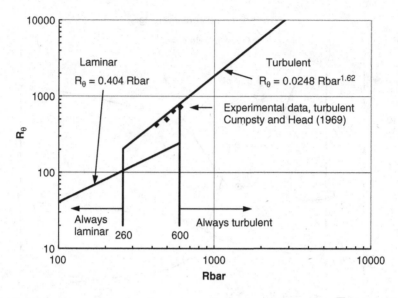

Figure 8.6.5   Dependence of boundary-layer state and momentum-thickness on $\overline{R}$ for the infinite-span attachment-line boundary layer

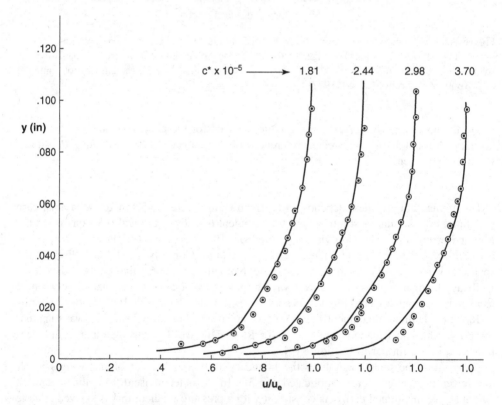

Figure 8.6.6   Spanwise velocity profiles in a turbulent attachment-line boundary layer. Calculations by the method of McLean (1977) and experimental measurements of Cumpsty and Head (1969)

Moving closer to the wall and into the boundary layer, we find essentially the same pressure gradient, since the boundary layer is thin and the pressure doesn't change much across its thickness. But the velocity magnitude is smaller than it is outside the boundary layer, so that the same pressure gradient produces greater streamline curvature. The result is *pressure-driven crossflow* of the type we discussed in Section 4.1.2, with cross-flow velocity directions as illustrated in Figure 8.6.7.

Typical outer-flow streamlines and surface streamlines (skin-friction lines) for a wing upper surface are shown in Figure 8.6.8. Note that the behavior of the outer-flow streamlines is generally what we would expect based on simple sweep theory. Over the forward part, where the pressure coefficient is generally negative and the velocity magnitude is higher than freestream, the streamlines are oriented generally inboard. In the aft pressure-recovery region, they turn toward the outboard direction, and the skin-friction lines follow suit by turning outboard even more strongly. On this particular wing, significant outboard turning of the skin-friction lines begins at a shock near midchord. We'll consider this flow further below in connection with Figures 8.6.10–8.6.11.

Because of the kind of general outboard turning of the flow seen in Figure 8.6.8, swept wings have a reputation for having problems with "spanwise flow" in the boundary layer. A common impression seems to be that boundary-layer fluid typically migrates from inboard to outboard on the aft part of the surface and "piles up" outboard, causing earlier separation than would otherwise occur. This reputation overrates the problem. Skin-friction lines like those in Figure 8.6.8 don't typically indicate much long-distance migration of fluid prior to separation. It is only after separation has developed that fluid can migrate long distances spanwise, as in Figure 5.2.13b. "Boundary-layer fences," which are intended to inhibit spanwise flow, can be beneficial, but usually only after separation has begun to move forward.

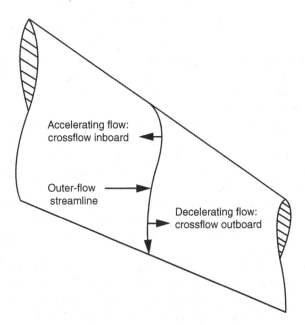

Figure 8.6.7 Typical cross-flow directions in the 3D boundary layer on the upper surface of a swept wing

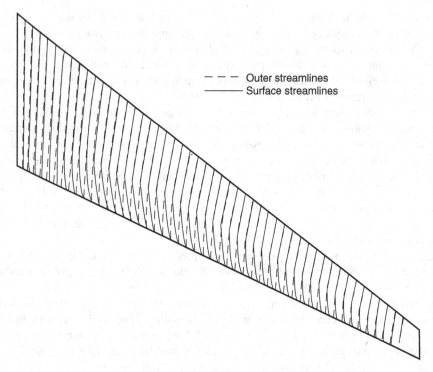

Figure 8.6.8 Typical outer-flow streamlines and surface streamlines (shear-stress lines) on the upper surface of a swept wing. From 3D boundary-layer calculations by M.D. Murray using the method of McLean (1977), for wing shape derived from the RAE 2822 airfoil

As in any general 3D boundary-layer flow, separation is indicated by the surface-streamline pattern according to the *region-of-origin* concept we discussed in Section 4.1.4. The usual kind of separation line on a swept wing is a surface streamline dividing a region in which the surface streamlines come from the general upstream direction from one in which they come from the general downstream direction. Surface streamlines coming from upstream often originate from the leading-edge attachment line, but they can also effectively originate from a reattachment following a separation. Figure 8.6.9 figuratively illustrates two common varieties of separation on a swept wing: separation with reattachment (either a laminar separation bubble with turbulent reattachment, or a turbulent separation bubble under a shock), and separation without reattachment, in which the surface streamlines downstream come from the trailing edge. Remember that separation in 3D is not generally accompanied by zero magnitude of the skin friction and that at a separation line on a swept wing, the skin friction usually has a significant spanwise component.

We can use zero C_f as a criterion for separation on swept-wings, but we must use the component of C_f perpendicular to the trailing edge, and we must keep in mind that what it can tell us is limited. In infinite-span-swept-wing flow, where any separation line must lie straight along a line of constant x/c, the component of C_f perpendicular to the constant-x/c lines must be zero at separation, regardless of where on the chord the separation occurs. In 3D swept-wing flow, strictly speaking, the zero-perpendicular-C_f criterion works only

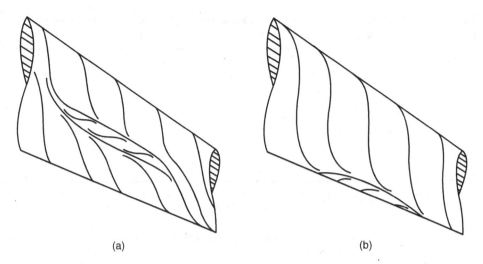

Figure 8.6.9 Two types of separation on a swept wing. (a) Separation with reattachment (either laminar separation with turbulent reattachment, or turbulent separation and reattachment under a shock). (b) Separation without reattachment

for incipient separation at the trailing edge (i.e., separation that is on the verge of moving forward from the trailing edge). In this situation, perpendicular C_f is zero at one isolated point and positive everywhere else on the trailing edge, as we noted in Section 5.2.3.2, in connection with Figure 5.2.13b.

A zero-perpendicular-C_f criterion (relative to constant-x/c lines) is sometimes used to identify separation that has moved forward from the trailing edge, but it is error-prone. In Figure 5.2.13b we saw that the region of negative perpendicular C_f at the trailing edge fails to identify the entire separation bubble (there is a portion of the separation bubble where perpendicular C_f is positive). If we apply the criterion ahead of the trailing edge, the opposite can also happen, and we can be fooled by regions of negative perpendicular C_f that are not part of the bubble, as you can imagine if you look just upstream of the separation line on the inboard side of the bubble in Figure 5.2.13b. If a bubble is sufficiently long and narrow near the trailing edge so that the flow is locally like infinite-span-wing flow, these discrepancies can be minor, and zero perpendicular C_f can be a reasonable separation criterion. But to be rigorous in general, locating a separation line on the surface requires mapping out the skin-friction direction field and choosing a line that satisfies our region-of-origin definition of 3D separation.

In some general ways, the response of a 3D swept-wing boundary layer to pressure gradients is similar to that of a 2D boundary layer on an unswept airfoil. For example, a favorable pressure gradient increases the magnitude of the skin friction and slows or even reverses the growth of boundary-layer thickness, and an adverse pressure gradient does the opposite. But quantitatively, the boundary layer on a 3D swept wing has a response to pressure gradients that is very different from that of a 2D boundary layer. To understand the sources of these differences and to get a feel for their magnitude, we'll compare the results of boundary-layer calculations that were carried out for nominally the same flow situation but used equation sets that included or excluded various effects. These equation

Table 8.6.1 Equation sets used in calculations comparing different ways of calculating the boundary layer on a swept wing (calculated results shown in Figures 8.6.10–8.6.12)

Equation set 1	General 3D boundary-layer equations
Equation set 2	Boundary-layer equations for infinite-span with sweep and taper, as described in Section 4.3.5 (uses the actual surface pressures from a single longitudinal cut across the wing, but assumes the isobar pattern is "conical" even if the real pattern is not, and uses the conical coordinate system of Figure 4.3.12a)
Equation set 3	Boundary-layer equations for infinite-span with sweep but without taper, as described in Section 4.3.5 (uses actual surface pressures from a single cut as in 2 above, but assumes the isobars are all swept at the midchord sweep angle, and uses the perpendicular coordinate system of Figure 4.3.11)
Equation set 4	2D boundary-layer equations applied to the "perpendicular-chordwise" velocity component only (uses actual surface pressures as in 2 and 3 above, converted to 2D by Equation 8.6.4, and the chord length converted to 2D by Equation 8.6.1, all done with the midchord sweep angle)
Equation set 5	2D boundary-layer equations applied to the "streamwise" velocity component only (uses the surface pressures from a single cut as in 2, 3, and 4 above with no sweep adjustment to the pressures or the chord length)

sets are listed in the Table 8.6.1 in order of decreasing applicability to a general 3D flow and therefore decreasing fidelity.

A wing shape was specially designed by M. D. Murray for these comparisons. The objective was to produce a pressure distribution similar to what might be seen on the upper surface of the outboard wing of a swept-wing transport airplane at a transonic cruise condition, with a nearly conical isobar pattern for which equation set 2 should provide a good approximation for the boundary-layer development. The planform is trapezoidal, with a quarter-chord sweep of $35°$, taper ratio 0.20, and aspect ratio 9. The initial wing shape was lofted with the RAE 2822 airfoil, for which we saw calculated 2D pressure distributions in Figure 7.4.27b. The airfoil was used as the streamwise cross section, thinned according to simple sweep theory (Equation 8.6.2) for the sweep of the midchord line. The camber and twist were then inverse-designed to produce the desired nearly conical isobar pattern on the upper surface over the mid-to-outboard portion of the span, with a weak shock just aft of midchord, as shown in the isobar plot in Figure 8.6.10. Calculated outer streamlines and surface streamlines were shown in Figure 8.6.8 and display a pattern typical of this kind of swept-wing flow. Freestream Mach number was 0.85, the Reynolds number was typical of full-scale flight, and the boundary layer was assumed to be turbulent, starting on the attachment line.

The design and the basic 3D flow solution were calculated by an automated viscous-inviscid interaction procedure that couples the transonic full-potential method of Jameson and Caughey (1977) with the finite-difference boundary-layer method of McLean (1977), solving the general 3D boundary-layer equations with an algebraic eddy-viscosity turbulence model. The basic solution thus used boundary-layer equation set 1 above, with the potential-flow velocity vectors at the displacement surface serving as the outer-flow boundary condition. The solution procedure cycled between the potential flow and the boundary layer, updating the wing shape by the displacement thickness calculated by the

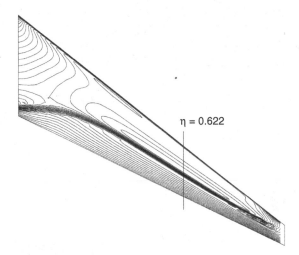

$\eta = 0.622$

Figure 8.6.10 Calculated pressure distribution for a swept-wing shape designed by M.D. Murray specially for comparing different levels of fidelity in predicting the development of a swept-wing boundary layer

boundary-layer calculations, and stopping when a converged shape and flow solution were reached. Then the final inviscid-flow pressures at the displacement surface from the basic solution at the span station $\eta = 0.622$ were used to define boundary conditions for the comparison boundary-layer calculations using equation sets 2, 3, 4, and 5. Chord Reynolds number at this span station was 31.6 million and the sectional lift coefficient was 0.53.

Figure 8.6.11 shows the pressure coefficient C_p, plus the perpendicular skin friction coefficient C_{fperp}, the perpendicular momentum thickness θ_{perp}, and the surface-cross-flow angle β_{sw} (angle between the surface-shear-stress direction and the direction of the local outer flow, positive when the surface-shear-stress direction is outboard relative to the local outer flow direction). The latter three are sensitive indicators of pressure-gradient effects.

Through the shock and as the trailing edge is approached, the combined effects of adverse pressure gradients and sweep are clearly seen: C_{fperp} decreases, and there is strong outboard turning, reflected in positive (outboard) and rapidly increasing β_{sw}.

Because the isobars in this case line up well with the constant-percent-chord lines, the assumptions of the infinite-span-with-taper theory (equation set 2) should be nearly satisfied, and, indeed, we see in Figure 8.6.11 that the results from equation set 2 agree closely with those from the full 3D equations. If taper is neglected and sweep is matched only at midchord (equation set 3), the predictions don't match the 3D results nearly as well. In particular, β_{sw} at the trailing edge is about 15° larger for equation set 3, and C_{fperp} is much closer to zero. Both of these are indications that equation set 3 is predicting flow closer to separation at the trailing edge. (Note that C_{fperp} would be zero at the trailing edge at the onset of separation of the type illustrated in Figure 8.6.9b.)

Thus neglecting taper in this case leads us to overpredict the effects of the adverse pressure gradient aft of the shock. We can interpret this result physically in terms of the perpendicular component of the local outer-flow velocity. When taper is neglected, the sweep at the trailing edge is higher, and the perpendicular outer velocity is lower. This

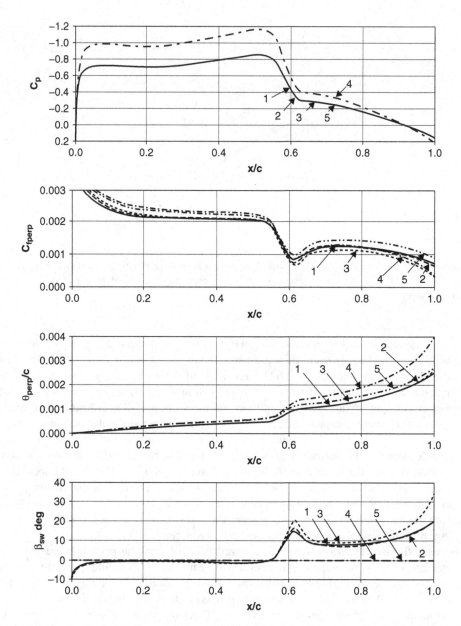

Figure 8.6.11 Comparison of boundary-layer predictions for the swept wing of Figure 8.6.10, using the equation sets from Table 8.6.1. Calculations by M.D. Murray using the method of McLean (1977). (1) General 3D boundary layer (basic solution). (2) Infinite span with sweep and taper. (3) Infinite span with sweep, using sweep angle at midchord. (4) 2D boundary-layer equations applied to the chordwise (perpendicular) component only. (5) 2D boundary-layer equations applied to the local streamwise component

greater reduction of the perpendicular outer velocity corresponds to a larger adverse pressure gradient in the perpendicular direction, which puts more "stress" on the boundary layer (i.e., brings it closer to separation).

Now consider equation sets 4 and 5, in which we use 2D planar boundary-layer equations with different boundary conditions. First, note that neither of these 2D calculations accounts for any outer-flow curvature, so that the surface-cross-flow angle is zero. Next, note that 2D equations applied to the local streamwise component (equation set 5) underestimate the combined effects of the adverse pressure gradient and sweep on C_{fperp} downstream of the shock, compared with equation set 3. Note also that equation set 5 underpredicts the growth of θ_{perp} through the shock and downstream, though this is offset by the fact that θ_{perp} is too high just ahead of the shock.

2D equations applied to the perpendicular component (equation set 4) predict the effects of pressure gradient on C_{fperp} fairly well, but overpredict the effects on θ_{perp}. Recall the *laminar independence principle* from Section 4.3.5, according to which the chordwise (perpendicular) component of a laminar, incompressible infinite-span boundary-layer flow can be solved for independently of the spanwise component. According to this principle, solving the 2D equations in the perpendicular direction properly accounts for the combined effects of sweep and pressure gradient, in the laminar, incompressible case. In the present turbulent, compressible case, solving 2D equations in the perpendicular direction compensates for the effects of sweep on C_{fperp}, but overcompensates for the effects on θ_{perp}.

Nash and Tseng (1971) used boundary-layer calculations to make comparisons similar to our comparisons of equation sets 3, 4, and 5, thus investigating the effects of sweep on wing boundary layers, but not the effects of taper. They used a typical subsonic airfoil pressure distribution with the suction peak near the leading edge and no shock. In their comparisons, equation set 4 came closer to equation set 3 than in our comparisons, and equation set 5 did worse than in our comparisons. These differences are probably due to the difference in the character of the airfoil pressure distributions. Nash and Tseng also proposed an adjustment to their turbulence model that made the predictions using equation set 4 match those using equation set 3 more closely.

So calculations with different boundary-layer equation sets have established that both sweep and taper, when combined with pressure gradients, have substantial effects on the development of a swept-wing boundary layer. Thus the temptation to use a simple 2D streamwise analysis to estimate boundary-layer quantities on a swept wing should always be resisted. For a given streamwise pressure distribution, a 2D streamwise analysis that ignores sweep can seriously underestimate the combined effects of sweep and pressure gradients on the integral thicknesses and the perpendicular skin friction.

And, of course, if perpendicular skin friction isn't predicted correctly, separation won't be predicted correctly. Recall that $C_{fperp} = 0$ is the threshold for infinite-span swept-wing separation, and note in Figure 8.6.11 that C_{fperp} at the trailing edge is farther from zero for the 2D streamwise analysis (equation set 5) than for the infinite-span-with-sweep analysis (equation set 3). This trend is typical, and it becomes more pronounced in flows that actually separate: If a swept wing and an unswept wing have the same streamwise pressure distribution, and the flow on the unswept wing separates somewhere on the chord, the flow on the swept wing will generally separate significantly earlier. Thus if a 2D streamwise boundary-layer analysis is applied to a swept wing, it will generally fail to predict separation soon enough and will overestimate the maximum-lift capability.

To illustrate this, M.D. Murray made comparison calculations (method of McLean, 1977) for one of the idealized pressure distributions that Smith (1975) used to study separation in 2D in Figure 7.4.13. Smith's canonical pressure distribution for $m = 1/2$, $x_o = 0.25$ was converted to a conventional C_p distribution, assuming $C_{pTE} = 0.20$ as was done for the families of idealized conventional pressure distributions in Figure 7.4.14. The result is shown as the dashed line in Figure 8.6.12. Calculating the boundary layer for this case as a 2D flow (corresponding to the "2D perpendicular" option, equation set 4) yields a predicted separation point in reasonable agreement with Smith's result, as should be expected, given the similarity of the eddy-viscosity models used in the two codes. Using simple sweep theory to convert this pressure distribution to 35° sweep yields the solid curve. Using this swept pressure distribution as input to an infinite-span-swept-wing analysis (equation set 3) leads to a predicted separation point very close to the predicted 2D separation point. Using the swept pressure distribution as input to a 2D streamwise analysis (equation set 5) leads to a predicted separation point much farther downstream. This underscores how unsuitable a 2D streamwise analysis will generally be for predicting separation on a swept wing.

These are the kinds of conclusions we can draw regarding the generic effects of sweep and taper, by comparing calculations with other calculations, but how do the best of such calculations compare with the real world? First, consider what we ought to expect, based on limitations of turbulence modeling. Turbulence models that are currently used in routine calculations of swept-wing flows generally assume that the eddy viscosity (see Section 3.7) is

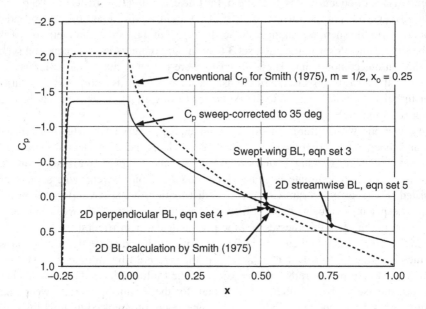

Figure 8.6.12 Assumed pressure distributions and calculated separation points for boundary-layer calculations illustrating the effect of sweep on separation. The baseline 2D case is Smith's (1975) canonical pressure distribution from Figure 7.4.13 for $m = 1/2$, $x_o = 0.25$ ft. The other calculations were made by M. D. Murray using the method of McLean (1977), and used the same equation sets 3, 4, and 5 that were used for the comparisons in Figure 8.6.11. The 2D streamwise analysis (equation set 5) predicts separation far downstream of the separation predicted by the swept-wing analysis

isotropic, that is that it acts the same in all directions. However, experimental measurements have indicated that the isotropic assumption is seriously violated in swept flows with adverse pressure gradients. For example, in the measurements of van den Berg and Elsenaar (1972) in a boundary-layer test rig (not a real "wing" geometry) that approximated infinite-span-swept-wing conditions, the cross-flow eddy viscosity was typically less than half the streamwise eddy viscosity in a strong adverse pressure gradient typical of the aft upper surface of a swept wing. Here, "cross-flow" and "streamwise" were defined relative to the flow direction at the boundary-layer edge, the same convention we used in discussing 3D velocity profiles in Section 4.1.2.

How much difference should this make? We can get some insight from boundary-layer calculations by the author (McLean, 1977) for the van den Berg and Elsenaar experiment. One set of calculations used an isotropic eddy viscosity model, and another used a crude model in which the cross-flow eddy viscosity was set to 0.4 times the streamwise eddy viscosity, which is nonisotropic to roughly the same degree as indicated by the experimental measurements. The calculated C_f magnitude and minimum flow angle β_1 relative to the spanlines are shown in Figure 8.6.13. A swept separation line is assumed to be located where β_1 goes to zero, and the nonisotropic calculation agrees much better with the separation location inferred in this way from the measurements. Velocity magnitude and direction

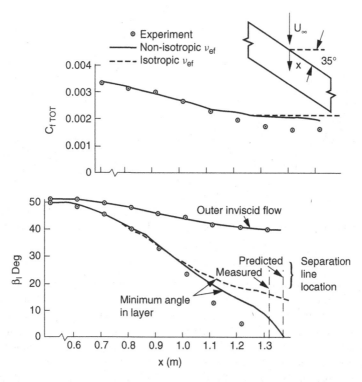

Figure 8.6.13 C_f magnitude and minimum flow angle β_1 relative to the spanlines for the "swept wing" experiment of van den Berg and Elsenaar (1972), compared with 2.5D boundary-layer calculations by the author (McLean, 1977)

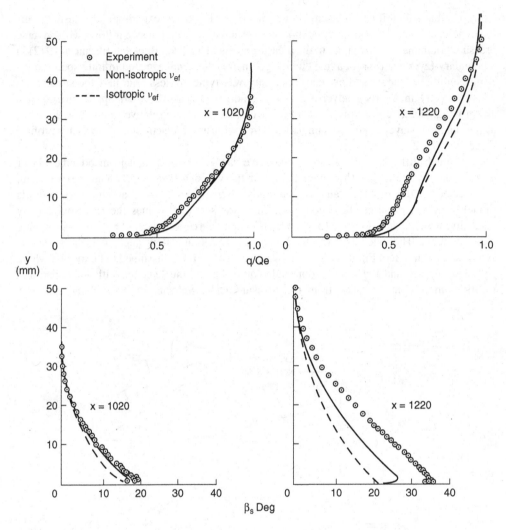

Figure 8.6.14 Velocity magnitude and direction profiles for the "swept wing" experiment of van den Berg and Elsenaar (1972), compared with 2.5D boundary-layer calculations by the author (McLean, 1977)

(β_S relative to the outer-flow direction) profiles are compared in Figure 8.6.14. Here also, the nonisotropic model moves the calculations in the right direction, just not enough to match the measurements very well. The main conclusion from these comparisons is that the decreased cross-flow eddy viscosity measured in the experiment can have strong effects on the development of a swept-wing boundary layer in an adverse pressure gradient, especially as it approaches separation. Thus for the flow on the aft upper surface of a swept wing, we should not expect high accuracy from CFD calculations using the usual isotropic eddy-viscosity models.

Now let's look at a real 3D wing case. As of this writing, I have seen only one example with comparisons of detailed velocity profiles, a comparison of RANS calculations with the experimental measurements of Brebner and Wyatt (1961) on a 45° swept wing in low-speed flow. The RANS code is CFL3D (Biedron and Rumsey, 1998) with the SST turbulence model of Menter (1994), in which the eddy viscosity is isotropic. Figure 8.6.15 shows the calculated surface streamlines (shear-stress lines), and Figure 8.6.16 compares calculated and measured velocity magnitude and direction profiles at three stations along the chord at 50% semispan on the upper surface. The calculated shear-stress lines indicate that the flow is approaching separation at the trailing edge. Note that because of the higher sweep, separation involves less turning of the flow than it did in the van den Berg and Elsenaar flow. The calculated velocity magnitude and direction profiles agree much better with the data than the isotropic boundary-layer calculations did for the van den Berg and Elsenaar flow in Figure 8.6.14.

There are a couple of possible reasons for the better-than-expected agreement for the Brebner and Wyatt flow. The first is that the SST model is a modern two-equation model that does better in predicting adverse-pressure-gradient effects in 2D flows than does the algebraic eddy viscosity used in the old boundary-layer calculations, and some of this improvement should carry over to 3D. Another is that there is less flow turning than in the van den Berg and Elsenaar flow, though the difference is small. A more important reason is probably that the adverse pressure gradient in the Brebner and Wyatt flow occupies practically the whole chord and is therefore relatively gentler. In the van den Berg and Elsenaar flow, there was some distance of flat-plate boundary-layer growth ahead of the start of the adverse gradient, so that the boundary layer was thicker at the start of the recovery, or in airfoil terms, the effective recovery-point location was somewhere aft on the chord (recall our discussion of the importance of the recovery-point location in

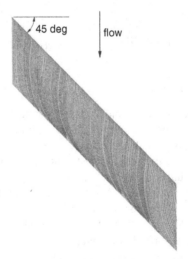

Figure 8.6.15 Surface-streamline pattern calculated by CFL3D with the SST turbulence model for the experimental swept-wing case of Brebner and Wyatt (1961). (Calculation by N.J. Yu. Plot by T.J. Kao)

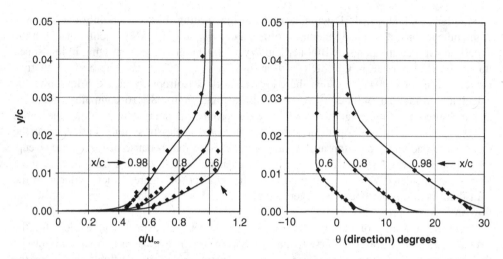

Figure 8.6.16 Comparison of velocity profiles calculated by CFL3D with the SST turbulence model with the experimental measurements of Brebner and Wyatt (1961). (Calculation by N.J. Yu)

connection with maximum lift in Section 7.4.3). The van den Berg and Elsenaar flow should thus be a more difficult test case for the turbulence models. It is also probably closer to being representative of the flow on the swept wing of a modern transport airplane than the Brebner and Wyatt flow is.

Though the predictions were better than expected for the Brebner and Wyatt flow, we've seen that an isotropic eddy-viscosity model poses a potentially serious limitation in CFD calculations of the development of the swept-wing boundary layer in an adverse pressure gradient, which is precisely where the boundary layer has its biggest impact on profile drag and on separation behavior. This shortcoming of the turbulence models could significantly degrade the prediction of all aspects of swept-wing performance.

8.6.3   Shock/Boundary-Layer Interaction on Swept Wings

In Section 7.4.8, we discussed shock/boundary-layer interaction in transonic flow around a 2D airfoil. We noted that this interaction thickens the boundary layer and increases profile drag, and that the boundary-layer thickening produces a "viscous wedge" effect that smears the shock pressure rise and modifies the downstream condition. Instead of looking like a normal shock, the pressure jump is close to that of an oblique shock with the maximum turning angle, for which the downstream Mach number is typically just below sonic. We also discussed shock-induced separation and noted that as the upstream Mach number M_1 increases, a separation bubble appears at $M_1 \approx 1.30$, followed by massive separation at $M_1 \approx 1.40$, largely independent of Reynolds number.

Sweep complicates this picture surprisingly little. Of course, if the shock is swept relative to the local upstream flow, it will turn the flow slightly in the lateral direction and produce cross-flow in the boundary layer, even if there was none to start with. But the turning and the cross-flow are just reflections of the fact that sweep introduces a velocity component parallel to the shock, in the lateral direction, and the presence of this parallel component has relatively little effect on the flow development that we noted in 2D. If we look only at the velocity component perpendicular to the shock's footprint on the surface, we should see something close to what we saw in the 2D case, at least qualitatively.

The physical argument for this relies on our observations regarding infinite-span swept-wing boundary layers in Section 8.6.2. Our general swept wing flow may not be much like an infinite span wing at all: The isobar pattern may be highly three-dimensional, and the shock may not line up with the constant x/c lines. But in the neighborhood of the shock, the infinite-span simplifications should apply very well. Within the region of the shock pressure rise, the pressure gradient perpendicular to the shock footprint is much, much stronger than the gradient in the parallel direction, and the general sweep of the isobars matches the sweep of the shock footprint very closely. To apply the infinite-span simplifications, all we have to do is take the sweep of our local "infinite-span wing" to be the local sweep of the shock footprint.

Recall that in laminar, constant-property flow on an infinite-span swept wing the perpendicular ("chordwise") velocity component develops completely independently of the parallel ("spanwise") component, according to what we called the *laminar independence principle*. In turbulent flow, the independence principle doesn't hold exactly because the Reynolds stress in the chordwise direction is affected by the flow in the spanwise direction. But the calculations compared in Figure 8.6.12 indicated that this effect is not large, and that the chordwise velocity profile is not strongly affected by the presence of the spanwise flow. Applying this observation to the region of the shock pressure rise, we can argue that the velocity component perpendicular to the shock footprint should develop close to the way it would if the parallel velocity component weren't there (i.e., the way it would in a 2D case with the same incoming perpendicular velocity profile). Then given that the gradients of all flow quantities in the parallel direction are much smaller than those in the perpendicular direction, the displacement thickness should be close to what it would be in 2D, and the smearing of the pressure rise and even the separation behavior should also be close.

The above line of argument served as the basis for a semi-empirical model for swept-wing shock/boundary-layer interaction for use in CFD calculations by viscous-inviscid coupling (McLean and Matoi, 1985).

8.6.4 Laminar-to-Turbulent Transition on Swept Wings

In Section 4.4.1, we discussed the general physical mechanisms and the associated theories of laminar-to-turbulent transition in boundary layers. The familiar pattern is for the boundary layer to start laminar at the nose of a body or the leading edge of an airfoil and to transition to turbulent after some boundary-layer growth has taken place. Transition may be caused by unstable growth of small disturbances or by direct introduction of large disturbances by the outer flow or by interaction of the flow with surface imperfections. The boundary layer on the leading edge of a swept wing doesn't always follow this simple pattern. Because of the spanwise flow along the leading edge, which we discussed in Section 8.6.2, the attachment-line boundary layer itself can be turbulent. In the usual case of an aft-swept wing, the "upstream" initial condition for the attachment-line boundary layer is at the wing-fuselage junction, and the fuselage boundary layer at the junction is usually turbulent. Thus the attachment-line boundary layer often gets a turbulent start (sometimes referred to as *leading-edge contamination*) and remains turbulent as far outboard as conditions will support turbulence, which may or may not be the full span of the leading edge. It is not unusual for the turbulent attachment-line boundary layer to transition from turbulent to laminar somewhere along the span, as the leading-edge radius and the resulting Reynolds number of the flow decrease.

On a forward-swept wing, the attachment-line boundary layer can get a laminar start at the tip. As the leading-edge radius grows larger inboard, the boundary layer can become thick enough to become unstable to small disturbances, shortly after which it would transition to turbulent.

So we see that the attachment-line boundary layer can undergo transition of either the laminar-to-turbulent or turbulent-to-laminar variety. Two thresholds thus play roles in determining the state of the boundary layer: one with increasing Reynolds number at which the laminar boundary layer becomes unstable and one with decreasing Reynolds number at which the turbulent boundary layer becomes unable to sustain turbulence (see Poll, 1979, 1989). Given that the attachment-line boundary layer is almost always practically identical locally to what it would be on an infinite-span attachment line, the two thresholds will generally be very close to what we saw in Figure 8.6.5. Under infinite-span conditions, the linear-instability threshold for the laminar boundary layer occurs at $\overline{R} = 600$, above which the boundary layer will always be turbulent. The threshold below which the boundary layer cannot sustain turbulence, and must therefore be laminar, is at $\overline{R} = 260$. Between 260 and 600, the boundary layer can be either laminar or turbulent, depending on conditions upstream.

This is the situation for an attachment-line boundary layer on a surface without area suction. If area suction is applied along the attachment line, R_θ will be reduced below the $R_\theta(\overline{R})$ curve shown in Figure 8.6.5. In the case with suction, the laminar-turbulent thresholds remain very close to the same R_θ values as for no suction, not the same $\overline{R}$ values. Thus a more general way to characterize the thresholds, valid with and without suction, is to say that the boundary layer is always laminar for $R_\theta < 100$, and it cannot be laminar for $R_\theta > 240$.

Several different devices have been demonstrated that can divert turbulent flow from the attachment line and give the boundary layer a laminar start. The *Gaster bump* (Gaster, 1965), shown in Figure 8.6.17, stagnates and sheds the turbulent boundary layer coming from inboard and starts a new laminar boundary layer at the stagnation point on the nose of the bump. Gaster's experiments showed the bump to be effective only up to nominal $\overline{R}$ values around 420, presumably because the outboard part of the bump has increased sweep and larger local $\overline{R}$. Seyfang (1987) tested an assortment of other devices at an $\overline{R}$ value of about 360. Figure 8.6.18 shows the minimum sizes, relative to leading-edge radius, at which they were effective. Whether they would be effective at higher values of $\overline{R}$ is not known.

Of course, the attachment-line boundary layer provides the starting condition for the boundary layer on the rest of the wing. On parts of the span where the attachment-line boundary layer is turbulent, the boundary layer off of the attachment line will also be turbulent, unless *relaminarization* occurs. In relaminarization, the boundary-layer turbulence dies out away from the attachment line, which generally requires that the chordwise acceleration must become much larger off of the attachment line than it was on the attachment line itself (Launder and Jones, 1969). This tends to happen only under high-lift conditions in which the attachment line is on the lower surface, well off of the leading edge, in which case the stronger acceleration required for relaminarization takes place as the flow rounds the leading edge in going from the lower surface to the upper surface. Relaminarization generally cannot occur when the attachment line is close to the leading edge, as it usually is

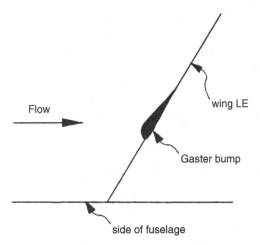

Figure 8.6.17 The Gaster bump (Gaster, 1965), a device for decontaminating the attachment line and giving the boundary layer a laminar start

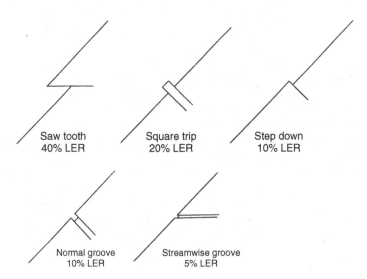

Figure 8.6.18 Other devices for decontaminating the attachment line and giving the boundary layer a laminar start, tested by Seyfang (1987). Numbers refer to the minimum sizes at which the devices were found to be effective, relative to LER (leading-edge radius)

under cruise conditions. In cruise, a turbulent attachment line generally leads to a turbulent boundary layer downstream.

A long run of laminar flow for cruise-drag reduction thus requires a laminar attachment line. Unless suction is applied on the attachment line itself, this requires that $\overline{R}$ be kept below 600, as an ideal upper limit. But practical limits are lower. For values of $\overline{R}$ above 260 without suction, a Gaster bump or other decontamination device will usually be required, and the Gaster bump has been shown to work only up to $\overline{R} = 420$. Then, even if the Gaster bump is successful, the laminar attachment-line boundary layer outboard will be vulnerable to unintended tripping by surface imperfections or contamination as long as $\overline{R}$ is above 260. Robust design for a laminar attachment line thus requires that $\overline{R}$ be kept below 260 or that suction be applied to thin the attachment-line boundary layer and thus to keep R_θ below 100.

When the attachment line is laminar, transition to turbulence off of the attachment line follows the general outline we discussed in Section 4.4.1: receptivity to disturbances from the outer flow or surface imperfections, linear growth of small disturbances, and nonlinear breakdown. The small disturbances can be either traveling TS (Tollmien-Schlichting) waves or stationary CF vortices. (Or, if the freestream turbulence level is high, as in some wind-tunnel situations, traveling CF vortices may come into play.) In the linear range, TS and CF disturbances grow independently of each other. For purposes of predicting transition, the TS and CF growth rates can therefore be computed separately, and transition identified when the accumulated growth of either passes a threshold e^n. In practice, it is advisable to apply the variable-n-factor method we discussed in Section 4.4.1, in which the values used are determined by comparison with experiments with flow environments and levels of surface quality comparable to the target application.

We see from the above that the possibility of transition by CF instability is another factor that makes designing a swept wing for laminar flow different from designing a 2D airfoil for laminar flow. As we noted in Section 4.4.1, the growth of CF disturbances requires a substantial amount of mean cross flow of the kind illustrated in Figure 4.1.7b. This in turn requires a pressure gradient. In applications of laminar flow to swept wings, CF tends to be critical only in the region of favorable gradient near the leading edge, and it determines how large a leading-edge radius and how much leading-edge sweep can be present without provoking early transition. Natural laminar flow (NLF) without surface suction is feasible on transport airplane wings, but only with leading-edge sweep angles somewhat lower than is currently typical of all-turbulent wings. If the sweep is higher than this, or the airplane is very large, laminar flow control by suction (see Joslin, 1998) would generally be required to prevent early transition by CF growth. CF disturbances tend to have high receptivity to disturbances caused by surface imperfections (distributed or discrete 3D roughness), and as a result, maintaining laminar flow around a swept leading edge can require very high levels of surface quality and very low levels of surface contamination, such as by insects (see Saric, Reed, and White, 2003).

8.6.5 Relating a Swept, Tapered Wing to a 2D Airfoil

Before 3D transonic CFD became routinely available, design technologies for swept wings were typically developed first on 2D airfoils. Today, 2D airfoil development is not nearly so important, but it is still used for early explorations of new ideas. So, what can we say about the correspondence of the performance of a swept, tapered wing to that of an "equivalent" 2D, unswept airfoil? In Section 8.6.2, we saw that turbulent boundary layers in the two cases behave in a qualitatively similar way, but that there are significant quantitative differences. Thus trends in performance should be similar, but the quantitative accuracy of comparisons will be limited.

And boundary-layer behavior is not the only limitation on the equivalence. In the invis-cid part of the flow, precise equivalence between 2D and 3D holds only in the case of infinite span with no taper (simple sweep theory). There is no rigorous way to take the effects of taper into account in a 2D context, as there is with the effects of sweep. Lock (1964) proposed that an approximate kind of inviscid equivalence requires the 2D airfoil to duplicate the perpendicular-Mach-number distribution on the surface of the 3D wing, that is, the component of Mach number perpendicular to the constant-percent-chord lines. The usefulness of this recipe in practice, however, has proved to be limited. If the freestream Mach number is chosen to correspond to the sweep at the quarter-chord or midchord of an aft-swept wing, the 3D perpendicular Mach number at the trailing edge is so high that the only way to match it on the 2D airfoil is to have an open (squared-off) trailing edge. The equivalence between a wing with a sharp trailing edge and an airfoil with an open trailing edge would always be questionable, especially with regard to viscous drag.

So, while 2D airfoil development may be useful for assessing general trends, its quantitative equivalence to a swept, tapered wing is limited.

8.6.6   Tailoring of the Inboard Part of a Swept Wing

The inboard part of a swept wing poses a design problem that is more 3D in nature than that of the outboard wing. The center plane of symmetry and the fuselage combine to restrict spanwise flow, so that the bound vortex lines of an aft-swept wing tend to unsweep and move aft, as shown in Figure 8.6.19. The resulting aft movement of the inboard lift loading is unfavorable for profile drag and can increase shock drag. It can also increase trim drag because the aft movement of the load requires a larger download on the horizontal tail for trim. To counteract this, the airfoil shapes of the inboard wing are usually specially tailored to move the loading forward. An additional consideration is that under transonic conditions the inboard wing can affect the pressure distribution and shock position far outboard.

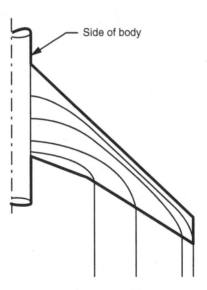

Figure 8.6.19 Illustration of the tendency of the bound vortex lines on the inboard part of a swept wing to unsweep and move aft

9

Theoretical Idealizations Revisited

In the preceding chapters, we looked at a wide range of aerodynamic phenomena, and we saw numerous examples of how theories based on conceptual models and simplifying assumptions allow us to construct explanations and make predictions without having to solve the full equations of motion. In this chapter, we'll take another look at the theoretical landscape, this time from the point of view of the models rather than the phenomena, taking a kind of inventory of the conceptual tools at our disposal. This will lead us to cover some of the ground we've already covered, but this time for the purpose of putting it all in a broad theoretical perspective, rather than the more phenomenological one we've taken sofar.

In Section 9.1, we'll look at the quantitative theories, categorizing them according to the type of model and what was done to simplify the equations. In Section 9.2, we'll explore the conceptual side, the tools available to us for doing Mental Fluid Dynamics (MFD).

9.1 Approximations Grouped According to how the Equations were Modified

The need for ad hoc models and simplifying assumptions was originally driven by the computational intractability of the full equations. This need is not as great now as it once was, but it is still there. Computers and solution algorithms have progressed to the point that solutions to the full NS (Navier-Stokes) equations for laminar flow can be calculated routinely, even for complicated 3D problems. Practical problems, however, almost always involve turbulent flow. In Section 3.7, we noted that Direct Numerical Simulation (DNS) of the turbulent motions in practical turbulent flows is beyond our reach and will be for decades to come. Thus for practical predictions of flows with turbulence, we must settle for the compromised accuracy that comes with turbulence modeling. And in many situations, further retreats from the full NS equations are useful as well.

Of course, with all retreats from the "full" equations come reductions in physical fidelity and range of applicability, and here we'll try to list them for a wide range of different types of theories. Though the losses in fidelity are easy to identify in general terms, they are not usually that easy to quantify. Theoretical arguments can sometimes determine the

Understanding Aerodynamics: Arguing from the Real Physics, First Edition. Doug McLean.
Images and Text: Copyright © 2013 Boeing. All Rights Reserved. Published 2013 by John Wiley & Sons, Ltd.

order of the error incurred by an approximation (how the error should be expected to grow as a function of some small parameter, for example). But the practical significance of the error can generally be determined only by comparison with more-accurate theories or with experimental measurements. Errors can usually be expected to depend strongly on the particular situation, and knowing what to expect in a given case generally requires experience with closely related cases.

To get a big-picture perspective on the landscape of theoretical approximations, we'll group them in tables of the major examples, categorizing them according to the types of things that were done to "simplify" the equations. We'll also identify what flow physics was assumed in addition to basic equations, the range of applicability of the resulting theory, and the loss of fidelity, if any.

9.1.1 Reduced Temporal and/or Spatial Resolution

For turbulent flows at practical Reynolds numbers, full temporal and spatial resolution of the flow is simply not feasible, and we must resort to time-averaging of the turbulent motions, combined with turbulence modeling. Then there are some situations in which even time-averaged spatial resolution of the distributions of flow quantities isn't needed, and the global view provided by control-volume analysis suffices. These two major types of approximation are summarized in Table 9.1.1.

9.1.2 Simplified Theories Based on Neglecting Something Small

In this class of approximations, some weak effect is either neglected altogether, or is accounted for only in some approximate way, often just to first order. For flows at high Reynolds numbers, the effects of viscosity can be either neglected or approximated. For flows at very low Reynolds numbers, the effects of inertia can be treated in the same ways. And if a body is very slender, the flow disturbances it causes can be linearized. Approximations of this type are summarized in Table 9.1.2.

9.1.3 Reductions in Dimensions

In many situations, the variation of the flow in time or in one or two of the spatial dimensions is so slow that it can be ignored or approximated. And in some of these situations, such as the flow along a very slender streamtube of a larger flow, or the local flow through a shock, the simplifying assumptions become practically exact in the limit, and the resulting theories are highly accurate. Approximations of this general type are summarized in Table 9.1.3.

9.1.4 Simplified Theories Based on Ad hoc Flow Models

In our final category, we have theories based on ad hoc models for the structure of the flow field. Such models don't have to reflect reality with great fidelity, but only well enough to scale in roughly the right way and to yield results that are accurate enough to be useful. The main examples of such theories are summarized in Table 9.1.4.

Table 9.1.1 Reduced spatial and/or temporal resolution

How equations are simplified	What modeling is added	Range of applicability	Loss of fidelity
Reynolds averaging with turbulence modeling			
Reynolds averaging: equations are time-averaged over time scales of the turbulent motions URANS: Only large-scale unsteadiness is computed; everything else is modeled LES and DES: Larger turbulent scales are computed; smaller scales are modeled. Computational grid is the "filter" that decides what scales are "small" RANS: Steady mean flow; turbulence is modeled	Turbulence modeling: turbulent transport of momentum and heat is modeled empirically Multiple PDEs, single PDE, or algebraic Direct Reynolds-stress modeling Eddy-viscosity modeling (often assumed isotropic)	Class of flows for which the turbulence model was calibrated	Can be reasonably accurate for simple boundary layers and free shear layers, but not generally very accurate for flows with significant curvature effects or other complications
Control-volume analysis			
Equations are integrated over some macroscopic volume of interest. Convection and transport terms then need to be considered only at the bounding surfaces	None	Generally applicable, but useful only when detailed distributions of flow quantities within the control volume and at the boundaries are not important	Difficult to apply accurately in situations where flow development is strongly influenced by detailed distributions of flow quantities at the boundaries

DES = detached-eddy simulation; LES = Large-eddy simulation; PDE = partial-differential field equation; RANS = Reynolds-averaged Navier-Stokes; URANS = unsteady Reynolds-averaged Navier-Stokes.

Table 9.1.2 Simplified theories based on neglecting something small

How equations are simplified	What modeling is added	Range of applicability	Loss of fidelity
Thin-layer RANS			
For 3D flow, viscous and turbulent diffusion terms are assumed small for one or two of the coordinate directions and are omitted. Mixed derivative terms are usually also omitted. Combinations of omitted and kept terms differ from code to code and depend on the type of grid. With diffusion represented in only one or two coordinate directions, the user must set up the grid so that the direction "normal" to any surface on which a boundary layer forms has the viscous terms appropriately represented	Turbulence model for RANS	Many external and internal flows at high Reynolds number, in which molecular and turbulent diffusion in one or more directions are small compared to diffusion in the other direction(s)	For many external and internal flows at high R, loss of fidelity is slight. In some cases, fidelity is actually improved through improved numerical accuracy because practical grids are often too coarse to represent the diffusion terms accurately in all directions. Examples in which thin-layer calculations were probably more accurate than full RANS for wing-body flows were seen in Mani et al. (2010)
Viscous/inviscid zonal decomposition for high R (1/R is assumed small)			
Outer flow is assumed governed by Euler equations or a potential equation	Turbulence model for RANS or RABL, or integral closure relations for integral BL	Many external and internal flows at high R, in which part of the flow away from the walls behaves as if it were inviscid	If BL approximation is assumed, pressure in the viscous region is not modeled accurately, and drag cannot be calculated by pressure integration

How equations are simplified	What modeling is added	Range of applicability	Loss of fidelity
Viscous/inviscid zonal decomposition for high R (1/R is assumed small) *(continued from previous page)*			
	Inner flow is assumed governed by RANS, RABL, or integral BL equations	Regions of separated flow may cause convergence problems for the interaction algorithm, and even if convergence is achieved, physical fidelity may be poor	Complicated behavior in viscous flow is not accurately predicted by turbulence modeling
Inviscid flow (Euler equations)			
Molecular and turbulent transport (viscosity and thermal conductivity) terms are dropped	None. Shocks can be captured by the numerical scheme, without need for explicit shock relations	Subsonic and supersonic flows at high R in which viscosity has little effect on surface pressures	Leaves viscous effects (skin friction and drag) unpredicted, and in many cases can badly miss the surface pressures (e.g., most transonic flows)
Inviscid flow (full potential equation)			
Same as Euler equations above, and flow is assumed irrotational so that velocity can be represented as the gradient of a potential	Same as Euler equations above	Same as Euler equations above, except that the vorticity generated by shocks is not represented	Leaves viscous effects unpredicted, as with Euler Shock drag from solution itself is incorrect, though it can be calculated after the fact by means of the Oswatitsch formula
Inviscid flow (incompressible)			
Same as full potential above, except that flow is assumed incompressible. Equation reduces to Laplace's equation	None	Flows with low Mach number everywhere in the field	Leaves viscous effects unpredicted, the importance of which depends on the situation

(continued overleaf)

Table 9.1.2 *(continued)*

How equations are simplified	What modeling is added	Range of applicability	Loss of fidelity
Inviscid flow (linearized for small perturbations)			
Flow is assumed uniform with small perturbations superimposed. Terms higher than first order in the perturbations are dropped. BCs are applied on a line or plane instead of the actual body surface	None	Accurate only for very slender bodies, wings, and airfoils	Not very accurate for most practical body and airfoil shapes
Slender viscous flows (BL or RABL)			
Molecular and turbulent transport terms in the streamwise direction are dropped All convective and transport terms in the normal-direction momentum equation are dropped	Turbulence modeling for RABL	Flows in which rates of change parallel to the flow are much smaller than rates of change perpendicular to the flow	Pressure in the viscous region is not modeled accurately Complicated behavior in viscous flow is not accurately predicted by turbulence modeling
Flows at very low R (Stokes flow)			
All convective terms are dropped, and only viscous and pressure terms remain Flow generally assumed incompressible	None	Flows with $R \ll 1$	No loss in fidelity if $R \ll 1$

How equations are simplified	What modeling is added	Range of applicability	Loss of fidelity
Oseen approximation for low R (Stokes flow with first-order approximation to convective effects)			
Same as for Stokes flow, but with first-order approximation to convective terms	None	Flows with $R \approx 1$	Very little loss of fidelity for $R \approx 1$
Lubrication theory			
Stokes flow for a 2D channel with very slowly varying depth	None	Lubrication flows with $R \ll 1$	Very little loss of fidelity for $R \ll 1$
Hele-Shaw flow (3D channel flow at very low R in a channel with small, constant depth)			
Same as for Stokes flow, plus channel is assumed to have a very small and constant depth. Result is that the velocity profile between the walls is parabolic, and that in the other two dimensions, the maximum velocity vector behaves like the velocity vector in a 2D potential flow obeying Laplace's equation. Hele-Shaw flows can thus be used as physical analog demonstrations of simple potential-flow patterns, as in flows 1–5 in Van Dyke (1982) depth	None	Shallow, constant-depth channel at very low R	Fidelity not very high for channels of practical depth

BC = boundary condition; RABL = Reynolds-averaged boundary-layer equation.

Table 9.1.3 Reductions in dimensions

How equations are simplified	What modeling is added	Range of applicability	Loss of fidelity
Steady flow			
All $\partial/\partial t$ terms are dropped. This simplification can be applied to any set of flow equations from full NS to incompressible potential flow	None	Flows that are nominally independent of time over sufficiently long periods	Typically little loss of fidelity in flows that are nominally steady
2D flow			
One spatial dimension is dropped. This simplification can be applied to any set of flow equations from full NS to incompressible potential flow	None	Ideally applicable only to flows that are really independent of the dropped dimension. Real flows are rarely very close to meeting the requirements	Ideally there is no loss of fidelity, but in comparison with real flows, the loss is often substantial
1D flow			
Two spatial dimensions are dropped	If flow is viscous, the effect is modeled as a 1D body force proportional to an empirical "friction factor." Likewise, if there is heat transfer, it must be modeled empirically	Flows in slender channels in which flow quantities are independent of the perpendicular coordinates. A streamtube in a 2D or 3D flow satisfies the requirement in the limit of small cross-section	For adiabatic inviscid flow there is ideally little loss of fidelity. In the presence of significant viscous effects or heat transfer, the loss is usually substantial
Simple sweep theory			
Flow is assumed inviscid (see Euler and full potential equations above) and invariant along the span of the wing, as it would be if the span were infinite	None	Inviscid flow over a wing of infinite span (can be swept) with spanwise-uniform far field boundary conditions	There is no loss of fidelity in the ideal situation assumed, but in comparison with viscous flows or flows over wings of finite span, the loss is often substantial

How equations are simplified	What modeling is added	Range of applicability	Loss of fidelity
2.5D boundary-layer theory			
Boundary-layer flow on a swept wing is assumed invariant in the spanwise direction, or, if the wing is tapered, the flow is assumed to vary spanwise according to a simple power-law scaling	Power-law spanwise scaling of boundary-layer thickness to account for wing taper	Boundary-layer flows on high-aspect-ratio swept wings with taper for which the isobars line up with the constant-percent-chord lines	Loss of fidelity is small if the taper is moderate, and the conical-isobar assumption is met
Fully developed duct or pipe flow			
Streamwise derivatives of everything but pressure are dropped, so that flow depends on only two coordinates in the duct cross-section, and flow is usually assumed incompressible	Turbulence modeling if flow is turbulent	Incompressible flows far from the entry or exit of a long, slender duct of constant cross-section	If flow is really fully developed, no fidelity is lost other than that associated with turbulence modeling
Fully developed flow in circular pipes			
Same as the more-general case above, plus flow is axisymmetric and depends only on r	Turbulence modeling if flow is turbulent	Incompressible flows far from the entry or exit of a long, slender circular pipe of constant diameter	If flow is really fully developed, no fidelity is lost other than that associated with turbulence modeling
Normal-shock relations (steady flow)			
Flow is assumed 1D, and conservation laws and thermodynamic relations are applied to derive the jump conditions	None	Steady flow through a normal shock	Jump conditions are very accurate. Detailed distributions through the shock are not predicted
Oblique-shock relations (steady flow)			
A Galilean transformation is used to add a tangential velocity to the 1D normal-shock relations	None	Steady flow through an oblique shock	Same as for normal-shock theory above
Shock relations for unsteady flow			
A Galilean transformation is used with steady-flow shock relations (normal or oblique) to account for motion of the shock	None	Flow through a moving shock	Same as for normal-shock theory above

Table 9.1.4 Simplified theories based on ad hoc flow models

How equations are simplified	What modeling is added	Range of applicability	Loss of fidelity
Lifting-line models for lifting surfaces			
Straight vortex filaments are assumed to stream straight back from the trailing edge of the lifting surface, and velocities affecting the wing are calculated by Biot-Savart law	Sections of the wing are assumed to respond to the local flow as if they were 2D airfoils	High-aspect-ratio wings at moderate lift loadings	Rollup of the vortex wake is not accounted, so that velocities far behind the wing are not well predicted
Trefftz-plane model for lift and induced drag of lifting surfaces			
Same wake model as for lifting-line models above, and velocities far downstream are used to infer lift and induced drag from momentum conservation in control-volume form	None	High-aspect-ratio wings at moderate lift loadings	Here fidelity is more questionable than for the lifting-line theories, since far field velocities are accurate only in the light-loading limit
Nikolski's (1959) model for effect of a fuselage			
Wake model is similar to lifting-line and Trefftz-plane theories above, except that trailing vortices are assumed to follow the contracting inviscid flow past the aft fuselage	Trefftz-plane lift is assumed to give the total lift. Difference between total lift and the lift on the exposed wing is assumed to be the fuselage carry-through lift	High-aspect-ratio wings at moderate lift loadings	Here fidelity is more questionable than for the lifting-line theories, since far field velocities are accurate only in the light-loading limit
Averaged actuator-disc (propeller) and turning-cascade models			
Effects of individual blades or vanes are averaged, and jump conditions for the changes in flow direction, speed, and, if appropriate, the total pressure are derived from the equations in control-volume form	Theoretical or empirical relationships to model the effects of blades	Applicable in the limit of a large number of closely spaced blades	Disturbances due to individual blades are not predicted

9.1.5 Qualitative Anomalies and Other Consequences of Approximations

The simplified theories listed in Tables 9.1.1–9.1.4 incur a variety of losses in physical fidelity relative to the full NS equations. But to keep this in perspective, we should recall that the NS equations themselves are not exact. The continuum representation that they embody is only an approximation to the integrated behavior of the molecules that make up the gas.

As a result, even the full NS equations produce inaccurate predictions when the assumptions underlying them are violated. For example, the NS equations don't predict the correct distributions of flow quantities through a shock because the very high gradients violate the assumption of local thermodynamic near-equilibrium. However, because the equations embody all the relevant conservation laws, the jump conditions across a shock are correctly represented, even though the internal details of the shock are not.

Solutions to the NS equations also display physically anomalous behavior when boundary conditions change too suddenly. For example, when a solid surface has a sharp convex corner, such as at a sharp trailing edge of an airfoil, the surface shear stress predicted by the NS equations goes to infinity (singularly) at the corner. This happens on such a small scale that the grids used in most NS calculations don't begin to resolve the singularity or even to hint at its presence. The presence of the singularity has nothing to do with the breakdown of the continuum assumptions at very small scales, but of course the assumptions must break down close to the singularity.

Approximating the effects of viscosity through boundary-layer theory introduces another kind of possible nonphysical singularity. When direct-mode solutions to the 2D boundary-layer equations predict separation from a smooth wall, the skin friction goes through zero with infinite slope, the Goldstein singularity depicted in Figure 4.2.11. In direct-mode solutions to the 3D equations, the skin friction perpendicular to the separation line displays the same singularity.

Neglecting viscosity can introduce qualitative anomalies in flows over surfaces with sharp corners. In a solution to the incompressible potential-flow equation, a sharp concave corner shallower than 90° produces a singular point of zero velocity with infinite velocity gradient on both the upstream and downstream sides (Figure 3.10.2c), and a sharp convex corner causes a singularity with infinite velocity (Figure 3.10.2b).

The situation at a convex corner is more complicated if the flow is compressible. In an otherwise subsonic flow governed by the Euler equations, attached flow around a sharp convex corner cannot produce infinite velocity, as it would in incompressible flow. It would, however, cause an acceleration to a sonic condition at the corner, with a Prandtl-Meyer expansion fan (see Shapiro, 1953) centered at the corner, producing a jump to supersonic speed on the downstream side. The corner would thus have a singularity with infinite acceleration, not infinite velocity, and downstream of the corner the supersonic flow would be terminated by a shock. This kind of attached-flow solution can exist only if the turning angle of the corner does not exceed the maximum Prandtl-Meyer turning angle of 130.5°. For typical airfoil shapes with sharp trailing edges, the turning angle around the trailing edge is too large (typically in excess of 170°), and the zero-lift flow pattern of Figure 7.1.3a is therefore impossible. This raises the interesting theoretical possibility that even an inviscid airfoil flow, provided it is compressible, would still have to satisfy the Kutta condition, as in Figure 7.1.3b. And even if the turning angle is less than 130.5°, the existence of an attached-flow solution is not guaranteed. It would depend on how much additional subsonic pressure recovery the flow faces downstream of the shock. If the shock terminating the supersonic bubble is strong enough, the resulting total-pressure deficit downstream could make it impossible for the flow to negotiate the rest of the pressure recovery. Finally, even when an attached-flow solution with a fan and a shock exists, it is only a theoretical possibility. In the real, viscous world, the boundary layer will either effectively fair over the corner, if it is shallow enough, or separate, eliminating the need for acceleration to supersonic flow.

In 3D numerical solutions of the Euler equations, a sharp convex corner can produce separation of the open, free-shear-layer type. An example would be the leading and trailing edges of a sharp-edged delta wing. In these situations, numerical dissipation has the same qualitative effect that real viscosity would have in bringing about the separation and enforcing a kind of de-facto Kutta condition. The mechanics of the resulting free shear layer are discussed in Section 10.4.3 and illustrated in Figure 10.4.1.

Free shear layers or shocks can also be modeled in solutions to inviscid equations as artificial discontinuities (artificial surfaces across which discontinuities are allowed), though even a shock has nonzero thickness in the real world. We discussed modeling shocks as discontinuities in Section 3.11.2. When such a shock discontinuity intersects a solid surface with convex curvature, the pressure and velocity distributions downstream of the shock are singular, the Zierep singularity that we discussed in Section 7.4.8.

A discontinuity tangent to the flow (a *slip surface*) can be used to model a physical shear layer such as can occur downstream of an intersection of shocks of different strengths. And as we discussed in Section 8.1.2, the shear layer in the wake of a 3D lifting surface is often modeled as a slip surface, or *vortex sheet*. Ideally, such a vortex sheet should be allowed to deform as required, so as not to have flow across it. In simplified models of lifting-surface flows, however, this requirement is usually allowed to be violated. An undeformed shape is assumed for the vortex sheet, and flow is allowed to pass through the sheet, as in the lifting-line theory discussed in Section 8.2.3, for example, and in many of the panel methods for lifting 3D potential flow that we'll discuss in Section 10.2.3.

9.2 Some Tools of MFD (Mental Fluid Dynamics)

In thinking about any aerodynamic flow for purposes of basic understanding or mental prediction, idealized models can be very helpful, even if they don't apply with great accuracy to the situation at hand. Even a model that is only a crude match for the flow in question can provide valuable insight. Of course care must be exercised in applying this kind of analysis, to avoid stretching an analogy too far or falling victim to any of the common fallacies.

The physical and theoretical bases for many such models were discussed in earlier chapters. In this section, we'll revisit some of these models, both as a summary and with an eye toward the sorts of flow situations in which they can be useful.

9.2.1 Simple Conceptual Models for Thinking about Velocity Fields

There are numerous ways that simple conceptual models can be invoked in thinking about the flow field a body will produce. Here we list some and briefly discuss how they apply.

9.2.1.1   Thinking of a Lifting Surface as Pushing the Flow in the Direction Opposite to the Lift Force

When a lifting surface such as a wing, tail surface, or a vortex generator (VG) produces lift, it pushes the flow between the tips (or between the root and the tip, if we are talking about a half-wing mounted on a wall) in the direction opposite to the lift force. This fact enables a quick mental determination of the direction of rotation of the tip vortices and the directions of the "induced" flow elsewhere in the field, given that the cross-stream field must look qualitatively like Figure 8.1.2.

This kind of mental modeling makes it easy to remember design guidelines for VG installations, such as how counter-rotating pairs should be grouped (Figure 4.5.2b) to delay vortex lifting, and how co-rotating VGs on a swept wing should be toed trailing-edge inboard to counter the outboard-directed boundary-layer cross-flow upstream of separation, as discussed in Section 4.5.2.

9.2.1.2   Reflection as a Model for the Effects of a Solid Surface

An easy way to model the effects of a solid surface such as the ground near a body is to place a reflection of the body on the other side of the surface. For an airfoil in 2D, it may suffice to image only the bound vortex as we did in Section 7.4.9. For wings in 3D, we must include the trailing vortex wake as we did in Sections 8.2.4, 8.3.9, and 8.5.1-2.

But lift is not the only effect that can benefit from being modeled by reflection. The volume displacement effect of a streamlined body can be modeled by a distribution of sources and sinks, which can be reflected if there is a nearby solid surface. The flow around a low-slung streamlined ground vehicle is strongly influenced by the ground, and the shape of such a vehicle can generally benefit from being designed in the presence of its reflection.

9.2.1.3   Flow Directions from Simple Sweep Theory

A quick estimate of the flow direction at the edge of the boundary layer at some location on a swept wing can be made if you know the local pressure coefficient and the local sweep, that is, the sweep of the local constant-percent-chord line. Use isentropic relations to calculate the local velocity-magnitude ratio V/U_∞. Then draw a vector diagram like Figure 8.6.3, using $U_\parallel = U_\infty \sin \Lambda$. On a wing with taper, simple sweep theory applied in this way is of course not exact, but it is a reasonable first estimate.

9.2.1.4   Identifying Separation in 3D Flows

Identifying separation in 3D is more difficult than in 2D because the skin-friction field in a 3D flow offers no purely local criterion corresponding to zero C_f at the separation point in 2D flow. In 3D, C_f is generally zero only at isolated nodal points of separation and is nonzero elsewhere on a line of 3D separation. Identifying separation in 3D thus depends more on the skin-friction direction than on the magnitude. It requires knowledge of the skin-friction direction field, defined with enough spatial resolution so that the separation line can be identified according to the region-of-origin concept described in Section 4.1.4 and illustrated in Figure 4.1.16. Some of the kinds of patterns that might have to be deciphered are illustrated in Figures 5.2.13 and 8.6.9.

9.2.1.5 Flow over a Slope Discontinuity (Crease) or Curvature Discontinuity in a Body Contour

Practical body shapes often have creases, either concave or convex, along which there is a discontinuity in the slope of the surface (or the direction of the surface normal). If the flow is parallel to the crease, the crease should have little effect except for changes in boundary-layer physics, which will be significant only if the crease is strongly concave or convex. But if the flow has a component perpendicular to the crease, there will be an effect on the pressure field. In the near field of the crease, the same assumptions behind simple sweep theory for a swept wing should hold, and the effect on the pressure should depend only on the perpendicular component of velocity. Then a good starting point for understanding the flow is the conformal-mapping solutions for the 2D potential flow past concave and convex corners illustrated in Figure 3.10.2b–d. These solutions are singular, with zero velocity and infinite velocity gradient if the corner is concave and infinite velocity if it is convex. The singularities are examples of the kinds of nonphysical anomalies that we discussed in Section 9.1.5. Of course such singularities don't materialize in the real world, but it is helpful to know that the inviscid version of the flow would see a large pressure disturbance at the crease. The boundary layer in a real flow tempers this behavior and smears out the pressure disturbance. If the angle change is large enough, the crease can cause boundary-layer separation, either at a convex corner, as in Figure 4.1.3c, or upstream (in the cross-flow sense) of a concave corner, in a situation analogous to Figure 4.1.3b (Imagine a crease at the bottom of the dip).

A similar situation occurs when there is a curve in the surface along which there is a discontinuity in surface curvature, as when a flat (zero curvature) portion of the surface abruptly adjoins a curved portion. This situation also produces a singularity in potential flow, one at which the velocity gradient is logarithmically infinite. In real life, a viscous boundary layer smears out the discontinuity, but the tendency toward an almost step change in velocity can sometimes actually be seen. This is why good aerodynamic design usually targets continuous surface curvature as well as slope.

9.2.1.6 A Source as the Far Field Effect of a Body with Viscous Drag

Viscous dissipation in the boundary layer and near-wake of a body produces a total-pressure deficit in the wake flow, and there is thus a velocity deficit in the wake. In simple situations where the viscous wake is not part of the core of a strong vortex wake, the velocity deficit is relative to the adjacent outer inviscid flow, as in Figure 6.1.3. In the wake vortex of a lifting wing, there is usually a jet in the core, as in Figure 8.1.9, and the velocity deficit due to viscosity is then relative to a stronger jet that would be there in the absence of viscosity. In either case, the presence of this deficit means that some mass flux that would otherwise be present in the space occupied by the wake must be compensated by greater mass flux outside the wake. It's the same mechanism as the displacement effect of a boundary layer (Section 4.2.3), extended to include the displacement effect of the viscous wake downstream of the tail of the body. To model the effects of this displacement on the outer flow, we can construct an equivalent inviscid flow that has the same outer-flow behavior as the real flow by placing sources and sinks on the body surface. These would be positive (sources) where the displacement thickness is growing and negative (sinks) where it is decreasing, as when the near wake "necks down" just aft of the tail (see Figure 5.3.4). Outside the near field, the displacement thickness of a viscous wake is generally positive, reflecting the fact that the source effect (growth) has dominated over the sink effect and that the net effect is that of a source. Far from the body and outside the viscous wake, the net effect is as if a single source were placed on or near the body.

9.2.1.7 A Vortex as the Far Field Effect of a Lifting Airfoil or Wing

According to the Kutta-Joukowski theorem (Section 7.2), the lift of any high-aspect-ratio lifting surface must be accompanied by circulation and bound vorticity. In the far field, only the net vorticity (the total, integrated over the chord of the surface) matters. In 2D, the far field thus looks like uniform flow with a vortex superimposed, with the vortex usually assumed to be located at the quarter-chord as in Figure 7.2.1d. If there is viscous drag, there would also be a source component as described in the previous item. In 3D, the bound vortex cannot end at the wingtips, and a horseshoe vortex system, including a trailing vortex wake, must be assumed, as in Figure 8.1.3.

9.2.1.8 The Kinematics of Vorticity

For thinking about the kinematics of flows with distributed vorticity, several ideas from Chapter 3 can be helpful.

The first applies very locally, and that is that the signature of vorticity in the deviations in velocity in the neighborhood of a point can be expressed as a solid-body rotation with angular velocity $\omega/2$ (*Helmholtz's first theorem*, see Section 3.3.5). If you've convinced yourself that there's no rotation, then you've probably shown that there's no vorticity.

Since most of the flows we're interested in are at reasonably high Reynolds numbers, vorticity diffuses slowly, and *Helmholtz's third and fourth theorems* (see Section 3.8.3) provide helpful guidance, even though they apply strictly only to inviscid flow. The third tells us that a vortex tube tends to be convected with the flow and that the only migration of vorticity into or out of the tube will happen slowly through viscous diffusion. The fourth tells us that the strength of a vortex tube tends to remain nearly constant, changing only slowly through diffusion.

In many examples of physical shear layers, vorticity is concentrated in a relatively thin sheet. An effective conceptual model for such flows is an ideal vortex sheet whose strength is equal to the velocity jump across the shear layer, as in Figure 3.3.8a. Remember that the velocity jump and the vorticity vector are perpendicular to each other, as in Figure 3.3.8c.

9.2.2 Thinking about Viscous and Shock Drag

Viscous drag generally has both a skin-friction part and a pressure part (the "viscous pressure drag"). For a change in the shape of a body to change the drag, it must change either one of these or both. The skin-friction part usually behaves in a straightforward way in that increasing the wetted area usually increases the skin-friction drag. The viscous pressure drag is usually not so simple. In Section 6.1.6, we discussed how misleading it can be to think of the viscous pressure drag in terms of the local surface pressures. For thinking about viscous pressure drag, there are two key concepts to keep in mind:

1. d'Alembert's paradox, that a body in inviscid flow without shocks or vortex shedding has zero drag. This is relevant because when the shape of a body is changed, much of the resulting change in the surface pressure distribution often originates in the outer inviscid flow, and unless there is a change in the lift distribution and induced drag, such changes inherently cause no change in pressure drag. For the viscous pressure drag to change, there must be a change in the displacement effect of the boundary layer and wake, which is difficult to estimate mentally. For intuitive purposes, it is often more helpful to think in terms of:
2. The fact that the total viscous drag is proportional to the volume integral of the viscous dissipation rate in the field, as discussed in Section 6.1.4.6. So an increase in the total viscous drag requires an increase in the integrated dissipation rate.

Thus if you want to predict or understand a change in the viscous drag, avoid thinking in terms of the surface pressure distribution and ask instead "where's the dissipation?"

Likewise, thinking about shock drag in terms of the surface pressure distribution is risky, especially in transonic flow. The pressure increments associated with shock drag are typically spread very diffusely over the body surface, and again it is more helpful to think in terms of dissipation in the field, in this case the Oswatitsch formula we discussed in Section 6.1.4.8.

9.2.3 Thinking about Induced Drag

Questions often arise as to how an increase in span, the addition of a tip device, or a change in spanload will affect induced drag. Thinking about such questions requires special care because the physics of induced drag is not easy to deal with intuitively and because there are such entrenched popular misconceptions about it. Not only is induced drag intuitively difficult, it is theoretically uncertain. In Section 6.1.3 we observed that induced drag cannot be defined precisely as a separate part of the drag except in terms of idealized approximate theories, and in Section 8.3.4 we noted that Trefftz-plane theory is the only practical option for inferring an approximate value for the induced drag from the geometry and spanload of a wing. Of course we can't do Trefftz-plane calculations in our heads, so for purposes of mental estimation we must rely on known results. Two rules of thumb that are easy to remember and provide a basis for useful estimating are:

1. For fixed lift and airspeed, the induced drag of a planar wing goes as the inverse square of the span (Equation 8.3.1) and
2. The effectiveness of a vertical winglet is the same as that of a horizontal span extension with about 55% of the Trefftz-plane span of the winglet (Figure 8.4.9).

These are ideal-induced-drag results, valid for spanloads optimized for minimum induced drag. Results for other tip configurations were shown in Figures 8.4.6 and 8.4.7.

When you encounter claims regarding the induced-drag performance of a change to a wingtip configuration, you should always ask yourself, "are the claims consistent with the Trefftz-plane-theory rules of thumb? If not, is there a plausible reason for a significant deviation from the theory?" This is a high hurdle. No tip configuration I'm aware of has ever conclusively demonstrated performance better than what Trefftz-plane theory would predict.

9.2.4   A Catalog of Fallacies

9.2.4.1   One-Way Causation of Pressures or Forces

According to Newton's second law, the direct cause of any acceleration, or change in the velocity vector, of a fluid parcel must be a net force applied to the parcel. In the effectively inviscid flow outside of boundary layers and wakes, a pressure gradient is the only source of such forces. Thus we can say that the pressure field causes the accelerations in the velocity field (though some of the explanations of lift we discussed in Section 7.3.1 imply causation in the other direction). But as we noted in Section 3.5, it is a fallacy to see the causation as going only one way. The cause-and-effect relationship between the pressure field and the velocity field is mutual, or circular. The pressure gradient causes the accelerations, and the accelerations sustain the pressure gradients. Likewise, the cause-and-effect relationship between the flow field and the integrated pressure force on a body is mutual.

9.2.4.2   Vortices as Agents of Causation

In Section 3.3.9, we discussed the Biot-Savart law, which allows us to infer part of the velocity field from the vorticity distribution. In the Biot-Savart law, the vorticity is the "input," and the "induced velocity" is the "output," which makes it all too easy to think of the vorticity at one location as "causing" the velocity at another. Such thinking is of course fallacious. There is no action-at-a-distance going on in ordinary fluid mechanics, and there is no way that vorticity can cause anything to happen remotely.

9.2.4.3   Trying to Influence a Global Flow Field by Tinkering with Vorticity Locally

This one is related to the previous item in that a vortex cannot physically influence what happens at another location. But there is more to it than that. Not only is it not possible to influence things remotely just by changing vorticity locally, it is not possible to significantly change the integrated vorticity in a local region by tinkering with the vorticity in a local streamtube. If you change the vorticity in one streamtube, compensating vorticity must appear adjacent to it, by kinematic necessity. This is why the erroneous wingtip device ideas discussed in Section 8.4.1 don't work. Another example, from Section 6.1.10, is that it's not possible for a propeller to "induce" circumferential velocities outside of its own slipstream.

9.2.4.4   Surface "Streamlines" (Skin-Friction Lines) as Indicators of Flow Direction

The surface "streamlines" visible in oil-flow photos like Figures 4.1.11 and 4.5.1b give the impression that they indicate the general direction of the flow over the part of the body in question. This impression is often misleading. Flow direction can change drastically within a short distance off the surface in a 3D boundary layer. The oil streaks indicate only the direction of the surface shear stress. A short distance off the surface the flow direction can be very different.

9.2.4.5   Zero C_f as the Universal On-the-Surface Signature of Separation

In steady 2D flow, the skin-friction coefficient C_f must go through zero at a separation point. In 3D, however, the magnitude of C_f is zero only at isolated singular points and is nonzero at typical points along a separation line. Identifying a separation line in 3D therefore requires looking at the global pattern of the skin-friction direction field and applying the region-of-origin concept as described in Section 4.1.4.

9.2.4.6   The Boundary-Layer Displacement Surface as an Effective "Solid Wall"

In Section 4.2.3 we defined the concept of a boundary-layer displacement surface as the effective location of the surface for an equivalent outer inviscid flow matching the actual flow far from the body. However, the idea that a displacement surface represents an effective "solid wall" is correct only in a limited way. A displacement surface is defined by a particular boundary-layer flow field consistent with a particular outer flow. The displacement surface so defined represents an effective solid wall only for that particular flow. If something is done to change the flow situation, the original displacement surface is no longer an effective solid wall. For example, if the displacement surface is defined for the wall boundary layers of a wind tunnel with an empty test section, it represents an effective solid wall only when the test section is empty. It is not an effective solid wall for the flow around a model placed in the test section.

9.2.4.7   Pressure is Constant across a Boundary Layer

One of the assumptions of first-order boundary-layer theory (Section 4.2.1) is to ignore the pressure gradient normal to the surface and thus to assume that the pressure is constant in the normal direction. It is therefore natural to adopt this assumption into our intuition and to think of the pressure as constant across the thickness of a boundary layer. This is not so bad if the wall is flat, but it isn't very accurate if the wall has pronounced curvature. In the real world, there must be a normal pressure gradient even within the boundary layer, consistent with the local mean velocity and streamline curvature.

9.2.4.8   The "Laminar Sublayer"

In the older literature, the sublayer of a turbulent boundary layer was referred to as the "laminar sublayer," and many commentators at the time apparently thought of the flow in the sublayer as actually being laminar. The flow in the sublayer is like laminar flow only in the sense that the Reynolds stress vanishes as the wall is approached, and the shear stress becomes all viscous, as we discussed in Section 4.4.2. Otherwise, in terms of the presence of substantial velocity fluctuations, the flow in the sublayer is anything but laminar, as can be seen in flow visualizations such as Figure 2.1f and in DNS calculations. Thus it is much more accurate to refer to the sublayer as the *viscous sublayer*.

9.2.4.9   Pressure Drag or Thrust of a Portion of a Complete Body

We sometimes like to assess the aerodynamic performance of a vehicle by looking at the drag-producing and thrust-producing parts separately. Such an assessment is always ambiguous because it requires making an arbitrary decision as to how to divide up the surface of the vehicle. Furthermore, we found in Section 6.1.3 that dividing the surface of a complete (closed) body into portions raises serious questions about the meaning of the pressure "drag" or "thrust" on each portion. The integrated pressure force on a portion of a body generally contains "spurious" contributions that are not properly seen as either drag or thrust.

9.2.4.10   Reducing Pressure Drag by Increasing Pressure on Aft-Facing Surfaces

In Sections 6.1.6 and 9.2.2 we discussed how misleading it can be to think of pressure drag in terms of the local surface pressures on parts of the surface and to think that the pressures can be manipulated in an essentially inviscid way by design of the surface shapes so as to influence the pressure drag. These kinds of changes to the surface pressure distribution usually produce little or no change in the viscous pressure drag.

9.2.4.11   Inviscid "Interference" Drag

As we noted in Section 6.1.9, the mutual inviscid buoyancy effect that occurs when two bodies with volume are "flown" in proximity is not properly considered interference drag. In a shock-free inviscid flow, such effects would always add up to zero for the total configuration. The total drag can change only if one or more of the drag-producing mechanisms in the field is affected. Changes to induced drag and shock drag are the only interference mechanisms that are effectively "inviscid" in the near field and can change total drag.

9.2.4.12   Base Drag and Hoerner's Jet Pump

In Section 6.1.7, we noted that the concept of "base drag" of bodies with squared-off tails is ambiguous at best. We also discussed Hoerner's correlation showing that the pressure on the blunt base of a body becomes less low as the body fineness ratio increases, and concluded that this is more likely a potential-flow effect than a result of Hoerner's proposed "jet-pump" mechanism in which the separating boundary layer "insulates" the wake from the jet-pump effect of the outer flow.

9.2.4.13   Explanations of 2D Airfoil Lift

There are many misconceptions in wide circulation in physical explanations of lift. These are discussed in detail in Section 7.3.1 and summarized in Section 7.3.3 under "popular misconceptions."

9.2.4.14   The "Dependence" of Induced Drag on Aspect Ratio

In Sections 8.3.1 and 8.3.4, we found that in both lifting-line theory and Trefftz-plane theory the induced drag depends on the span of a wing, but not on the chord or area. However, when we nondimensionalize induced drag in the usual way, dividing by qS so as to be consistent with the way we nondimensionalize other forces, aspect ratio appears in the formula for the induce-drag coefficient. This gives the false impression that induced drag depends on aspect ratio.

9.2.4.15 2D Biplane Lift Reduction as a Loss of "Efficiency"

In Section 8.3.12, we discussed how biplane pairs of airfoils incur a loss of lift at constant angle of attack, but we found that the loss of sectional "efficiency" is not due to the lift loss but is due instead to an induced-thickness effect. The loss in maximum sectional L/D is much smaller than what would be inferred from the lift loss.

9.2.4.16 Integrated Downward Momentum in the Trefftz-Plane "due to" the Trailing Vortices

A classical analysis of the flux of vertical momentum in the Trefftz plane behind a lifting wing looked at it in terms of velocities "induced" by the trailing vortices, which were assumed to be infinitely long. The result, Equation 8.5.2, indicated an integrated downward momentum corresponding to the lift. However, this turns out to be wrong because the double integral that was evaluated is nonconvergent. Assuming the vortices are finite in length, and taking the limit as the length goes to infinity yields the correct answer of zero integrated vertical momentum "induced" by the trailing vortices. There is an integrated vertical momentum in planes behind the wing, but it is in the velocities "induced" by the bound and starting vortices, not the trailing vortices. Details, including the effects of a ground plane, are discussed in Section 8.5.3.

9.2.4.17 An Expectation of High Accuracy in Calculations That Depend on Turbulence Modeling

In Section 3.7, we discussed the nature of turbulence and the strategies we have available for modeling it in computational fluid dynamics (CFD) calculations. Given the complexity and non-locality of turbulent motions and the way that models must gloss over much of this complexity, we should not expect high accuracy in calculations using turbulence models.

10

Modeling Aerodynamic Flows in Computational Fluid Dynamics

Since the late 1960s, the capabilities of computers and computational fluid dynamics (CFD) codes, and our knowledge of how to use them, have all grown tremendously. Over that time, CFD-based analysis, design, and optimization methods have revolutionized the practice of aerodynamics.

In some applications (e.g., the design of the cruise configuration of a transport airplane), knowledgeable use of CFD can now routinely produce good designs that don't need further refinement in the wind tunnel, only verification. For such applications, the role of wind-tunnel testing has changed from the screening of many candidate designs to the detailed evaluation of just a few. For applications where the geometry and/or the flow physics are more complicated (e.g., design of high-lift systems, or the analysis of off-design flight conditions), the wind tunnel still plays a major role in design development, but it will diminish as CFD capabilities advance.

CFD offers decided advantages over the wind tunnel in several ways. With CFD, we can explore a much larger number of design geometry variations than it is practical to test in the wind tunnel. And of course CFD can be run without interference from wind-tunnel walls or model supports. CFD can model full-scale Reynolds numbers, which in wind-tunnel testing is possible only in very expensive cryogenic pressure tunnels. With some additional computational effort and coupling with a structural model, CFD can simulate something closer to the real aeroelastically deflected shape of a vehicle than is possible in the wind tunnel, where it is usually possible at best to match the shape at only one flight condition. Last but not least, CFD has design and optimization capabilities that are now essential to our design practices.

With these advantages come costs, however. Defining the surface geometry and generating the grid for a calculation can represent a significant investment in engineering time. For 3D Reynolds-Averaged Navier-Stokes (RANS) calculations requiring high spatial resolution, computer time is still a significant cost, though it is diminishing rapidly as computers and algorithms improve. Then there is the personnel cost: Effective use of CFD is a specialized skill that requires time and practice to develop. Effective users must often spend a good part of their careers learning the mechanics of lofting surface geometry, generating grids, and running codes as well as developing the judgment needed for effective use of the results.

Judgment is needed because the simulation of reality that CFD can provide is usually far from perfect. A fundamental physical limitation on accuracy comes from turbulence modeling, which we discussed in general in Section 3.7 and for swept wings in particular in Section 8.6.2. Inadequate grid resolution also often degrades accuracy. A user may inadvertently use a grid that is too coarse, or, in many cases the densest grid one can afford to use isn't dense enough to provide the resolution needed for an accurate simulation.

The rapid advances in CFD capabilities in recent years have made it all too easy to forget that grid resolution can still be such a limitation. Capabilities that were on the cutting edge not long ago are now available at little cost. But no matter how much the capabilities advance, it seems that there are always applications of engineering interest that are at or beyond practical grid-resolution limits. Thus, for the foreseeable future, there will be a market for the most advanced codes available. Such capabilities at the cutting edge will always be costly and will be available only to those with deep pockets.

These are some of the costs and limitations inherent in CFD. Something else we should be aware of is a major psychological factor that influences how CFD is used and how the results are perceived. This is the tendency toward an overly optimistic perception of the accuracy of CFD among the entire community of code developers, code users, and customers (those who use or rely on the results). Much of this is just basic human nature, as I know from my own experience. When we see any sort of theory that has a reasonable physical basis and predicts the right trends, we have a tendency to forget the limitations and give the theory more credence than it deserves. In general, resisting this tendency requires vigilance. In the case of CFD, however, the problem is more serious than it was for "simpler" theories, and we tend not to be vigilant enough. A major contributor to this problem is the digital nature of the CFD "medium." CFD results are now almost universally viewed using advanced digital graphics, and the effects are powerful. These days it is common to see a complicated flowfield, predicted with all the right general features, and displayed in glorious detail that looks like the real thing. Results viewed in this way take on an air of authority out of proportion to their accuracy. In this regard, modern CFD is a very seductive thing.

With this general overconfidence comes a tendency, usually unintentional, to downplay the limitations of CFD's physical fidelity, to show and publish biased samples of results, and thus to oversell the codes' capabilities. Overconfidence induces us to think that the best comparisons with experimental data are the most representative and that the not-so-good comparisons probably reflect problems with things other than the code, such as shortcomings in the experiments. Of course experimental data are never perfect, but the CFD is at fault more often than its believers would like to think. The user community seems to be almost as prone to this kind of subtle self-deception as the developers are.

This tendency toward overconfidence, combined with the "high-tech" aura surrounding CFD, leads to a kind of overuse, in the sense that CFD is often used in situations where other options would be more appropriate. In many cases where a simpler theory or even a simple desktop experiment would provide an answer with less effort, or even a better answer, there is a temptation to use CFD anyway. A similar syndrome is in play in the overuse of other expensive tools such as the wind tunnel: the urge to hit a problem with the biggest hammer available. The tendency is understandable. If something goes wrong later in the program, you're less likely to be criticized if you used the "best" tool available.

So CFD is far from perfect, its capabilities are often overrated, and it can mislead us if we let it. But it is such a powerful tool that there is no way that the modern practice of aerodynamics could get along without it. In the rest of this chapter, we'll take a broad overview of CFD from a perspective similar to that of the rest of the book; that is, we'll focus on the basic physics and conceptual issues that CFD raises. We'll look at how CFD works, its strengths and weaknesses, and at some of the things users need to know to be effective.

10.1 Basic Definitions

- **CFD**: Using computer codes to generate numerical solutions to equations of motion for fluid flow, for purposes of making qualitative or quantitative predictions of flow behavior, or designing aerodynamic shapes that produce desired flow characteristics.
- **CFD analysis**: Computing the flow field produced by a given aerodynamic shape at a given flow condition. Outputs typically include flow quantities throughout the field, pressures and shear stresses at bounding surfaces, and integrated forces and moments on bounding surfaces.
- **CFD design or optimization**: This is similar to analysis in that the same flow equations are solved, but different in that a new ("designed") surface geometry is one of the outputs. Surface pressures or some other objectives, such as drag, are imposed, along with a starting ("seed") surface geometry.
- **Preprocessing**: Generation of geometry definitions for surfaces and in some cases the generation of the computational grid.
- **Postprocessing**: Generation of force and moment data for performance or loads analysis, and solution visualization (plots of quantities on surfaces or in the field).

10.2 The Major Classes of CFD Codes and Their Applications

Practically any fluid-flow model that is amenable to numerical solution can become the basis for a CFD code. From the full Navier-Stokes (NS) equations to the simplest lifting-line model, the possibilities run the gamut of the theoretical approximations summarized in Chapter 9. In this section, we'll look at the main categories of CFD codes and the applications for which they're currently suited. In the future, the range of things that can be considered practical will of course expand.

10.2.1 Navier-Stokes Methods

10.2.1.1 Direct Numerical Simulation (DNS)

The full unsteady NS equations are solved without a turbulence model, generally for the purpose of studying some aspect of turbulence physics. Even if the mean flow is 2D, a 3D unsteady calculation is required because turbulence is always 3D and unsteady, and spatial and temporal resolution must be fine enough to resolve the smallest details of the turbulence. This kind of simulation is so computationally demanding that it is generally limited to incompressible flow, simple flow geometries, and Reynolds numbers much lower than those typical of practical applications.

The typical Direct Numerical Simulation (DNS) calculation is still a major undertaking, and applications of DNS have been limited to fundamental physics studies. Turbulent boundary layers, wakes and other shear layers, and laminar-to-turbulent transition have been investigated in DNS studies. A DNS solution can provide much greater detail in a given flow than it is possible to record in an experiment. For some aspects of turbulence structure, especially for quantities like pressure-velocity correlations or vorticity fluctuations that are difficult or impossible to measure experimentally, DNS has provided much of our knowledge. As an example, Figure 10.2.1 shows instantaneous isocontours of turbulence quantities in the wall region of a flat-plate boundary layer, calculated by Robinson, Kline, and Spalart (1988). Compare the streaky structure visible here with that of the experimental visualization in Figure 2.1f.

10.2.1.2 Large-Eddy Simulation (LES)

The full unsteady NS equations are solved so as to resolve the larger-scale motions of the turbulence that can be resolved by the computational grid, while the effects of sub-grid-scale motions are modeled by a turbulence model. Sophisticated algorithm technology is required to blend the calculation smoothly and unobtrusively between the resolved scales and the modeled effects (see Sagaut and Germano, 2002). Large-Eddy Simulation (LES) can handle much more complicated geometries and higher Reynolds numbers than DNS, and it is therefore used for some practical applications in addition to fundamental physics studies. Still, practical applications are extremely demanding computationally, and LES is used only sparingly. An example of instantaneous quantities from a DNS solution was seen in the isovorticity surfaces in the wake of a circular cylinder in Figure 5.3.3a.

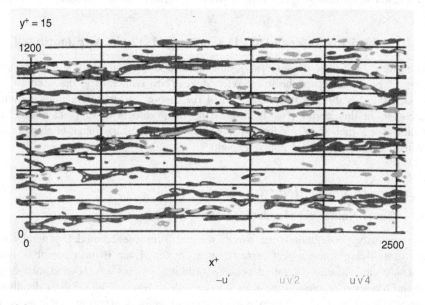

Figure 10.2.1 An example of the kinds of turbulence physics that can be simulated in DNS calculations: instantaneous contours of constant u' and $u'v'$ in a flat-plate boundary layer. Calculation by Robinson, Kline, and Spalart (1988). (Used with permission from Spalart)

10.2.1.3 Detached-Eddy Simulation (DES)

This is similar to LES, except that the large-scale motions that are resolved by the solution are limited to separated parts of the flow, and turbulence modeling is applied for everything else. The algorithm issues are discussed in Spalart (2009). Results of a Detached-Eddy Simulation (DES) calculation for the flow around a circular cylinder were shown in Figure 5.3.3. For practical applications, DES is much more economical than LES, but it is still very demanding computationally. DES is sparingly used in studies of noise generated by separated flow.

10.2.1.4 Unsteady Reynolds-Averaged Navier-Stokes (URANS)

The full unsteady NS equations are solved with a turbulence model with the intention of resolving unsteadiness only on length and time scales much longer that those associated with "turbulence." The grids and time steps can therefore be much coarser than for either LES or DES, and the calculations are thus much less expensive. However, Unsteady Reynolds-Averaged Navier-Stokes (URANS) is still expensive enough that applications are much less common than those of RANS. URANS has obvious applicability in cases where flow unsteadiness is forced by changes in the onset flow or by unsteady motion of the body. It can also simulate some kinds of spontaneous unsteadiness that arise from within a flow with steady boundary conditions, as, for example, buffeting associated with unsteady movement of separation, but this capability is limited. The damping imposed on such motions by the turbulence model is not generally realistic, with the result that the motions are not often realistically predicted.

10.2.1.5 Reynolds-Averaged Navier-Stokes

The full RANS equations with a turbulence model are solved with the intention of predicting a steady mean flow. The unsteady equations are usually used, and the solution is advanced in time from some initial state until a steady state is reached. Because only the final steady solution is of interest, not the time history, the solution is advanced in a nonphysical pseudo-time in which the time step varies form point to point in the domain, taking advantage of the fact that numerical stability allows longer time steps in some places than in others. Pseudo-time-stepping is generally combined with other convergence-acceleration schemes (see various articles in VKI, 1995). This allows a steady solution to be reached much faster than it could be if time-accurate URANS were used.

By their nature, RANS codes seek steady solutions to the equations. But what happens if the real flow under the conditions in question is unsteady on length and time scales longer than those of ordinary boundary-layer turbulence? That depends on the details of the particular solution algorithm. But before we look at what the codes actually do, let's speculate briefly about what we'd like them to do.

Ideally we might think we'd like a code to predict the onset of real unsteadiness by failing to reach a steady solution. Even as an ideal, however, this is problematic: At what level of unsteadiness would we like the code to fail? Besides, practically speaking, we'd really like the code to continue giving usable results in situations where the real flow has significant unsteadiness, as, for example, at maximum lift of an airfoil or wing. How accurate or useful the results would be under such conditions would have to be evaluated (This kind of evaluation is part of what is referred to as "CFD validation," which we'll discuss further in Section 10.5), but having a code "quit" early is a practical nuisance.

Of course what we'd like the codes to do and what they actually do are two different things. The behavior of RANS codes in situations where the real flow is unsteady varies widely depending on the details of the solution algorithm and on the particular flow situation. It ranges from failing near the onset of real unsteadiness to plowing onward and getting steady solutions well past the onset. The Boeing GGNS code, for example, produces steady solutions for 2D airfoils at 90° angle of attack (see Allmaras *et al.*, 2009). What such solutions mean, physically speaking, is an interesting question.

A common assumption or option in many RANS codes is to solve *Thin-Layer Navier-Stokes* equations in which the viscous and turbulent diffusion terms have been omitted for one or two of the coordinate directions, as described in Table 9.1.2. This is akin to a boundary-layer-theory assumption in that diffusion is neglected in one or two directions parallel to viscous shear layers, but it is unlike first-order boundary-layer theory in that the pressure variation perpendicular to shear layers is still accounted for. In principle, the thin-layer assumption entails some loss of fidelity, but it can be offset by an improvement in numerical accuracy. Grids used in 3D calculations are often not fine enough to resolve the viscous terms in directions parallel to solid surfaces, with the result that when the full NS equations are solved, numerical dissipation produces exaggerated surface-parallel diffusion, leading to nonphysical effects. For example, some participants in a NASA drag-prediction workshop (see Mani *et al.*, 2010) observed that a 3D separation in a wing-body junction was unrealistically large in their full NS solutions but not in their thin-layer solutions.

RANS codes account for an overwhelming portion of all NS calculations in practical applications. The most natural applications are in *analysis,* in which the body geometry and onset flow conditions are given as inputs, and the flowfield quantities are outputs. RANS is also used in *inverse design,* where the objective is to determine the body shape that produces a particular pressure distribution, and in *optimization,* in which there can be several kinds and combinations of objectives. But design and optimization are not the natural forte of RANS codes in their current form.

There is currently no direct way to solve the inverse design problem in RANS. One way of attacking inverse design is by the ad hoc CDISC scheme, in which the body shape is updated based on a formula that uses surface curvature and the mismatch between the current pressure solution and the target (see Campbell and Smith, 1987). It works after a fashion, but it is cumbersome. The other approach is called "modal" inverse design, which uses a number of shape-change *modes* whose amplitudes are to be determined, as we discussed in Section 7.4.10 in connection with 2D airfoil design. Doing modal inverse design with a RANS code requires using an external optimizer to determine the modal amplitudes so as to minimize some measure of the deviation between the actual pressure distribution and the target.

Doing other kinds of optimization (e.g., minimizing drag) using RANS codes is currently very expensive and time consuming. In the usual approach, an external optimizer is used to guide the optimization, and the RANS code is run many times to evaluate design candidates, thus providing *function evaluations,* usually a large number of them in the course of a typical optimization. External optimizers can be of either the *global* type, which can in principle find a global optimum, or the *gradient-based* type, which can find only a local optimum but can home in on it very precisely. Sometimes both types are used in sequence, a global optimizer to identify the neighborhood of the global optimum, followed by a gradient-based optimizer to home in closely on the optimum. LeDoux *et al.* (2004) describe a system that can use both types.

Gradient-based optimizers of course require derivative information, that is, rates of change of the objective with respect to the *design variables* (shape-change modes, for example). If the RANS code is used only to provide function evaluations, these derivatives must be calculated by finite differences between flow solutions, and each derivative thus requires an additional flow solution. When finite-difference derivatives are used, applications are typically limited to small numbers of design variables and *single-point optimization* (optimization at a single flow condition). An alternative approach for obtaining derivative information is to use a suitably modified version of a CFD code to solve an adjoint problem (see Elliot and Peraire, 1996). A single adjoint solution can provide derivatives for any number of design variables simultaneously, thus greatly reducing the number of flow solutions required.

Most practical aerodynamic optimization problems really require *multipoint optimization,* in which the objective function takes into account more than one operating condition (more than one lift coefficient or Mach number, for example). This is currently prohibitively expensive and time consuming to do in RANS with finite-difference derivatives, and is still quite expensive even if adjoint derivatives are used because a separate adjoint solution is required for each operating condition. Most multipoint aerodynamic optimization is currently done with coupled viscous/inviscid codes (Section 10.2.2), which incur much lower expense and flow time than RANS.

10.2.2 Coupled Viscous/Inviscid Methods

Coupled viscous/inviscid codes take advantage of the basic inner/outer structure of many high-Reynolds flows that we discussed in Section 3.6 and in the introduction to Chapter 4. The boundary-layer equations are used in the inner region, inviscid equations in the outer region, and the two solutions are matched through the displacement thickness or surface transpiration, as discussed in Section 4.2.3. Such methods are nearly always limited to steady flows and have been implemented in both 2D and 3D.

The main advantage of such methods is that they require roughly an order of magnitude less computing than RANS. They are thus more practical than RANS for multipoint aerodynamic optimization or modal inverse design. A further advantage of viscous/inviscid coupling is that the inviscid portion of the method can represent small changes in surface shape by transpiration from the surface instead of through the modification of the grid that is required in RANS methods. TRANAIR (Melvin *et al.*, 1999) uses this approach to reduce the number of times the grid must be modified and thus to speed up modal design and optimization in 3D. The 2D airfoil codes XFOIL and MSES (Drela, 1989, 1993) do not use transpiration, but they do provide the option of solving the inverse-design problem directly, that is, defining the new surface shape grid point-by-grid point and not relying on shape-change modes.

For flows that are attached all the way to the rear of a body or have only shallow separated regions, the physical fidelity of coupled viscous/inviscid codes is comparable to that of RANS. Coupled viscous/inviscid codes typically fail to converge for flows with massive separation, but RANS also has problems in such flows. Because of the limitations of the boundary-layer equations, coupled viscous/inviscid codes miss viscous-flow details such as junction flows (e.g., necklace-vortex separation).

The inviscid component of a coupled code can be any method that models the body geometry with sufficient fidelity and produces a sufficiently realistic and smooth prediction of the pressure or velocity distribution on the surface (an Euler, full-potential, or panel method, as discussed in Section 10.2.3). The boundary-layer component can be either an integral method or a field (differential) method. In earlier codes, coupling was often just sequential iteration back and forth between an inviscid module and a boundary-layer module, usually with some under-relaxation of the displacement thickness or surface transpiration to reduce oscillations. Convergence was not very robust, and flows with separation were typically off limits. More-recent codes solve the inviscid and boundary-layer equations in more simultaneous ways and use convergence-acceleration schemes such as Newton's method (see Drela and Giles, 1986). These newer methods can often converge for flows with some separation.

Coupled viscous/inviscid codes are available for a variety of 2D and 3D applications. For low-speed airfoils XFOIL (Drela, 1989) is probably the most versatile in that it handles laminar separation bubbles with reasonable accuracy. For single-element and multielement airfoils and transonic flow, MSES (Drela, 1993) offers a full range of analysis, design, and multipoint optimization options. For 3D subsonic, transonic, and supersonic flow (potential-flow plus integral boundary layer), TRANAIR (Melvin et al., 1999) handles analysis and a broad range of design and optimization applications.

10.2.3 Inviscid Methods

Codes for inviscid flows cover a wide range in physical fidelity from the full Euler equations to incompressible potential flow with small perturbations. At any of these levels, the flow can be steady or unsteady, though codes for steady flow are by far the most common. Inviscid methods by themselves generally apply only to flows for which neglecting viscous effects is justified, and many entail further restrictions, as we'll see next. Inviscid codes encompass field methods for the nonlinear full-potential or Euler equations, and panel methods for linear potential flow.

10.2.3.1 Euler Methods

The Euler equations are just the NS equations with the viscous and heat-conduction terms omitted. The equations apply to flows at any Mach number, as long as the equation of state that is used remains valid, and viscous and heat-conduction effects are negligible.

Discontinuities (shocks and slip surfaces) can appear in solutions to the Euler equations. In numerical solutions, these are generally "captured" by the numerical method and are therefore "smeared" over distances of at least several grid cells. Shocks numerically captured in this way tend to approximate the "real" discontinuous ones that would occur in exact solutions to the equations. Smeared slip surfaces tend to emanate from sharp edges, where the artificial dissipation associated with the numerical method causes the flow to separate much like a viscous flow would. In effect, artificial dissipation enforces something like the Kutta condition that we discussed in connection with the potential flow around airfoils in

Section 7.1. The smeared slip surface that emanates from a sharp edge is thus similar to a real physical shear layer, except that its diffusion (smearing) is artificial and not realistically predicted. We'll consider these issues further in Section 10.4.3.

The Euler equations are applicable to flows with or without vorticity, but given the lack of viscosity, there are only limited ways that vorticity can arise. It must either be fed into the domain through the inflow boundary conditions, be generated by passage of the flow through a shock of nonuniform strength, or result from "separation" from a sharp edge. The equations themselves have no mechanism for vorticity to diffuse, only for it to be convected with the flow, but in numerical solutions there will be some diffusion due to artificial dissipation.

A major advantage of Euler over RANS is that Euler doesn't need to resolve boundary layers and therefore doesn't require the dense grids that RANS requires in such regions. Grid generation is therefore easier, and computing times are much shorter. However, Euler methods by themselves have limited applications. Transonic airfoil and wing flows, for example, are quite sensitive to the displacement effect of the boundary layer, and accurate simulation requires RANS or coupled viscous/inviscid methods. In Figure 7.4.29, we saw an example of the large effects viscosity can have on the pressure distribution and shock location on a transonic airfoil. On the other hand, in some supersonic flows the boundary layer remains much thinner than in typical transonic flows, and thus supersonic flows can be reasonable applications for Euler methods. Some flows involving separation from sharp edges, such as the leading-edge-vortex flow over a delta wing, can also be appropriate applications for Euler, though assuring adequate grid resolution of the separated shear layer is not always easy.

10.2.3.2   Nonlinear Potential (Full-Potential) Methods

In the full-potential equation, the velocity is represented as the gradient of a potential, as it is in incompressible potential flow, but the "full" equation is nonlinear because it effectively incorporates the "exact" 1D is entropic relation (Equation 3.11.3) for the density as a function of the local Mach number (see Garabedian and Korn, 1971). Solutions therefore cannot be constructed by superposition as in the panel methods, and a field grid is required.

Slip surfaces cannot be "captured" as they are in Euler solutions, so that to model the vortex wake of a 3D wing in full-potential, there must be a grid surface behind the wing, on which a wake boundary condition is applied. We'll discuss what this means in terms of the numerics and physical modeling in Sections 10.3 and 10.4. Shocks can be captured in full-potential as they are in Euler. The pressure jump across a shock is reasonably predicted, as long as the shock is not too strong, but the shock drag imposed on the body surface is erroneous, and shock drag must be determined by means of the Oswatitsch formula applied in the field. We'll discuss this issue in some detail in Section 10.4.1.

Full-potential methods are applicable to subsonic, transonic, and supersonic flows up to low supersonic Mach numbers. Strong shocks are not accurately represented, but the weak shocks typical of many transonic and supersonic cruise applications are adequately predicted. Like Euler methods, full-potential methods by themselves have only limited applications. In many of the applications where we might like to use such methods, we really should include viscous effects, either through a coupled viscous/inviscid code or RANS.

10.2.3.3 Linear Potential (Panel) Methods

Panel methods solve either the incompressible potential-flow equation (Laplace's equation 3.10.2) or one of the versions applicable to compressible flow with small disturbances (see Liepmann and Roshko, 1957, Section 10.2). Because the potential equation is linear, solutions can be constructed by superposition. In panel methods, the basic solutions that are superimposed are those for flows around elementary singularities that are generally distributed on *panels* placed on the contours of the body and on other surfaces where the flow simulation requires a discontinuity in the inviscid flowfield. The singularities automatically satisfy the potential equation, so the equation itself isn't explicitly used. Instead, a surface integral equation that represents the boundary conditions is solved for the strengths of the singularities. No field grid is required. This is an example of a general class of methods referred to as *boundary integral methods*.

Unlike Euler and full-potential, panel methods do not allow shocks. Slip surfaces are allowed and can be used to model shear layers such as boundaries of propeller slipstreams or vortex wakes of lifting surfaces in 3D. Propeller discs can also be modeled in an average sense (averaged over the disc, ignoring the discrete propeller blades) by surfaces of discontinuity that produce the appropriate velocity jumps. When surfaces of discontinuity divide the flow domain into regions with different total-pressure, as in the case of a propeller disc and slipstream boundary, special care is needed to ensure that the formulation is consistent. All such surfaces of discontinuity must be *paneled,* so that the velocity jumps across them can be "produced" by the appropriate singularities.

Of course a "real" slip surface is a stream surface across which there is no flow, and to model one with a paneled surface ideally requires solving for the location and shape of the paneled surface along with the distribution of singularity strength, which is a nonlinear problem. Some 3D panel codes have the option to determine the shapes of slip surfaces by iteration in a process called "wake relaxation," but the process rarely converges and must usually be truncated after a few iterations. One reason for this is that vortex wakes tend to roll up at their edges, and when they are modeled as thin sheets they form such tight coils that no reasonable paneling can resolve them.

Given the difficulty of iterating for real wake shapes, most panel calculations are done with *fixed wakes* that are fixed in space and must therefore generally allow flow through them. What this means to the physical modeling is discussed in Section 10.4.3.

Historically, panel methods provided the first practical CFD capabilities for complex 3D configurations, and they are still useful for some applications, such as providing rough estimates of the effectiveness of high-lift systems. But linear compressibility has a very limited range of applicability, and, as with Euler and full-potential, panel methods without viscous coupling have limited applications.

10.2.3.4 Lifting-Surface Methods

Lifting-surface methods are essentially 3D panel methods that model only the camber lines of lifting surfaces, not the thickness. Vortex wakes must of course be paneled, just as in general panel methods.

Relative to general panel methods, physical fidelity and range of applicability are reduced. In return, lifting-surface methods provide ease of geometry definition, and short computing times. Versatility is also a strong suit, as many lifting-surface codes have options for inverse design (designing the camber shape to produce a given load distribution) and optimization, in addition to analysis. Because they don't model the thickness of lifting surfaces, and therefore predict only load distributions, not realistic surface-pressure distributions, lifting-surface methods are not suitable for viscous/inviscid coupling.

10.2.4   Standalone Boundary-Layer Codes

In the early days of boundary-layer CFD in the 1960s and early 1970s, most boundary-layer calculations were done in a "one-shot" mode. The outer-flow boundary conditions for the boundary layer were taken from a single inviscid calculation (usually potential) or from pressures measured in an experiment, and the boundary-layer code was run once, without coupling to account for the effect of the boundary layer on the outer flow. A standalone boundary-layer code was then a multipurpose tool that could take inputs from a variety of sources.

Now, most viscous-flow CFD is done with coupled viscous/inviscid coupling codes or with RANS, and there are fewer applications for standalone boundary-layer calculations. One of the few remaining applications is to provide the mean-velocity field (boundary-layer velocity profiles) for laminar-flow stability calculations, as done, for example, by the eMalik code (Malik, 1992).

10.3   Basic Characteristics of Numerical Solution Schemes

There are many different kinds of numerical methods in use in CFD codes. In this section, we'll look at some of their basic characteristics, concentrating more on the operational and practical issues than the mathematical.

10.3.1   Discretization

The flow problems we wish to solve with CFD methods are continuous problems in that they involve flow quantities and boundary conditions that are at least piecewise continuous. To attack such problems numerically requires discretizing the spatial domain in some way such that solution quantities need to be determined at only a finite number of locations. Here, we'll look very briefly at some of the ways that a continuous flow problem can be converted into a discrete problem that can be solved numerically. The objective in such methods is generally to define the discrete problem such that its solution *spatially converges* to a solution of the continuous problem as the discretization is refined. We'll discuss such convergence further in Section 10.3.3.

In panel methods for linear potential equations (Section 10.2.3), a grid of *panels* is defined on all body surfaces exposed to the flow and on other surfaces where the flow field is required to be discontinuous, such as vortex wakes. Solution quantities defined on these panels are parameters defining the distribution of singularity strength to be solved for. The number of parameters per panel depends on the order of the mathematical representation of the singularity distribution. When there is more than one parameter per panel, continuity conditions between panels constrain the parameters so that there is only one independent parameter to be determined per panel.

The boundary conditions in a panel method are evaluated and imposed at a single *control point* on each panel. For a non-through-flow condition on a solid surface, for example, the component of velocity in the direction of the panel normal vector is evaluated at the control point. The boundary value that is to be constrained on each panel is influenced by the singularity parameters on all the panels through a matrix of influence coefficients. Imposing the boundary conditions on all the panels results in a set of linear algebraic equations to be solved for the unknown independent singularity parameters. In principle, no iteration is required because the problem is linear, but for large numbers of unknowns, an iterative method of solving the system is often the fastest. Once the values of the singularity parameters are known, other influence coefficients are used to calculate flow outputs such as velocity vectors and pressures at the control points. Calculating flow quantities at points other than the control points requires calculating influence coefficients for each point where the solution is to be evaluated.

In methods for any of the nonlinear field equations (NS, Euler, full-potential, or boundary-layer), the spatial domain must be filled with a *grid* consisting either of *points* or of *cells* and their bounding surfaces. In cell-based schemes, the flow variables can be defined and stored at either the cell centers or the vertices. The equations can be used in differential form (finite-difference methods) or control-volume form (finite-volume methods), and within each of these categories there are a variety of ways the equations can be discretized. Different methods represent flow quantities to different orders and converge to the exact solution at different rates with grid refinement. We'll discuss grids and spatial convergence further in Sections 10.3.2 and 10.3.3.

10.3.2 Spatial Field Grids

A spatial field grid for solving nonlinear equations typically consists of a space-filling *volume grid*, and a grid surface that lies on the surface of the body, the *surface grid*.

Grids can be either *structured* or *unstructured*. The simplest kind of structured grid is one that can be morphed into a rectangular mesh with ordered rows, columns, and planes (in 3D) of points or cells. Such a grid can be thought of as defining a curvilinear coordinate system in which locations are defined by their coordinates a in *grid parameter space*. Figure 10.3.1a shows a 2D example, a grid wrapped around an airfoil in an arrangement called a *C-grid*, with grid parameters denoted by s and t. An unstructured grid usually consists of triangular or tetrahedral cells not arranged in ordered rows, as shown in Figure 10.3.1b for a 2D airfoil. An unstructured grid doesn't have a multidimensional grid parameter space.

A grid can consist of a *single block,* as in Figure 10.3.1, or an arrangement of multiple blocks. The blocks of a *multiblock* grid can cover the spatial domain in a non-overlapping fashion, in which each point in the domain is included in only one grid block, and adjacent blocks *abut* each other precisely, as in Figure 10.3.2a. An alternative is that grid blocks can *overlap,* as in Figure 10.3.2b. In the case of abutting grid blocks, special conditions must be applied across block boundaries to ensure proper continuity of the flow solution. In overlapping grids, continuity of the solution is enforced by interpolation algorithms that can be quite complicated to implement.

Structured grids can be morphed around body contours in many different ways, which raises interesting issues of grid topology. A sharp corner or crease in the body contour is called a *lost corner* if part of a single grid block wraps around it as in Figure 10.3.3a. Solution accuracy tends to suffer in the neighborhood of a lost corner because of the high

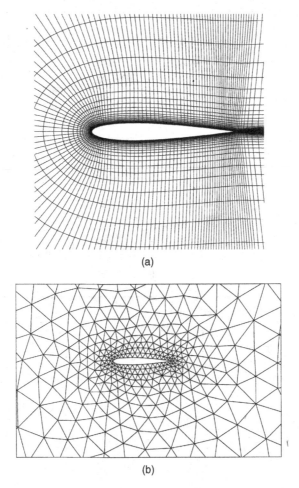

(a)

(b)

Figure 10.3.1 Examples of *structured* and *unstructured* grids for a 2D airfoil. (a) A single block of structured grid wrapped around the airfoil in an arrangement called a *C-grid*. (b) An unstructured grid of triangular elements around the same airfoil.

distortion of the grid. A physical corner is more naturally represented if the grid parameter space also has a corner, so that the surfaces on opposite sides of the corner represent different directions in parameter space. For a concave corner in the computational domain, this can be done with the corner of a single grid block, as in Figure 10.3.3b. A convex corner must generally be represented by a corner between grid blocks, as in Figure 10.3.3c.

Even with the flexibility afforded by multiple grid blocks, it is not always easy to avoid lost corners, and they are tolerated in many multiblock grids. For example, the *H-grid* topology is often used on wings and airfoils, with lost corners at the leading and trailing edges, as in Figure 10.3.4. For complex configurations, especially in 3D, overlapping grids and unstructured grids offer great flexibility and are seeing increasing use.

When a viscous boundary layer is part of the flow solution, the grid must be able to adequately resolve the details of the boundary-layer velocity profile. For a turbulent

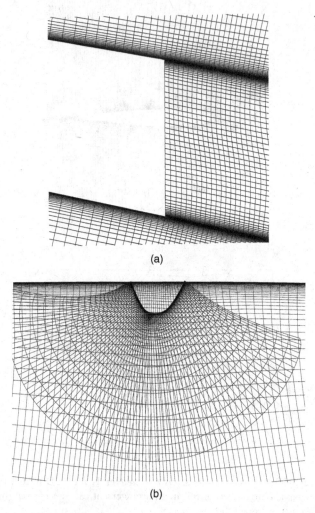

(a)

(b)

Figure 10.3.2 Examples of multiblock grids. (a) Non-overlapping. This is an extreme blowup of a 2D grid around the blunt trailing edge of an airfoil, where a separate grid block has been used to resolve the region behind the trailing edge. (Plot by N. J. Yu.) (b) Overlapping. This view shows two grid surfaces extending downward from the surface of a wing in a 3D volume grid. The one with the straight upper edge is part of the general grid around just the wing and abuts the wing lower surface, and the one with the curved boundary is part of a collar grid to resolve a small fairing protruding from the wing surface. (Plot by E. R. Setiawan.)

boundary layer, this requires resolving the viscous sublayer. In practice, codes often provide an assessment of how well this requirement was satisfied after the calculation is completed, by displaying the thickness of the first grid interval off the wall in turbulent-boundary-layer wall units (units of y^+ as defined by Equation 4.4.4). A good value for this thickness is in the neighborhood of one or two. This of course means having a normal-direction grid

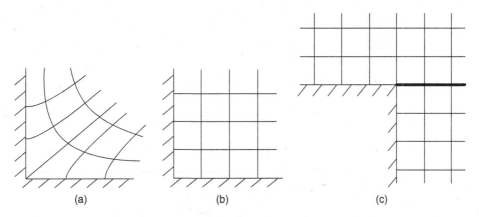

Figure 10.3.3 Grid topology at a sharp corner of a boundary. (a) A lost corner where part of a single grid block wraps around. (b) A corner of a single grid block fitted to a concave corner. (c) A corner between grid blocks fitted to a convex corner.

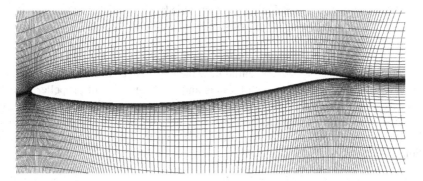

Figure 10.3.4 An *H-grid* for an airfoil, with a 90° lost corner at the leading edge and a subtle lost corner at the trailing edge

spacing at the wall that is very small compared to the boundary-layer thickness, which has two serious consequences, if unreasonably large numbers of grid points are to be avoided:

1. The spacing along the surface must be much larger than the near-wall normal-direction spacing, making for grid cells with very high aspect ratios, which can degrade accuracy or cause numerical instability.
2. The normal-direction spacing must stretch away from the surface, quickly increasing the spacing outside the viscous sublayer. Strong stretching also degrades accuracy.

An appreciation for the typical velocity gradient that must be resolved at the bottom of a turbulent boundary layer can be gotten from the velocity profiles plotted in Figure 4.4.5. And the situation becomes increasingly difficult as Reynolds number increases. A typical grid for resolving a turbulent boundary layer is illustrated in Figure 10.3.5.

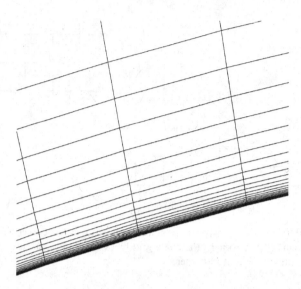

Figure 10.3.5 Typical grid for resolving a turbulent boundary layer, including the viscous sublayer. This is a crop of a small portion of the airfoil C-grid in Figure. 10.3.1a

Recall that in panel methods for potential flow the paneled geometry has to include known shear layers such as the vortex wakes of wings and the boundaries of propeller slipstreams. The same requirement applies to methods for the nonlinear full-potential equation: the "surface grid" must include surfaces in the field where any discontinuity, such as a vortex wake, is to appear, and the flow solver must be set up to apply the appropriate boundary condition there.

10.3.3 Grid Resolution and Grid Convergence

Recall from Section 10.3.1 that our objective is to have the solution to the discrete problem *spatially converge* to a solution of the continuous problem as the discretization is refined. The discrepancy between a discrete solution and a continuous solution is called the *spatial truncation error*, and it should decrease as the grid is refined. In the limit of small grid spacing, the truncation error in principle should vanish in proportion to the grid spacing to some power called the *order of convergence*, which can generally be deduced by an analysis of the formulas used in discretizing the equations. I said "in principle" because in practice a given computer carries only so many digits in its arithmetic calculations, and before spatial truncation error reaches zero, *numerical round-off error* will begin to increase.

In practice, we don't generally know the continuous solution (if we did, we wouldn't need to be carrying out the discrete calculation), and we therefore don't know the actual spatial truncation error. But we can estimate it, based on the grid spacing and the gradients in the discrete solution. Then there are various ways of averaging the estimated error to define different kinds of *error norms*. A rapidly developing technology called *adaptive gridding* uses such information to automatically and selectively refine the grid and to achieve a minimum level of error for a given number of grid points.

Whether or not adaptive gridding is used, there are practical limits to the grid resolution that can be used, imposed sometimes by the amount of computer memory available, but more often by the amount of computing time one is willing or can afford to spend on the solution. In practice, the degree of spatial convergence that is achieved by the finest grid the user decides to use varies greatly, depending on the details of the flow solution and on the user's skill in setting the problem up, particularly with regard to gridding.

Following the better known gridding rules doesn't always guarantee a level of accuracy the user would like to see. For example, meeting the requirement for a small value of y^+ at the first grid point off the wall doesn't guarantee that the grid is adequate to resolve the whole boundary layer. In many grids, the stretching factor away from the wall is so severe that the grid becomes too coarse to resolve the outer part of the boundary layer accurately. Providing grid to adequately resolve wakes and other shear layers off of the body is difficult when their locations are not known a priori. The same thing applies to the resolution of shocks. Viscous effects that involve flow gradients in directions other than surface-normal can also be a source of difficulty, as grid spacings in these other directions are often chosen with only inviscid effects in mind, something we'll touch on again in Section 10.4.2. These are all areas in which robust adaptive gridding could be very helpful.

10.3.4 Solving the Equations, and Iterative Convergence

Solving the equations in a CFD code involves different processes depending on the equations and the particular solution scheme:

1. Panel methods for potential flow involve the solution of a set of linear algebraic equations, which in principle requires no iteration, only the inversion of a matrix.
2. In unsteady flow problems, the solution must be marched forward in time from a known initial condition. At each new time step, the solution at each point in the grid is calculated based on values on a small stencil of surrounding points at one or more previous time steps.
3. Nonlinear steady-flow problems are generally solved by time marching, line relaxation, or global relaxation with Newton's method. All of these involve repetition, often generically referred to as *iteration*. Time-marching schemes use the unsteady equations and march the solution forward in time until it settles to a steady state. Line-relaxation methods typically start from an initial guess and then repeatedly sweep through the grid line-by-line, improving the solution iteration-by-iteration until it satisfies the discrete equations to within some tolerance. Global relaxation schemes use Newton's method to update the solution for the entire grid at once, a step that generally requires inverting a matrix.
4. Viscous/inviscid coupling methods in the early days iterated back and forth between the inviscid and viscous flow analyses. In newer methods, the coupling is done simultaneously with the iteration for the nonlinear flow solution, and Newton's method is often used to update the solution at each iteration.

Time-marching methods use numerical schemes that are either *explicit* or *implicit*. In explicit schemes, the solution at each new time step is explicitly calculated point-by-point in the spatial grid, with the spatial derivatives in the equations calculated from values at neighboring points that are already known from previous time steps. In implicit schemes, the spatial derivatives are represented in terms of values at the new time step. Thus in an implicit method, the new solution at each point depends on a set of implicit relations that must be solved simultaneously, either by matrix inversion or iteration. Explicit methods require less computation per time step than implicit methods but have the disadvantage that numerical stability imposes limits on the size of the time step that can be taken. For implicit methods, on the other hand, there is no stability limit on the time step.

The maximum allowable time step for an explicit method depends on local conditions in the grid and generally varies from point to point. For overall stability in an explicit *time-accurate* unsteady-flow calculation, the time step used globally must not exceed the shortest of the local maximum time steps. When only a final steady-flow solution is sought, the solution need not be time-accurate, and nonphysical *pseudo-time* marching is often used, in which the time step varies from point to point according the local stability criterion, rather than the global one. The time history in such a calculation is not realistic, but the same final steady state is reached, generally in far fewer time steps than a time-accurate calculation (a calculation with constant global time steps) would have required.

Settling down to a final steady-state solution, whether by time marching, pseudo-time marching, or relaxation, is called *convergence*. Of course, the changes in solution quantities from one iteration to the next decrease as convergence is approached, but small changes are not a reliable indicator of convergence. The real measure of convergence is how well the discrete equations for the steady problem are satisfied. Take any one of the steady discrete equations and arrange it with all the terms on one side. You now have an expression whose value should be zero for a solution to the steady problem. At any given point in the grid, at any given iteration of the solution procedure, the discrepancy from zero is called the *residual*. The goal of the iterative solution process is to drive the residuals to zero for all of the equations for every point in the grid.

How well real CFD codes meet this goal in practice varies widely. Convergence is typically assessed according to the maximum residual or an average, usually rms. Many of the earlier RANS codes can reduce the average residual only 3 or 4 orders of magnitude, at which point they "stagnate." (The residual stops decreasing.) The better new codes can drive the residual to machine zero. Remember, however, that no matter how low the residual is, the residual measures only how well the discrete equations are satisfied. A low residual says nothing about how well the discrete solution approximates the continuous solution (i.e., a low residual says nothing about the spatial truncation error).

10.4   Physical Modeling in CFD

How well a CFD code models the physics of a particular flow depends on the applicability of the equations and boundary conditions to the flow in question. In this section, we'll address some of these physical modeling issues.

10.4.1   Compressibility and Shocks

As we pointed out in Section 3.11, there are small-disturbance regimes, both subsonic and supersonic, but not transonic, in which the effects of compressibility can be linearized. However, this kind of treatment of compressibility has very limited ranges of applicability and is no longer used much. Predicting real compressibility effects in most applications requires nonlinear equations: NS, Euler, or full-potential.

In transonic and supersonic flows, shocks nearly always appear. In principle, a solution to the NS equations predicts the correct jump conditions across a shock and nearly the correct shock thickness, but the detailed distributions across the shock are not accurate because the simple representation of the viscosity in the NS formulation is inadequate, as we saw in Section 3.1. In practice, this is never an issue because grids are never fine enough to produce spatially converged solutions through shocks. With practical grid densities, there is always numerical dissipation such that a shock is smeared over at least several grid intervals and is therefore orders of magnitude thicker than a real shock. Fortunately, the jump conditions across the shock can still be accurately predicted because they are independent of the details of the dissipative process, whether real or numerical. Of course the jump conditions lose accuracy if the degree of artificial shock smearing is such that the smearing encroaches significantly on flow gradients upstream or downstream.

Shocks can also be approximated in numerical solutions to inviscid equations, either by *shock fitting* or *shock capturing*. Shock fitting explicitly addresses the mathematical problem of representing shocks as discontinuities in solutions to multidimensional inviscid equations: The geometry of the surface of discontinuity is explicitly determined as part of the solution process, and the appropriate jump conditions are explicitly imposed. Shock capturing avoids the problem of having to explicitly define the shock surface. Given the right kind of numerical dissipation, a smeared approximation to the shock forms naturally within the numerical grid. Although the smeared structure is nonphysical, the conservation properties embodied in the equations result in the correct jump conditions across the shock without requiring explicit imposition of jump conditions. As with a shock in an NS solution, the jump conditions lose accuracy if the degree of shock smearing is such that the shock encroaches significantly on flow gradients upstream or downstream. Shock capturing is by far the more common of the two strategies because it is much simpler to implement, especially in 3D.

When a shock appears in an Euler solution, artificial dissipation produces the same loss of total-pressure as would occur in a real shock or a numerically smeared NS shock. In the full-potential equations, the flow is assumed irrotational, and the vorticity and loss of total pressure that appear downstream of a real shock are not allowed. As a result, when a shock is modeled as a discontinuity or captured in the grid in a potential-flow solution, it is not possible to enforce all of the conservation laws across the shock. Either conservation of mass or conservation of momentum, or a combination of the two, must be waived. Most full-potential methods use differencing schemes that capture shocks automatically in such a way that mass is conserved, and conservation of momentum is waived. Such differencing schemes are referred to as conservative. The spurious momentum produced at a shock in a conservative scheme is mirrored in a nonzero contribution to the pressure drag on the body. In the weak-shock limit, this pressure drag is equal to the "real" shock drag that would have occurred in a solution to the Euler equations, but as shock strength increases, it becomes seriously in error. A formula that quantifies this error was derived by Steger and Baldwin (1971).

The error is significant for anything but the very weakest of shocks. Thus a reasonable prediction of shock drag in a potential-flow method requires an approach other than surface-pressure integration. The only real choice is to identify streamtubes that pass through the shock in the flowfield, apply either the Oswatitsch formula (Equation 6.1.11) or Tognaccini's (2003) higher order formula to each one, and to integrate (sum) the results. This is not a simple thing to do, and algorithms that can do it robustly can be quite complicated. It requires accurately determining the location and the upstream flow condition in each streamtube where the flow enters the shock, even though the shock is smeared in the streamwise direction over several grid cells. And the upstream Mach number that is required is the shock-perpendicular Mach number, which means that the direction of the shock normal must be accurately determined. The shock location on adjacent streamtubes often doesn't define a smooth enough "shock surface" to support an accurate determination of the normal direction. Instead, the shock normal direction is usually assumed to coincide with the direction of the pressure-gradient vector, which is usually well-behaved.

Heat-transfer effects can be modeled in solutions to the NS equations or the boundary-layer equations, provided the appropriate temperature or heat-flux boundary conditions are specified, and the turbulence model is extended to calculate the turbulent heat transport.

10.4.2 Viscous Effects and Turbulence

As I claimed in Chapter 2 and we discussed further in Section 3.1, the continuum formulation embodied in the "full" NS equations represents molecular viscous stresses and heat fluxes in a way that is nearly exact, for nearly any practical purpose. Most practical applications in aerodynamics, however, involve turbulent flow, for which we must resort to RANS equations with turbulence modeling, as we saw in Section 3.7. This entails a substantial loss in physical fidelity, and for swept wings (Section 8.6.2) and many other practical applications we must settle for CFD predictions that are far from perfect. Turbulence modeling also increases computing time and can reduce the robustness of convergence of many numerical schemes. And as we saw in Sections 10.3.2 and 10.3.3, turbulent boundary layers impose difficult demands on the grid, requiring very small spacing near the wall to resolve the viscous sublayer.

Other complicating factors that impact the fidelity of viscous-flow CFD are separated flow, laminar-to-turbulent transition, and laminar separation bubbles. The fidelity of turbulence models, not perfect even for attached boundary layers, deteriorates further in separated regions. Flows with long runs of laminar boundary layer can be very sensitive to the location of transition. Even if the CFD code has a built-in transition prediction, such flows are difficult to predict accurately because transition is sensitive to environmental factors: surface quality (part of the "environment" seen by the flow), freestream turbulence, and acoustic disturbances. Even if such factors are known, the sensitivity of transition to them is often not. And most current viscous-flow codes don't have built-in transition prediction anyway. Laminar separation bubbles with turbulent reattachment are handled by very few CFD codes, the MIT MSES (Drela, 1993) code being one of the few. Laminar separation bubbles play an important role in stall behavior in many inlet and airfoil flows, and the inability to handle them is one of the more serious shortcomings of current RANS codes.

The "full" NS equations provide for all of the components of the molecular viscous stress that are likely to be of any consequence in any flow of a simple fluid in the continuum regime. Of course predicting the turbulent counterparts of these stresses accurately is problematic, but the "full" RANS equations at least have provisions for all of them. That said, there are many practical situations in which only some of them are significant, and we needn't account for them all. The *thin-layer NS* equations neglect the streamwise viscous (and turbulent) diffusion terms and all terms involving cross derivatives, thus saving computing time and reducing artificial dissipation, allowing the use of grids that are too coarse to resolve the neglected terms accurately. Many RANS codes have the option of running "full" or "thin-layer." Finally, coupled inviscid/viscous methods make use of the *boundary-layer approximation* (Section 4.2), in which only the viscous (and turbulent) transport normal to the surface is accounted for, and the normal-direction momentum equation is omitted, which is adequate in many flows in which the boundary layer remains thin.

10.4.3 Separated Shear Layers and Vortex Wakes

Separated shear layers and vortex wakes are naturally captured in solutions to the NS equations. The fundamental limitation on the accuracy of prediction is the fidelity of the turbulence modeling, though in practice grid resolution is often just as serious a limitation, especially in 3D.

Surprisingly, shear layers (vortex wakes) emanating from sharp edges can also arise in numerical solutions to the Euler equations. First, note that the Euler equations themselves, before they are discretized, allow contact discontinuities (slip surfaces across which there is zero normal velocity and a jump in tangential velocity) as weak solutions to the equations. Such slip surfaces are the Euler-equations analog to the vortex sheets we saw in idealized models of lifting flows around 3D wings (Sections 8.1.2 and 8.3.4). Thus in principle, a shear layer can be modeled in the Euler equations as a discontinuity and fitted, just as a shock can be. But in practice, shear-layer fitting is even less practical than shock fitting because shear layers often have edges (as would come from the tips of a lifting wing) where they tend to roll up into tight spirals, as we saw in Section 8.1.2. Such rolled-up shapes are all but impossible to deal with by shear-layer fitting. Therefore, just as with shocks, it is much more common to allow smeared versions of vortex wakes to be captured in the numerical grid.

Shear layers captured in Euler solutions have interesting features, as analyzed and described out by Powell *et al.* (1987). Like real shear layers, they are of course characterized by a "jump," across the layer, in the velocity vector. And, like real shear layers, they tend not to support much of a pressure jump across the layer. Thus when an Euler shear layer is wet on both sides by fluid from the freestream, there tends to be not much of a jump across it in velocity magnitude, only in direction. The velocity component perpendicular to the velocity jump tends to be nearly constant through the layer, as illustrated in Figure 10.4.1. There is thus a deficit in velocity magnitude inside the layer, which is accompanied by a deficit in total-pressure, since the static pressure varies relatively little.

When a shear layer or wake is captured in an NS solution or an Euler solution, inadequate grid resolution often degrades accuracy because the location of the shear layer is not usually known a priori, and it is usually not practical to provide sufficient grid density to resolve a shear layer throughout the region where one might be expected to appear.

In panel-method codes based on potential-flow theory, the wakes of lifting surfaces must be paneled. In Section 10.2.3, we discussed wake "relaxation" schemes that attempt to determine a wake configuration that is also a stream surface (no through flow). These schemes, however, are not typically robust because they come up against the same problem we discussed above in connection with shear-layer fitting in Euler solutions: Vortex wakes

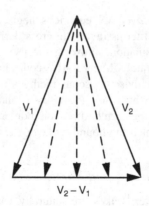

Figure 10.4.1 Illustration of how the velocity vector varies through a shear layer captured in a solution to the Euler equations, as explained by Powell *et al.* (1987). V_1 and V_2 are the velocity vectors on the two sides of the shear layer, $V_2 - V_1$ is the velocity jump across the layer, and the dashed vectors are velocities at intermediate levels within the layer. The component perpendicular to $V_2 - V_1$ is nearly constant, and at intermediate levels there are therefore deficits in velocity magnitude and total-pressure

tend to roll up into tight spirals. In a panel method, no reasonable panel spacing would be able to resolve a tightly wound spiral.

Thus it is much more common to fix the wake in space and model it in a way that is equivalent to the assumptions of Trefftz-plane theory. A wake sheet is defined that extends back from the trailing edge of each lifting surface, usually in such a way that its cross-section in a Trefftz-plane view remains constant. In surface panel methods, the sheet is paneled with doublet singularities whose strengths are distributed so as maintain a velocity jump across the sheet that is equivalent to having vortex lines oriented straight back, at least in plan view. A typical wake paneling is shown in Figure 10.4.2. In finite-difference field methods for the full potential equation, the sheet is identified with a particular grid surface, and a velocity-jump boundary condition is imposed across the sheet that has the same effect as the singularity distribution used in panel methods.

In both panel methods and field methods for 3D potential flow, the spanwise distribution of effective shed vortex strength is determined as part of the flow solution so that the upper- and lower-surface pressures are equal at the trailing edge, which is equivalent to imposing the Kutta condition for a 2D airfoil (Section 7.1). For high-aspect-ratio wings, the distribution of effective shed vortex strength works out to be consistent with the spanload on the lifting surface, just as in Trefftz-plane theory.

A potential-flow wake-vortex sheet defined in this way has approximately the right distribution of vortex strength where it leaves the trailing edge, but it reflects reality less accurately downstream because in a real wake the vortex lines migrate generally outboard, and the sheet rolls up at the outer edge. Well downstream of the wing, some details of the overall flowfield, such as the downwash field at the location of an aft tail, will not be accurately predicted as a result.

In Section 8.3.4, we discussed how a wake sheet in which the vortex lines are aligned in the freestream direction is not force-free, but is drag-free. We noted that this doesn't guarantee that the induced drag calculated by Trefftz-plane theory based on the distribution

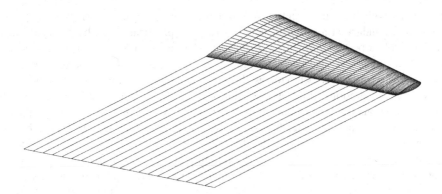

Figure 10.4.2 Typical paneling of a wake sheet behind a lifting surface in a potential-flow panel method. (Plot by M.F. Smith.)

of vortex strength in the wake is correct, but only that in a shock-free potential flow it is consistent with the pressure drag on the configuration. Still, short of a "relaxed" stream-surface wake, a drag-free wake with vortex lines in the freestream direction would seem to be the best option we have. A drawback in practice is that if solutions are desired at more than one angle of attack, a different wake definition is needed for each one. Of course it is easier to define just one wake anchored to the configuration and to allow it to tilt relative to freestream when the configuration is analyzed at different angles of attack. But how much is accuracy degraded if we do so?

Our discussion so far might lead one to think that tilting a potential-flow wake sheet up or down in this way would be a very bad thing to do, because then the wake would not be drag-free. However, counterintuitive as it might seem, a wake that is tilted by a modest amount will usually have very little effect on the results that usually matter. Tilting the wake up or down will usually have almost no visible effect on plots of pressure distributions and very little effect on the integrated lift and moments. The spanload shape will also typically be changed very little, and as a result, the induced drag calculated by Trefftz-plane theory will not be significantly changed. The Trefftz-plane drag calculation gives the total induced drag on the entire paneled configuration, including the wake sheet. So presumably when a wake is tilted and takes on a drag force, the integrated pressure drag on the configuration surfaces will change by roughly the same amount in the opposite direction. But this usually involves subtle changes to the pressure distributions that are practically invisible on a pressure-distribution plot and have almost no effect on lift and moments. So as long as we follow the usual practice of inferring induced drag from the Trefftz-plane calculation and not from pressure integration, a tilted wake will have little effect on the results we actually use.

10.4.4   The Farfield

In most applications in external aerodynamics, the flow approaches a uniform condition in the farfield. When we simulate such a flow in a panel method, this condition is automatically enforced by the superposition of a uniform flow and the singularities that are used to simulate the body and its wake. The contributions from the singularities die off with distance in just the right way to simulate a free-air flow in an infinite domain.

In methods with field grids, getting the farfield "right" is a more complicated problem. For an outer boundary at a finite distance, there is no practical set of boundary conditions that can provide a perfect simulation of flow in an infinite domain, and we must generally settle for something less than perfect. A minimum requirement should be that the solution converges toward the right behavior as the outer boundary is moved farther away. For a given set of flow equations (full-potential, Euler, or NS), there are several choices of farfield boundary conditions that are acceptable in this sense. For boundary conditions of this kind, the farther away we put the outer boundary, the less impact imperfections in the boundary conditions will have. But the bigger the domain gets, the more grid points or cells we will have to use. So regarding the impact of the farfield boundary conditions, there is a trade to be worked between fidelity and cost, just as there is regarding spatial truncation error.

What the "right behavior" is at the outer boundary is complicated by the fact that vortex wakes associated with lift in 3D persist over long distances downstream (forever, in the case of a gridded wake in full-potential). We must therefore allow some strong flow disturbances to pass through the downstream boundary. However, even if we didn't have these disturbances, enforcing uniform flow at the farfield boundary wouldn't be an option because it requires more boundary conditions than we're allowed to specify in any of our equation sets (potential flow, Euler, or NS).

In full-potential methods, because there is only one unknown, only one boundary condition can be imposed. Thus only one component of the velocity vector can be specified, and the others must be allowed to be determined by the solution. Specifying uniform flow is not allowed because it would require specifying two components in 2D or three components in 3D.

The situation with the Euler or NS equations is more complicated. Recall from Section 3.4.4 that we have six unknowns in Euler, and eight in NS. The combinations we are allowed to specify depend on whether the flow is entering or leaving the domain, whether the flow is subsonic or supersonic, and whether the equations are Euler or NS. These issues are discussed in Hirsch (2007).

For internal flows in ducts, including solid-wall wind tunnels, the appropriate boundary condition at the wall for any equation set would be no through-flow (normal component of velocity equals zero). For the NS equations, a no-slip condition and a thermal boundary condition would also be required. Wind tunnels with holes or slots in the walls are more difficult. The holes or slots can be modeled individually, but conditions that are easy to impose on the flow through them cannot always be relied on to model the situation accurately. Modeling of slotted walls is discussed, for RANS by Bosnyakov *et al.* (2008), and for viscous/inviscid coupling by Krynytzky (2004). There are also models available that area-average the effects of the holes or slots, but with some further loss of fidelity, and these have been largely superseded by models that treat the slots or holes as discrete. Internal flows also require boundary conditions upstream and downstream, the usual combination being stagnation temperature and pressure upstream, and static pressure downstream.

10.4.5 Predicting Drag

In previous chapters, we discussed many aspects of the physics of drag: The drag of streamlined and bluff bodies in general in Section 6.1, the profile drag of airfoils in Section 7.4.2, the shock drag of airfoils in Section 7.4.8, and lift-induced drag in Section 8.3. A common thread in all of this is that drag depends on subtle flow details that CFD cannot be expected to predict perfectly. Turbulence modeling, with all its imperfections, will always compromise the prediction of viscous drag, even in attached flows. Separated flow is not predicted well by CFD unless extraordinary measures are taken (DNS, LES, or DES), and this compromises prediction of drag in any flow involving separation. Shocks and vortex wakes must be captured in grids that are often not as fine as they should be, compromising the prediction of shock drag and lift-induced drag.

To these difficulties we must add that drag is particularly sensitive to any lack of iterative or spatial convergence in the flow solution. A lack of spatial convergence in particular shows up as artificial dissipation, producing spurious drag that looks like additional viscous drag or shock drag.

Although CFD drag predictions are generally far from perfect, there is a widespread assumption that predicted drag increments due to small changes in the configuration or the flow conditions are much more accurate than predicted absolute levels. There is some justification for this, in that the increments we are usually interested in are small fractions of the total drag, and the error in an increment is therefore likely to be smaller than the error in the absolute level. But there is no other theoretical reason to expect an advantage in dealing in increments, and we're depending on the assumption that the bias in the absolute level doesn't change so much between the two calculations as to swamp the increment we're looking for. For this assumption to be justified, we must be diligent about making the two calculations as comparable as possible, avoiding unnecessary changes to the grid, and minimizing truncation errors and residuals.

CFD can certainly produce useful drag predictions. We just need to remember that they are not infallible.

10.4.6   Propulsion Effects

In CFD modeling of full airplane configurations it is not usually practical to resolve propeller blades or the internal details of turbojet or turbofan engines. Such systems are usually modeled by inflow and outflow surfaces in the grid, where boundary conditions can be applied to simulate the time-averaged effects of the propulsion device. Boundary conditions used for this purpose can be fixed or can be formulated to respond in some realistic way to the surrounding flow conditions, so as to model the response of the device. This kind of modeling of propulsion is similar to the basic actuator disc theory we discussed in Section 6.1.10 in connection with the physics of propellers, but it is often extended to include compressibility and heat addition.

10.5   CFD Validation?

"CFD validation" is a term that is heard frequently among users of CFD codes in industrial environments, among their customers, and in the engineering literature. But what does it mean? If we take the words literally, there are two relevant definitions to consider (Webster's, 1976):

- **Validate**: To confirm the validity, and
- **Valid**: Seen to be in agreement with the facts or to be logically sound.

According to these definitions, successful CFD validation requires that the code in question be found to agree with "the facts," which I would take to mean experimental measurements covering the whole range of flows for which the code is to be used. However, for most practical applications, CFD codes tend not to agree with the facts as well as we would like, over as wide a range of flows as we would like, and we must conclude that CFD validation in any literal sense is usually an unattainable goal.

This is not to suggest that the enterprise we call "CFD validation" is hopeless, just that it is misnamed. It can actually serve a very necessary purpose. When we compare a code against a wide range of test cases, we learn how physically accurate the solutions are likely to be and what kinds of biases to expect in particular situations. Then when we use the code in an application, we have a basis for making uncertainty estimates and applying adjustments to the results. Looked at in this light, the purpose that CFD testing serves is really more *calibration* than "validation."

Of course the knowledge gained in CFD testing helps us in our application work only if we remember to actually use it. Even when we know the limitations of a code, we have a natural tendency to forget them and to be overly optimistic about the accuracy of the results, something we noted in the introduction to this chapter.

10.6   Integrated Forces and the Components of Drag

In principle, determining total integrated forces and moments from a solution of the NS equations is simply a matter of carrying out the integrations over the body surface called for in Equation 5.4.1 for the lift, Equation 6.1.1 for the drag, and corresponding integrals for moments. For NS codes, these forces and moments from surface-pressure and surface-shear integration are the primary force outputs.

But in the case of the pressure drag, surface-pressure integration yields only the total and doesn't determine the "components": Shock drag, induced drag, and the viscous contribution to pressure drag. Recall from Section 6.1.3 that these drag "components" are not rigorously separable, and that they can be determined only through the appropriate theoretical idealizations. Estimating shock drag requires finding the shocks in the flowfield, identifying streamtubes that pass through them, applying the Oswatitsch formula (Equation 6.1.11) to each one, and integrating (summing) the results, as we discussed in Section 10.4.1. Estimating induced drag requires a Trefftz-plane calculation (Equation 8.3.10).

Other ways of integrating flowfield quantities can also be used with solutions to the NS equations. Vorticity and total-pressure data on a single cross-plane downstream of the configuration can be used, as described in Section 6.1.4.7, or more-general formulas involving volume integration of the viscous dissipation can be used, as described by Tognaccini (2003). An advantage claimed for the latter method is that drag determined in this way converges faster with grid refinement than does drag determined by surface-pressure integration.

For viscous/inviscid interaction methods, surface-pressure integration doesn't work at all for determining the pressure drag. Instead, it requires building up estimates of the "components": shock drag and induced drag as described above, and viscous profile drag from one of the variants of the Squire-Young formula described in Section 6.1.4.5, applied either at the wing trailing edge or somewhere in the viscous wake downstream.

When propulsion inlets and exhaust streams are represented in CFD calculations, special care is required to ensure that thrust/drag bookkeeping is done in a physically consistent way.

Drela (2009) discusses the physical issues that arise when propulsion effects are present in the flowfield.

10.7   Solution Visualization

A CFD field solution defines the flowfield everywhere and can therefore tell us a great deal more about a flow than just the integrated forces. On solid surfaces, there are surface distributions of pressure, temperature, and shear stress as well as patterns of skin-friction lines. Away from the surfaces there are volume distributions of scalar quantities: pressure, total-pressure, density, temperature, turbulence kinetic energy, and eddy viscosity. And there are volume distributions of vector quantities: velocity and vorticity. The velocity field defines streamlines, streaklines, particle paths, and timelines; and the vorticity field defines vortex lines. Understanding such multidimensional information requires graphic visualization, which can range from simple 2D Cartesian plots to elaborate 3D perspective presentations, both still and animated. In this section, we'll look at how the various sorts of displays serve different purposes in illuminating different aspects of a flowfield.

Graphic visualization is the most powerful tool we have for gaining any intuitive understanding of complex quantitative information, and the computer has greatly expanded the range of visual displays that it is practical to produce. But before we get into the specifics of applications in aerodynamics, it's worth reflecting on the general nature of the graphic medium and how it limits what we can actually do.

First, the nature of the visual sense itself imposes limitations on our ability to view real 3D information. Our binocular vision provides only very low resolution in the depth dimension and works only at short distances. So for quantitative purposes, we see little more than a 2D projection of what's in front of us. Much of what we perceive about depth comes from prior knowledge and context: We know the tree is closer to us than the mountain because we know something about the relative sizes of trees and mountains and because the tree blocks part of the view of the mountain. So blockage of one object by another helps us with depth perception, but it is also a limitation: Along any given direction from our eyes, we can usually see only one thing at a time. Unless an object in our field of view is nearly transparent, it blocks our view of anything behind it.

Next are the limitations imposed by the physical media we typically use. We have no practical alternative to viewing graphics on computer screens and paper surfaces, which are basically 2D displays that take no advantage of our binocular vision. A depth dimension can be conveyed only through perspective, shading, and context. As a result, displays that depict 3D geometry generally incur some degree of spatial ambiguity.

A 2D display medium offers us only five basic graphical constructs for displaying spatial distributions of quantitative data:

1. A Cartesian plot of a scalar quantity versus one coordinate.
2. A contour plot of a scalar quantity versus two coordinates. A contour plot can be plotted in the 2D plane, or it can be constructed on a curved surface in space, such as the surface of a body, and viewed in perspective.
3. A shaded isosurface showing how a single value of a scalar quantity is distributed in space in three dimensions.

4. An array of arrows showing the distribution of a two-component vector in two spatial
   coordinates.
5. Space curves, viewed either in 2D or 3D, representing a direction field.

We've used examples of all of these to illustrate flowfield phenomena in earlier chapters.
Now let's look at how we can best represent different aspects of flowfields, using these
constructs.

In 3D flow, the overall pattern of pressure on the surface of a body is most clearly
visualized in an isobar contour plot using either colors or lines, as in Figure 10.7.1. In this
plot of the pressure distributions on the upper surface of a swept wing at two different
angles of attack, the changes in the position and the sweep of the shock are visible at a
glance. However, subtle details of the local chordwise pressure distributions, such as the
magnitude of the shock pressure rise, the slope of the "rooftop" ahead of the shock, and
whether or not there is a re-expansion aft of the shock are not easy to see in a contour plot
and are visualized better in Cartesian pressure plots for station cuts along the span, like
airfoil C_p plots, of which there are several examples in Section 7.4. Putting data for more
than one spanwise station on a single plot tends to be more confusing than helpful, and it is
usually better to show multiple stations as an array of plots, as in Figure 10.7.2. This kind
of display shows subtle details with much greater pressure resolution than a contour plot,
but it doesn't provide nearly the same feel for the overall pattern.

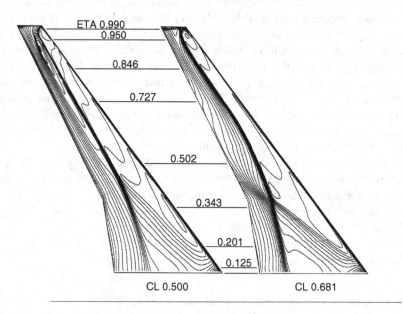

Figure 10.7.1   Isobar contour plots of the pressure distribution on the upper surface of a swept wing
at two different angles of attack. The shock is visible as the concentration of contours traversing most
of the span. Eta values indicate spanwise stations for which Cartesian plots of the pressure distribution
are shown in Figure 10.7.2. CFL3D solutions and plots by B.J. Rider. Wing geometry is the Common
Research Model developed by Vassberg *et al.* (2008)

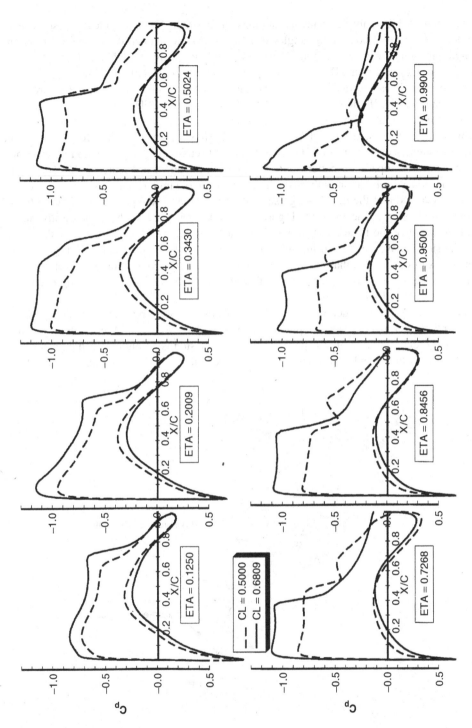

Figure 10.7.2 Cartesian plots of the pressure distributions at stations along the span of the wing shown in Figure 10.7.1. Such a plot shows subtle local details of the pressure distribution not easily seen in the contour plot

Though showing multiple spanwise stations on a single C_p plot typically results in a jumble, a sequence of conditions (angles of attack or Mach numbers) at a single station on a single plot tends to form a "family" of curves that are easily readable and can be quite informative. Examples for 2D airfoils were shown in Figures 7.4.26 and 7.4.27.

Skin-friction lines are often used to visualize the distribution of shear-stress direction on body surfaces in 3D flow. Skin-friction lines plotted from CFD solutions are analogous to the oil-flow visualization pictures from experiments, and they have the same strengths and weaknesses that we discussed in Section 4.1.2. They are useful mainly for showing the surface signatures of the kinds of 3D separation we discussed in Chapter 5, and an example of a complicated separation pattern on a swept wing is shown in Figure 10.7.3. Note that the shock-induced separation with reattachment looks very similar to the pattern sketched in Figure 8.6.9a and that the trailing-edge separation is similar to the pattern in Figure 8.6.9b.

As pointed out in connection with Figure 4.1.11, skin-friction lines have the disadvantage that they can give the misleading impression that they represent the general flow direction over the surface. We must keep in mind that skin-friction lines represent only the surface-shear-stress direction, and that the flow direction can be very different only a short distance off the surface, as seen, for example, in the flow direction plotted for a swept-wing boundary layer in Figure 8.6.13.

Visualizing 3D flow structure out in the field poses interesting challenges because of the essentially 2D nature of the graphic constructs we have available. We've already noted that in a 2D representation of 3D space the depth dimension is ambiguous, which can make interpretation difficult and error-prone. Animation in which the 3D configuration represented

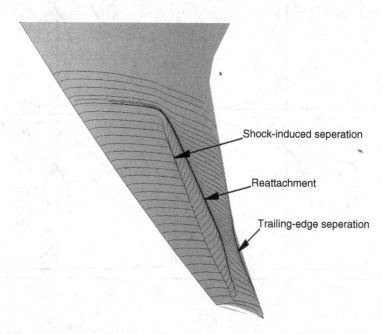

Figure 10.7.3 A plot of computed skin-friction lines on a swept wing. RANS calculation and plot by N.J. Yu

by the graphic rotates continuously, or a screen display in which the viewer can rotate the configuration at will, can make things much easier to understand.

It is often effective to show distributions of flow quantities on planar cuts in the field, where the cuts are displayed as part of a perspective view of a 3D configuration. In Figure 10.7.4, isobars are plotted in a longitudinal cut in the flow around a wing, clearly showing the shock that caused the separation visible in Figure 10.7.3. Of course, the relatively low pressure resolution of an isobar plot is still a disadvantage, as it was in the case of surface pressures.

Streamlines, streaklines, or particle paths displayed as space curves in a perspective view of the field can also be quite useful. The effectiveness of such a plot can vary widely, depending on how many curves are plotted and whether they are distributed effectively, aspects that are controlled by the code user. Each curve must be defined by the coordinates of a point through which it passes, and graphics software generally requires these to be chosen by the user. Some skill and trial and error are usually required to make an effective streamline plot. A plot needs to have a sufficient number of curves, as curves too sparsely spaced can miss important details of the flow. And the curves need to be positioned so as to mark the details that are of interest, particularly vortices, as we'll see next.

Vorticity is often a key to understanding the structure of a flow, and concentrated vortices in particular are indicators of interesting things happening in the flow. In steady flows, concentrated vortices can be seen in 3D streamline plots, but visualizing a vortex this way is a hit-or-miss proposition if one doesn't know ahead of time where the vortex will form. Figure 10.7.5 illustrates this problem using the tip vortex behind a wingtip as an example. In (a), the streamlines are released too far inboard and fail to show evidence of the vortex. In (b), the streamlines are released close enough so that some of them orbit the vortex, indicating its location.

There are several kinds of displays that can make a vortex visible, but depend less on prior knowledge of the location. Figure 10.7.6 shows three examples using flow quantities in a cross-stream plane behind a wingtip, in the same flowfield as Figures 10.7.3–10.7.5. In (a), streamlines constructed from velocity vectors projected in the cut clearly orbit the vortex and make it visible. For this kind of plot to work, however, the cutting plane must be

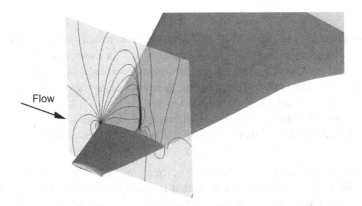

Figure 10.7.4   Isobars in a longitudinal cut of a 3D flow around a wing, seen in a perspective view. Same flowfield as in Figure 10.7.3. RANS calculation by N.J. Yu. Plot by A.M. Malone

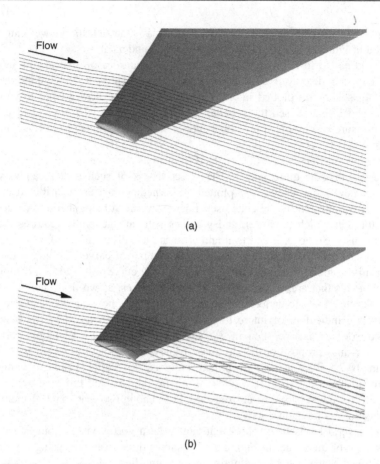

(a)

(b)

Figure 10.7.5 Streamlines in the flowfield around a wing indicating the location of the tip vortex. Same flowfield as in Figures 10.7.3 and 10.7.4. RANS calculation by N.J. Yu. Plots by A.M. Malone. (a) Streamlines released too far inboard of the tip fail to mark the vortex. (b) Streamlines released close enough to the vortex orbit the vortex, marking its location

roughly perpendicular to the vortex. In (b), contours of constant vorticity magnitude mark much of the vortex sheet behind the trailing edge, not just the concentrated vortex rolling up at its edge. In (c), contours of constant pressure clearly indicate the location of the vortex core and don't show the vortex sheet inboard. A drawback to these three types of display is that a weak vortex would likely be difficult to see.

Other approaches to visualizing vortices without prior knowledge of their locations have been proposed. Saderjoen *et al.* (1998) identify two main classes of telltales:

1. *Point-based* flow quantities, of which the *vorticity magnitude* $\omega = |\nabla \times \mathbf{V}|$ is the most widely familiar but has the disadvantage that it doesn't distinguish between shear layers and concentrated vortices, as we saw in Figure 10.7.6b. Because concentrated vortices in steady flows must generally be aligned with the prevailing local flow direction, the *normalized helicity* $\mathbf{V} \cdot \omega / (|\mathbf{V}| \cdot |\omega|)$ may be a better choice than the vorticity magnitude.

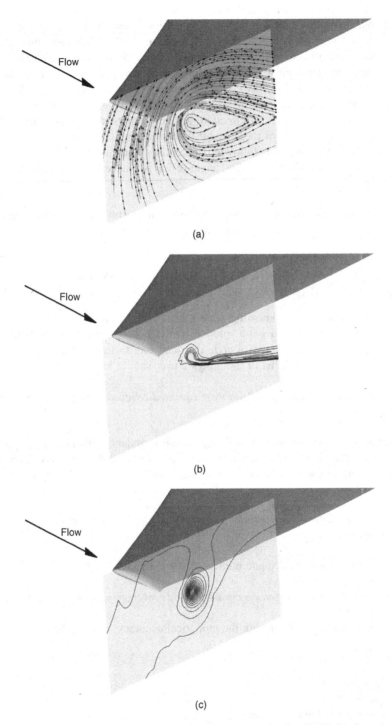

Figure 10.7.6 Visualizing the tip vortex in a cross-stream cut in the field behind a wingtip. Same flow field as in Figures 10.7.3–10.7.5. RANS calculation by N.J. Yu. Plots by A.M. Malone. (a) Streamlines from velocity vectors projected in the cut. (b) Contours of constant vorticity magnitude in the cut. (c) Contours of constant pressure in the cut.

2. *Curve-based* quantities based on nonlocal properties of the streamline pattern. The basic idea was put forward by Robinson (1991): "[A] vortex exists when instantaneous streamlines mapped into a plane normal to the vortex core exhibit a roughly circular or spiral pattern, when viewed from a frame of reference moving with the center of the vortex core." Saderjoen *et al.* (1998) give two examples of automatic vortex telltales based on this idea, both applied to projections of the streamlines into a plane. One involves calculating the locations of centers of curvature at many points along many streamlines and then calculating the density of these center points in the plane, called the *curvature center density*. The other involves following streamlines and calculating the *winding angle*. A vortex is identified when a winding angle of 2π is reached, and the end point is near the starting point.

Complicated unsteady vortex structures in wakes and turbulent boundary layers can be clearly visualized in terms of isovorticity surfaces, as in Figures 5.3.3a and 10.2.1.

10.8   Things a User Should Know about a CFD Code before Running it

Running a CFD code without knowing what's in it is risky. It is too easy to make poor choices of input parameters and get a solution that is not as accurate as it should be, and not understanding the limitations of a code can lead one to misinterpret the results. This section provides a brief check list of the things a user should know.

What equations does the code solve? Do these equations adequately model the physics of the problem?

This includes not just the basic equations, but the turbulence model and the provisions, if any, for modeling laminar-to-turbulent transition. These considerations determine the applicability of the method to the problem at hand, and the physical fidelity of the results.

What order of spatial convergence does the numerical method provide, and how fine a grid resolution can be used on your particular problem? Can adequate grid density be concentrated where it is needed?

These determine whether adequate numerical accuracy can be obtained.

What provisions does the code have for monitoring grid convergence and iterative convergence?

These are important for monitoring the numerical accuracy of the solution.

Where did the code come from?

- In-house developed?
- Vendor supplied?
- Vendor supplied and in-house modified?

This affects where you may have to go to get questions answered knowledgeably.

What is the code's state of development?

- *Experimental?*
- *Well developed and documented, but as a code only?*
- *Well developed and documented as a code and integrated into the process in which you are using it?*

The code's status in this regard will affect:

- The reliability of execution and convergence.
- The user-friendliness of the inputs and outputs.
- The degree of automation of pre- and post-processing steps.

Has it been used for similar applications, and how good were the results?

If this is a pioneering application, be prepared to spend considerable time in trial-and-error adjustment of input parameters and in assessing the quality of the results.

Are consultation and support available?

If not, arriving at good results could prove to be very time consuming.

In light of all of the above, is the code really applicable to the problem at hand?

Index

Acceleration, convective, **17** (2nd term in Eqn 3.2.1), **39**, 103
Ackeret propulsion, **225–6**, 240–1
Actuator-disc theory, **227–9**, 239, 240, 480, 515
Adiabatic wall, 120–1, **158**, **159–62**
Advance ratio, **229–30**, **235–8**
Adverse pressure gradient:
    effect on transition, **132**, **135**, 341
    effect on velocity profile, **84**, 104, 138–9, 140–2, 149, 457–64
    role in separation, **84–5**, 87, **93–5**, 150, 276–9, **323–7**, **459–60**
Aileron, 297
Airfoil, 307
    design, 311, 338, 348, **352**, 355
    drag, 200, **316–19**, 338
    in ground effect, 350–2
    laminar-flow, 338–41
    linear theory, 307–15
    low-Reynolds-number, 341
    maximum lift, 319–29
    shape function, 355–7
    slotted, 329–35
    stall, 296–7, **319–29**
    supercritical, 344
    transonic, 342–50
Angle of attack, **287**, **296–7**, 306–7, **309–10**
Aspect ratio, **386**, 391, 395
Attachment, flow, **80–2**, 124–5, **169–74**, 275–80, 448–52

Attachment line, 105–6, 125–8, **171–4**, **448–52**, 465–8
Axisymmetric boundary layer, **123–5**, 126
Axisymmetric flow, 123–5, **199–202**, 227

Base drag, 210–212
    Hoerner's jet pump, **211–12**, 489
Basic and additional spanloads, 376–9
Bernoulli equation, **58–60**, 168,
Bernoulli lift, **270–5**, **281–2**, 298–9
Biot-Savart law, **31–3**, 389, 392–3, **413**, 480, 487
Biplane drag, 409
Blade-element theory, 234–39
Blowing for separation prevention, 155–7
Body forces, 32, **35**, 37, 41, 47, 56, 58
Boltzmann equations, 13
Bound vorticity, **265–9**, 305, **363–4**, 380–2, 429–31
Boundary conditions, 15
    boundary-layer, 92–3, **103**, 109, 110–13
    farfield, 513–14
    flow tangency (no-through-flow), 111–12, **308**, 382, 401, 429
    free-stream, 513–14
    general flows, **37**, 381–2
    linearized, 308
    no-slip, **15–16**, 37, 79, 83, 103, 140, 193
    thermal, 158–9

Understanding Aerodynamics: Arguing from the Real Physics, First Edition. Doug McLean.
Images and Text: Copyright © 2013 Boeing. All Rights Reserved. Published 2013 by John Wiley & Sons, Ltd.

Boundary layer:
  attachment-line, 105–6, 125–8, **171–4**,
    448–52
  axisymmetric, **123–5**, 126
  CFD codes, 501
  compressible, 158–62
  concept of, 47–8, 57, 63–4, **79,**
    99–100
  cross-flow, 85
    angle, **86–7**, 457–8
    eddy viscosity, 461–2
    effect on displacement, 92
    effect on vortices from VGs, 154–5
    generation by pressure gradient,
      85–90, **105–7**, 124–6
    instability, **134**, 468
    pressure-driven, **85–7**, 453
    separation, 96–7
    shear-driven, 85
    velocity profile, 85–9
  displacement effect, **90–3, 110–113**,
    196, 207, 311–14, 344–7
  equations, 100–108
  flat plate, **117–21**, 242, 248, 318
  independence principle, **129**, 459, 465
  infinite-span-wing theory, 128–30
  integral momentum equation, **108–10,**
    125–8
  laminar flat-plate, 117–20
  laminar similarity, 121–3
  laminar skin friction, **117–21**, 251, **338**
  laminar, **82–3**, 101–2, **117–18**,
    121,123, 338–41, 465–8
  momentum thickness, **108–10**, 127,
    200–1, 451–2
  plane-of-symmetry, 125–8
  propulsion, **225–6**, 240–1
  separation, **93–9**, 102, 115–16,
    139–40, **150–7**, 169–76, 210,
    319–35
  separation, shock-induced, **347–9**,
    554–5
  shape factor, **108–9**, 161–2
  shock interaction, **345–9, 464–5**
  similarity solutions, 121–3
  suction for laminar flow, **135**, 251
  suction for separation control, **157,**
    240–1
  surface roughness, **162, 212–22,**
    243–250
  temperature profile, 158–62
  theory, **79, 99–108**
  thermal, 158–62
  thickness, **83**, 243–4
  three-dimensional (3D), 80
    equations, **104–8**, 109–10
    displacement thickness, 112–13
    on swept and tapered wings,
      128–30, 449–64
    separation, 96–9
    turbulent prediction, 147–50
    velocity profiles, 85–90
  transition, **130–8**, 338–341, 465–468
  trip, **132**, 341
  turbulent, 82–3, 94–6, **117–21**, 123,
    138–50
    inner layer, 138–141
    law of the wake, 145–7
    law of the wall, **142–5, 215**, 245–6
    outer layer, **142–3, 145–7**
    overlap layer, 143–5
    shear stress, **49–55**, 139–41
    similarity, 123
    skin friction, **117–21**, 162, **212–222,**
      243–50
    viscous sublayer, 95–6, **139–44,**
      217–22, 488
  velocity profile, 46, **82–90**
    approaching separation, 96–7
    effect of shape on growth, 109
    in adverse pressure gradient, 104
    initial, 101
    shape factor, 108–9
    turbulent, 139–47
    with similarity, 121–3
    3D, in arbitrary coordinate system,
      105
Boussinesq eddy viscosity, 54
Branch cut, 262
Buffer layer, 140–1
Buffeting, 48, **322**, 347, 495
Bulk viscosity, 44, 59

Buoyancy force, **189–90**, **195**, 207, 223, 489
Bypass transition, 131–2

Camber line, 266, **309–10**, 354–5
Canard effect on induced drag, 404–9
Canonical pressure coefficient, **323–8**, 332–3
Cascade, 335–8
Causation, one-way, **9–10**, 41, 270, 299, 487
Cause and effect, **40–3**, 270
Chord line, **310**, 354, 377
Circulation, **25**, 56–8, 233, 262–3, **265–9**, 303
Circulation preservation, 56–8
Clauser parameter (β), 123
Clauser plot, 144–5
Coanda effect, 275–80
Coherent turbulent structures, 48, **138–9**
Coles' law of the wake, 145–7
Collapse of vortex wake, **369–70**, 436–43
Compliant wall, 252
Compressible boundary layer, 158–62
Compressible flow, 59–60, 63, **70–7**
Compressibility, 59–60, **63**, 66, **70–7**, 342–50
Compressor blade, 335–7
Computational fluid dynamics (CFD), **4**, **491–3**
   basic definitions, 493
   methods, classes of, 493–501
   physical modeling, 508–15
   solution schemes, numerical, 501–8
   solution visualization, 517–24
   validation, 515–16
Conduction, thermal, **14**, 45, **62**, **159**
Conformal mapping, **67–70**, 122, **264**, 484
Connectivity of a domain, 184
Conservation laws:
   energy, **36–7**, 197–8
   mass, 34–5
   momentum, **35–6**, 198–202
Constitutive relations, 37
Continuity equation, 34–5
Continuum formulation, **13–16**, 43–8

Control of separation, **150–7**, **319–35**
Control of transition, **251**, **338–41**, **465–8**
Control surface (boundary of control volume), **18**, 198–203, 431–5
Control surface (on airfoil), 297–8
Control-volume analysis, **18**, 197–203, 303, 336–7, 387–8, 431–5, 473
Convection, 14, 34, **38–40**, 114, 159
Convective acceleration terms, **17** (2nd term in Eqn 3.2.1), **39**, 50, 62, 64–5, **103–4**
Convective heat transfer, 158–9
Convergence, CFD:
   iterative, 507–8
   spatial, 501–2, **506–7**
Convergence, flow, **88–90**, 97–8, 116, 124, **125–8**
Correlations, empirical, 242–3
Couette flow, 46, 220
Critical condition (sonic or choked), **73**, 342–4
Critical Mach number, 342–4
Critical point, topological, 176–185
Critical Reynolds number:
   for drag behavior, 210
   for instability growth, 133–5
   for surface roughness, 213–14
Critical roughness height, 217–18
Crocco energy integral, 159
Crocco's theorem, **60**, 366
Cross-flow, boundary-layer, 85
   angle, **86–7**, 457–8
   eddy viscosity, 461–2
   effect on displacement, 92
   effect on vortices from VGs, 154–5
   generation by pressure gradient, **85–90**, **105–7**, 124–6
   instability, **134**, 468
   pressure-driven, **85–7**, 453
   separation, 96–7
   shear-driven, 85
   velocity profile, 85–9
Crow instability, **369–71**, 437
Curl, **24**, 32–3, **44**, 47
Curvature discontinuity, 355, 484
Curvature vector, 106–7

Curvature, coordinate, 100, 104–5
Curvature, streamline, 85–6, **106–7**
Curvature, wall, 100, 104–5
Cylinder, circular, 19, 21, **67–8**, 102,
    165–6

d'Alembert's paradox, **192**, 195, 207, 316,
    486
Damping factor, van Driest, 140
Decay of vortex wake, **369–70**, 436–43
Defect, velocity:
    boundary layer, **123**, **143**
    wake, 186, 188
Deformation of a fluid element, 24–5
Density, 13–14, **34–5**
Detached shock, 71–2
Detached-eddy simulation (DES), 495
Diffuser, **96**, 155, 337
Dimensionless parameters, 61–3
Direct numerical simulation (DNS), 493–4
Discontinuity:
    in body slope or curvature, 355, 484
    shock (idealized), 39, **74–7**, 509
    vortex sheet (idealized), **29–30**, 500,
        506, 511
Discretization, 501–2
Displacement effect, **90–3**, **110–13**,
    188–9, 311–16, 344–7
Displacement thickness, **91–3**, **110–13**
Dissipation, turbulent, 37, **49**
Dissipation, viscous, **45**, 49, 62, 159,
    192–3, **202**, 204–7
Disturbance amplitude, 131–8
Disturbance frequency, 133–8
Disturbance, small, **70**, **131–7**, 307–11,
    379–80, 500, 508
Divergence, flow, 23, **88–90**, 92, 125–8,
    152–4, 165, 172
Drag, 191–2
    crisis, 210
    of a sphere, 162, **209–10**
    polar, **317–19**, **397–8**
    prediction, **241–50**, **514–15**
    reduction, laminar flow, 251, **338–41**,
        465–8
    reduction, turbulent, 251–8

rise, transonic, 344–5
Drag categories:
    base, 210–12
    excrescence, 224–5
    induced, **362**, **385–409**
    interference, 222–5
    inviscid "interference", 195, 222–3, **489**
    parasite, 210
    pressure, **194–6**, **207–10**, 316–19,
        344–5, 489
    profile, **316–19**, **344–5**
    shock, **203–4**, **344–5**
    skin friction, **194**, 204–5, **208**, **317–19**
    slot, 212
    surface roughness, **162**, **212–22**,
        243–250
    viscous, 189–90, **192–3**, **196**, **199–202**
Dumping velocity, trailing edge, **332–4**,
    337
Dynamic pressure, 64–5
Dynamic similarity, 61–6

Eddy viscosity, **53–5**, **95**, **140–1**, **147–9**,
    246–8, **460–64**
Eigenvalues (laminar stability theory), 133
Elementary particles, 5–8
Elevator, 297
Elliptic equations, 37–8, 101
Elliptic spanload, **394–5**, 420
Emmons spot, 6, 138
Empirical drag correlations, 242–3
Endplate, 411, **420–2**
Energy:
    conservation of, **36–7**, 197–8
    internal, 45, **59**, 63
Engine inlet, 27–8, **239–40**
Enstrophy, 202
Enthalpy, total, **59–60**, 62–3, 75, **160**,
    197–8
Equation of state, **14–15**, 37
Equations of motion, 33–40
Equilibrium turbulent boundary layer, 123
Equivalent sand-grain roughness, **214**, 217
Euler equations, 474–5, **498–9**, 509,
    511–12
Eulerian formulation, **16–17**, 34, **38–40**

Excrescence drag, 224–5
  magnification factor, 224
Exhaust nozzle, **73**, 226, 240, 241
Explicit finite difference method, 507–8

Fallacy:
  2D biplane lift reduction as loss of
    "efficiency", **410–11**, 490
  base drag and Hoerner's jet pump,
    210–12, 489
  boundary-layer displacement surface as
    effective wall, **92–3**, 488
  constant pressure in boundary layer, 488
  dependence of induced drag on aspect
    ratio, **386**, **394**, 489–90
  expectation of high accuracy from
    turbulence modeling, **53**, 462, 490
  explanations of 2D airfoil lift, **269–70**,
    299–302, 489
  induction, **32–3**, 487
  integrated downward momentum in
    Trefftz plane, **433**, 490
  inviscid "interference" drag, 195,
    222–3, **489**
  laminar sublayer, **140**, **218**, 488
  one-way causation, **9–10**, 41, 270, 299,
    487
  pressure drag or thrust on a portion of a
    body, **194–5**, 489
  reducing pressure drag, **207–8**, 489
  surface streamlines as indicators of flow
    direction, **90**, 487–8
  tinkering with vorticity, **411–14**, 487
  zero Cf as universal signature of
    separation, **97**, **454–5**, 488
Farfield boundary conditions, 513–14
Favorable pressure gradient:
  effect on transition, **132**, **135**, **338**, 468
  effect on velocity profile, **83–4**, 109,
    121, 138–9
First law of thermodynamics, 36–7
Flap, 297
  Krueger, 330
  slotted, 297, **329–35**
Flat-plate boundary layer, **117–21**, 242,
    248, 318

Flow tangency boundary condition,
    111–12, **308**, 382, 401, 429
Flow visualization:
  by streaklines, 19
  by timelines, 22–3
Fluctuating part, 50
Fluid, definition of, 43
Form factor, profile-drag, 242–3, **318–19**
Free-stream boundary condition, 513–14
Frequency, disturbance, 133–8
Friction velocity, **142**, 245
Fuel mass, 225
Full-potential equations, 456, 498, **499**
Fully developed duct flow, 46, **479**
Fully rough wall, 213–17
Fuselage carry-through lift, 402–4

Galilean invariance, 53, **55–6**, 288
Genus of a region, 177
Goldschmied body, 240–1
Goldstein loading for propeller, 234
Goldstein singularity, **115–16**, 481
Görtler vortices, 134
Green's theorem, 23
Grid:
  CFD, 502–7
  convergence, 506–7
  resolution, 506–7
Ground effect:
  on induced drag, 401
  on lift in 2D, 350–2
  on lift in 3D, 382–4

Heat:
  conduction, **14**, 45, **62**, 159
  convective transfer, 158–9
  viscous dissipation into, **45**, 49, 62,
    159, 192–3, **202**, 204–7
Hele-Shaw flow, 477
Helmholtz vortex theorems:
  1st, 24
  2nd, **27**
  3rd and 4th, 58
Hiemenz flow, 122
Horseshoe vortex, **138–9**, **363–4**
Hybrid laminar-flow control, 251

Hydraulically smooth wall, **213–14**, 217
Hydrostatic pressure, **14**, 35, 189
Hyperbolic differential equation, **37–8**,
    101, **113**

Ideal (perfect) gas, **15**, **37**, 59–60, 62, 66,
    73, 75
Implicit finite difference method, 507–8
Incompressible flow, 66–70
Incompressible potential flow, 67–70, 500
Independence principle, swept-wing
    boundary layer, **129**, 459, 465
Index of a region, 180
Induced drag, **362**, **385–409**
    biplane, 409
    effect of fuselage, 402–4
    effect of tail or canard, 404–9
    in ground effect, 401
    minimum (ideal), 394–6
    polar, 397–8
    reduction, 411–27
    scaling, 385–6
    theory of Spreiter and Sacks, 389–90
    theory, momentum balance, 386–9
    theory, Trefftz-plane, 391–3
Induced efficiency, propeller, 235
Induction:
    Biot-Savart law, **31–3**, 389, 392–3,
        413, 480, 487
    fallacy, **32–3**, 487
Inlet vortex, 27–8
Inlet, engine, 27–8, **239–40**
Inner layer, turbulent, 138–141
Integral momentum equation, **108–10**,
    125–8
Interference drag, 222–5
Intermittency (turbulence), **138**, 147
Internal energy, 45, **59**, 63
Inverse design, 311, 338, **353–4**, 496, 497
Inverse mode, boundary layer, **103**, **109**,
    110, 116
Inviscid flow:
    Euler equations, 474–5, **498–9**, 509,
        511–12
    full-potential equations, 456, 498, **499**
    incompressible potential flow, 67–70,
        500

    linearized for small perturbations,
        308–9, **379–80**
Irrotationality, **25**, **26**, **47–8**, **56–7**, 60, 67
Isentropic-flow relations, 71–4
Isobar, 166, 306–7, **518**
Isotropic eddy viscosity, **54**, 149, 461–4
Iterative convergence, 507–8

Jet propulsion, 239–40
Junction (necklace) vortex, 150–1

Kelvin's circulation theorem, 57
Kinematic description, 18
Kinematics of vorticity, **24–5**, **26–33**
Kinetic energy of turbulence, 37, **49**
Kinetic theory of gasses, 8, **15**, **44**
Krueger flap, 330
Kutta condition, **263–4**, 481–2
Kutta-Joukowski theorem, **265–9**, 303,
    380–1

Lagrangian formulation, **16–17**, **34–35**, 40
Laminar boundary layer, **82–3**, 101–2,
    117–18, 121,123, 338–41, 465–8
Laminar flat-plate boundary layer, 117–20
Laminar flow, natural (NLF), **251**, 338–41
Laminar skin friction, **117–21**, 251, **338**
Laminar sublayer, **140**, **218**, 488
Laminar-boundary-layer similarity, 121–3
Laminar-flow airfoil, 338–41
Laminar-flow control (LFC), 251
Laminar-flow stability theory, 132–8
Laminar-to-turbulent transition, **131–8**,
    338–341, **465–8**
Laplace's equation, **67**, 475, 477, 500
Large-eddy simulation (LES), 261, **494**
Law of the wake, 145–7
Law of the wall, **142–5**, **215**, 245–6
Lift, 259, **286**
    Bernoulli, **270–5**, **281–2**, 298–9
    fuselage carry-through, 402–4
    in ground effect, 2D, 350–2
    in ground effect, 3D, 382–4
    manifestations in atmosphere, 427–43
    maximum:
        limited by 3D effects, 384

limited by viscous effects,
    multi-element, 329–35
limited by viscous effects,
    single-element, 319–29
physical explanations, **269–70,**
    299–302, 489
reaction, 281–2
Lifting-line theory, **380–2**, 385, 393, 480
Lifting-surface theory, 379–80
Linearized airfoil theory, 307–15
Linearized boundary condition, 308
Linearized for small perturbations, **308–9,**
    379–80
Logarithmic law, turbulent, 142–5
    effect of roughness, **215–17**, 245–8
Longer path length, **270–2**, 298–9
Lubrication theory, 477

Mach number, 62–3
Mach number, critical, 342–4
Magnification factor, excrescence drag,
    224
Mapping, conformal, **67–70**, 122, **264**, 484
Marching solution of boundary-layer
    equations, **101**, **105**, 116
Mass, conservation of, 34–5
Matched asymptotic expansions, method
    of, 100, 381
Maximum lift:
    limited by 3D effects, 384
    limited by viscous effects,
        multi-element, 329–35
    limited by viscous effects,
        single-element, 319–29
Mean flow, **48–9**, 437
Mean free path, **14–16**, **44**, 70
Meme, 286
Minimum (ideal) induced drag, 394–6
Mixing length, 53–5
Models, turbulence, 48–55
Molecular motion, 9, **13–15**, 16, 34, 44,
    48–53, 164, 289–90
Moment of momentum, 227
Momentum:
    area, 200
    coefficient, jet blowing, 156

conservation of, **35–6**, 198–202
integral equation, boundary layer,
    108–10, **125–8**
thickness, boundary layer, **108–10**, 127,
    200–1, 451–2
Moody chart, 213

Natural laminar flow (NLF), **251**, 338–41
Natural transition, 131–8
Navier-Stokes (NS) equations, **13–18,**
    35–6, **44–8**, 493–7
Neutral-stability curves, 135–6
Newton's laws:
    second law, **35**, 41, 67, **289**, **292–5**
    third law, **35**, 190, **288**, 298
Newton's theory of lift, 260
Newtonian fluid, 14
Newtonian worldview, 11
No-slip condition, **15–16**, 37, 79, 83, 103,
    140, 193
Nodal-point singularity, 171
Nonuniqueness of lift curve, **321–2**, 334
Normal pressure gradient, **276**, **278**, 281,
    282, 488
Normal-force coefficient, 395
    (Figure 8.3.7)
Normal-shock relations, 74–5
Nozzle, exhaust, **73**, 226, 240, 241

Oblique shock, **75–6**, 345–7, 464, 479
Obstacle effect, **164–8**, 272, 299
One-dimensional flow assumption, **71**, 478
Origin, region of, as definition of
    separation, **97–9**, 180–2, 454–5
Orr-Sommerfeld equation, 135
Oseen theory, 477
Oswald efficiency factor, 396–7
Oswatitsch shock-drag formula, **203–4,**
    475, 486, 499, 510, 516
Outer layer, turbulent, **142–3**, **145–7**
Overlap layer, turbulent, 143–5

Parabolic differential equation, 101–2
Parallel-flow assumption, 135
Parasite drag, 210
Perfect (ideal) gas, **15**, **37**, 59–60, 62, 66,
    73, 75

Perturbation, singular, 64
Perturbation, small, **64**, **308–9**, 379, 435,
    476
Pinching, streamtube, **165–7**, **272–3**,
    299
Pitching moment, **310–11**, 404, **446**
Plane-of-symmetry flow, 125–8
Plausible falsehood, von Karman's, 285
Plume, propulsion, 225–6
Potential flow, **67–70**, 474–8, 481,
    499–500
  equation, 67
  potential function, 67
  sink, 67–8
  source, 67–8
  vortex, **67–8**, 233, 303, 368
    (Figure 8.1.8a)
Power loading coefficient, propeller, 234
    (Equation 6.1.20)
Prandtl boundary-layer theory, **79**, **99–108**
Prandtl lifting-line theory, **380–2**, 385,
    393, 480
Prandtl mixing length, 54
Prandtl number, **62–3**, **159–61**
Prandtl propeller induced-flow factors, 234
Pressure, 35–6
  coefficient, canonical, **323–8**, 332–3
  drag, **194–6**, **207–10**, 316–19, 344–5,
    489
  dynamic, 64–5
  footprint on ground, 429–31
  gradient:
    effect on transition, **132**, **135**, **338**,
        341, 468
    effect on velocity profile, **82**-**7**, 104,
        138–9, 140–2, 149, 457–64
    role in general momentum balance,
        36, 46–7, **62**, **83–4**, **108–9**
    role in separation, **84–5**, 87, **93–5**,
        96–9, 150, 276–9, **323–7**,
        459–60
  recovery, **95–6**, **98–9**, 168, **323–9**, 334
  static, **35–6**, 49
  total, 59–60
Pressure-driven cross-flow, **85–7**, 453
Profile drag, **316–19**, **344–5**

Propeller:
  actuator-disc theory, **227–9**, 239, 240,
      480, 515
  advance ratio, **229–30**, **235–8**
  blade-element theory, 234–39
  efficiency, 230–9
  induced efficiency, 235
  power loading coefficient, 234
      (Equation 6.1.20)
  Prandtl induced-flow factors, 234
  thrust loading, 234–9
  torque coefficient, 234–9
  torque, 229–36
Propulsion, 191
  Ackeret, **225–6**, 240–1
  boundary-layer, **225–6**, 240–1
  plume, 225–6
Protrusion height (riblets), 255–7
Pseudo-Lagrangian viewpoint, 40

Reaction lift, 281–2
Reattachment, **80–2**, **320–2**, 347–8
Receptivity theory, 132
Recovery factor, 160–1
Recovery temperature, 160–1
Recovery, pressure, **95–6**, **98–9**, 168,
    323–9, 334
Reflection in a solid surface, 93 (Figure
    4.1.13), 305, **483**
Region-of-origin definition of separation,
    97–9, 180–2, 454–5
Relaminarization, 466–7
Reversal, flow, **93–7**, 276, 334
Reynolds averaging, 50–1
Reynolds number, 62–3
Reynolds stresses, 49–55, 139–41
Reynolds-averaged Navier-Stokes
    equations (RANS), 261, **495–7**
Riblets, 252–8
Ring wing, 239
Rollup, vortex, 208, 232, **364–6**
Roughness Reynolds number, 215
Roughness, equivalent sand-grain height,
    214
Roughness, surface, **162**, **212–22**,
    243–250

Saddle-point singularity, **171**, 174
Sailboat analogy for winglets, 415–6
Sand-grain roughness, 212–17
Second law of thermodynamics, **45**, **75**
Separation, 3D, 96–9
Separation, boundary-layer, **93–9**, 102,
      115–16, 139–40, **150–7**, 169–76,
      210, **319–35**
Separation, shock-induced, **347–9**, 554–5
Shape factor, boundary-layer, **108–9**,
      161–2
Shear layer:
    idealized as vortex sheet, **29–30**, 364–5
    in inviscid CFD solutions, 498, 500,
      511–13
    physical, 29–30 (Figure 3.3.8b)
Shear stress, 35–6
    laminar, 35–6, **43–7**
    turbulent, **49–55**, 139–41
Shear-driven cross-flow, 85
Shock:
    bow, 71–2
    capturing, 509
    detached, 71–2
    discontinuity (idealized), 39, **74–7**, 509
    drag, **203–4**, **344–5**
    fitting, 509
    normal, **74–5**, 345–7
    oblique, **75–6**, 345–7, 464, 479
    Oswatitsch drag formula, **203–4**, 475,
      486, 499, 510, 516
    unsteady, 75
Shock/boundary-layer interaction, **345–9**,
    464–5
Similarity:
    dynamic, 61–6
    parameter, laminar boundary layer,
      121–2
    parameter, turbulent boundary layer,
      123
    solutions, boundary-layer, 121–3
Simple fluid, 43
Simple sweep theory, 444
    angle of attack, 446–7
    flow directions, 448–9
    force coefficient formulas, 446

Mach number, 446
Sin-series spanloads, 398–400
Singular perturbation problem, 64
Singularity:
    Goldstein, **115–16**, 481
    nodal point, 171
    saddle point, **171**, 174
Sink, potential-flow, 67–8
Skin friction, 83
    drag, **194**, 204–5, **208**, **317–19**
    effect on momentum balance, 109
    laminar, **117–21**, 251, **338**
    reduction, turbulent, 251–8
    surface lines, **88–90**, **97–9**, 150–1,
      176–183, **453–5**
    turbulent, **117–21**, 162, **212–222**,
      243–50
    use in empirical drag correlations,
      242–3
Slat, 297–8
Slip velocity:
    apparent, in turbulent flow, 142–3
    molecular, **15–16**, 252
Slope discontinuity, 355, 484
Slot drag, 212
Slot effect, 330–35
Slotted flap, 297, **329–35**
Small disturbance, **70**, **131–7**, 307–11,
    379–80, 500, 508
Small perturbation, **64**, **308–9**, 379, 435,
    476
Source, potential-flow, 67–8
Spalart-Allmaras turbulence model, 55
Span-efficiency factor, 396–7
Spanload:
    basic and additional, 376–9
    sin-series, 398–400
Spatial field grid, 502–6
    convergence, 501–2, **506–7**
    resolution, 503–7
Specific heat, 15, **62–3**, **159**
Speed of sound, **62–3**, 67, **73**
Sphere drag, 162, **209–10**
Spoiler, 298
Squire-Young formula, 201–2
Stability theory (laminar-flow), 132–8

Stability, numerical, 508
Stagnation conditions, 59
    temperature, 160–1
Stall, airfoil, 296–7, **319–29**
Static pressure, **35–6**, 49
Stator, 335–7
Steady flow, **17**, **19–22**, **58–60**
Stokes flow, **209–10**, 476–7
Stokes's theorem, **25**, 29, 56, 58, 223, 265,
    364, 439
Streaklines, 18–21
Stream function, 22
Streamline coordinates, 105
Streamline curvature, 106–7
Streamline topology, 168–85
Streamline, **18–22**, 59–60, 91–2, **111–12**
Streamlined body, **150**, 169–70, **192**, 196,
    207–8
Streamtube:
    definition, 19–22
    pinching, **165–7**, **272–3**, 299
Stress tensor, **18**, **35**, **44**
Sublayer, "laminar" or viscous, 95–6,
    139–44, **217–22**, 488
Suction, boundary-layer separation control
    by, **157**, 240–1
Suction, laminar flow control (LFC), **135**,
    251
Supercritical airfoil, 344
Surface roughness, **162**, **212–22**, 243–250
Surface streamlines, **88–90**, **97–9**, 150–1,
    176–183, **453–5**
Surface tension, 279–80
Sweep theory, simple, 444
    angle of attack, 446–7
    flow directions, 448–9
    force coefficient formulas, 446
    Mach number, 446
Swept wing, 444
    attachment-line boundary layer, 105–6,
        125–8, **171–4**, **448–52**
    boundary layer independence principle,
        129, 459, 465
    inboard tailoring, 469
    infinite-span boundary-layer theory,
        128–30

laminar-to-turbulent transition, 465–8
    related to a 2D airfoil, 468–9
    separation criterion, 454–5
    shock/boundary-layer interaction,
        464–5
    turbulent attachment line, 449–52
    turbulent boundary layer, 147–9,
        455–64
    vortex generator installation, 154–5
Swirl (in propeller flows), **233**, 235–6

Taper, effect on wing boundary layer,
    129–30, **455–9**
Temperature, recovery, 160–1
Temperature, total, 160–1
Thermal boundary condition, 158–9
Thermal boundary layer, 158–62
Thermal conduction, **14**, 45, **62**, **159**
Thermodynamic equilibrium, **14**, **44**
Thin-layer Navier-Stokes equations, 474,
    496
Three-dimensional boundary layers, 80
    equations, **104–8**, 109–10
    displacement thickness, 112–13
    on swept and tapered wings,
        128–30, **449–64**
    separation, 96–9
    turbulent prediction, 147–50
    velocity profiles, 85–90
Three-dimensional wings, 359
Thrust loading, propeller, 234–9
Time-averaging:
    of molecular motion, **13–14**, 16,
        48–49, 50–2
    of turbulent motion, 48–49, **50–2**
Timelines, 22–3
Tip vortices, **232–3**, **364–71**, **411–15**,
    436–44
Tollmien-Schlichting waves, 6,
    133–8
Topology, streamline, 168–85
Torque coefficient, 234–9
Total enthalpy, **59–60**, 62–3, 75, **160**,
    197–8
Total pressure, 59–60
Total temperature, 160–1

Trailing-edge dumping-velocity elevation, **332–4**, 337
Transition to turbulence, **130–8**, 338–341, 465–468
Transonic airfoil, 342–50
Transpiration to represent displacement effect, 111
Transport properties, 15
Transverse curvature, effect on axisymmetric boundary layer, 123–4
Traveling wave disturbance, **134**, 468
Trefftz plane, **198–9**, **391**
   integrated downward momentum in, 431–5
   integrated pressure in, 435–6
   theory, **391–6**, 402–11, 420–1, 480, 486–7, 489–90
Trip, boundary layer, **132**, 341
Triple-deck theory, 116
Turbine blade, 335–7
Turbofan and turbojet, 239–40
Turbulence:
   coherent structures, 48, **138–9**
   eddy viscosity, **53–5**, **95**, **140–1**, **147–9**, 246–8, **460–64**
   intermittency, **138**, 147
   kinetic energy, 37, **49**
   mixing length, 53–5
   modeling, 48–55
   Reynolds averaging of, 50–1
Turbulent boundary layer, 82–3, 94–6, 117–21, 123, **138–50**
   inner layer, 138–141
   outer layer, **142–3**, **145–7**
   overlap layer, 143–5
   shear stress, **49–55**, 139–41
   similarity, 123
   skin friction, **117–21**, 162, **212–222**, 243–50
   viscous sublayer, 95–6, **139–44**, **217–22**, 488
Turbulent dissipation, 37, **49**
Turbulent shear stress, **49–55**, 139–41
Turning vane, 335–6

Unsteady blowing (separation control), 157

Validation, CFD, 515–16
Van Driest damping factor, 140
Velocity defect:
   boundary layer, 123, 143
   wake, 186, 188
Velocity potential, 67
Velocity profile, boundary-layer, 46, **82–90**
   3D, in arbitrary coordinate system, 105
   approaching separation, 96–7
   effect of shape on growth, 109
   in adverse pressure gradient, 104
   initial, 101
   shape factor, 108–9
   turbulent, 139–47
   with similarity, 121–3
Vertical momentum in atmosphere, **427–9**, **431–5**
Viscosity, 43–8
Viscous dissipation, **45**, 49, 62, 159, 192–3, **202**, 204–7
Viscous drag, 189–90, **192–3**, **196**, **199–202**
Viscous sublayer, 95–6, **139–44**, **217–22**, 488
Viscous/inviscid interaction methods, **110**, 261, **497–8**
Visualization:
   of CFD solutions, 517–24
   of flow, by streaklines, 19
   of flow, by timelines, 22–3
von Karman constant, 143
von Karman momentum integral equation, 108–9
von Karman plausible falsehood, 285
von Karman vortex street, 21
Vortex:
   decay and collapse, **369–70**, 436–43
   filament, 27–9
   generators, 150–5
   generators, sub-boundary-layer, **150**, 155
   induction fallacy, **32–3**, 487
   line, 26–9
   rollup, 208, 232, **364–6**
   shedding wake sheets from wings, **362**, 375

Vortex: (*continued*)
   sheet, 28–30
   stretching, 26–7
   unsteady wake shedding, 187–9
   wake myths, 411–14
   wake of a wing, 362–71
   wake reconnection, 439–43
Vortices, tip, **232–3**, **364–71**, **411–15**,
   436–44
Vorticity, 24
   Biot-Savart law, **31–3**, 389, 392–3,
      413, 480, 487
   bound, of a lifting airfoil or wing,
      265–9, 305, **363–4**, 380–2,
      429–31
   budget of a boundary layer, 113–14
   equation (creation, convection,
      diffusion, stretching), **47**, 58,
      113–14
   interaction with a boundary, 27–8
   kinematics of, **24–5**, **26–33**
   relation to circulation (Stokes's
      theorem), **25**, 29, 56, 58, 223,
      265, 364, 439
   stretching, **26–7**, 47, 58

Wake:
   law of the, 145–7
   momentum area, 200
   reversal, 334
   viscous, **186–9**, 199–202
   vortex, 186, **362–71**, 411–14
Wall jet, **155–7**, **331**
Wall, law of the, **142–5**, **215**, 245–6
Water-faucet demonstration, 279–80
Wave number, 133
Wedge flows, **69**, **121–3**
Wing structural weight, **250**, 399, 409–10,
   418, **420**, 422
Winggrid tip device, 397
Winglet, **395–6**, **411–12**, **415–18**, **422**,
   425–6
Wingtip device, 411–23
Wingtip vortex, **232–3**, **364–71**, **411–15**,
   436–44

Zierep singularity, 346–7

Note: For index entries listing more than one instance, the more important instances are indicated by bold page numbers